Photorefractive Materials and Their Applications II

Survey of Applications

Edited by P. Günter and J.-P. Huignard

With 209 Figures

Springer-Verlag Berlin Heidelberg GmbH

Professor Dr. *Peter Günter*

Institut für Quantenelektronik, ETH Hönggerberg
CH-8093 Zürich, Switzerland

Dr. *Jean-Pierre Huignard*

THOMSON-CSF, Laboratoire Central de Recherches, Domaine de Corbeville, B.P. 10
F-91401 Orsay Cedex, France

ISBN 978-3-662-31244-5   ISBN 978-3-540-39135-7 (eBook)
DOI 10.1007/978-3-540-39135-7

© Springer-Verlag Berlin Heidelberg 1989

Originally published by Springer-Verlag Berlin Heidelberg New York in 1989.
Softcover reprint of the hardcover 1st edition 1989

2154/3150-543210 – Printed on acid-free paper

Topics in Applied Physics Volume 62

Topics in Applied Physics Founded by Helmut K. V. Lotsch

Volumes 1–56 are listed on the back inside cover

Foreword

It is now well established that a unique feature of coherent optical beams is their capability to transmit, to process and to interconnect in parallel a large number of high bandwidth information channels. However, although a considerable potential exists for these techniques, their development depends critically on the availability of nonlinear optical materials that work at high speed and low incident optical power. At present, these requirements are stimulating a great deal of research in materials science and are challenging existing technologies, in particular high speed electronics. This volume devoted to photorefractive crystals and their applications presents a detailed account of materials and device issues. Starting from the large index nonlinearities observed in several photorefractive crystals, the authors present and analyse a variety of new applications based on all-optical interactions between the incident laser beams. Photorefractive crystals permit beam interactions with a large gain at low power levels. A remarkable consequence is the demonstration of complex signal processing functions using the amplification and feedback concepts for performing tasks comparable to electronics, but now on optical wavefronts containing spatial information. These concepts are clearly outlined and developed in the different chapters of this book which cover theory and experiments relevant to image amplification, phase conjugation, self-induced optical resonators and associative memories. This collection of articles provides a broad survey of the current research on the applications of photorefractive crystals. We hope that, together with the significant results already demonstrated, it will stimulate further research into the optimization of photorefractive crystals as well as the investigation of new effects and the invention of new devices and device architectures. With the advent of other types of nonlinear optical materials such as organic materials and semiconductors, photorefractive crystals will undoubtly play a role in the future development of high performance laser systems and optical processors.

Paris, March 1988   *E. Spitz*

Preface

The objective of this second volume on photorefractive materials is to focus on the applications of the effects associated with these materials. Since the discovery of the basic phenomena in different electro-optic crystals, an increasing amount of work is now devoted to the applications of photorefractive nonlinearities. Thus it is hoped that the following chapters will help to present an overview of the field with a unified treatment of the beam interactions. The main applications to be covered in this book include dynamic holography, phase conjugation, wavefront amplification, oscillators and spatial light modulators. The photorefractive crystals have already demonstrated very promising capabilities since efficient beam interactions and large gain coefficients are now achieved with low power lasers. Further optimization of the materials should stimulate a growing interest in photorefractive nonlinearities for device applications in optical information processing and laser wavefront control.

Most of the applications considered in this volume are based on beam interference effects that spatially modulate the refractive index of the electro-optic crystals. In contrast to photographic materials, the photoinduced gratings in photorefractive materials adapt in real time to any change of the spatial properties of the incident recording beams. In photorefractive crystals, the main parameters which govern the amplitude of the photoinduced index modulation are: the applied electric field, the grating period, and the electro-optic coefficient. The first volume detailed the influence of each of these parameters according to the physical mechanisms (drift and diffusion of the photocarriers) involved in the grating build-up and erasure. The applications presented here derive from the specific properties of the photorefractive dynamic gratings. Firstly, photorefractive crystals enable one to perform holographic experiments with the advantage of real time operation with a reusable photosensitive support. Some crystals have already shown writing-erasure energies comparable to silver halide media. Several examples of parallel image processing operations and of holographic interferometry are discussed extensively in different chapters of this book, along with other highly developed applications based on the beam interactions during grating writing in the nonlinear crystal.

A related phenomenon is the energy transfer from the pump beam to a low intensity signal beam. This remarkable effect is characterized by an exponential gain coefficient Γ, which, under optimized recording conditions, ranges from $\Gamma = 5$ to $\Gamma = 15 \, \mathrm{cm}^{-1}$ in common photorefractive crystals. Consequently, an efficient probe beam amplification is achieved with most of the materials considered here. Moreover, such a large two-wave-mixing gain also leads to an amplified phase conjugate beam when using a four-wave-mixing interaction with two antiparallel pump beams. Numerous applications are presented by the authors of this second volume. They are mainly oriented towards image amplification, parallel optical logic, spatial light modulation, phase distortion compensation, associative memories, laser beam steering and laser beam combining. All these experiments are clearly explained, as are the physical mechanisms involved in the beam interactions and the corresponding properties of the chosen photorefractive crystal. Several outstanding experimental results illustrate the capabilities of the photorefractive nonlinearities while the parameters which limit the device performances are also outlined.

Last but not least, these amplifiers permit the realization of a new type of oscillator through the addition of an optical feedback. The oscillation starts on the light induced scattering generated by the incident pump on the photorefractive ring and phase conjugate oscillators. Preliminary applications to laser beam combining and nonlinear image processing are convincingly demonstrated. The real advantage of photorefractive nonlinearities over other nonlinear mechanisms relies on the fact that such interactions are achieved with a few tens of milliwatts of laser power and with millisecond response times.

We hope that this volume, by presenting a detailed survey of the wide range of topics outlined above, will provide a useful source of reference on the applications of photorefractive materials, as well as a stimulus for new work in the field. With the advent of new photorefractive crystals with optimized performances it is likely that this class of nonlinear materials will play an important role in the design of future high performance optoelectronic and laser systems.

Zürich and Orsay,   *P. Günter*
January 1989   *J.-P. Huignard*

Contents

Contributors

Cressman, Paul J.
Eastman Kodak Company, 901 Elmgrove Road, Rochester, NY 14650, USA

Cronin-Golomb, Mark
Electro-Optics Technology Center, Tufts University, Medford, MA 02155, USA

Feinberg, Jack
Department of Physics, University of Southern California,
Los Angeles, CA 90089-0484, USA

Fischer, Baruch
Department of Electrical Engineering, Technion, Haifa, Israel

Günter, Peter
Institut für Quantenelektronik, ETH Hönggerberg,
CH-8093 Zürich, Switzerland

Holman, Robert L.
Battelle Columbus Laboratory, 505 King Avenue, Columbus, OH 43201, USA

Huignard, Jean-Pierre
THOMSON-CSF, Laboratoire Central de Recherches, Domaine de Corbeville,
B.P. 10, F-91401 Orsay Cedex, France

Johnson, Richard V.
Department of Electrical Engineering, University of Southern California,
University Park MC-0483 Los Angeles, CA 90089-0483, USA

Khomenko, Anatolii V.
A.F. Ioffe Physico-Technical Institute of the USSR Academy of Sciences,
SU-194021 Leningrad, USSR

Kwong, Sze-Keung
Ortel Cooperation, 2015 W. Chestnut Street, Alhambra, CA 91803, USA

MacDonald, Kenneth R.
Institute of Optics, University of Rochester, Rochester, NY 14627, USA

Marrakchi, Abdellatif
Bell Communications Research, 331 Newman Springs Road,
NVC 3X 115, Red Bank, NJ 07701-7020, USA

Odoulov, Serguey G.
Institute of Physics of the Ukrainian Academy of Sciences, Prospekt Nauki,
SU-252650 Kiev 28, USSR

Petrov, Mikhail P.
A.F. Ioffe Physico-Technical Institute of the USSR Academy of Sciences,
SU-194021 Leningrad, USSR

Psaltis, Demetri
California Institute of Technology, Department of Electrical Engineering,
Pasadena, CA 91125, USA

Soskin, Marat S.
Institute of Physics of the Ukrainian Academy of Sciences, Prospekt Nauki,
SU-252650 Kiev 28, USSR

Tanguay, Armand R., Jr.
Optical Materials and Devices Laboratory, 523 Seaver Science Center,
University of Southern California, University Park MC-0483,
Los Angeles, CA 90089-0483, USA

Verber, Carl M.
School of Electrical Engineering, Georgia Institute of Technology,
Atlanta, GA 30332, USA

White, Jeffrey O.
Hughes Research Laboratories, 3011 Malibu Canyon Road,
Malibu, CA 90265, USA

Wood, Van E.
Columbus Division, Battelle Memorial Institute, 505 King Avenue,
Columbus, OH 43201-2693, USA

Yariv, Amnon
Department of Applied Physics, California Institute of Technology,
Pasadena, CA 91125, USA

Yu, Jeffrey W.
Department of Electrical Engineering, California Institute of Technology,
Pasadena, CA 91125, USA

1. Introduction

Jean-Pierre Huignard and Peter Günter

The development of reliable laser sources in the late 1960s has contributed a great deal to the growing interest in the applications of coherent optical techniques. It is now recognized that coherent optics offers significant advantages for signal processing with a high degree of parallelism and permits information to be transmitted at very high rates using either free-space or guided-wave propagation. Moreover, the concept of recording the amplitude and phase of spatially and temporally modulated wavefronts has opened new applications for optical holography. Obviously, the implementation of these concepts for performing processing tasks requires nonlinear materials allowing efficient and high speed interaction between the incident optical beams. For these reasons, materials which exhibit a reversible index change under a low power laser beam illumination are of particular importance. Such requirements on the nonlinear optical properties of materials are met in electro-optic crystals due to the presence of photorefractive effects. The index nonlinearity arises from a charge redistribution under nonuniform illumination causing a semi-permanent change in the crystal refractive index through the linear electro-optic effect. These photoinduced index variations are erased by flooding the crystal with uniform illumination. Since its discovery in $LiNbO_3$, the effect has been observed in a variety of electro-optic materials and a great deal of effort has been devoted to investigations of the microscopic mechanisms responsible for the charge transport. The physics of the effect and the experiments on material properties were presented in the first volume TAP Vol. 61, on the photorefractive effect and materials.

The aim of this second volume is therefore to develop the applications of the photorefractive effect and this includes image storage, information processing, phase conjugation and nonlinear wave mixing. All these applications require materials that possess high photorefractive sensitivity, speed, spatial resolution and diffraction efficiency. To date, no material displays all of these ideal features, but several extremely attractive characteristics have already been achieved in existing crystals such as $LiNbO_3$, $KNbO_3$, $BaTiO_3$, $Bi_{12}(Si,Ge, Ti)O_{20}$, GaAs .... It was recognized early, that the photorefractive effect can be applied to phase volume holographic recording and high density optical memories based on selective reconstruction of images superimposed in the same crystal. This type of optical memory had no practical impact because of the concurrent development of semiconductor technologies. However, the concept has become important again for the new field of all-optical associative memories. Besides this, different applications were developed which utilize dynamic holographic recording for performing holographic interferometry, image convolution and correlation, pattern recognition, parallel optical logic, photolithography and reconfigurable optical interconnections. An evident advantage of the photorefractive crystals used for these experiments is that no development of the recording medium is needed. Moreover, ultrahigh speed is not usually required since high information rates are already achieved by the high degree of parallelism of the input image. In these holographic imaging experiments, the crystal response times typically range from milliseconds ($Bi_{12}SiO_{20}$, $KNbO_3$: Fe^{2+}, GaAs ...) to seconds ($BaTiO_3$, $LiNbO_3$) for an incident recording beam intensity of $1\,W\,cm^{-2}$. However, operation at nanosecond speed is also possible when recording with a high power blue-green pulsed laser.

In other investigations, the origin of the energy redistribution between the two recording beams was identified. The effect arises from the self-diffraction of the incident beams by the dynamic phase volume grating whose photoinduced index modulation is spatially shifted with respect to the fringe pattern. The physics of the photorefractive effect is such that the index modulation is often $\frac{\pi}{2}$ phase shifted and, therefore, one of the beams is coherently amplified at the expense of the other. Thus, the photorefractive crystal may be regarded as a nonlinear device allowing parametric amplification of coherent signal waves containing spatial information. The consequences for coherent optical signal processing are important since large gain coefficients for an incident probe beam are attained in wave mixing experiments. The two wave mixing interaction is applied to image amplification by energy transfer from the pump beam and high reflectivity phase conjugate wavefront generation is obtained by degenerate or nearly degenerate four wave mixing. Phase conjugation, a new and important branch of coherent optics, permits restoration of phase distorted wavefronts after beam propagation through the atmosphere and in laser resonators. By feeding the amplified signal back into the photorefractive crystal, new types of ring and phase-conjugate oscillators have been demonstrated. Many other very attractive nonlinear features have been investigated including self-pumped phase conjugation, laser beam cleanup and coherent coupling of different laser beams. These applications as well as the physics of the beam interactions with the photorefractive nonlinear medium are detailed by the authors of this second volume. The last two chapters deal with another important device, the spatial light modulator, whose function is to convert an incoherent input image into a coherent image carried by a readout laser beam. This optoelectronic transducer is used for real time data input in a digital or analog optical processor.

In summary, the applications of the photorefractive effects have expanded tremendously in the last few years in different directions. To gain full benefit from the exceptional capabilities of coherent optics for signal processing, further extensive research work on the nonlinear optical materials is necessary. Photorefractive nonlinearities in electro-optic crystals have already shown their unique potential for image processing, phase conjugation and wavefront amplification with low power laser beams. These applications should stimulate further efforts to improve the crystal characteristics, in particular the two key parameters, namely the amplitude of the nonlinearity and the operating speed. It is our hope that this survey of the applications of photorefractive crystals will contribute to the optimization of existing materials and to the discovery of new physical effects and applications.

2. Amplification, Oscillation, and Light-Induced Scattering in Photorefractive Crystals

Serguey G. Odoulov and Marat S. Soskin

With 25 Figures

Photorefractive crystals (PRC), which make possible an efficient steady-state energy transfer from one or several reference beams into an object (or phase-conjugate) beam [2.1–3], are attracting particular attention as media for dynamic recording and transformation of light beams. Investigations in the last decade have revealed some new nonlinear effects involving the so-called non-local response, which has greatly extended the scope of nonlinear interactions enabling a signal beam to be amplified even in a strictly frequency degenerate case. The newly discovered nonlinearities provide exponential gains on the order of 10 or even $100\,cm^{-1}$, which many times exceed the gains attainable with widely used laser-active media such as ruby or neodimium glasses; it is this feature that enables one to obtain enormous (hundred- and thousandfold) amplification of the signal beam in PRC samples a few millimeter thick [2.4, 5].

This chapter presents an analysis of the physical processes of light amplification in PRC, a description of optical oscillation [2.6] and of light-induced scattering [2.7]. It demonstrates that frequency-degenerate optical oscillators are inherently a striking analogue of ordinary lasers and, in many respects, the light-induced scattering resembles the enhanced luminescence of active media in laser oscillators and amplifiers.

2.1 Fundamentals of Light Amplification Due to Quasi-Degenerate Four-Wave Mixing

2.1.1 Beam Coupling on Dynamic Gratings: A New Type of Coherent Light Amplification

An optical amplifier is a device where the coherent light beam intensity at the output exceeds that at the input. A quantum amplifier generally contains an active medium where a population inversion is produced by external pumping and the amplification results from stimulated emission processes. The coupling of interacting beams in dynamic hologram recording in nonlinear media can be utilized to develop a novel type of optical amplifier, a holographic amplifier whose action is based on classical interference and diffraction phenomena as well as on a light-induced change of the refractive index.

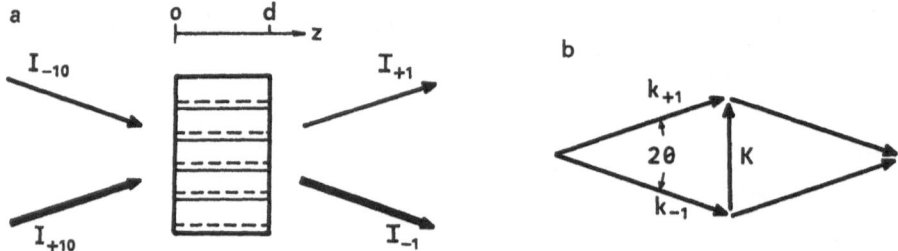

Fig. 2.1. Two-beam interaction in nonlinear media: (a) wave propagation (schematic); (b) interaction vector diagram. Solid lines in sample represent index change maxima; dashed lines are intensity maxima

The holographic amplifier functions as follows. A signal beam of intensity I_{-10} and a coherent pump beam (or beams) I_{+10} intersect in a nonlinear recording sample. The interference pattern of the beams records a dynamic grating (Fig. 2.1). For the amplifier to operate, energy transfer from the pump beam into the signal beam should be provided (Sect. 2.1.2). The amplification process proceeds as follows: the increase in signal beam intensity I_{-1} leads to an increase in the recording pattern contrast, $\sim (I_{-1}I_{+1})^{1/2}$, thus increasing the dynamic grating amplitude Δn and efficiency η, which in turn results in a further amplification of I_{-1}. The resulting amplification is nonlinear, and at $I_{-1} \ll I_{+1}$ it may sometimes be exponential[1]. The resemblance between ordinary and holographic amplifiers has been convincingly proved by the development of coherently pumped holographic oscillators (Sect. 2.6).

The development of holographic amplifiers has opened up new possibilities within coherent optics. First, amplification with the necessary response speed of complex light beams carrying a great amount of information has been made feasible. Second, it has become possible, by appropriately selecting the pump beam parameters and interaction scheme, to control the spatial spectrum of amplified light beams, including the possibility of phase conjugation (PC).

2.1.2 Spatial Shift of Recording Patterns and Holographic Gratings as Prerequisite for Energy Transfer

For noncentrosymmetric materials (which include most PRC) the nonlinear response may be nonlocal, i.e. the dynamic grating arising in a nonlinear medium is spatially shifted in the direction of the grating wavevector $K = k_{+1} - k_{-1}$ so that its extreme do not coincide with those of the steady-state light fringes[2]. A typical nonlocal response due to diffusion nonlinearity (Sect. 2.2.1) is:

---

[1] At great I_{-10}, saturation of amplification ensues due to depletion of the pump beam intensity, as in ordinary amplifiers.

[2] The term "nonlocal response", widely used already in special literature on PRC, calls for some clarification. The term "asymmetric nonlocal response" would be more consistent to emphasize the unidirectional spatial shift of the grating being recorded. A large number of symmetric nonlocal responses (e.g. thermal nolinearity) exist, and these involve no shift of the induced grating.

$$\Delta n(x) \propto [I(x)]^{-1}\{d[I(x)]/dx\} \quad , \tag{2.1}$$

where the first Fourier component $\Delta n_1(x)$, with the same spatial frequency as the light fringes $I(x)$, is $\frac{\pi}{2}$ out of phase with respect to the fringes, the shift direction depending on the sign of refractive index change Δn [2.1, 8].

In the case of a nonlocal response, a steady-state energy transfer is possible and reaches its maximum at a $\frac{\pi}{2}$ phase shift between grating and light fringes [2.2]. This fact is qualitatively explained by a $\frac{\pi}{2}$ phase shift between the zero- and the first-order of diffraction on a phase grating. At a two-beam diffraction on a spatially shifted grating (nonlocal response) an additional $\pm \frac{\pi}{2}$ phase shift of each diffracted beam appears, its sign depending on that of the spatial shift. Hence, for the direction of one transmitted beam, both its components, the zero-order beam and the first-order diffracted beam, will be in phase, whereas for the second one they will be π out of phase, i.e. the interference in the former case will be constructive, and in the latter case, destructive.

The study of diffusion-nonlinearity beam coupling in photorefractive LiNbO$_3$ crystals [2.1, 2] resulted in the formulation of a general principle for holography: energy transfer between two beams interacting in a volume phase grating is possible if and only if the interference pattern of the beams is out of phase with respect to the grating by $\Phi \neq n\pi$, where n is an integer. This assertion holds for both two-beam diffraction on static gratings (where no energy transfer occurs either with equal-intensity beams at a phase shift of $\Phi = 0$ or at a zero phase shift averaged over the grating [2.9]) and light-induced dynamic holograms.

This assertion is further proved (by contradiction) by a complete absence of any steady-state two-beam energy transfer between two light beams recording a hologram in a medium with local nonlinear response [2.10] when the first Fourier component of the refraction index, $\Delta n_1(x)$, is strictly proportional to the intensity distribution in light fringes:

$$\Delta n_1(x) \propto I(x) \quad . \tag{2.2}$$

As a consequence, the isophase surfaces of the light fringes and of the grating exactly coincide in any layer of the nonlinear sample so that no phase shift exists between them.

Fortunately, for the simultaneous self-diffraction of several light beams, the phase shift required for energy transfer may arise automatically also in locally nonlinear media. Such is the case, e.g., for various parametric interactions where the energy transfer between any pair of light beams results from diffraction at the grating recorded by another pair of beams which also participate in the interaction. A good example is the degenerate backward four-wave mixing (BFWM), widely used at present for phase conjugation [2.11, 12]. Two collinear pump beams 1 and 2 (Fig. 2.2) enter a nonlinear sample from opposite sides. The signal wave 4 forms an angle of 2θ with one of the pump beams. Let pump beam 1 and signal beam 4 record a transmitting phase grating. Diffraction of pump beam 2 on the grating gives rise to beam 3, which travels backwards

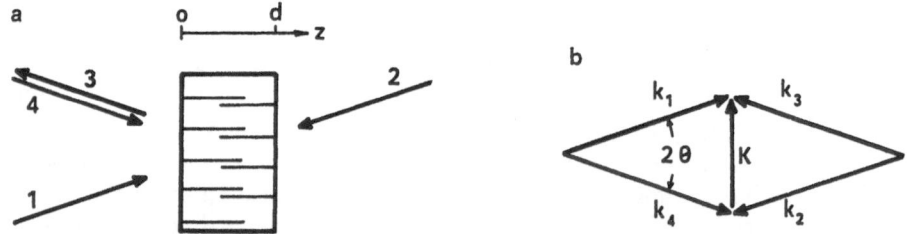

Fig. 2.2. Backward four-wave interaction in nonlinear media: (a) wave propagation (schematic); (b) interaction vector diagram. Solid lines in sample — index change maxima coinciding with intensity maxima. Note spatial shift of grating recorded by beams *1* and *4* with respect to that recorded by beams *2* and *3*

with respect to beam 4. The wavevector difference $K = k_1 - k_4$ is exactly the same as $K = k_3 - k_2$, i.e., the grating recorded by beams 2 and 3 is identically spatially oriented to the initial grating (1–4) [2.12], but is always out of phase by a quarter of the fringe spacing $\Lambda = 2\pi/|K|$ with respect to the latter, because beam 3 arising due to diffraction is always $\frac{\pi}{2}$ out of phase relative to its zero order (beam 2). Thus, the grating arising in the sample consists of two components that are $\frac{\pi}{2}$ shifted with respect to each other, and the energy transfer may occur due to diffraction of one pair of beams on the grating produced by the other pair [2.13]. Note that here the spatial shift automatically provides a simultaneous transfer of energy from pump beams 1, 2 into both signal beam 4 and the generated beam 3.

A similar explanation applies to various other types of phase-matched parametric interaction, resulting in a net amplification of the signal beam (Sect. 2.3).

Direct two-beam coupling and four-wave parametric interaction give quite different amplification characteristics. Beam coupling in the nonlocal-response nonlinear media resembles various stimulated processes (like stimulated Brillouin or stimulated Raman scattering, etc.). Amplification of the weak probe beam is exponential in this case, as distinct from parametric mixing where the probe intensity grows as the square of the crystal thickness (Sect. 2.3.1). The direction of energy transfer in two-beam coupling depends on the sign of the coupling constant (amplification or attenuation), whereas for parametric interaction in local response media the probe beam is amplified regardless of the sign of the coupling constant. These distinguishing features of the two amplification processes determine the characteristics of optical oscillators utilizing nonlinear media with local and nonlocal response.

2.1.3 Holographic Amplification of Quasi-Degenerate Frequency Beams

Until now, only interacting beams with exactly coinciding temporal frequencies and steady-state (saturation) recording where the interaction time greatly exceeds all characteristic relaxation times τ of the nonlinear recording medium

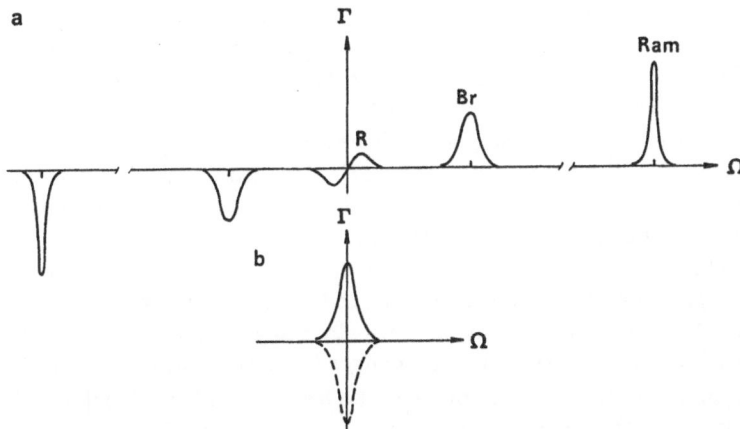

Fig. 2.3. Gain frequency spectrum for **(a)** ordinary stimulated processes (Rayleigh, Brillouin, and Raman scattering) and **(b)** stimulated scattering due to nonlocal nonlinearity in PRC. In **(b)** both signal beam amplification (solid line) and attenuation (dashed line) are possible in the same medium, the sign of the gain depending on beam orientation with respect to the polar axis or on its polarization

have been considered. However, the principle of shifted gratings is also applicable to elucidating the energy transfer between stationary beams with frequency detuning $\Omega = \omega_1 - \omega_2 \simeq \tau^{-1}$ or, similarly, between modulated beams with pulse duration $\Delta t \gtrsim \tau$.

Two beams with slightly different frequencies produce, in a nonlinear medium, a moving interference pattern which, with a local-response medium, at any given moment records in it a phase grating with the same refractive index distribution and at the next moment shifts away from this grating to record a new one. Due to the inertia of the nonlinear response, the recorded grating does not disappear instantaneously, but exists at least for relaxation time τ, with the result that even for a local-response medium a steady-state spatial lag of the grating behind the moving fringes appears and hence the energy transfer becomes possible[3].

Nondegenerate interaction of light beams with $\Omega = \omega_1 - \omega_2 > \tau^{-1}$ has been studied in detail in nonlinear optics [2.14, 15]. Figure 2.3 shows a calculated spectrum of exponential gain for a weak probe beam with a strong pump beam in a third-order nonlinear medium (local response). Resonant lines of amplification or absorption for Raman (Ram) and Brillouin (Br) stimulated scattering and the central line of Rayleigh (R) scattering are indicated. In agreement with conclusions of dynamic holography, for a strict degeneracy, $\omega_1 = \omega_2$, no gain exists in a local-response medium.

---

[3] Another, generally equivalent, means of producing the energy transfer is to displace the recording medium at a certain speed in the direction of grating wavevector K in the course of recording [2.16].

Investigations of PRC as recording media for dynamic holography revealed processes with a nonlocal nonlinear response, giving rise to an additional central resonance in the gain spectrum. This gain (Fig. 2.3b) occurs only in noncentrosymmetric media; the sign of energy transfer depends either on the probe beam orientation with respect to the polar axis and pump beam [2.1] or on the polarization of the probe and pump beams [2.17]. Note that such a behaviour does not exist in local-response media, where a 180° sample rotation about any crystal axis does not change the physical situation.

For nonlocal-response media, a frequency detuning from an exact resonance only decouples the interacting beams to reduce the gain (refer, e.g., to the experiment in [2.18a]). This is fairly obvious, since any frequency change offsets the phase shift between the grating and fringes from its optimal value of $\frac{\pi}{2}$. (In a four-wave mixing arrangement, however, the frequency offset $|\pm \Omega| \sim \tau^{-1}$ can sometimes improve the phase conjugate reflectivity [2.18b]).

2.2 Steady-State Frequency Degenerate Amplification in Two-Beam Coupling

2.2.1 Amplification Due to Diffusion Nonlinearity

Diffusion nonlinearity in noncentrosymmetric crystals was historically the first and the most studied mechanism of nonlocal nonlinearity, discovered in PRC [2.1]. Diffusive redistribution of the initial profile of mobile photoexcited carriers leads to the formation of a grating of space charge ϱ_{sc} whose extrema coincide with those of the intensity distribution in light fringes. According to the Poisson equation, the periodic space charge field modulating the refractive index is spatially shifted with respect to ϱ_{sc} and consequently also relative to the intensity distribution, with the result that the recorded phase grating in linear electro-optic (Pockels effect) crystals is also shifted with respect to the light fringes.

The steady-state intensities of the interacting beams are [2.2]:

$$I_{\pm 1} = I_0[1 + \beta_0^{\pm 1} \exp(\pm \Gamma z)]^{-1} \quad , \tag{2.3}$$

where gain $\Gamma = d^{-1} \ln (I_{-1}I_{+10}/I_{-10}I_{+1})$ depends on crystal parameters and experimental conditions:

$$\Gamma = \frac{2\pi^2 r_{33} n_e^4 kT \sin 4\theta [1 + (r_{13}n_o^2/r_{33}n_e^2) \tan^2 \theta]}{\lambda^2 e(1 + \sigma_d/\sigma_p)(1 + l_s^2/\Lambda^2)} \quad . \tag{2.4}$$

Here 2θ is the full angle of beam intersection inside the sample; e is the carrier charge; r_{ij} are electro-optic tensor components; n_o, n_e are refractive indices for the ordinary and the extraordinary wave respectively; k is the Boltzmann constant; T is the absolute temperature; σ_d and σ_p are the dark conductivity and the photoconductivity respectively; l_s is the space charge screening length; and $\beta_0 = I_{-10}/I_{+10}$ is the intensity ratio of incident beams.

Equation (2.4) for Γ is valid for the interaction of two extraordinary beams propagating symmetrically at $\pm\Theta$ to the (100) or (010) face of $3m$ or $4m$ point group crystals. A more general expression for asymmetric interaction is given by *Feinberg* in [2.19] and in Chap. 5 of the present volume. For polar crystals, the direction of propagation of interacting beams with respect to the polar axis is an essential parameter. The fringe spacing dependence of the gain, $\Gamma = \Gamma(\Lambda)$, is affected by both the interaction angle 2θ (since $\sin\theta = \lambda/2\Lambda$) and the correction factor $(1 + l_{\mathrm{s}}^2/\Lambda^2)$, which takes into account the finiteness of screening length l_{s}.

It should be emphasized that (2.3) describes the entire evolution of interacting beams without any limitation on coupling strength (Γz) or on intensity ratio β_0. For a weak probe beam the amplification is exponential:

$$I_{-1} \cong I_{-10} \exp(\Gamma z) \quad .$$

For large Γz, however, the signal beam saturates at $I_{-1} = (I_{-10} + I_{+10})$, i.e. the total incident intensity can be transferred into the acceptor beam. The range of Γz where the intensities of the donor and acceptor beams are comparable corresponds to the saturation regime in ordinary quantum amplifiers.

The same beam can be either amplified or attenuated depending on its direction with respect to the other beam and the crystal c-axis. So the gain may be positive or negative.

In accordance with (2.1) the amplitude of the refractive index change[4] is independent of the light intensity and depends on the intensity distribution alone. That is why the expression for Γ does not include the light intensity (the latter appears only in $\sigma_{\mathrm{p}} \gtrsim \sigma_{\mathrm{d}}$; for most practical cases, $\sigma_{\mathrm{d}}/\sigma_{\mathrm{p}} \ll 1$).

Equation (2.4) is in a good agreement with experimental data for all the known crystals (e.g., Fig. 2.4). At low spatial frequencies, where $l_{\mathrm{s}}/\Lambda \ll 1$, the

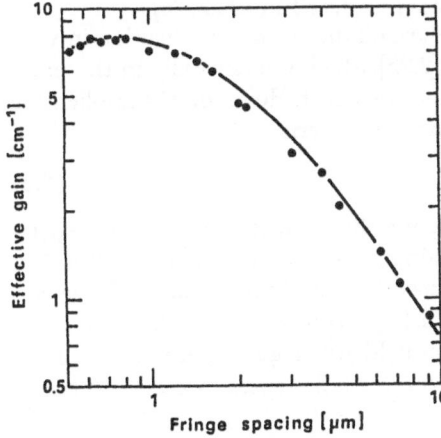

Fig. 2.4. Grating spacing dependence of effective exponential gain $\Gamma_1 = \Gamma - \alpha$ for $Ba_2NaNb_5O_{15}$ crystal. Solid line − results of calculation from (2.4) at $l_{\mathrm{s}} = 0.7\,\mu m$

---

[4] In holographic amplifiers, the refractive index modulation plays the same role as does the population inversion in ordinary laser amplifiers.

gain increases with decreasing Λ. In this approximation there are no free parameters in the theory, and Γ can be calculated with known r, n, λ, θ, and T. This stems from the fact that at a given temperature the diffusion field depends solely on the grating spacing Λ, i.e. on θ and λ. For small Λ, the best agreement between experimental points and theoretical calculations (solid line) is attained by varying the screening length l_s, which in turn yields the concentration of trapping centers [2.2]:

$$N_D^+ = \left(\frac{4\pi^2 \varepsilon \varepsilon_0 kT}{e^2 l_s^2}\right)^{1/2} \quad .$$

In particular, for $Ba_2NaNb_5O_{15}$ samples $l_s \cong 0.7\,\mu m$ and $N_D^+ = 5 \times 10^{16}\,cm^{-3}$ [2.20].

From (2.4) it follows that the sign of the gain depends on that of the photoexcited carriers. This sensitivity of the energy transfer to the sign of free carriers was first used by *Krätzig* and *Orlowski* [2.21] to prove that the mechanism of photoconductivity in $LiNbO_3$ differs for visible and ultraviolet light (electrons in visible, holes in UV).

The diffusion-nonlinearity gain is proportional to the carrier distribution gradient, and therefore the interaction of two counterpropagating beams with the smallest possible fringe spacing, $\Lambda_R = \lambda/2n$, is the most advantageous for beam coupling. Due to the PRC anisotropy, however, the actual electro-optic constant is generally lower than in the transmission geometry (e.g., for $LiNbO_3$ the ratio r_{33}/r_{13} is about 3), which reduces the advantage of using the reflection geometry. Obtaining the highest gain with the reflection geometry also calls for the use of materials where the screening length is much smaller than the fringe spacing, $l_s \ll \Lambda_R$. Experiments with $LiNbO_3$ demonstrated a considerable gain increase for the reflection geometry [2.22]. Specific features of reflection interactions have been analyzed in detail in [2.23, 24].

Table 2.1 summarizes experimental data on diffusion amplification in various materials, including PLZT ceramics [2.25] which are normally in the para-electric phase and display no linear electro-optic effect. However, the application of an external electric field E_0 gives rise to linear term

$$\Delta n \propto E_0 E_{sc} \quad , \tag{2.5}$$

where E_{sc} is the space charge field modulating the refractive index. The experiments demonstrated the possibility of attaining gains as high as $\Gamma \simeq 100\,cm^{-1}$.

Many PRC clearly exhibit effects caused by "hot" unthermalized free carriers, e.g. the bulk photovoltaic effect [2.26, 27]. "Hot" carriers may show up in a so-called "hot diffusion", where diffusion field E_D^h is governed not by sample temperature T, but by photoexcited electron (or hole) energy excess $\Delta \mathcal{E}$ before thermalization:

$$E_D^h = |K|(\Delta \mathcal{E}/e) \quad .$$

Table 2.1. Diffusion-type amplification in PRC

Crystal	Γ ($\Lambda=1\mu m$), (Γ_{max}) [cm^{-1}]	Experimental conditions: d [mm], (β_0)	λ, [μm]	l_s [μm]	$N_D \times 10^{16}$, [cm^{-3}]	Ref.
LiNbO$_3$ [a]	4 (12)	4 (0.0001)	0.441	0.4		2.2
LiNbO$_3$: Fe [b]	– (20)	– (0.16)	0.488	–	–	2.22
KNbO$_3$	1 (1)	3.3 (1)	0.515	–	0.59	2.42
BSO	0.4	10 (0.01)	0.515	0.7	1	2.42
BaTiO$_3$ [c]	13 (15)	2.2 (1)	0.515	0.68	0.7	2.101
SBN : Ce [d]	– (100)	1.48 (10^{-6})	0.441	–	–	2.7
SBN : Ce [e]	30 (50)	1.2 (–)	0.488	–	–	2.90
Ba$_2$NaNb$_5$O$_{15}$	7 (8)	4 (0.01)	0.488	0.7	5	2.20
PLZT	30 (100)	0.26 (1)	0.441	–	–	2.25
GaAs [f]	0.35 (0.4)	4 (0.1)	1.06	0.8	0.13	2.33

[a] Nominally pure sample (transmission grating)
[b] Data from experiment on nonlinear reflection of (001) sample
[c] Coupling of ordinary waves in symmetric scheme
[d] Data from scattering experiment
[e] Data from oscillation experiment
[f] response time 20 μs as distinct from all other cited PRC with slow nonlinear response

At $\Delta\mathcal{E} \gg kT$ a considerable increase in the gain may be expected [2.28]. Experimental difficulties in detecting this process stem from other strong photovoltaic-type nonlinearities that always accompany it.

Processes in which coherent light beams interact in PRC may be considerably affected by an additional recording of an amplitude dynamic grating [2.29]. Since the space charge is formed because of a charge transfer between impurity centres, its formation is always more or less associated with a spatial modulation of absorption. In LiNbO$_3$: Fe, for instance, in illuminated regions, Fe^{2+} ions strongly absorbing at $\lambda = 0.5\,\mu$m are converted into the nonabsorbing state Fe^{3+} (in dark regions the opposite occurs). A specific feature of PRC is that their amplitude gratings are always $\frac{\pi}{2}$ shifted with respect to their phase gratings (the absorptivity decrease is proportional to the space charge, while the field modulating the refractive index is, according to the Poisson equation, proportional to the integral of the charge distribution).

For a local response an additional shifted amplitude grating cannot give rise to beam coupling. For a nonlocal response, there is a difference between amplification (and attenuation) factors measured for each of the beams separately. The exponential gain for the acceptor beam can be increased by selecting the sign of the nonlinearity, but it should be pointed out that no gain in its intensity over that in a nonabsorptive phase medium can be attained. A dynamic theory of self-diffraction, including the effects of amplitude gratings, is presented in [2.30].

The great majority of studies of diffusion amplification have involved the use of broad-band ferroelectric and electro-optic crystals (LiNbO$_3$, KNbO$_3$,

LiTaO₃, BSO, BGO) whose dielectric relaxation time for illumination by conventional gas lasers is relatively long, e.g. fractions of a second and more. A considerable improvement in the sensitivity and response time can be expected by using semiconductors as the active media, thus giving a higher mobility compared to ferroelectrics, and linear electro-optic effect, such as that found in II–VI compounds [2.31]. The first report of a unidirectional energy transfer for a dynamic holographic recording was for a CdS crystal [2.32]. Photorefraction in semiconductors has been convincingly demonstrated most recently by the examples of GaAs [2.33] and InP [2.33, 34] crystals. This subject is discussed in detail in Chap. 6 of Vol. 61 of Topics in Applied Physics.

Some of the experimental results have not yet been fully explained. The theory [2.2] predicts the Γ value to be completely independent of the intensity ratio β_0 of the two beams. However, after the present authors' study on LiNbO₃, where a twofold decline in Γ over a β_0 variation from 0.001 to 10 was observed [2.2], qualitatively similar relationships have also been detected for KNbO₃ [2.35], BSO [2.36], Ba₂NaNb₅O₁₅ [2.20], and BaTiO₃ [2.37].

Another effect observed in BSO crystals and PLZT ceramics is a strong dependence of the experimentally measured Γ on the sample thickness. In BSO, for example, increasing the thickness from $d = 0.127$ cm to $d = 1$ cm reduces Γ from 12 to 7 cm⁻¹ [2.38]. The explanation of both discrepancies with the theory should probably be based on allowing for the thickness dependence of the light-induced scattering of a strong beam into noise components [2.39].

2.2.2 Amplification at Large Transport Lengths

The nonlinear response of a PRC in a moderate external electric field ($E_0 \ll E_q$) is local. Due to the unidirectional transport of photoexcited carriers, the space charge distribution maxima are $\frac{\pi}{2}$ out of phase with respect to the light fringes, and the space charge field, according to the Poisson equation, is in turn $\frac{\pi}{2}$ shifted with respect to the maxima of the space charge itself, i.e. it is either in phase or in antiphase with the light fringes.

For strong external fields ($E_0 \gtrsim E_q$) the situation changes radically: the spatial distribution of photoexcited carriers becomes virtually uniform. The space charge maxima then coincide with those of ionized donors, i.e. with the light fringes, whereas the space charge field is phase shifted, so that the nonlinear response is nonlocal.

The gain in this case is expressed as follows

$$\Gamma = \frac{2\pi n^3 r_{33} E_q \cos\theta \cos 2\theta}{\lambda(1 + \sigma_d/\sigma_p)} \quad , \tag{2.6}$$

where $E_q = N_D^+ e\Lambda/2\pi\varepsilon\varepsilon_0$ is the limiting space charge field. The sign of Γ is independent of the external electric field direction, but, as in the diffusion case, depends (through E_q) on the sign of mobile charges.

This type of response to the electric field was first observed in KNbO₃ crystals by *Krumins* and *Günter* [2.40, 41] and then also in BSO crystals [2.36]. These experiments are described in detail in [2.42].

2.2.3 Amplification Due to Circular Photovoltaic Currents

The third presently known nonlocal response is the photovoltaic response with $\frac{\pi}{2}$ shift of the current discribution with respect to interference pattern. It arises in some PRC which belong to gyrotropic symmetry classes and therefore allow the appearance of so-called circular photovoltaic currents, i.e. currents that can be excited only by circularly polarized light and whose directions depend on the direction of polarization rotation [2.27]:

$$J_i = i\beta^a_{ijk}\tilde{E}_j\tilde{E}^*_k = i\gamma_{im}[\tilde{E}\tilde{E}^*]_m \quad . \tag{2.7}$$

Here $\tilde{E}$ is the electric field amplitude of the light wave, β^a_{ijk} is the anti-symmetric component of the third-rank photovoltaic tensor (which is sometimes written in another form as the component of second-rank tensor γ_{im} similar to the gyration tensor).

The orthogonally polarized light beams incident on a crystal at a certain angle give rise to a stationary grating of spatially modulated currents (Fig. 2.5), whose extrema are $\frac{\lambda}{4}$ shifted with respect to those of the incident beams[5]; hence, the circular photovoltaic current records a shifted grating.

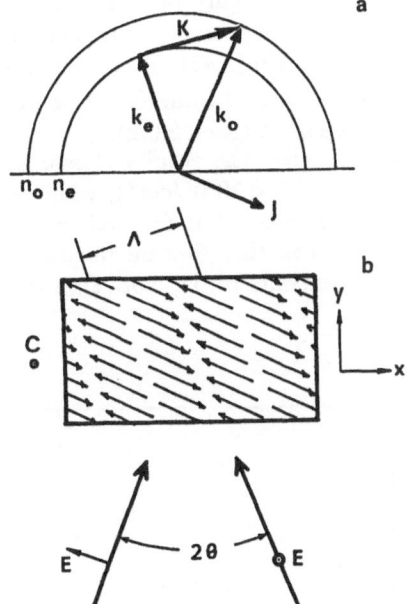

Fig. 2.5. (a) vector diagram of ordinary and extraordinary wave interaction in birefringent PRC; (b) spatial distribution of spatially oscillating current in sample

---

[5] For orthogonally polarized waves with no intensity modulation, the definition of light fringe "maxima" becomes ambiguous and calls for explanation. Since the result of two-beam interaction on a dynamic hologram depends ultimately on the phase relation between the zero-order wave and the diffracted wave propagating in the same direction (Sect. 2.1.1), isophase surfaces with a phase difference of $2p\pi$ (where p is an integer) should be regarded as the interference pattern "maxima".

The gain Γ for this mechanism is as follows [2.17]:

$$\Gamma = 2\pi n_o^2 n_e r_{51}(se_o)^2 \beta_{15}^a (\lambda \kappa \cos \theta)^{-1} \quad , \tag{2.8}$$

where $s = |k|/k$ is the grating unit vector; $\kappa = \sigma_p/I$ is the specific photo-conductivity and e_o is the ordinary wave unit vector. The factor $(se_o)^2$ allows for the difference in current propagation and grating vector directions, tending to zero for small intersection angles or for counterpropagating beams, where the grating wavevector is perpendicular to transverse axes. In both cases the grating wavevector is essentially parallel to the crystal optical axis, whereas the circular photovoltaic current propagates in a plane normal to this.

The main distinguishing feature of this amplification mechanism is that, here, the energy transfer direction is conditioned by the polarization of interacting beams rather than by the orientation of crystal polar axes and is unambiguously determined by the sign of the product of the corresponding electro-optic constant via the antisymmetric component of the photovoltaic tensor $(r_{ijk}\beta_{lmn}^a)$. When the product is negative, energy is transferred from the extraordinary to the ordinary wave, and vice versa.

Extraordinary wave amplification in $LiNbO_3$: Fe crystals for various argon laser lines $(\beta_{131}^a < 0)$ has been measured experimentally (Fig. 2.6a). The frequency spectrum of β_{131}^a calculated with (2.8) is roughly proportional to the absorption spectra of iron-doped samples. Ordinary wave amplification $(\beta_{131}^a > 0)$ was noted for a similar material, $LiTaO_3$: Cu [2.43]. The gains are within $10\,\mathrm{cm}^{-1}$, while the Γ value, which can be extracted from experimental data on oscillation [2.44], is much higher: $\Gamma \simeq 100\,\mathrm{cm}^{-1}$ (Sect. 2.6.2).

The two experiments differ in the power density: the results shown in Fig. 2.6 have been obtained with unfocused beams $(I_0 \gtrsim 10\,\mathrm{W/cm}^2)$, whereas the oscillator has been pumped by the same laser beam, but focused in the sample by a $F = 25\,\mathrm{cm}$ lens $(I_0 \simeq 10^3\,\mathrm{W/cm}^2)$. It can therefore be suggested that the discrepancy stems from the intensity dependence of β^a. Note that the

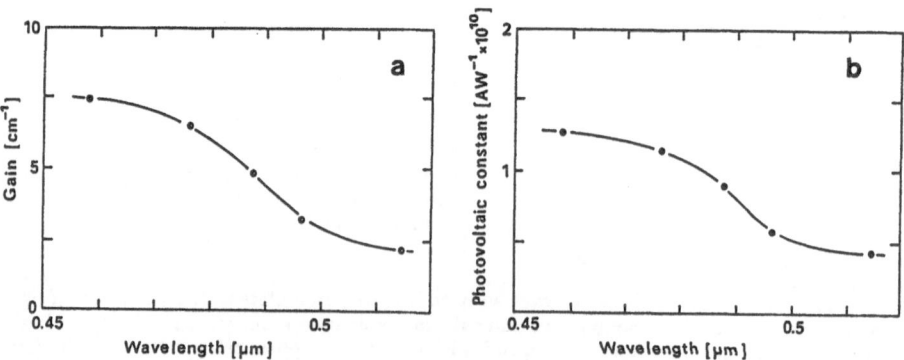

Fig. 2.6. (a) gain spectrum for extraordinary wave in $LiNbO_3$: Fe 0.02 wt. % sample: (b) antisymmetric component $|\beta_{131}|$ of the photovoltaic tensor calculated from (a)

$\beta = \beta(I)$ dependence has also been observed for the linear photovoltaic effect [2.45], but it was not so strong.

Circular photovoltaic currents have been found in many other materials [2.46–48], in particular in sillenite crystals [2.48] which show no natural birefringence. However, because of the high conductivity of these materials, the effective compensating photovoltaic field

$$E_{\mathrm{phv}} = -j^{\mathrm{phv}}/\sigma_{\mathrm{p}} \quad . \tag{2.9}$$

turns out to be negligible and brings about no appreciable photorefraction. The situation could perhaps be improved by cooling the crystal down to liquid nitrogen temperature or introducing appropriate carrier trapping centres.

2.3 Steady-State Amplification Due to Degenerate Parametric Interactions

Energy transfer, apart from the forward two-beam transfer described in Sect. 2.2, can also result from simultaneous interaction of several (three or four) beams. Each beam pair records its own grating which is spatially shifted relative to that recorded by the other beam pair (Sect. 2.1.2). The phase-matching condition here is exactly the same as the condition for a simultaneous Bragg diffraction of different light waves on a phase grating resulting from the interaction.

Four-beam interactions are often called parametric, since any two beams, when recording a grating, modulate the interaction parameter and the phase of the grating that appears upon interaction of the other beam pair. A steady-state energy transfer due to a parametric four-beam interaction is accomplished for both local and nonlocal responses. In the latter case the interaction result depends on the two-beam energy transfer direction: at a diffusion-nonlinear degenerate backward four-wave mixing (BFWM) the transmitted signal beam may be either amplified or attenuated [2.49]. It should be emphasized that, in contrast to well-known three-wave parametric processes (second-order nonlinearity $\chi^{(2)}$ [2.50]), the parametric interactions under consideration here are degenerate four-wave mixing processes based on third-order nonlinearity $\chi^{(3)}$.

Parametric interactions are of particular practical interest since they enable not only amplification of the signal beam, but also generation of a beam phase-conjugated with the latter.

2.3.1 Degenerate Backward Four-Wave Mixing

Degenerate BFWM is now extensively used to conjugate laser beam wavefrons [2.11, 12] (Fig. 2.2). It differs favourably from other parametric interactions in its extremely large signal beam acceptance angle: for strictly counterpropagating pump beams of an equal intensity, the phase-matching condition

$$k_4 + k_3 = k_1 + k_2 \tag{2.10}$$

imposes no restrictions at all on the direction of signal beam 4.

The signal beam output intensity in the undepleted pump approximation for transmission gratings is [2.3]:

$$\frac{I_4(d)}{I_4(0)} = \left| \frac{\exp(-\gamma d/2)}{\cosh[(\gamma d/2) - \ln r/2]} \right|^2 \frac{(1+r)^2}{4r} \quad , \tag{2.11}$$

where r is the ratio of pump beam intensities and $\gamma = i\omega \Delta n \exp(i\Phi)(2c \cos \theta)^{-1}$.

For local-response media ($\Phi = 0$ or π) at $r = 1$ and for small phase modulation γd :

$$I_4(d)/I_4(0) \cong 1 + (|\gamma|d/2)^2 \quad , \tag{2.12}$$

i.e. the signal beam is amplified as the square of the phase modulation, whatever the crystal orientation.

For nonlocal-response media ($\Phi = \pm \frac{\pi}{2}$):

$$I_4(d)/I_4(0) \cong 1 - \gamma d \quad , \tag{2.13}$$

i.e. the amplification varies linearly with the phase modulation and therefore changes into attenuation when the crystal axis direction is reversed. Signal beam amplification and attenuation in the four-wave interaction was first observed in a nominally pure lithium niobate crystal [2.49] with a diffusion-type response.

This sensitivity to the sign of γ is qualitatively accounted for by the fact that energy transfer due to a direct interaction of two recording beams is allowed for nonlocal-response media even in the first approximation ($\Gamma = 2\gamma$), whereas for local-response media it only appears in the second approximation, $\propto (\gamma d/2)^2$ as a consequence of multi-beam parametric interaction.

It should be pointed out that all the preceding considerations are valid only for a relatively weak interaction ($\gamma d \ll 1$). If this condition is not met, then higher-order terms, whose contribution may turn out to be decisive, should be retained in (2.13).

From (2.11) it follows that at a local response the maximum signal beam amplification is attained at $r = 1$, i.e. for a symmetric pumping, whereas for a nonlocal response the optimum is an ordinary two-beam interaction, i.e. $r = 0$, (Fig. 2.7), where the linear addition to the intensity of the amplified beam is twice that in (2.13), i.e. $2\gamma d$.

Figure 2.8 shows calculated dependences of the amplified signal beam intensity on the coupling strength for two-wave interaction in a nonlocal-response medium and for four-wave mixing in a local-response medium for a strong depletion of pump beams. As can be seen, in the first case the pump beam intensity can be fully transferred to the signal beam. For FWM the intensity of each pump beam is transferred to amplified signal or to phase conjugate beam. As a result in this case too the whole pump beam intensity can be totally transferred to the signal and phase-conjugate beams.

The difficulty in BFWM of simultaneously amplifying the oppositely directed signal and conjugated beams, which is due to the opposite signs of their wavevector projections on the polar axis, typical for diffusion-type response

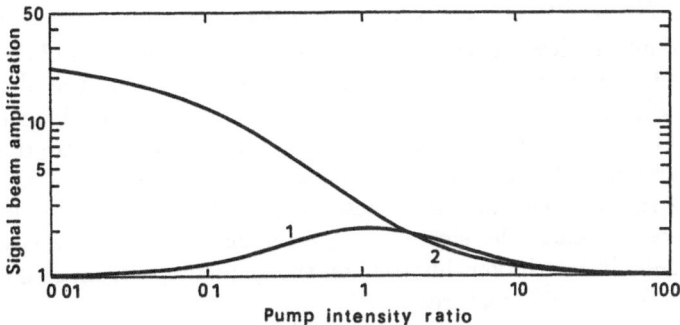

Fig. 2.7. Signal beam amplification γ_0 versus pump beam intensity ratio r for local *(1)* and nonlocal (diffusion-type) *(2)* nonlinear response

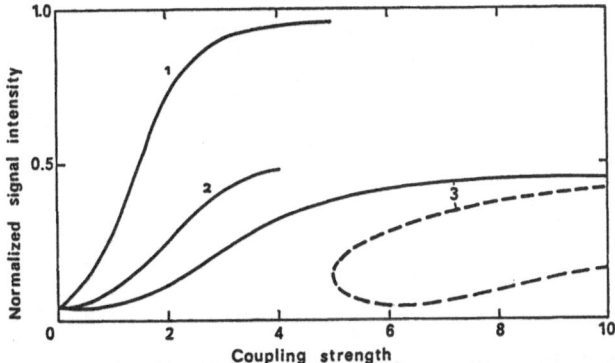

Fig. 2.8. Coupling strength dependence of output signal wave intensity normalized to total intensity of pump beams for *(1)* two-beam interaction in nonlocal-response medium and degenerate *(2)* forward and *(3)* backward four-wave mixing in local-response medium

media, can in some cases be overcome by employing setups which rotate the polarization of one beam pair with respect to the other. In BSO (and similar) crystals cut so that the holographic grating field is oriented along $[1\bar{1}0]$, the refractive index changes for the two polarizations are of opposite sign. This allows one to spatially superimpose (and thereby amplify) the gratings recorded by each pair of interacting beams, the signal beam intensity in the nondepleted pump approximation taking the form of [2.51]:

$$\frac{I_4(d)}{I_4(0)} = \left| \frac{\Gamma d}{4 - \Gamma d} \right|^2 . \tag{2.14}$$

This expression turns out to be valid also for nonlocal photovoltaic nonlinearity due to circular currents [2.52]. Transmission gratings recorded by two pairs of interacting beams are also superimposed here. A distinguishing feature of the four-wave mixing (FWM) with such a nonlinearity is that the maximum gain is attained at equal pump beam intensities.

2.3.2 Degenerate Noncoplanar Forward Four-Wave Mixing

This type of FFWM interaction is attained when the two pump beams and the signal beam, incident on one side of nonlinear medium, lie on the surface of a right circular cone whose vertex angle equals that between the pump beams [2.53, 54] (Fig. 2.9). The interaction gives rise to beam 3 which lies opposite the signal beam with respect to the cone axis and whose wavefront is conjugated with respect to the signal beam, and the weak signal beam is amplified. In contrast to the preceding case, the signal beam direction is rather strongly restricted, which results in restrictions imposed on the angular spectrum of the signal beam and in other specific features.

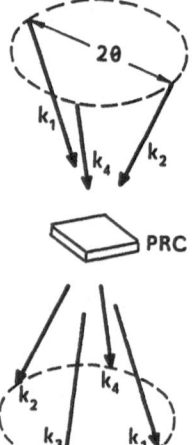

Fig. 2.9. Schematic illustration of forward four-wave mixing

The possibility of amplification is, as in other parametric processes, associated here with the $\frac{4}{4}$ spatial shift of the gratings recorded by two pairs of interacting beams. The phase-matching condition for this interaction is the same as (2.10).

The intensity of a signal beam leaving a local-response nonlinear medium at $I_3(0) \ll I_1(0)$ is expressed by [2.53]:

$$I_3(d) = I_3(0)\left[1 + \frac{16r}{16r - (1+r)^2}\frac{\sinh^2(\sqrt{3/4}\gamma d)}{1+B}\right] \quad , \tag{2.15}$$

where:

$$B = \frac{32I_3(0)r\cosh^2(\sqrt{3/4}\gamma d)}{I_1(0)(1+r)[16r - (1+r)^2][1 - 2\sqrt{1 - 4r/(1+r)^2}/3]} \quad ;$$

$$\gamma = \frac{\omega\Delta n}{2c\cos\theta} \quad .$$

For a weak interaction ($B \ll 1$) the signal beam intensity increases in proportion to $\sinh^2(\sqrt{3/4}\gamma d)$, i.e. as the square of the phase modulation, as for other parametric processes. At large phase modulations the signal beam intensity reaches its maximum value

$$I_3(d) \cong I_3(0) + \tfrac{1}{2}I_1(0)(1+r)\frac{1 - 2\sqrt{1 - 4r/(1+r)^2}}{3} \quad , \qquad (2.16)$$

i.e. it approaches the pumping intensity at equal pump beam intensities. Figure 2.8 shows the calculated dependence (2.15) for $I_4(0)/I_1(0) = 0.2$ and $r = 1$. Up to $\gamma d \simeq 5$ the approximate solution shows a fair agreement with the rigorous one from [2.53]. The calculation of Fig. 2.8 took into account that only transmission (or only reflection) gratings can be recorded in PRC and that γ has the same meaning as in (2.11–13), but can assume only real values corresponding to a local response.

The data shown in Fig. 2.8 indicate that the FFWM is more efficient than the BFWM: the same gain is attained in the former case at a lower coupling strength.

FFWM has been experimentally accomplished with LiNbO3 crystals [2.53, 54]. A typical time variation of the signal beam is shown in Fig. 2.10. As seen, its maximum value is seven times its initial one.

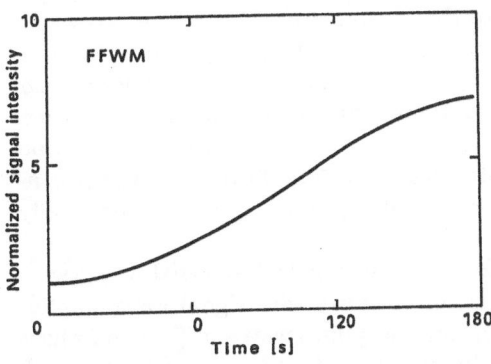

Fig. 2.10. Time dependence of output signal intensity for FFWM in LiNbO$_3$: Fe 0.02 wt. % ($\lambda = 0.44\,\mu$m; $I_1 = I_2 = 10^{-3}$ W/cm^2; $I_4 = 0.01 I_1$)

2.3.3 Degenerate Coplanar Forward Four-Wave Mixing for Orthogonally Polarized Beams

The symmetry properties of PRC of the $3m$, $4m$, 23, etc. classes permit anisotropic diffraction [2.55], in which the diffracted beam turns out to be polarized orthogonally to the reconstructing one. With birefringent crystals, the moduli of wavevectors of differently polarized beams are not equal, which makes it possible to meet the phase-matching condition (2.10), leaving the beams in one plane (coplanar interactions).

Further prerequisites for amplification are the feasibility of grating recording by each pair of orthogonally polarized beams (e.g., k_1 and k_4, k_2 and k_3) (Sect. 2.2.3) and the possibility of diffraction of one beam from the grating recorded by another pair of beams (in our example, k_2 from the grating $k_1 - k_4$ and k_1 from the grating $k_2 - k_3$).

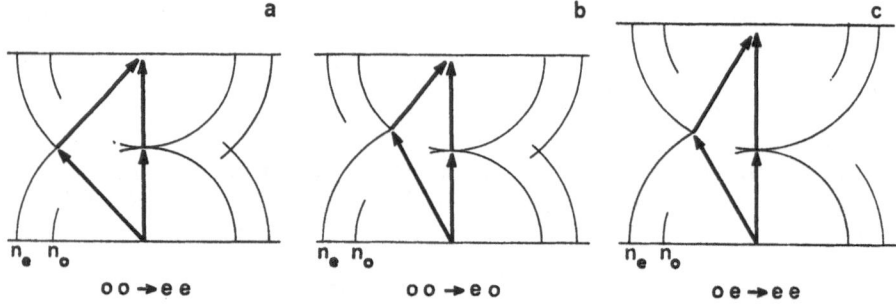

Fig. 2.11. Vector diagrams for FFWM in LiTaO$_3$ ($n_e > n_o$)

A typical example is the oo-ee interaction [2.43] in LiTaO$_3$: Cu crystals (Fig. 2.11a). Ordinary wave 1 is normally incident (for simplicity) upon the sample surface and extraordinary signal wave 4 is incident at an angle

$$\theta \simeq \sin^{-1} \sqrt{n_e^2 - n_o^2} \ , \tag{2.17}$$

both waves being in a plane normal to the crystal optical axis. The two waves give rise to a spatially oscillating photovoltaic current which generates a space charge grating with wavevector $\mathbf{K} = \mathbf{k}_o - \mathbf{k}_e$ shown in the figure. Since the linear photovoltaic effect and Pockels effect tensors are of a similar structure, there will always exist a suitable electro-optic constant which will provide diffraction from the grating recorded by the photovoltaic currents. In the case under consideration, the diffraction is due to component r_{51} which couples orthogonally polarized waves [2.55].

A wave with a conjugated wavefront, symmetric to the signal wave (with respect to the pump wave), emerges simultaneously with signal wave amplification. The grating recorded by the signal and pump waves is $\frac{4}{4}$ out of phase relative to the grating recorded by the conjugated and pump waves, which makes possible a parametric amplification.

A calculation [2.56] using the undepleted pump approximation shows the scattered light intensity in this case to be:

$$I_s = I_{s0} \exp(\Delta G d) \cosh^2(G_0 d) \ , \tag{2.18}$$

where I_{s0} is the initial noise scattering intensity,

$$\Delta G \propto \beta_{131}^a \sin^2 \varphi \ , \quad \text{and} \tag{2.19}$$

$$G_0 \propto \sqrt{(\beta_{131}^c)^2 + (\beta_{131}^a)^2} \sin^2 \varphi \ . \tag{2.20}$$

As seen, (2.18) contains an exponent allowing for the direct beam coupling due to the nonlocal photovoltaic nonlinearity (Sect. 2.2.3) as well as a hyperbolic cosine characterizing the parametric process.

A direct beam coupling may, depending on the sign of β_{131}^{a}, either amplify or attenuate the scattered wave, whereas a parametric process is insensitive to the sign of the nonlinearity and invariably amplifies the wave.

In LiTaO$_3$: Cu crystals a direct beam coupling attenuates the extraordinary wave ($\beta_{131}^{a} < 0$), and therefore the oo-ee scattering (Fig. 2.11a) observed in this material is evidence that symmetric component β_{131}^{c} of the photovoltaic tensor greatly exceeds antisymmetric one, β_{131}^{a}.

In negative birefringent crystals ($n_o > n_e$), other three-beam parametric interactions of the oo-oe, eo and oe-ee types (Fig. 2.11b,c) are also possible, their phase-matching angles being determined by the conditions:

$$\theta_2 \simeq \sin^{-1}\sqrt{n_e^2 - (n_o + n_e)^2/2} \quad ; \tag{2.21}$$

$$\theta_3 \simeq \sin^{-1}\left(\sqrt{10n_o^2 n_e^2 - 9n_o^4} - n_e^4/4n_o\right) \quad . \tag{2.22}$$

The possibility of recording a grating with the same wavevector using both a pair of identically polarized waves and a pair of orthogonally polarized waves is a prerequisite for these processes. As the linear photovoltaic current due to β_{333} propagates along the z-axis, and the spatially oscillating currents due to β_{131}, along transverse axes, the maximum scattered light intensity for processes (2.21, 22) corresponds to an azimuth angle $\varphi \simeq 45°$.

2.4 Quasi-Degenerate Two-Beam Interaction

For the nondegenerate case a direct two-beam coupling in the inertial local-response medium becomes possible (Sect. 2.1.3). A frequency difference is attained either by phase-modulating one of the beams [2.16, 57, 58] or by displacing the nonlinear medium in the course of recording. The phenomena also include a transient energy transfer between beams of different intensity in a local-response medium.

2.4.1 Amplification of Difference-Frequency Waves

The gain in a quasi-degenerate interaction for a third-order (local) nonlinear medium with relaxation time τ (Fig. 2.3a) takes the form [2.59]:

$$\Gamma = \left(\frac{\omega}{\sqrt{\varepsilon_0}c}\right)\mathrm{Im}\{\delta\varepsilon_{\mathrm{eff}}\} = \frac{4\pi\Delta n(\Omega)\Omega\tau}{(1 + \Omega^2\tau^2)\lambda} \quad , \tag{2.23}$$

where $\Omega = \omega_1 - \omega_2$ is the frequency difference between interacting waves. With increasing Ω, $\Phi \to \frac{\pi}{2}$, but at the same time Δn and hence also Γ drop to zero. $\Gamma(\Omega)$ assumes a maximum value at $\Omega_{\mathrm{opt}} = \tau^{-1}$:

$$\Gamma^{\mathrm{max}} = 2\pi\Delta n(\Omega_{\mathrm{opt}})/\lambda \quad . \tag{2.24}$$

As seen, the exponential gain is independent of the spatial frequency of the recorded grating. At Ω_{opt}, the optimum interference pattern displacement speed is automatically attained:

$$v_{\mathrm{opt}} = \Omega_{\mathrm{opt}}/|\boldsymbol{K}| = (|\boldsymbol{K}|\tau)^{-1} \quad , \qquad\qquad (2.25)$$

which is a function of $|\boldsymbol{K}|$. For a wide signal beam spectrum it is preferable to modulate the phase of the reference wave [2.16]; this allows v_{opt} to be attained for all the partial gratings simultaneously. With amplification due to displacement of the active medium, the displacement speed should be selected from condition (2.25) for the mean carrier spatial frequency of the hologram.

The energy transfer direction, i.e. the sign of Γ, depends on the properties of the medium [on the sign of $\Delta n(\Omega)$] and also on the direction (towards higher or towards lower frequencies) of the pump beam detuning (i.e. on the sign of Ω).

Amplification in a quasi-degenerate interaction has been experimentally investigated by *Huignard* [2.38, 57, 60]. A local response of BSO and BGO crystals is attained by applying relatively weak external fields $E \ll E_q \simeq 42\,\mathrm{kV/cm}$ (at $\Lambda = 10\,\mu\mathrm{m}$). An exponential gain of the signal beam, linear in applied field E_0, was observed up to $E_0 \simeq 8\,\mathrm{kV/cm}$, after which the increase rate slowed down. A gain $\Gamma \simeq 2.2\,\mathrm{cm}^{-1}$ was attained. Amplification in the absence of the electric field at the same displacement speeds was negligible. Optimization of the interaction conditions can result in a considerable increase of the gain to as much as $\Gamma \simeq 12\,\mathrm{cm}^{-1}$ [2.38].

This result confirms the general assertion that the optimum shift between the grating and the light fringes should be of $\frac{\pi}{2}$ because $E_0 > E_q$ (Sect. 2.2.2), thus any technique which introduces an additional artificial shift will only attenuate the two beam energy transfer. But if the initial shift is of $0°$ (field $E_0 \gtrsim 7\,\mathrm{kV/cm}$ is applied), then increasing the frequency difference produces an increase in Φ and thereby enhances the effect.

2.4.2 Transient Energy Transfer

Self-diffraction of two light beams of dissimilar intensities (but of identical frequencies) in a local-response medium results in dissimilar nonlinear changes of their phases, i.e. the effective refractive indices for the weak and for the strong wave turn out to be different.

The transient process of recording consists of three distinct stages. In the initial stage ($t = 0$) (Fig. 2.12a) the grating efficiency is as yet so low that nonlinear phase changes are insignificant, and the vector of the interference pattern $\boldsymbol{K}$ is, as in a free space, perpendicular to the bisector of the angle between the beams.

As the grating is being recorded, at $t \simeq \tau$ diffracted beams arise and rotation of isophase surfaces inside the medium occurs (Fig. 2.12b). The grating lags behind the interference pattern giving rise to a spatial mismatch between the grating and the fringes, which increases with the medium depth. With

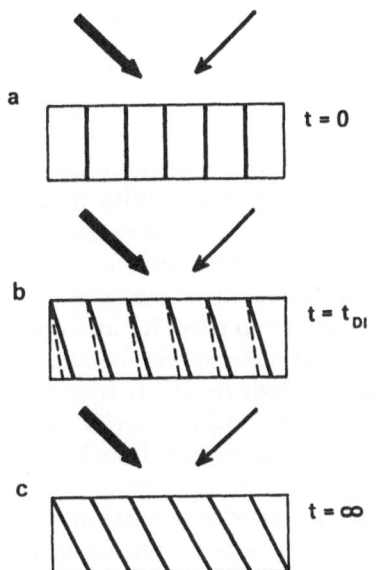

Fig. 2.12a–c. Time development of grating and light fringe positions in inertial local-response medium: **(a)** $t = 0$: there is no grating, and fringes are parallel to bisector of interacting beams; **(b)** $t \simeq \tau_{DI}$: both fringes and grating are tilted with some relative spatial mismatch; **(c)** $t \gg \tau_{DI}$: grating extrema and fringes are exactly superposed and tilted with respect to incident beam bisector

$\Delta n > 0$ the fringes rotate towards the positive projection of the strong beam wavevector (Fig. 2.12b,c), and with $\Delta n < 0$, towards the positive projection of the weak beam wavevector, but the grating shift with respect to the fringes always coincides with the positive projection of the weak beam wavevector. As a result, the energy transfer at the initial stage, whatever the sign of the nonlinearity, always proceeds from the more intense to the less intense beam. At large phase modulations the energy transfer may become oscillatory.

At $t \to \infty$ and under steady-state conditions the inclination angles of iso-hase planes of both the fringes and the grating become equal so that all energy transfer ceases (Fig. 2.12c).

Transient energy transfer in ferroelectrics was first discovered in lithium niobate placed in an external field [2.61]. As predicted by the theory, the weak beam was always amplified, regardless of the direction of the applied field; the effect increased with the square of the field; and at $t \gg \tau$ the energy transfer ceased. It was noted that at the initial stage of recording the increase in the acceptor beam intensity was proportional to its initial intensity and to the square of the exposure time. This experiment spurred the development of theory [2.62], which resulted in the derivation of an expression for the initial energy-transfer kinetics:

$$I_{\pm 1} = I_{\pm 10} \mp \left(\frac{2\pi d r_{33} n_e^3 E_0}{\lambda \cos \theta} \right)^2 \frac{1 - \beta_0}{1 + \beta_0} \frac{t^2 I_{+10} I_{-10}}{\tau^2 I_0} \quad . \tag{2.26}$$

A qualitative analysis of the set of equations describing the transient energy transfer also made it possible to explain the oscillatory nature of the intensity redistribution at large phase modulations and to estimate the period of the intensity oscillation [2.62]:

$$T = \frac{8\tau\lambda\cos\theta}{dr_{33}n_e^3 E_0} \ .$$ (2.27)

The maximum transient energy transfer was attained with iron-droped lithium niobate crystals with a photovoltaic nonlinearity [2.63]. With 0.4 cm thick samples the weak beam was amplified by a factor of over 2500; essentially the entire energy of the strong beam was pumped into the acceptor beam up to the initial intensity ratio of $1:1000$ (Fig. 2.13). It should be pointed out that the approximate expression (2.26) predicting the maximum effect for an initial beam intensity ratio $\beta_0 = 1 : 3$ is not valid for these extreme circumstances [2.61]. Such a strong shift of the maximum towards high β_0 stems from the tremendous phase modulation typical for doped crystals, which can amount to tens of π ($E_{\mathrm{pgv}} \cong 100\,\mathrm{kV/cm}$). This suggestion was substantiated by a numerical simulation of the experiment [2.64].

Further theoretical studies of the transient energy transfer were conducted by *Solymar* [2.65, 66] who developed a different mathematical technique to solve time dependent equations.

Employing the transient energy transfer to amplify real beams and images requires an analysis of its frequency-contrast and intensity characteristics. As in parametric interactions, the amplification here is not exponential [refer to (2.26)]. The relative increase in the weak beam intensity, $\Delta I_{-1}/I_{-10}$, is independent of both spatial frequency and light intensity, and, provided the donor beam is much stronger than any spectral component of the acceptor beam, also of the beam intensity ratio. Then, with (2.26) applicable, a transient energy-transfer-based amplifier is ideal since it introduces no distortions in the distribution of amplitudes of individual spectral components $\Delta I_{-1}(K) \sim I_{-1}(K)$, or in the phase ratio between the individual components (all the components have

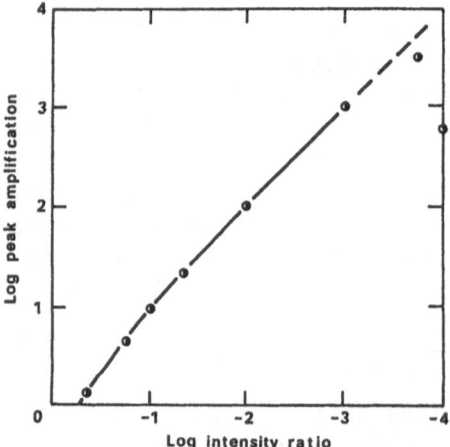

Fig. 2.13. Transient peak amplification γ_0' versus interacting beam intensity ratio β_0, in LiNbO$_3$: Fe 0.03 wt. % ($\lambda = 0.44\,\mu$m; $I_0 = 10^{-3}$ W/cm^2)

the same addition in the absolute value of the phase) [2.62]. At the same time the nonlinear dependence of the transient energy transfer on beam intensity ratio β_0 (2.26) allows it to be employed for image processing, e.g. for filtering out the low-frequency components and for improving image contrast [2.67].

Transient energy transfer with violation of quasi-neutrality in the external field was analyzed in [2.68].

2.5 Light-Induced Scattering

Quite a number of stimulated scattering processes in nonlinear optics differ from an ordinary scattering in that the number of scattering centres or the scattering mode is a function of the light intensity. Prerequisites for nonlinear scattering are a high enough amplification and an initiating noise. Several different nonlinear light scattering processes have been observed in PRC [2.7, 2.69–78]; they are readily apparent because of the considerable gain provided by the respective amplification processes (Sects. 2.2–4). Under steady-state conditions more than half the intensity of the light beam incident on a sample is often transformed into scattered radiation [2.7, 70, 78].

Three likely sources of coherent noise have been mentioned in literature: (i) finite angular spectrum of recording beams [2.69]; (ii) optical imperfections of the sample; and (iii) fluctuations of the photovoltaic constant and hence also of the gain, stemming from a nonuniform doping of the sample [2.79, 80]. The third cause is typical for doped samples where photorefraction is brought about by the photovoltaic effect whose constant fluctuates due to a nonuniform distribution of the dopant throughout the sample [2.26, 27]. Therefore, even for a uniform illumination of the sample, space charge field macro-nonuniformities arise, which cannot be optically "erased" (in contrast to ordinary gratings recorded in the same crystal through a nonuniform light intensity distribution), and the resulting crystal refractive index changes show up as static rather than dynamic nonuniformities.

Scattering phenomena deserve close attention since, firstly, amplification mechanisms existing in PRC are capable of providing such great coupling strengths across a few millimeter thick nonlinear layer ($\Gamma d \gtrsim 20$) that the intensity of scattered beams becomes comparable with that of the incident one and, secondly, it is the coherent scattered radiation that is the agent initiating oscillation on dynamic gratings (Sect. 2.6).

2.5.1 Asymmetric Scattering in Crystals with Considerable Diffusion-Type Amplification

Because of the anisotropy of photorefractive crystals themselves, of initiating scattering, and of recording and reconstructing processes, the light-induced scattering in PRC is never isotropic, but is localized near certain planes or directions, Assuming the initiating noise scattering in the region of the most intense light-induced scattering to be isotropic, it may be expected that the angular structure of scattered light will be primarily conditioned by the angular spectrum of amplification. This supposition has been fairly well substantiated in studies of the light-induced scattering in materials displaying a considerable diffusion-type amplification [2.7, 69]. Thus, in SBN crystals, for which the estimated gain is positive and over $100 \, \text{cm}^{-1}$[6], helium-cadmium laser beam illumination produced scattered radiation that was concentrated in the positive direction of the polar axis. At $\lambda = 0.63 \, \mu\text{m}$ the gain, and with it also the scattered light intensity, declined, although an asymmetry of the scattering along the axis and counter-axis directions was still clearly visible (Fig. 2.14).

Similar scattering was later observed in (100) samples of barium titanate [2.69], which also offer high gains of up to $40 \, \text{cm}^{-1}$, and in PLZT ceramics with $\Gamma = 100 \, \text{cm}^{-1}$ under an external field of $10 \, \text{kV/cm}$ [2.25]. In the latter case the sign of Γ depends on the direction of the applied field, so that inverting the voltage changes the scattering pattern to its mirror image.

In the cases considered, both the exciting wave and the scattered waves were extraordinary waves in a crystal. However, in some cases the scattering pattern is greatly complicated by crystal birefringence and the resulting change in polarization of the light in the course of its propagation [2.72]. Thus, back-scattered radiation in (001) $\text{LiNbO}_3 : \text{Fe}$ samples, depending on the incident beam polarization, either has the shape of a bright cross or is localized over the surface of a cone [2.72] (Fig. 2.15). The angular distribution of amplification for this case is smooth, with no specific features [2.81], and differences in intensities of individual scattered light components stem from variations in the effective length of interaction in the crystal. For the brightest portions of the cross in

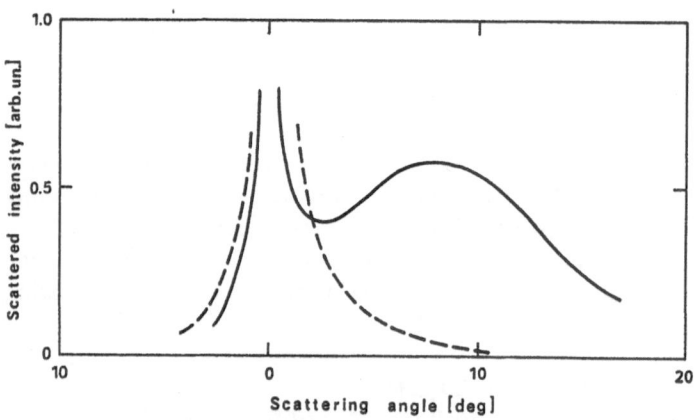

Fig. 2.14. Angular dependence of scattered light intensity in plane containing sample optic axis for SBN crystal for $\lambda = 0.44 \, \mu\text{m}$ (solid line) and $\lambda = 0.63 \, \mu\text{m}$ (dashed line)

---

[6] The gain $2\gamma_0'$ adopted in [2.7] differs from that used in the present paper in that it contains no angular dependence; the two values are related by the equation $\Gamma = 2\gamma_0' n \sin 4\theta$.

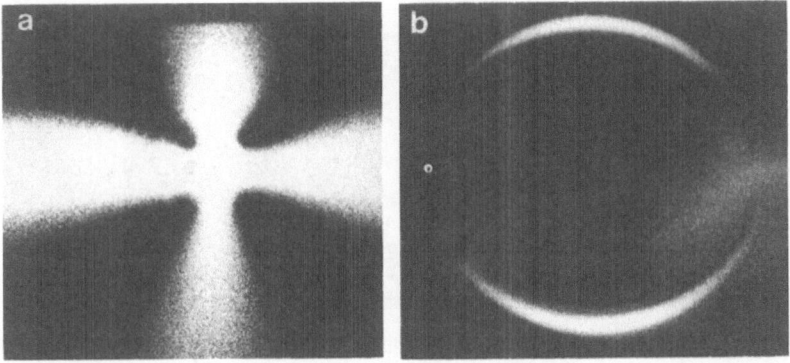

Fig. 2.15a,b. Far-field distribution of back-scattered light for (001) LiNbO$_3$: Fe 0.05 wt. % crystal with incident light polarized (a) in plane containing sample c-axis and (b) at 45° to this plane

Fig. 2.15, the electric field of the scattered waves oscillates in a plane coinciding with the plane of oscillation of the electric vector of the incident wave, and the effective interaction length coincides with the crystal thickness. However, for other directions, the scattered wave polarization is not identical everywhere to that of the incident wave, so that the effective interaction length, and hence also the coupling strength Γd, decline.

2.5.2 Polarization-Anisotropic Scattering by Photovoltaic Amplification

In LiNbO$_3$: Fe crystals a focused wave polarized normally to the optic axis gives rise to orthogonally polarized steady-state scattering (Fig. 2.16) which is predominantly localized in a plane perpendicular to the optic axis; its intensity drops to zero in the incident wave direction [2.70].

This type of scattering was first reported in [2.76] and subsequently explained in [2.70] as being associated with the excitation in the crystal of spatially oscillating photovoltaic currents induced by photovoltaic tensor component β^a_{131}. Owing to the existence of this component, holographic gratings in LiNbO$_3$ and LiTaO$_3$ crystals can be recorded by orthogonally polarized beams, i.e. by an ordinary and an extraordinary wave (Sect. 2.2.3).

Since in this experiment the excitation is effected by a wave preserving its polarization throughout the crystal, the spatial distribution of scattered light is governed by the angular spectrum of amplification and by the configuration of the pumped volume of the crystal (because of a limited pump beam diameter, different components of scattered light have differing lengths of interaction with the pump wave).

LiNbO₃ : Fe

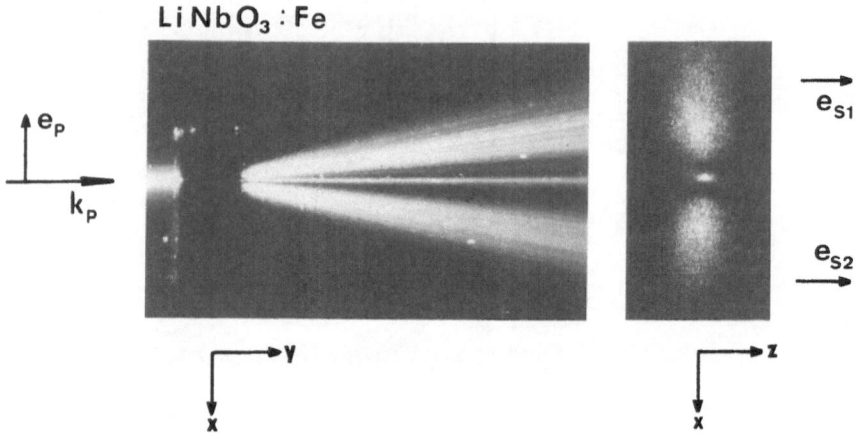

Fig. 2.16. Angular distribution of scattered light (a) in XOY-plane (scattered and transmitted beams are made visible in Rhodamine C dye solution) and (b) in XOZ-plane (far-field pattern)

2.5.3 Forward Polarization-Anisotropic Parametric Scattering

The parametric amplification discussed in Sect. 2.3.3 is the cause of associated scattering processes observed in copper- [2.43] and iron-doped [2.56] lithium tantalate crystals (Fig. 2.17).

Illumination of a LiTaO₃ : Cu crystal by a focused ordinary wave gave rise to extraordinary polarized scattered radiation localized on the surface of a cone whose angle is expressed by (2.17). Further experiments confirmed that for a parametric scattering the phase-matching cone angle is in fact conditioned by the birefringence of the material (which was varied by cooling the crystal down to $T = -60°$ C [2.43] and also by varying the melt composition at crystal growth). The scattered light intensity rose with the square of the sample thickness [2.43].

The azimuthal distribution of scattered light intensity corresponded to the calculated one (2.18). The maximum intensity occurred in the plane of propagation of spatially oscillating currents, i.e. in a plane perpendicular to the crystal's Z-axis. In the plane containing the Z-axis the scattered light intensity dropped to zero.

When the same LiTaO₃ : Cu samples were illuminated by a light wave polarized at 45° to the crystal's Z-axis, both the oo → ee and the oe → ee scattering were observed simultaneously (Fig. 2.17c). For such scattering to arise, both the spatially oscillating currents determined by β_{131} and the ordinary photovoltaic currents associated with β_{333} should be nonzero at the same time, and hence the spatial distribution of scattered light becomes a four-lobed one.

Finally, one more scattering type corresponding to the oo → oe interaction was successfully identified [2.56] in LiTaO₃ : Fe crystals (Fig. 2.17c).

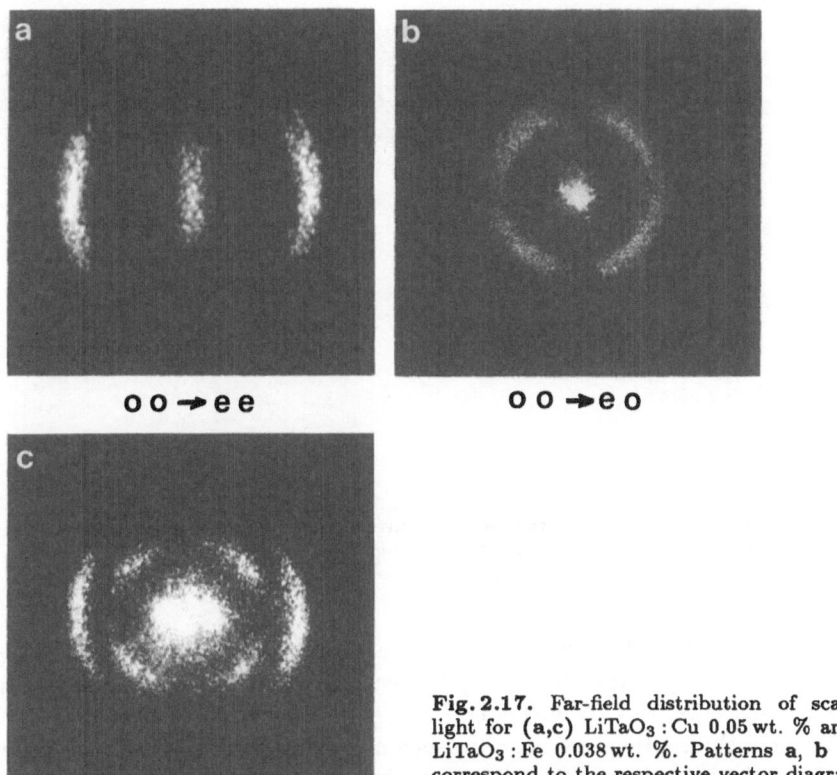

$$\mathbf{o}\,\mathbf{o} \rightarrow \mathbf{e}\,\mathbf{e}$$

$$\mathbf{o}\,\mathbf{o} \rightarrow \mathbf{e}\,\mathbf{o}$$

$$\mathbf{o}\,\mathbf{e} \rightarrow \mathbf{e}\,\mathbf{e}$$

Fig. 2.17. Far-field distribution of scattered light for **(a,c)** $LiTaO_3 : Cu$ 0.05 wt. % and **(b)** $LiTaO_3 : Fe$ 0.038 wt. %. Patterns a, b and c correspond to the respective vector diagrams in Fig. 2.11

Apart from the fact that increasing the number of possible phase-matched parametric interactions extends the potential of these crystals as light beam amplifiers, the presence or absence of one or another parametric process may provide information on the relation between components of the photovoltaic tensor.

2.5.4 Polychromatic Scattering in Nonlocal-Response Photorefractive Crystals

When a $LiNbO_3 : Fe$ crystal was exposed to a multifrequency argon laser beam exciting ordinary waves in the sample, an intense cone-type polarization-anisotropic scattering with polarization plane rotation arose [2.82]. The phasematching condition for ordinary parametric processes (Sect. 2.3.3) prohibits such a degenerate interaction for a negative crystal. It was therefore suggested that the cone-type scattering is accounted for in this case by a biharmonic process

$$k_1(\omega_1) - k_4(\omega_1) = k_2(\omega_2) - k_3(\omega_2) \quad . \tag{2.28}$$

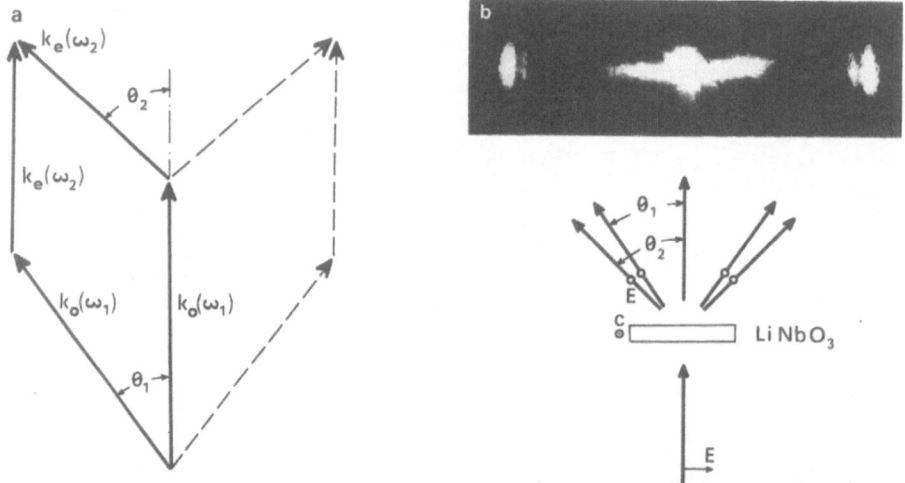

Fig. 2.18. Vector diagram of biharmonic four-wave mixing in $LiNbO_3$ **(a)** and far-field distribution of corresponding light-induced scattering **(b)**

This proposition was directly proved by the absence of cone-type scattering when the same sample was illuminated by even the most intense isolated argon laser lines ($0.488\,\mu m$ or $0.51\,\mu m$) (Fig. 2.18).

For monochromatic pumping a two-beam coupling with an energy transfer from the o- to the e-wave (Sect. 2.2.3), which provides the wide-angle light-induced scattering described in Sect. 2.5.2 (Fig. 2.15), is allowed in $LiNbO_3$: Fe crystals. An addition of the second wavelength into the pump beam results in suppression of the wide-angle scattering and in growth in intensity of the components which meet condition (2.28).

A calculation shows that the addition of the second component does not increase the gain at spatial frequencies meeting condition (2.28), but instead reduces the gain for all the remaining noise components because of the light-induced erasure by the non-Bragg pumping component.

Note that all the parametric scattering processes considered in Sect. 2.5.3 have their biharmonic analogs (e.g., $o_1 o_2 \rightarrow e_1 e_2$ for $LiTaO_3$: Cu crystals). Moreover, both amplification and scattering on "foreign" gratings, meeting conditions of the type (2.28), are possible for all the above-described multi-beam processes.

2.5.5 Transient Scattering in Local-Response Photorefractive Crystals

Preceding sections (2.5.1–4) have dealt with steady-state scattering of the type involved in various shifted grating recording processes. However, the scattered radiation intensity may also increase as a result of a transient energy transfer due to recording of unshifted gratings (Sect. 2.4.2), but in this case at $t = \infty$ the scattered light intensity should once again become equal to the intensity of the initiating noise radiation. The spatial distribution of scattered light intensity should in this case be symmetric in the positive and negative directions of the polar axis because the transient energy transfer varies as the square of the phase modulation.

Studies of the light-induced scattering of the extraordinary and ordinary waves in $LiNbO_3$: Fe crystals indicate a strong transient scattering. Nevertheless, in essentially all the experimental situations, the scattered light level at $t \to \infty$ remains much greater than the initiating noise intensity. Most explanations of this effect are based on a consideration of additional factors not taken into account by the transient energy transfer theory [2.61, 62, 65, 66]. However, even in the theoretical solutions for large phase modulations ($\Gamma d \gtrsim 20$), the donor and acceptor beam intensities start oscillating about certain mean values which are not equal to the initial values [2.64]. For $LiNbO_3$: Fe crystals with an iron dopant concentration of several tenths of a percent, the effective photovoltaic field is $\sim 10^5$ V/cm, and the phase modulation at thickness $d \cong 0.5$ cm corresponds to the above-discussed instability region.

The additional factors that may entail a steady-state energy transfer include diffusion of hot unthermalized electrons [2.28], simultaneous presence of both phase and amplitude spatial gratings in the crystal (due to the $Fe^{2+} \to Fe^{3+}$ conversion) [2.29, 30], as well as an optical "charging" of nonuniform dopant accumulations [2.79, 80], giving rise to a refractive index nonuniformity not erasable by light. Note that the two former mechanisms result in a scattering asymmetry in the positive and negative directions of the axis, whereas the latter one produces a symmetric scattering.

2.6 Oscillation in Photorefractive Crystals

Incorporating a positive feedback into a system where amplification exceeds losses results, by definition, in self-excitation and oscillation. The source of the initiating radiation, corresponding to luminescence in ordinary laser systems, is in this case scattering by inhomogeneities in the medium (the amplified scattered radiation is an analog to the superluminescence).

The process of oscillation development can be represented as follows. When the sample is illuminated by pump wave (or waves), every noise light beam acquires an increment proportional to its initial intensity: $I_{-1} - I_{-10} = I_{-10}(\Gamma d)$ for steady-state two-beam interactions; $I_4(d) - I_4(0) = I_4(0)(\pi \Delta n d/\lambda)^2$ for parametric amplification processes. Some of the initial noise radiation gets into high-Q modes of the cavity, i.e. it transforms into itself after a full round trip of the cavity. Due to the nonlinearity of the process, the noise diffraction gratings corresponding to high-Q modes develop at a faster rate than the remaining ones (the recording interference pattern contrast being higher for them than for the remaining gratings); correspondingly, in the noise radiation angular spectrum a peak appears, determined by the cavity axis direction.

This stage corresponds to the linear development stage in luminescence-centre lasers. It is terminated when the intensity of the radiation in high-Q modes becomes so high that the undepleted pump approximation is no longer applicable. During the nonlinear stage the oscillation intensity rises sharply, and under favourable conditions a major part of the pumping intensity is transferred into the oscillation intensity. At the same time the noise radiation becomes strongly suppressed as an intense oscillating beam results in a non-Bragg optical erasure of the majority of noise gratings.

Backward four-wave mixing nonlocal-response PRC oscillators are commonly treated as passive self-pumped PC mirrors [2.3] where their functions are limited to generating the oscillating beams playing an auxiliary part in

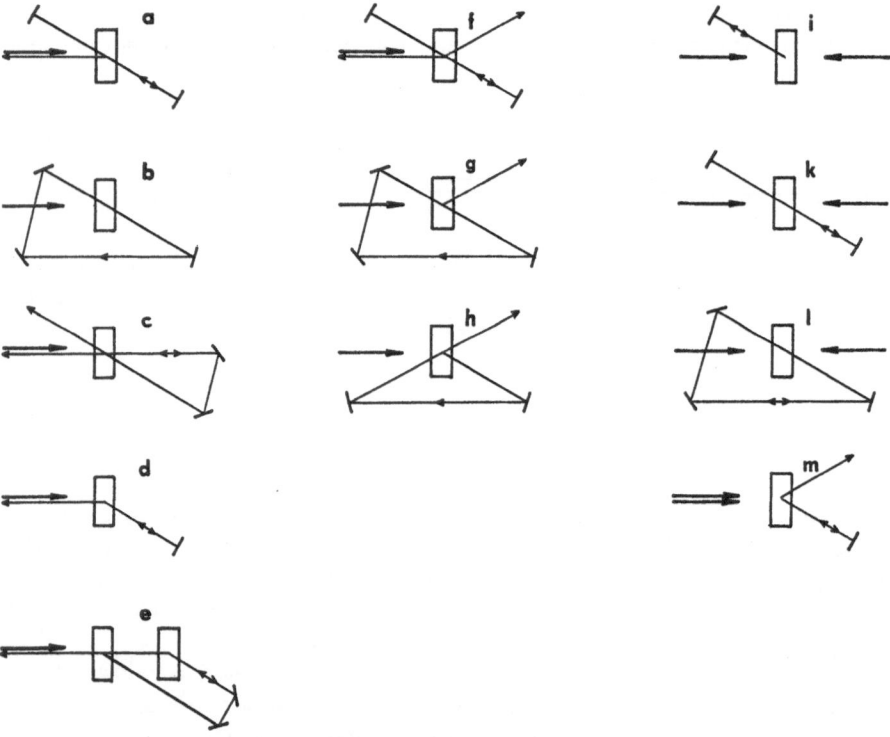

Fig. 2.19a–m. Some feasible configurations of PRC crystal-based optical oscillators. Oscillators (a–e) employ direct two-beam coupling and hence are possible only with nonzero nonlocal response media. Other configurations employing various parametric amplification processes can operate with either nonlocal or local response media. A single pump beam suffices to excite oscillation in configurations (a–h); configurations (i–m) require two pump beams (in (m) they propagate in plane normal to plane of figure). In configurations (f–k, m) the angle between cavity axis and pump wavevector is defined by the phase-matching condition; for remaining configurations this angle is not strictly limited

producing a wave phase-conjugated with respect to the incident one. This aspect is described in detail in Chap. 6 of the present volume. At the same time dynamic-grating oscillators deserve attention in their own right as a new type of frequency quasi-degenerate pumping lasers. Some introductory information on BFWM-PRC oscillators is presented below.

All the possible configurations of the oscillators involve the provision of a positive feedback by applying to the system input either some portion of the amplified signal beam or the fourth beam resulting from the interaction (Fig. 2.19). The sample's internal losses and cavity losses being neglected, the former type of oscillator has no threshold, and oscillation starts at any arbitrarily low amplification. The threshold condition for the latter type is that the intensity of the emerging fourth wave should exceed the initial intensity of the signal wave. Either linear or circular cavities may be used in both cases.

Another classification scheme is based on the number of pump beams required for oscillation. For all the nonlocal mechanisms where a two-beam amplification is possible, oscillation is effected with a single pump beam. Except for exotic configurations of semi-open cavities [2.19, 83, 84], for the case of diffusion nonlinearity, adding the second pump wave only impairs the effect: the oscillation threshold increases. A single pump beam is also sufficient for self-excitation of a parametric forward coplanar interaction oscillator (Sect. 2.3.3) since, there, the two pump beams are indistinguishable. Only for the backward four-wave interaction (Sect. 2.3.1) and noncoplanar forward interaction (Sect. 2.3.2) are two pump beams mandatory, any imbalance in their intensities increasing the oscillation threshold.

2.6.1 Diffusion-Type Amplification Oscillators

The first PRC oscillator was developed by *Hellwarth* and *Feinberg* [2.4], who used barium titanate and nonoptimum two-beam pumping with the configuration shown in Fig. 2.19i. The great amplification margin of this material allowed the most arbitrary and unlikely objects, such as kitchen utensils [2.85] to be used for providing the feedback. Many researchers [2.5, 6, 19, 83, 86–88] have subsequently developed a variety of oscillators (Fig. 2.19a–e) in BaTiO$_3$, strontium-barium niobate [2.89, 90], and barium-sodium niobate [2.19] crystals, all based on the same diffusion-type nonlinearity. The radiation emerging in the cavity often had a spatial structure corresponding to transverse modes of the optical cavity [2.5, 90] (Fig. 2.20).

Figure 2.21 shows the output characteristics of diffusion nonlinearity oscillators [2.91] as calculated by *Yariv* et al. [2.3]. As can be seen, the amplification asymmetry shows up distinctly also in the oscillating mode. For a coupling strength greatly exceeding the threshold value, the efficiency of the transmission grating tends to unity. In this case the entire pumping energy is transformed into a forward oscillating beam which, having been reflected from the cavity mirror, once again diffracts, practically without loss, into a wave which travels backward with respect to the pumping. Qualitatively, just this

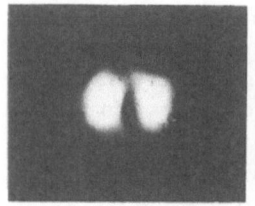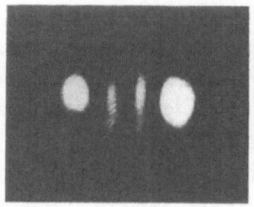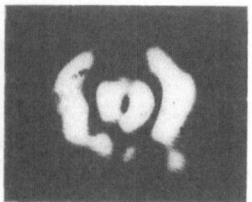

Fig. 2.20. SBN crystal oscillation transverse mode structure

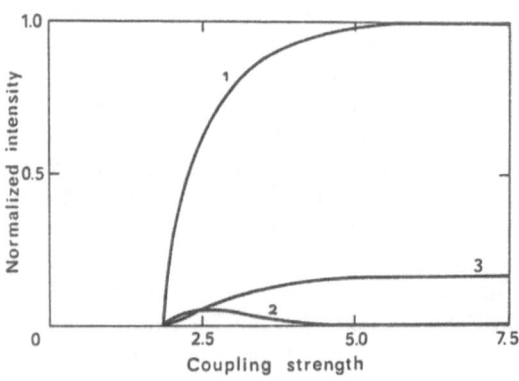

Fig. 2.21. Calculated coupling strength dependence of the intensities of *(1)* forward and *(2)* backward oscillating beams and of *(3)* phase-conjugate beam for diffusion-nonlinearity oscillator (strict frequency degenerate case). $R_1 = R_2 = 0.16$

type of dependence of output energy on gain was observed for a $Ba_2NaNb_5O_{15}$ crystal [2.20].

The oscillation threshold conditions for linear (Fig. 2.19a) and ring (Fig. 2.19b) cavities are close to those of traditional lasers:

$$(\Gamma d)_{th} = -\ln R_1 R_2 + 2\gamma_{int} \quad , \tag{2.29}$$

$$(\Gamma d)_{th} = -\ln R_1 R_2 R_3 + \gamma_{int} \quad , \tag{2.30}$$

where internal absorption, scattering, etc. losses of the nonlinear medium, γ_{int}, enter the threshold conditions with a factor of 2 for a linear cavity (2.29). This is because in every round trip of the cavity the radiation passes through the sample twice, whereas for a ring cavity it does so only once (2.30).

A decrease in the reflectivity of one of mirrors, R_1 or R_2, in (2.29, 30) to zero results in an infinite increase of the threshold. Calculations for the diffusion-type nonlinearity and experiments demonstrate, however, the possibility of oscillation from some threshold value of oscillation wave intensity in a semi-open linear cavity [2.84] and in a semi-open cavity with two interaction regions [2.19, 89]. Strictly speaking, these systems are not self-starting. To excite the oscillation a seed beam must be directed into the cavity. In practice a suitable angular component of light-induced scattering may often serve as the seed radiation, and real systems employing barium titanate crystals are self-starting. After some light-induced scattering development stage the intensity of an appropriate angular component may suffice to excite oscillation.

Threshold conditions for the configurations of Figs. 2.17e,f are as follows:

$$(\Gamma d/2)_{th} = \sqrt{1+R}\ln\left[\frac{\sqrt{1+R}-1}{\sqrt{1+R}+1}\right] + \gamma_{int} \quad , \tag{2.31}$$

$$(\Gamma d/2)_{th} = \frac{R+1}{R-1}\ln\left[\frac{R+1}{2R}\right] + \gamma_{int} \quad , \tag{2.32}$$

which at $R = 1$ gives $|\Gamma d|_{th} = 4.98$ and $|\Gamma d|_{th} = 2$, respectively, if γ_{int} is ignored.

With two pump beams (Fig. 2.19f) and for nonlocal diffusion-type nonlinearity both two-beam coupling and parametric FWM are involved in oscillation. The analysis shows that the threshold can be achieved for both positive and negative Γ by an appropriate choice of pump intensity ratio:

$$(\Gamma d)_{th} = 2\ln\left[\frac{\sqrt{Rr}\pm 1}{\sqrt{R}\mp r}\right] \quad . \tag{2.33}$$

The ambiguity stems from the fact that changing $r \rightarrow 1/r$ (inverting the beam intensity ratio) and inverting by 180° the crystal polar axis direction, i.e. $\Gamma \rightarrow -\Gamma$, while keeping the position of the ordinary cavity mirror unchanged, does not change the threshold. However, when the threshold is exceeded, the oscillation output characteristics for these two cases will be radically different, which agrees with the first observation by *Feinberg* and *Hellwarth* [2.85].

All the above-listed oscillating systems are based on recording of transmission volume phase gratings. A single implementation of a reflection-grating-based generator involved the use of a (001) LiNbO₃ : Fe crystal [2.92] where a diffusion-nonlinearity oscillation can be excited over a wide range of wavelengths, from a He-Cd to a He-Ne laser. With regard to the effect of crystal birefringence on amplification of noise gratings (discussed in Sect. 2.5.1) the optimum conditions for oscillation for the case when the cavity axis coincides with the crystal polar axis correspond to the excitation of an ordinary or an extraordinary wave.

2.6.2 Oscillators Based on Circular Photovoltaic Currents

Only a single oscillator of this type, employing an iron-doped LiNbO₃ : Fe crystal, is known so far. Irradiation of the sample, which had plane parallel faces with Fresnel reflection $R \cong 16\%$, by a focused wave corresponding to the ordinary wave gave rise to oscillation [2.44, 92]. The oscillating wave at the same as the pump wave frequency corresponds to the extraordinary wave.

Calculations [2.52] show that the oscillation threshold condition for such a "laser" coincides with that for a diffusion-nonlinearity laser [2.29] and that

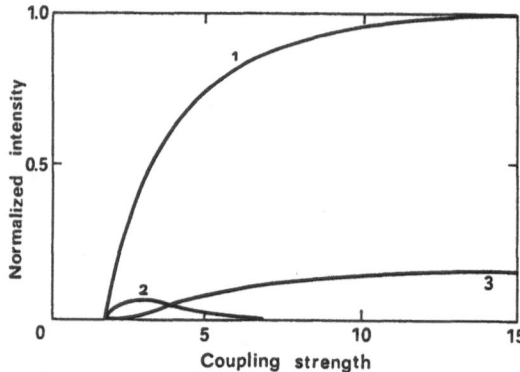

Fig. 2.22. Calculated coupling strength dependence of the intensities of *(1)* forward and *(2)* backward oscillating beams and of *(3)* phase-conjugate beam for photovoltaic nonlinearity oscillator (strict frequency degenerate case) with phase-conjugate beam polarization orthogonal to that of oscillating bemas. $R_1 = R_2 = 0.16$

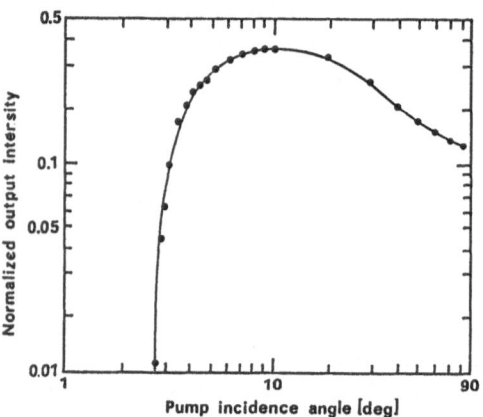

Fig. 2.23. Pump incidence angle dependence of forward oscillating beam intensity in $LiNbO_3 : Fe$ 0.02 wt. %, normalized to incident beam intensity

the dependence of output energy on coupling strength is similar for the two oscillators (Fig. 2.22 and Fig. 2.21).

In practice it is convenient to vary the coupling strength by directing the pump beam at varying angles with respect to the cavity axis (2.8). The dependence of oscillation output energy on the angle has been obtained experimentally (Fig. 2.23). With increasing angle the coupling strength grows sharply, and the oscillation output power rises accordingly, reaching 35 % of the incident wave intensity. As the angle of the pump beam incidence on the sample-interferometer increases, however, the overlap between pump beam and oscillating beam starts diminishing (this is clearly observed in photographs of the crystal end face where interacting beams are visible due to scattering on volume inhomogeneities of the crystal) (Fig. 2.24). As a result, beginning from some angle, the output power starts decreasing despite stabilization of the Γ value over this range of angles.

The coincidence of oscillation threshold conditions (2.29) with those for ordinary lasers makes it possible to employ standard procedures to estimate

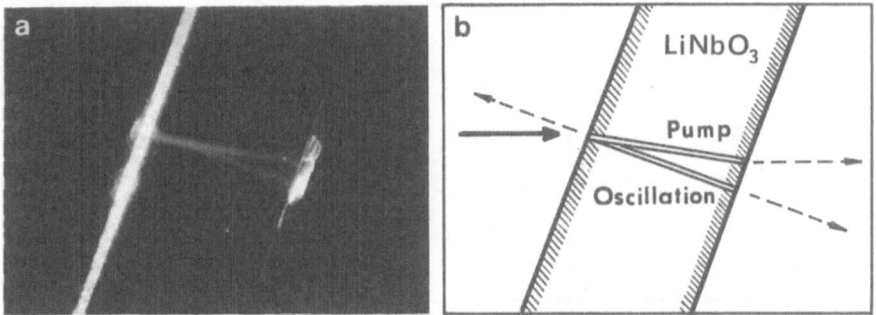

Fig. 2.24. Pump wave and oscillation wave traces inside oscillating $LiNbO_3$: Fe sample. Note the incomplete overlap of beams, which limits the interaction length

gains Γ and internal losses γ_{int}. The cavity losses were controlled by immersing the crystal into water, glycerol, and other liquids, which reduced the Fresnel reflection from sample end faces serving as mirrors. Values of $\Gamma \simeq 10^2\,\mathrm{cm}^{-1}$ and $\gamma_{int} \simeq 10\,\mathrm{cm}^{-1}$ were obtained.

The spectral dependence of Γ for the range covered by an argon laser turned out to be the same as that obtained from two-beam interaction experiments.

2.6.3 Other Oscillator Types

Oscillation in any laser is always effected at those high-Q modes of cavity for which the active medium provides the maximum amplification. In some specific PRC, such as BSO and similar crystals, the maximum gain corresponds not to a strictly frequency-degenerate case, but to some frequency shift of the signal beam relative to the pump beam (Fig. 2.3a).

Oscillation with a BSO crystal in a ring cavity was reported in [2.93]a. The oscillation frequency corresponded to the optimum gain and coincided with that calculated from (2.25). A mode close to the oscillating one was also obtained with BTO crystals using two counterpropagating orthogonally polarized pump beams [2.94, 95].

Much attention has been given in recent years to studying the properties of FWM oscillators with an additional active medium inside the cavity [2.96 98] and also to lasers with FWM oscillators as conjugator mirrors [2.99, 100]. This problem is dealt with in detail in Chap. 6 of the present volume, but it seems appropriate to note here some essential features of a laser, one of whose mirrors is a conjugator based on the FWM in a local-response medium (Fig. 2.19i). The dependence of output power on the coupling strength for such a laser, calculated within the framework of the model of [2.3], is shown in Fig. 2.25. The nonmonotonic nature of the dependence is immediately obvious, the solution being a number of separate branches with transitions between the branches being effected jumpwise; the coupling strength value corresponding to a jump depends on the prehistory, i.e. bistable oscillation modes may arise.

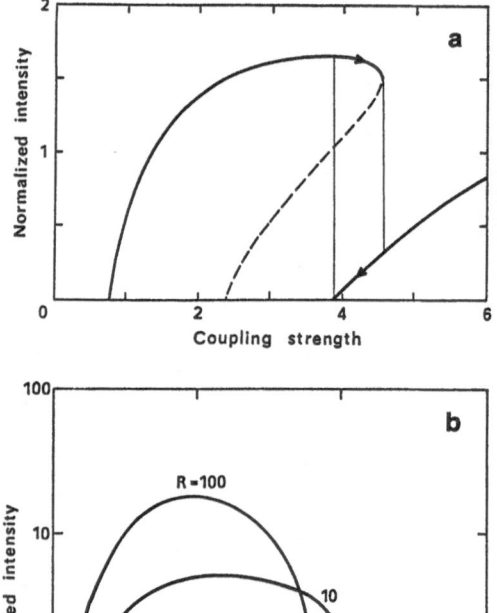

Fig. 2.25a,b. Coupling strength dependence of oscillating beam intensity normalized to pump beam intensity for local response medium: (a) cavity with high-reflection ordinary mirror; (b) amplifier between ordinary and phase-conjugate mirrors

Introducing an additional active medium into the cavity (Fig. 2.25b) is equivalent to increasing the reflectivity of an ordinary mirror to $R > 1$. Calculations for such a case demonstrate a natural decrease of the oscillation threshold and an increase of the output intensity, which may exceed that of the pump beams.

2.7 Conclusion

Investigations into the processes of recording light-induced gratings in PRC began with the brilliant studies of *Amodei* [2.1, 8] and in 16 years have seen many new developments and yielded a number of improtant results. The achievement of a novel-type degenerate oscillation on a dynamic grating, enabling the generation of phase-conjugated beams, may be rightfully regarded as the highlight of the story to date. The principal mechanisms of holographic amplification in PRC appear to have been established and their specific features and parameters studied. Finally, many types of lasers based on dynamic gratings in ferroelectrics (barium titanate, lithium niobate, etc.) have been developed.

At the same time a number of problems remain unsolved. These include: generation and amplification of complex beams; discovering and minimizing the main sources of detrimental losses; finding out the optimum pumping configurations; increasing the oscillation efficiency and attaining the maximum signal-to-noise ratios for these processes. By analogy with ordinary lasers, the free oscillation mode alone may be regarded as attained for holographic lasers. The possibility of obtaining other known modes, such as Q-switch generation, etc., may well become a reality.

For the most part one may assume that dynamic-grating amplifiers and oscillators are only at the beginning of a long and fruitful course of development which may be expected to yield important fundamental and applied results in the near future.

Acknowledgement. The authors wish to express their gratitude to B.G. Shilman and to Dr. A. Lahee for their expert advices in translating and editing of this paper.

References

2.1   D.L. Staebler, J.J. Amodei: J. Appl. Phys. **43**, 1043 (1972)
2.2   N. Kukhtarev, V. Markov, S. Odoulov, M. Soskin, V. Vinetski: Ferroelectrics, **22**, 946 and 961 (1979)
2.3   M. Cronin-Golomb, B. Fisher, A. Yariv: IEEE J. QE-20, 12 (1984)
2.4   R. Hellwarth, J. Feinberg: Opt. Lett. **5**, 519 (1980) and **6**, 257 (1981)
2.5   F. Laeri, T. Tschudi, J. Albers: Opt. Commun. **47**, 387 (1983)
2.6   J.O. White, M. Cronin-Golomb, B. Fisher, A. Yariv: Appl. Phys. Lett. **40**, 450 (1982)
2.7   V.V. Voronov, I.R. Dorosh, Yu. S. Kuzminov, N.V. Tkachenko: Kvant. Electron. **7**, 2313 (1980) [English transl.: Sov. J. Quantum Electron. **10**, 1346 (1980)]
2.8   J.J. Amodei: Appl. Phys. Lett. **18**, 22 (1971)
2.9   V. Kondilenko, V. Markov, S. Odoulov, M. Soskin: Opt. Acta **26**, 239 (1979)
2.10   B.Ya. Zieldovitch: Kratkie soobshchenija po fizike, No. 5, 20 (1970) (Lebedev Physical Institute Letters, Moscow, in Russian)
2.11   R. Hellwarth: J. Opt. Soc. Am. **67**, 1 (1977)
2.12   A. Yariv: Opt. Commun. **25**, 23 (1978)
2.13   M.S. Soskin, A.I. Khyzhniak: Kvant. Electron. **7**, 42 (1980) [English transl.: Sov. J. Quantum Electron. **10**, 21 (1980)]
2.14   R.J. Harrison, P.Y. Key, V.I. Little: Proc. R. Soc. (London), A334, 193 (1973)
2.15   B.Ya. Zieldovitch, I.I. Sobelman: Usp. Fiz. Nauk **101**, 3 (1970) [English transl.: Sov. Phys.-Usp. **13**, 307 (1970)]
2.16   Yu. Anan'ev: Kvant. Electron. **1**, 1669 (1974) [English transl.: Sov. J. Quantum Electron. **10**, 929 (1974)]
2.17   B.I. Sturman: Kvant. Electron. **7**, 483 (1980) [English transl. Sov. J. Quantum Electron. **10**, 276 (1980)]
2.18a Y.H. Ja: Opt. Quantum. Electron. **16**, 399 (1984)
2.18b K. MacDonald, J. Feinberg: Phys. Rev. Lett. **55**, 821 (1985)
2.19   J. Feinberg: Opt. Lett. **7**, 468 (1982)
2.20   S.G. Odoulov, O.I. Oleinik: Fiz. Tverd. Tela **27**, 3470 (1985) [English transl.: Sov. Phys. Solid State **27**, 2093 (1985)]
2.21   R. Orlovski, E. Kratzig: Solid State Commun. **27**, 1351 (1978)
2.22   I.F. Kanaev, V.K. Malinovski, B.I. Sturman: Zh. Eksp. Teor. Fiz. **74**, 1599 (1978) [English transl.: Sov. Phys.-JETP **47**, 834 (1978)]
2.23   Y.H. Ja: Opt. Quantum Electron. **14**, 547 (1982)

2.24 Y.H. Ja: Appl. Phys. B33, 51 (1984)
2.25 M.M. Butusov, N.V. Kukhtarev, A.E. Krumins, A.E. Knyazkov, A.V. Saikin: Ferro-electrics 45, 63 (1982)
2.26 A.M. Glass, D. von der Linde, T.K. Negran: Appl. Phys. Lett. 25, 233 (1974)
2.27 B. Sturman, V. Belinitcher: Usp. Fiz. Nauk 130, 415 (1980) [English transl.:Sov. Phys.-Usp. 23, 199 (1980)]
2.28 V.G. Brovkovitch, B.I. Sturman: Pis'ma Zh. Eksp. Teor. Fiz. 37, 464 (1983) [English transl.: Sov. Phys.-JETP Lett. 37, 550 (1983)]
2.29 A.V. Alekseev-Popov, A.V. Knyazkov, A.S. Saikin: Pis'ma Zh. Tekh. Fiz. 9, 1108 (1983) [English transl.: Sov. Tech. Phys. Lett. 9, 457 (1983)]
2.30 A.N. Knyazkov, M.N. Lobanov: Pis'ma Zh. Tekh. Fiz. 11, 882 (1985) [English transl.: Sov. Tech. Phys. Lett. 11, 365 (1985)]
2.31 V. Vinetski, N. Kukhtarev, S. Odoulov, M. Soskin: USSR Inventors Certificate No. 603276, published in Bulletin of Inventions No. 43 (1978) (in Russian)
2.32 V.Vinetski, N. Kukhtarev, E. Salkova, L. Sukhovertkova: Kvant. Electron. 7, 1191 (1980) [English transl.: Sov. J. Quantum Electron. 10, 684 (1980)]
2.33 M.B. Klein: Opt. Lett. 9, 385 (1985)
2.34 A.M. Glass, A.M. Johnson, D.H. Olson, W. Simpson, A.A. Ballman: Appl. Phys. Lett. 44, 948 (1984)
2.35 P. Günter: Proc. Electroopt./Laser Int. Conf. 76, ed. by H.G. Jerrard (IPC Science & Techn. Press Ltd. 1976) p. 121
2.36 A. Marrakchi, J.-P. Huignard, P. Gunter: Appl. Phys. 24, 131 (1981)
2.37 Sze-Keung Kwong, Young-Hoon Chung, M. Cronin-Golomb, A. Yariv: Opt. Lett. 10, 359 (1985)
2.38 P. Refregier, L. Solymar, H. Rajbenbach, J.-P. Huignard: Electron. Lett. 20, 656 (1984)
2.39 A. Knyazkov, M. Lobanov, A. Krumins, J.Seglins: Ferroelectrics 69, 81 (1986)
2.40 P. Günter, A. Krumins: Appl. Phys. 23, 199 (1978)
2.41 A.E. Krumins, P. Günter: Phys. Status Solidi A55, K185 (1979)
2.42 P. Günter: Phys. Reports 93, 199 (1982)
2.43 S. Odoulov, K. Belabaev, I. Kiseleva: Opt. Lett. 10, 31 (1985)
2.44 S. Odoulov: Kvant. Electron. 11, 529 (1984) [English transl.: Sov. J. Quantum Elec-tron. 14, 360 (1984)]
2.45 I. Kanaev, V. Malinovski: Fiz. Tverd. Tela 24, 2149 (1982) [English transl.: Sov. Phys. Solid State 24, 1223 (1982)]
2.46 V.M. Asnin, A.A. Bakun, A.M. Danishevski, E.L. Ivtchenko, G.E. Pikus: Pis'ma Zh. Eksp. Teor. Fiz. 28, 801 (1978) [English transl.: Sov. Phys.-JETP Lett. 28, 74 (1978)]
2.47 V.M. Fridkin: Izv. Akad. Nauk SSSR. Ser. Fiz. 47, 626 (1983) [English transl.: Bull. Acad. Sc. USSR, Phys. Ser. 47, 1 (1983) (Allerton Press)]
2.48 M.P. Petrov, A.A. Gratchev: Pis'ma Zh. Eksp. Teor. Fiz. 30, 18 (1978) [English transl.: Sov. Phys.-JETP Lett. 30, 15 (1978)]
2.49 S. Odoulov, N. Kukhtarev: Pis'ma Zh. Eksp. Teor. Fiz. 30, 6 (1979) [English transl.: Sov. Phys.-JETP Lett. 30, 4 (1979)]
2.50 F. Zernike, J.E. Midwinter: Applied Nonlinear Optics (J. Wiley, New York 1973)
2.51 S. Stepanov, M. Petrov: Opt. Commun. 53, 64 (1985)
2.52 A. Novikov, S. Odoulov, O. Oleinik, B. Sturman: 75, 295 (1987) Ferroelectrics
2.53 A. Khyzniak, Yu. Kutcherov, S. Lesnik, S. Odoulov, M. Soskin: J. Opt. Soc. Am. A1, 169 (1984)
2.54 V. Kondilenko, S. Odoulov, M. Soskin: Ferroelectr. Lett. 1, 16 (1983)
2.55 M.P. Petrov. S. Stepanov, A. Kamshilin: Ferroelectrics 21, 631 (1978)
2.56 K. Belabaev, I. Kiseleva, S. Odoulov, V. Obukhovski, R. Taratouta: Fiz. Tverd. Tela 28, No. 2 (1986) [English transl.: Sov. Phys. Solid State 28, 321 (1986)]
2.57 J.-P. Huignard, A. Marrakchi: Opt. Commun. 38, 249 (1981)
2.58 V. Kremenitski, S. Odoulov, M. Soskin: Pis'ma Zh. Tekh. Fiz. 6, 931 (1980) [English transl.: Sov. Tech. Phys. Lett. 6, 403 (1980)]
2.59 N. Tabiryan, B. Zeldovitch: Usp. Fiz. Nauk 147, 602 (1985) [English transl.: Sov. Phys.-Usp. 28, 1059 (1985)]
2.60 J.-P. Huignard, A. Marrakchi: Opt. Lett. 6, 622 (1981)
2.61 N. Kukhtarev, V. Markov, S. Odoulov: Opt. Commun. 23, 338 (1977)
2.62 N. Kukhtarev, V. Markov, S. Odoulov: Zh. Tekh. Fiz. 50, 1905 (1979) [English transl.: Sov. Phys.-Tech. Phys. 50, 1109 (1980)]

2.63 V.P. Kondilenko, V.B. Markov, S.G. Odoulov, M.S. Soskin: Ukr. Fiz. Zhurn. (in Russian) **23**, 2039 (1978)
2.64 N. Kukhtarev, T. Semenets: Kvant. Electron. **8**, 217 (1981) [English transl.: Sov. J. Quantum Electron. **11**, 130 (1981)]
2.65 L. Solymar, J.M. Heaton: Opt. Commun. **51**, 76 (1984)
2.66 L. Solymar, J.M. Heaton: Opt. Acta **32**, 397 (1985)
2.67 V. Markov, S. Odoulov: Kvant. Electron. **6**, 2236 (1979) [English transl.: Sov. J. Quantum Electron. **9**, 1310 (1979)]
2.68 P. Günter: Opt. Commun. **41**, 83 (1982)
2.69 J. Feinberg: J. Opt. Soc. Am. **72**, 46 (1982)
2.70 E.M. Avakyan, K.G. Belabaev, S.G. Odoulov: Fiz. Tverd. Tela **25**, 3274 (1983) [English transl.: Sov. Phys. Solid State **25**, 1887 (1983)]
2.71 R. Magnussen, T.K. Gaylord: Appl. Opt. **13**, 1545 (1972)
2.72 R. Grousson, S. Mallik, S. Odoulov: Opt. Commun. **51**, 342 (1984)
2.73 N.D. Khat'kov, S.M. Shandarov: Avtometrija (in Russian) No. 2, 61 (1983)
2.74 F. el Guibaly, L. Young: Ferroelectrics **46**, 201 (1983)
2.75 W. Philips, J.J. Amodei, D.L. Staebler: RCA Rev. **33**, 94 (1972)
2.76 E. Avakyan, S. Allahverdyan, K. Belabaev, V. Sarkisov, K. Tumanyan: Fiz. Tverd. Tela **20**, 2429 (1978) [English transl.: Sov. Phys. Solid State **20**, 1401 (1978)]
2.77 J. Kanaev, V. Malinovski, B. Sturman: Opt. Commun. **34**, 95 (1980)
2.78 E. Avakyan, K. Belabaev, I. Kiseleva, S. Odoulov: Ukr. Fiz. Zh. (in Russian) **29**, 790 (1984)
2.79 V.V. Obukhovski, A.V. Stojanov: Kvant. Electron. **12**, 563 (1985) [English transl.: Sov. J. Quantum Electron. **15**, 367 (1985)]
2.80 V.V. Obukhovski, A.V. Stojanov: Ukr. Fiz. Zh. (in Russian) **58**, 378 (1985)
2.81 V.L. Vinetski, N.V.Kukhtarev: Kvant. Electron. **5**, 405 (1978) [English transl.: Sov. J. Quantum Electron. **8**, 231 (1978)]
2.82 I.N. Kiseleva, V.V. Obukhovski, S.G. Odoulov, O.I. Oleinik: Ukr. Fiz. Zh. (in Russian) **31**, 1682 (1986)
2.83 J. Feinberg, K.R. MacDonald: J. Opt. Soc. Am. **73**, 548 (1983)
2.84 M. Cronin-Golomb, B. Fisher, J.O. White, A. Yariv: Appl. Phys. Lett. **41**, 689 (1982)
2.85 Laser Focus **17**, 118 (1981)
2.86 M. Cronin-Golomb, B. Fisher, J.O. White, A. Yariv: Appl. Phys. Lett. **41**, 219 (1982)
2.87 M. Cronin-Golomb, B. Fisher, J.O. White, A. Yariv: Appl. Phys. Lett. **42**, 919 (1983)
2.88 R.K. Jain, G. Dunning: Opt. Lett. **7**, 420 (1982)
2.89 B. Fisher, M. Cronin-Golomb, J.O. White, A. Yariv, R. Neurgaonkar: Appl. Phys. Lett. **40**, 863 (1982)
2.90 A. Krumins, Ya. Seglins, S. Odoulov, Yu. Kuzminov, N. Polozkov: Pis'ma Zh. Tekh. Fiz. **12**, 6 (1985) [English transl.: Sov. Tech. Phys. Lett. **12**, 2 (1985)]
2.91 S.G. Odoulov, L.G.Sukhoverkhova: Kvant. Electron. **11**, 575 (1985) [English transl.: Sov. J. Quantum Electron. **14**, 390 (1984)]
2.92 S.G. Odoulov, M.S. Soskin: Pis'ma Zh. Eksp. Teor. Fiz. **37**, 243 (1983) [English transl.: Sov. Phys.-JETP Lett. **37**, 289 (1983)]
2.93 H. Rajbenbach, J.-P. Huignard: Opt. Lett. **10**, 137 (1985)
2.94 S. Stepanov, M. Petrov, M. Krasinkova: Zh. Tekh. Fiz. **54**, 1233 (1984) [English transl.: Sov. Phys.-Tech. Phys. **29**, 703 (1984)]
2.95 S. Stepanov, M. Petrov: Opt. Acta **31**, 1335 (1984)
2.96 R.K. Jain, G. Dunning: Opt. Lett. **7**, 420 (1982)
2.97 G. Valley, G. Dunning: Opt. Lett. **9**, 513 (1984)
2.98 R.K. Jain, K. Stenersen: Opt. Lett. **9**, 546 (1984)
2.99 W.B. Whitten, J.M. Ramsay: Opt. Lett. **9**, 44 (1984)
2.100 J. Feinberg, G. Bacher: Opt. Lett. **9**, 420 (1984)
2.101 J. Feinberg, D. Heiman, A.R. Tanguay, Jr., R. Hellwarth: J. Appl. Phys. **51**, 1297 (1980)

3. Photorefractive Effects in Waveguides

Van E. Wood, Paul J. Cressman, Robert L. Holman, and Carl M. Verber

With 15 Figures

In optical waveguides, light beams are confined in one or two dimensions to regions only a few micrometers in width. The high optical intensities which are thus readily obtained make it relatively easy to observe effects of photorefractivity in waveguiding layers or channels. The inhibition of diffraction in waveguides, moreover, leads, particularly in channel guides, to effects which do not occur in bulk samples. In addition, waveguides are chemically and microstructurally different from the substrates on or into which they are grown. They can, therefore, provide clues to the microscopic origins and dynamics of photorefractive effects. In addition to these fundamental reasons for investigating photorefractivity in waveguides, the high light intensities, together with the efficient beam modulation possible using surface electrodes, make it easier to construct photorefractivity-based devices in waveguides than in the bulk; several such devices have been demonstrated [3.1]. However by far the main incentive for studying photorefractivity in waveguides is concern that it may contribute to unacceptable degradation or variability in device performance. In many of the applications presently envisioned for integrated optical devices – in optical communications, rotation sensing, and remote sensing, in particular – reliable unattended operation over long periods of time is mandatory. Most of the materials being considered for these devices are photorefractive; in any proposed device, the presence of photorefractive optical damage must be shown not to lead to deleterious effects at the wavelength and other conditions of operation. Alternatively, the device must be designed so that effects of the photorefractivity are adequately compensated.

3.1 Overview

In this chapter, we will provide an introduction, necessarily very brief, to the goals and techniques of integrated optics and to the methods of optical waveguide formation. Following a short survey of the history of our subject over the past decade, we will describe the methods by which waveguide photorefractivity may be observed and studied. We then try to provide a snapshot of the present state of knowledge concerning effects in various waveguide materials and waveguide configurations. This is currently quite an active area of research, and many new results have appeared since the beginning of 1985 when we first contemplated writing this review. While much progress has been made, there is

45

still much to be learned about the phenomenology and about the fundamental sources of waveguide photorefractivity, and the definitive treatise on the subject is still a long way from being written. In our final summary we list some open questions and areas that would benefit from further research.

3.1.1 Integrated Optics

The techniques and objectives of integrated optics are described in several textbooks and monographs [3.2–4] to which the reader is referred for details. In brief, a light beam may be confined by a phenomenon akin to total internal reflection to a region of higher refractive index than its surroundings. This region may be planar, confining the beam in one dimension only, or it may take the form of a channel or a ridge, confining the beam in two dimensions. We will use the term "channel" generically in this chapter to refer to any type of two-dimensional guiding structure. When the confining dimensions of the guide are sufficiently reduced, a monochromatic beam is constrained, as in microwave guides, to propagate in one of a small number of discrete modes of definite wave vector and polarization. Integrated optics seeks to exploit this situation in several ways:

- In electro-optically or acousto-optically responsive materials, a beam confined at the surface may interact very efficiently with electric fields or with surface acoustic waves produced by electrodes deposited on the surface. The absence of diffraction in the waveguiding dimensions simplifies device design, and the high interaction efficiency means the electrode structures can be very small, typically extending only a few millimeters in the direction of light propagation. Thus low-power, high-bandwidth modulators of phase, frequency, or amplitude can be constructed.
- Because of the small size of the components, a large number of them may be aligned on one small substrate to form, for example, an integrated optical-signal-processing or optical distribution system that will retain its alignment indefinitely and within which the beam cannot easily be broken or otherwise interfered with.
- The modal nature of the waveguided beams makes it possible to design devices that are not otherwise feasible in the optical regime. The foremost examples of this are switches and modulators [3.5, 6] based on evanescent wave coupling between channel waveguides.

The high optical power densities available in waveguides make it in some ways easier than in bulk configurations to observe and exploit nonlinear optical phenomena such as second-harmonic generation, multi-photon absorption and optical-intensity dependence of the refractive index. While useful devices based on these phenomena have not yet been demonstrated, this is an area of great current research interest, which the availability of semiconductor lasers of greatly increased power should further stimulate.

The permanent alignment of components claimed above as a possible advantage of integrated optical devices obviously will not accrue if photorefractive effects alter the optical path lengths in the waveguide. Thermo-optic effects may cause similar problems in some materials. These effects have to be considered in bulk devices as well, but again the high light intensity in integrated devices exacerbates their influence. Moreover, mechanical realignment of components is not generally feasible in a waveguide device. Nonetheless, device designers have until recently paid little attention to these phenomena. This attitude may be partially justified by awareness of the reduced photorefractive sensitivity of the most popular materials, such as lithium niobate, in the near-infrared wavelength regime where future devices are most likely to operate. In other materials, improvements in waveguide optical quality may have to be relied on, in conjunction with appropriate device and package design, to obtain useful devices.

The materials most widely used in integrated optics have not, of course, been selected for their freedom from such potentially deleterious phenomena, but on the basis of their known advantages. Chief among these are:

— capability for reproducible formation of good quality low-loss waveguides,
— good resistance to chemical attack and to environmental degradation,
— sizable electro-optic and/or acousto-optic interactions, and
— availability at reasonable cost.

Up to now three main strategies have been followed in materials research.

1. All-Semiconductor Optical Systems containing integrated elements now under development utilize III-V semiconductors for all the light sources and for most of the photodetectors. The ultimate in integration would be to combine these along with modulators, switches, or similar devices on a single substrate. Drive electronics might also be integrated on the same chip. The versatility and sophistication of III-V processing technology make such schemes feasible. Fairly good electro-optic modulators can be made using the III-V materials, and problems of coupling light into and out of the device might in some cases be completely eliminated. Some simple demonstrations of integration of source and modulator; sources, switch, and detectors; source and drive electronics; and detector and amplifier have recently been described [3.7, 8]. The main deficiency of this strategy up to now has been the high optical attenuation found in these materials. Recently, however, GaAs ridge-type waveguides with attenuations well below 1 dB/cm have been produced [3.9]. It should be recognized, though, that since light must be coupled into or out of many devices through optical fibers, monolithic integration of a source or detector with other components may not always offer an advantage. Investigations of photorefractive effects in bulk samples of III-V and II-VI semiconductors are discussed in [3.10].

2. Passive Waveguides. A particularly inexpensive way to form good quality waveguides is to use glass. In one version of this approach, a glass is sputtered onto an isolation layer on a silicon substrate, which may contain photodetectors or drive electronics. Alternatively, waveguides may be formed by ion-exchange in the glass or by deposition of high-index glass on a lower index substrate.

Since the glass waveguide is not electro-optically or acousto-optically sensitive, additional materials must be introduced to provide active components such as modulators or switches. For example, ZnO films have been sputtered onto glass waveguides to provide regions for acousto-optic modulation. Photorefractive effects are either not present or have not been investigated in this category of hybrid structures, so we shall have no more to say about them here.

3. Ferroelectric Oxides. Oxide ferroelectrics are among the best and most widely used materials for electro-optic and acousto-optic devices, both in bulk and in thin-film form. For the most part single crystals of these oxides have been used, but poled fine-grained polycrystals of PLZT have been used in electro-optic devices in both bulk [3.11] and waveguiding [3.12] configurations. Optical waveguides have also been made [3.13, 14] in some piezo-electric, but not ferroelectric, oxides such as $Bi_{12}GeO_{20}$. Photorefractive effects in these sillenite materials are discussed in detail elsewhere in these volumes [3.15].

With both the passive waveguides and the oxide materials, integration of sources, detectors, and electronics into compact devices must be accomplished in a hybrid, as distinguished from a monolithic, format.

By far the greatest amount of both fundamental and applied work in integrated optics has been done using crystalline lithium niobate, $LiNbO_3$. This is because it is one of the few materials possessing the four advantages listed above. It has been available because it has been grown in quantity for acoustic-wave devices, and, as we shall describe, methods have been developed for growing low-loss (less than 0.3 dB/cm) waveguides in this material. Before its widespread use in integrated optics, lithium niobate was studied extensively as a holographic-memory material [3.16, 17]. Probably more is known about its photorefractive behavior under a variety of circumstances than about any other material. Its photorefractive and other properties are summarized elsewhere in this series [3.15] and in a recent review [3.18]. For the present discussion, we remind the reader that $LiNbO_3$ is a trigonal ferroelectric crystal, transparent throughout the visible and near infrared, with a moderately large birefringence (~ 0.1) and a very high Curie temperature – around 1140° C in commercial material – close to the melting point of 1250° C. By poling near the Curie temperature, large single-domain plates can be produced. It is not a stoichiometric compound, but exists over a considerable range of lithium oxide deficiency [3.19–21]. Commercial crystals are grown by the Czochralski method from melts near the congruently melting composition at 48.6 mole percent $(Li_2O)_{0.5}$; for several reasons we shall not go into, they show enough variation from sample to sample to affect self- and impurity-diffusion coefficients. This variability, along with variations in waveguide processing technique, undoubtedly affects the waveguide microstructure and consequently its photorefractive sensitivity. Since most integrated optics development has emphasized lithium niobate, the preponderance of waveguide photorefractive work has used this material as well; consequently a major portion of our review is devoted to this one material. As we mentioned earlier, the possibility of having to contend with long-term photorefractive changes in LiNbO$_3$ devices has not greatly deterred device designers, and at this time several prototype integrated modulators and switches are being offered for sale [3.22, 23] and are being used in experimental systems [3.24–27].

Lithium tantalate, a very similar crystal to LiNbO$_3$, has not been as widely used in integrated optics although it is generally believed to be more than an order of magnitude less susceptible to photorefractive optical damage [3.28]. This relative neglect apparently results from the higher cost of the crystal and from the necessity, in many circumstances, of repoling after waveguide formation, since the Curie temperature of LiTaO$_3$, 660° C, is below the diffusion temperatures required.

3.1.2 Optical Waveguide Formation

In this section, we present a brief survey of the principal methods used to make optical waveguides, using lithium niobate as the primary example. The discussion will primarily concern planar waveguides. Channel waveguides are made by similar processes, except that photolithographic masking is used to restrict the extent of the diffusion and/or counter-diffusion sources. Ridge or rib waveguides are generally made by depositing planar layers and then etching or ion-milling the non-waveguide regions back to the substrate.

Out-Diffusion

LiNbO$_3$ loses lithium or lithium oxide copiously at temperatures above 950° C. This has been found to increase the extraordinary, but not the ordinary, refractive index in a layer at the crystal surface. Thus a region is formed which guides only the polarization corresponding to this index – the TE mode for light propagating perpendicular to the polar (Z, or crystallographic c) axis in an X- or Y-cut crystal plate, and the TM mode in a Z-cut sample. There is no strong crystal orientation dependence of the lithium out-diffusion rates [3.29]. The first out-diffused LiNbO$_3$ waveguides were made by heating in vacuum [3.30]; this tends to darken the crystals and they had to be subsequently heated in air to obtain low-loss guides. It was later found that heating in flowing air or oxygen worked about as well, and that out-diffusion in a controlled atmosphere [3.20, 31, 32] could be used to make the process more reproducible and to control the overall index change and diffusion depth. Out-diffused guides can also be made in LiTaO$_3$ [3.29]. The crucible out-diffusion technique, a controlled-atmosphere technique which has yielded the best quality single-mode out-diffused guides, is described in Sect. 3.4.2a.

In-Diffusion

The vast majority of lithium niobate waveguide devices are formed by diffusion of titanium from a film on the crystal surface. Many metals can be diffused into LiNbO$_3$ to form waveguides, but it has been found from the time of the earliest experiments [3.33] that titanium provided the best combination of large refractive-index change and low attenuation. Typically, to form a single-mode planar waveguide operating in the visible or near infrared, a layer of 20 to 25 nm of Ti is deposited on the crystal surface and is in-diffused at 950–1050° C for several hours in an atmosphere of slowly flowing air or oxygen which has been dampened by bubbling through hot water [3.34–36]. At the end of the diffusion period, the sample is cooled quickly to around 600° C to minimize the formation, while passing through a reentrant two-phase region [3.37], of scattering centers of $LiNb_3O_8$; then it is cooled slowly to room temperature. The exact nature of the compound or compounds providing the actual diffusion source at 1000° C is a subject of current research and some controversy [3.21, 38]. The titanium film reacts with oxygen and probably also with cations from the lithium niobate in the course of attaining the diffusion temperature [3.38]. Gravimetric experiments [3.38a] show that samples continue to gain weight during 5 h in-diffusion anneals at 1000° C. This indicates either that the diffusion source is not fully oxidized at this temperature or that the lithium niobate has become reduced through reaction with the Ti film and is gradually re-oxidizing during the in-diffusion period. Titanium dioxide, sputtered on [3.39] or formed by oxidizing a film of a metal-organic precursor [3.40], has also been used successfully as the diffusion source. In a variant of the usual method, part of the heating is carried out in argon rather than oxygen. This does not, however, appear ordinarily to prevent oxidation of the titanium.

At the in-diffusion temperature, $LiNbO_3$ naturally tends to lose lithium by out-diffusion. To produce a well-confined guide, this loss has to be inhibited or compensated. Carrying out the process in dampened oxygen was at one time thought to suppress lithium out-diffusion, but it now appears [3.36, 41] that it compensates for the Li loss by exchange with protons. Other methods of preventing out-diffusion involve heating in Li-rich atmospheres [3.35, 42]; for reasons not yet understood, these procedures have not proved as reliable as the damp-oxygen method. When carried out properly, the titanium-in-diffusion technique produces a layer with a Gaussian concentration profile 1–2 μm thick. The diffusion source is completely exhausted; so the surface requires little or no polishing to reduce surface scattering losses. The titanium in-diffusion raises both the ordinary and extraordinary refractive indices of the $LiNbO_3$; thus a so-called "single-mode" guide usually supports one mode of each polarization.

Ion Exchange

The damp-atmosphere processing method described above for the manufacture of $LiNbO_3$: Ti waveguides was originally adopted because it was believed to reduce the photorefractive sensitivity of the guides [3.34, 35, 43], rather than with the intention of compensating for the out-diffusion of Li. With the gradual realization that proton entry into the crystal might be involved in the compensation process, the concept arose of waveguide formation by intentional exchange of protons for lithium ions. The first such guides were made by *Jackel* et al. [3.44] by immersing $LiNbO_3$ plates in molten benzoic acid (C_6H_5COOH) at temperatures in the range 217–249° C for 1–24 hours. This procedure produces waveguide layers on X- or Z-faces with a large increase, up to 0.12, in extraordinary refractive index at the surface. Y-faces are etched by the acid. Other weak acids, organic and inorganic, can also be used, but appear to be somewhat less suitable. Strong acids [3.45] cause conversion of the surface to $HNbO_3$. The low processing temperatures involved are an advantage additional to the large index change. Waveguides made in pure benzoic acid, though, are subject to a number of problems, including high surface strains, leading to cracking and light scattering [3.46]; irregular or spatially variant mode-index changes [3.47] over periods of days or weeks; reduction in electro-optic [3.48, 49] or acousto-optic [3.50–54] efficiency; and strong photoconductive optical damage [3.49, 55] in moderate dc fields, e.g., 0.5–$1.5\,V/\mu m$ applied for 1 minute. Annealing the guides [3.44, 46a, 56, 57] may improve their stability somewhat, but a more satisfactory process [3.56–58] is addition of Li ions, in the form of lithium benzoate, to the melt, in order to reduce the ionic exchange rate. This also allows some control over the refractive index change in the guides and improves their transmission characteristics; guiding on the Y-face also becomes possible. It has been reported [3.54] that guides made by exchange in melts diluted this way do not suffer degradation of electro-optic response. The amount of dilution required is small, however – at about 4 % lithium benzoate in the melt, ion exchange stops completely – and the process is somewhat difficult to control. Ideally a sealed melt should be used, although even then effects of crystal stoichiometry and initial hydrogen content might be significant.

An alternative to the use of proton exchange alone is its application to a crystal that already contains a Ti-in-diffused waveguide [3.56, 59]. Mode index measurements on such so-called TIPE (for titanium in-diffused proton-exchanged) guides show that the hydrogen-induced increase in extraordinary index is accompanied by a substantial decrease (~ 0.04) in the ordinary index. The presence of the titanium slows down the ion-exchange process considerably, and also inhibits etching on the Y-faces. By using diluted benzoic acid melts, and by varying the other processing parameters, a wide variety of index profiles may be obtained. It is believed [3.46b, 3.54] that the titanium-in-diffusion and proton-exchange steps may produce opposite, partially compensating, strains in the waveguiding layer; so TIPE guides of optical quality superior to simple proton-exchanged guides should be attainable.

It was at one time thought [3.60] that other ions, such as Ag^+ and Tl^+, could be exchanged with Li^+ to form waveguides in $LiNbO_3$. It is now understood that these ions do not enter $LiNbO_3$ rapidly [3.61], and the effects observed were probably the result of hydrogen-containing impurities in the melts employed.

Other Methods; Other Materials

Other methods by which $LiNbO_3$ waveguides have been formed include sputtering [3.62], epitaxial growth from molten $LiNbO_3$ [3.63], liquid phase epitaxy [3.64], vapor-phase epitaxy [3.65], molecular-beam epitaxy [3.66], and metal-organic chemical vapor deposition [3.67]. In most cases, $LiTaO_3$ crystals were used as substrates; in one experiment [3.68] LiTaO$_3$ was sputtered onto glass. In general, these methods have produced waveguiding layers with much higher optical losses than the diffusion based methods discussed above. Since better and easier waveguide fabrication procedures have been available, these techniques have not drawn much interest, and there is no information available on photorefractivity in LiNbO$_3$ or LiTaO$_3$ guides made by any of these processes. These methods have been used, though, to make good optical quality films of other photorefractive materials. A few examples are discussed in Sect. 3.7. In an electro-optic material like LiNbO$_3$, a channel waveguide can be induced simply by applying a field to parallel strip electrodes placed on the crystal surface. This method has proved useful [3.69] in obtaining waveguides in materials like KNbO$_3$, where diffusive processes are very slow.

Most of the diffusive methods used to make waveguides in LiNbO$_3$ have also been tried in LiTaO$_3$, but less information is available on the quality and variability of the layers produced. In addition to titanium, niobium can be diffused into LiTaO$_3$ to make good waveguides. Since most of the waveguide formation procedures involve temperatures exceeding the Curie temperature (650° C) of LiTaO$_3$, repoling of the crystal while cooling is generally believed necessary to retain fully the electro-optic properties of the material [3.70], although there is evidence [3.71] of "memory" of the crystal poling conditions in samples which are not repoled. LiNbO$_3$ doped with around 5 % MgO has recently been advanced [3.72] as a material more resistant to photorefractive degradation than pure LiNbO$_3$. The bulk photovoltaic coefficient is not substantially reduced in this material; rather the photoconductivity is increased [3.73] by one or two orders of magnitude. The electro-optic coefficients are not markedly altered [3.74]. Waveguides have been made in the doped material by Ti in-diffusion [3.75] and by proton exchange [3.76].

3.2 Historical Sketch

In this section we provide a chronological resumé of some significant events in the rather brief history of our subject. Our purpose is not to assign historical credit although we apologize to anyone whose work we have overlooked; we simply want to assist the reader of earlier literature to place it in context with other developments in the understanding of photorefractivity and in optical waveguide technology. Naturally, we will refer to some topics that are not discussed in detail until later in the article.

The first waveguides in LiNbO$_3$ were voltage-induced guides demonstrated by *Channin* [3.77] in 1971. Aware of the possibility of photorefractive degradation of the guides, he attempted to avoid it by using an alternating applied field; he did not, in fact, observe any effects of photorefraction. After the introduction of Li-out-diffused guides by *Kaminow* and *Carruthers* [3.30] in 1973 and of metal-in-diffused guides the following year by *Hammer* and *Phillips* [3.78], *Noda* and co-workers [3.79], and *Schmidt* and *Kaminow* [3.33], many research groups prepared such guides. Doubtless many of them observed effects of photorefractivity in these guides, but the first published mention of such an effect appears to be *Schmidt* and *Kaminow*'s [3.80] comments on damage problems at 633 nm in some $LiNbO_3$: Ti waveguides fabricated for study of acousto-optic Bragg diffraction. The first report on intentional experimental study of the phenomenon is that of *Wood* et al. [3.71] on hologram formation in waveguides made by out-diffusion of Li from Fe-doped $LiNbO_3$ and by in-diffusion of Nb into $LiTaO_3$. The first integrated device using a photorefractive grating was the electro-optically controlled switch devised by *Verber* and co-workers [3.81] in 1975.

Channel and rib waveguides, confining the guided beam in two dimensions, had been made in many materials [3.82] prior to their introduction in $LiNbO_3$ and $LiTaO_3$. *Kaminow* et al. [3.83] described in 1974 an electro-optic modulator using a ridge waveguide produced by ion-beam etching an out-diffused crystal, while *Standley* and *Ramaswamy* [3.84] produced channel waveguides the same year by diffusing a stripe of Nb into $LiTaO_3$. The following year *Noda* and co-workers [3.39, 85] at NTT in Japan reported on a ridge-guide modulator in $LiTaO_3$: Cu and on a channel guide in $LiNbO_3$ by in-diffusion from a sputtered TiO_2 strip. Also in 1975, *Papuchon* et al. [3.6] at Thomson-CSF in France were first to publish a description of a device – the so-called COBRA electro-optic directional-coupler modulator – made by the now familiar process of depositing a Ti film, masking and etching to obtain the desired guide configuration, and then in-diffusing the titanium to form an arrangement of channel guides. The occurrence of troublesome photorefractive index changes in channel waveguide devices was noted in 1976 by *Burns* et al. [3.86], studying a branching waveguide modulator fabricated on Y-cut $LiNbO_3$; and by *Schmidt* and *Buhl* [3.87], who arranged 5 directional couplers on X-cut $LiNbO_3$ to form a 4×4 switching network. The latter authors noted a drift with time in coupling characteristics when dc voltages were applied to the network; similar effects were noted later in branching waveguide devices [3.88]. This drift, which places definite restrictions on device design, is largely attributable to a photoconductive contribution to optical damage. The first quantitative study of this effect, and of channel waveguide photorefractivity in general, was carried out by *Schmidt* et al. [3.89] in 1979.

Photorefractive effects in $LiNbO_3$ resulting from two-photon absorption without a real intermediate state were demonstrated in a bulk configuration by *von der Linde* et al. [3.90] in 1974. Similar effects in an out-diffused waveguide configuration were found by *Verber* et al. [3.91] two years later.

Most early experiments on waveguide photorefractivity involved hologram formation, although the effect of photorefraction on the shape of a single beam had been known since the pioneering bulk-crystal work of *Chen* [3.92]. The first report of photorefractive changes in a single guided beam was the description, in 1976, by *Barnoski* and *Lotspeich* [3.93] of "optical cleanup" in $LiTaO_3$: Ti planar guides. In 1977, *Holman* et al. [3.31] began the development of the single-beam method as a quantitative tool for observation of the kinetics of photorefraction in waveguides, and reported superior laser power-handling performance of out-diffused guides.

The relative short-term resistance to optical damage of waveguides formed in moist atmospheres was noticed and investigated by *Jackel* and co-workers [3.43] and by *Holman* and *Cressman* [3.94] in 1981. (That introduction of H or OH ions could affect the photorefractive sensitivity of LiNbO3 had already been demonstrated by *Smith* et al. [3.95] in 1968.) This research led to the development, also by Jackel, in the same year of the low-temperature technique of waveguide formation by proton exchange in weak acids, initially hydrated Mg(NO3)2 [3.45] and subsequently benzoic and other organic acids [3.44]. Proton exchange in previously titanium in-diffused guides was introduced in 1982 by *De Micheli* and co-workers [3.59] at the University of Nice, primarily for the purpose of improving the coupling efficiency for second harmonic generation. Short-term resistance, relative to that of titanium-in-diffused guides, to optical damage of proton-exchanged guides was noted by *Jackel* [3.96] and *Becker* [3.48] in 1983, while *Kopylov* et al. [3.97] and *Holman* et al. [3.98] found photorefractive changes in such guides in longer-term experiments carried out the same year. Relaxation effects in dc fields, similar to those seen in Ti-in-diffused guides, and long-term index instabilities in guides prepared in undiluted benzoic acid had been noted the previous year by *Wong* et al. [3.55] and by *Yi-Yan* [3.47], respectively.

Somewhat surprisingly, the phenomenon of photoinduced conversion of ordinary- to extraordinary-polarized guided waves was not discovered until 1981. *Zolotov* and colleagues [3.99] at the Lebedev Physics Institute, Moscow, gave the first description of the effect and provided a qualitative explanation of the relationship to photorefraction. The effect was also found independently by *Haus* and his students [3.100–102].

It had long been realized that the high optical intensities in waveguides would make it easier than in the bulk to observe photorefractive changes at long wavelengths. The first published mention of such changes at the laser-diode wavelength of $0.83\,\mu$m seems to be that of *Hammer* and *Neil* [3.103] in 1982, although measurable laser-diode damage to bulk LiNbO3 had been reported [3.104] as early as 1969. The first quantitative studies on photorefraction in devices designed to operate in this wavelength region were those of *Becker* [3.48] and of *Mueller* and *Garmire* [3.105] in 1983. Becker also mentioned effects at $1.3\,\mu$m. Data on the relatively small initial effects at this wavelength in a variety of devices were presented by *Bulmer* et al. [3.106] in 1984.

The reduction in or absence of observed photorefractive change when a light beam propagates along the polar axis of a ferroelectric is a phenomenon known since the early days of photorefractivity [3.107]. Its explicit observation in a waveguide configuration and the suggestion of using Z-axis propagation to mitigate optical damage in devices are due to *Holman* et al. [3.98], 1983. Devices using this orientation have been constructed by *Becker* [3.108], 1984, and *Thaniyavarn* [3.109], 1985. Photorefractive polarization changes in Z-oriented channel guides have recently been reported by *Sanford* and *Robinson* [3.110].

For additional detail on the development of the understanding of photore-fractivity and its consequences, the reader is referred to other articles in these volumes. For more information on the development of integrated optics, the textbooks listed earlier may be consulted.

3.3 Experimental Techniques

In this section, we describe the more commonly used experimental arrange-ments for producing, observing and measuring photorefractive changes in opti-cal waveguides. It has gradually been recognized that quantitative measurement of photorefractive sensitivity and its change with time in the waveguide envi-ronment is an enterprise fraught with many experimental and interpretational difficulties. It is premature to discuss these difficulties, though, until the ba-sic experiments are described; therefore, the major part of that discussion is deferred to the beginning of the next section.

The principal methods of studying photorefraction in planar waveguides have been (1) formation and readout of simple grating holograms, and (2) observation of changes in shape of a single beam propagating in the guide and of light scattered out of this beam. In channel guides, when coupling regions or branches are present, phenomena associated with photorefractive modification of coupling or beam-splitting ratios provide the most convenient method of investigation. We shall describe these methods in turn.

3.3.1 Hologram Formation

The most usual method of hologram formation in planar waveguides is that shown in Fig. 3.1a. Two coherent guided beams of the same polarization are made to cross within the waveguide and are subsequently coupled out. As in-dicated in the figure, the beams are symmetrically disposed relative to the normal to the polar axis; so an interference pattern is formed along the polar axis, and, given adequate photorefractive sensitivity at the wavelength used, a phase grating is produced in the same way as in the bulk configuration. It should be noted that the beams intersect in a diamond-shaped region, a cir-cumstance which may have a decided effect [3.111] on the diffraction efficiency of the resulting hologram. Most experimental work has been interpreted using the simpler *Kogelnik* [3.112] theory, in which the grating formation region is assumed to be bounded by two parallel planes. With an appropriate defini-tion of grating thickness [3.111] this is usually not a bad approximation, since deviations from this theory as a result of the shape of the beam intersection region do not become significant until the diffraction efficiency reaches 80 % or so. It is also possible effectively to confine the hologram to a region bounded as in Kogelnik's theory by limiting the area of photorefractive sensitivity. We demonstrated this possibility [3.113] by diffusing a rectangular film of iron into an outdiffused waveguide and forming a high-efficiency grating by intersecting two guided beams in this region.

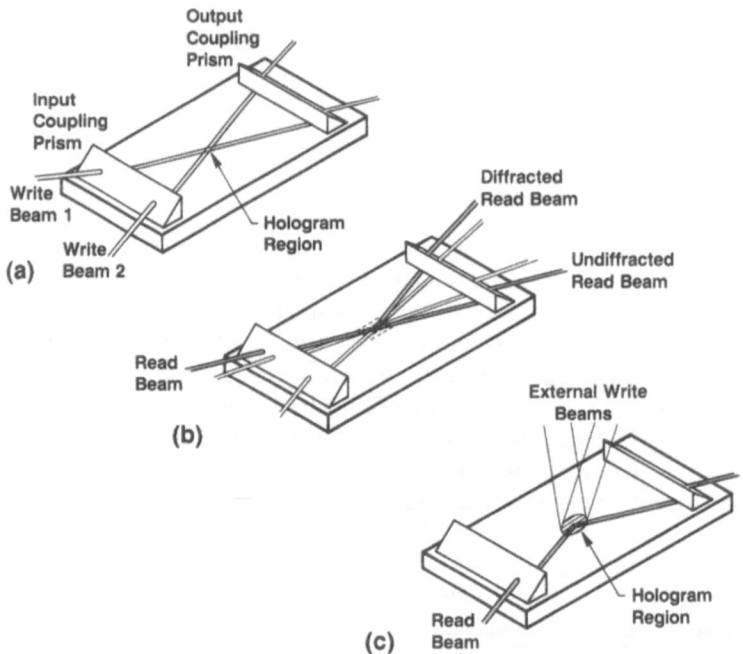

Fig. 3.1. Experimental arrangements for forming simple holograms in optical waveguides; (a) two guided beams intersecting in waveguide; (b) guided read beam of longer wavelength added; (c) two unguided beams intersecting in waveguide containing a guided read beam

To read out the hologram, any of the techniques used in the bulk may be employed. Most often, one of the writing beams is simply blocked momentarily. Alternatively, the phase of one writing beam can be briefly altered instead, in the type of experiment that has come to be known as degenerate two-wave mixing. These readout methods suffer the disadvantage that they alter the recording characteristics during the readout intervals – for instance, if one recording beam is blocked, the other tends to erase the hologram formed up to that time. The simplest way to avoid this problem, and to obtain a continuous record of the grating formation process, is to introduce, at the appropriate Bragg angle, an additional long-wavelength low-intensity read beam, as indicated in Fig. 3.1b.

As an alternative to forming the hologram with guided waves, it may be created using external beams, as shown in Fig. 3.1c, and only read out with a guided beam. Readout of guided-wave holograms with an external beam has also been used on occasion. It is also possible to form a waveguide hologram photorefractively by intersecting one guided and one unguided beam. We have used such an arrangement to make a grating coupler, for instance. When multiphoton processes for grating formation are considered, a variety of additional arrangements become possible. As the diagrams indicate, waveguide holograms have almost always been formed under open-circuit conditions; so build-up of an envelope field at the edges of the beam intersection region should influence the dynamics of hologram recording. Photorefractive changes of the type described in the following section should also affect the individual recording and readout beams; the possible effects on hologram formation and erasure have not received any experimental attention.

A phase grating in a waveguide constitutes a periodic perturbation of the dielectric permittivity that permits coupling among the different guided modes. Thus in a multimode guide it is possible both to record and to read out using beams of different modes [3.114]. For example, one might write a grating using guided modes 1 and 2, address it with mode 3, and observe the light diffracted into mode 4. Naturally, recording and diffraction efficiencies are highest when all the beams are in the same, and lowest-order, mode (unless there is something very peculiar about the distribution of photorefractive centers in the guide). The complications of intermodal scattering and the desire for the highest possible diffraction efficiency have led most experimenters to use single-mode guides.

3.3.2 Single Beam Methods

When a beam of wavelength suitable for inducing photorefractive damage propagates in a planar waveguide in lithium niobate or a similar polar material, photorefractively formed fields gradually distort the shape of the beam and diminish the total intensity within the original beam dimensions. When the polar axis lies in the waveguide plane, the light removed from the main beam is scattered towards the negative c-axis [3.115]. The interpretation of these observations is left for later discussion; in the present section we describe how this phenomenon can be used for the study of the dynamics of photorefractive change in waveguides. Gradual distortions of beam shape and accompanying changes in mode index are the most basic deleterious photorefractive changes that may be expected in planar guides; single-beam experiments thus provide a convenient and very direct method for assessing and comparing the suitability of different waveguide preparation procedures for devices that must survive long periods of unattended operation.

In the simplest experiments of this type, the intensity and shape of the output beam from a planar waveguide are measured as a function of input power and time. A schematic of the arrangement used [3.94, 115] is shown in Fig. 3.2. The beam from the laser is first coupled into the waveguide using a "tent" prism [3.116]. The use of this prism together with a prism-waveguide spacer $0.5\,\mu$m thick centered between the input and output coupling regions provides

— easy access to the reflected beam to enable determination of the input coupling efficiency,
— high coupling efficiencies because of the shaped or tapered coupling region,

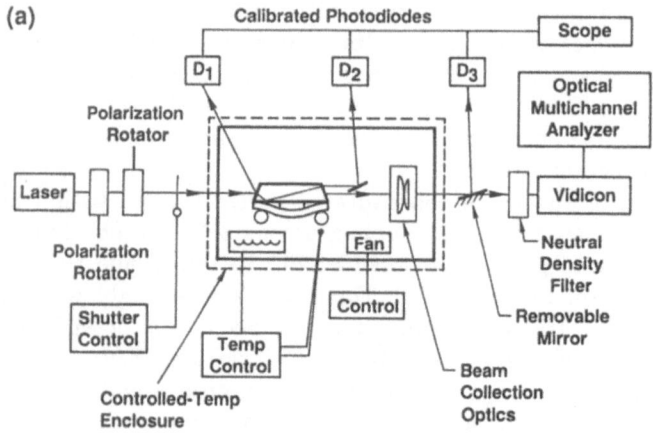

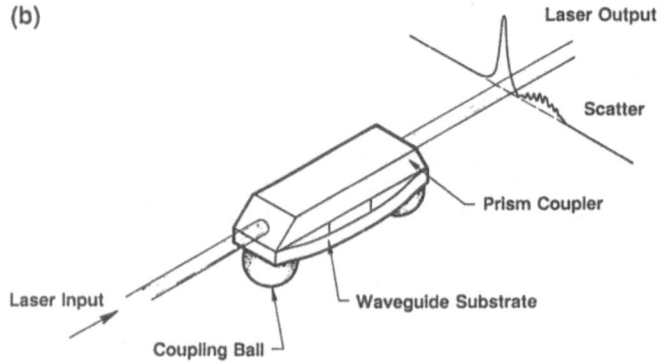

Fig. 3.2. Arrangement for measurement of optical damage in planar waveguides using a single guided beam: **(a)** block diagram; **(b)** detail of prism coupling system

- reproducible coupling, and
- ease of assembly.

Small bearings under the coupling region allow high-quality coupling spots to be formed readily at the desired positions. A pair of compound polarizing prisms are used to control the input intensity and the direction of polarization. Photodiodes continuously monitor the intensities of

- the beam reflected from the entrance of the prism coupler,
- the beam emerging from the output prism reflected from the coupling region, and
- the output beam.

The coupling angle depends on the indices of refraction of the prism and waveguide. Since these indices are temperature-dependent, the waveguide-prism unit is enclosed and maintained at a constant temperature $(30 \pm 0.1^\circ \text{ C})$ in order to provide maximum coupling. This allows stable measurements to be made for periods of over 200 hours. In some experiments, the waveguide output is recorded using an optical multichannel analyzer rather than a photodiode. The light scattered due to the photorefractive effect is sufficiently separated in angle from the main beam so that both beams can be observed independently. A condenser lens may be required to collect the light for recording purposes since the range of angles involved is quite large. For assessing the short time performance at high power and/or shorter wavelengths the photodiode used to detect the output beam is connected to an oscilloscope. In the majority of our experiments the light source has been the $0.633\,\mu\text{m}$ line of a $5\,\text{mW}$ He-Ne laser. The beam diameter is around $1\,\text{mm}$, and the initial beam profile is close to Gaussian. In all our experiments described here, the tent prism was of strontium titanate.

3.3.3 Channel Waveguide Devices

In Fig. 3.3, we show the optical waveguide arrangement in three of the most widely used fundamental structures for channel waveguide modulators and switches — the simple branch, the proximity directional coupler and the cross-channel switch. Electrodes are not indicated, since a variety of arrangements are possible and the details are not essential to the present discussion. While there has been little investigation of photorefractive effects in cross-channel devices [3.117], the other structures have been studied extensively. Integrated

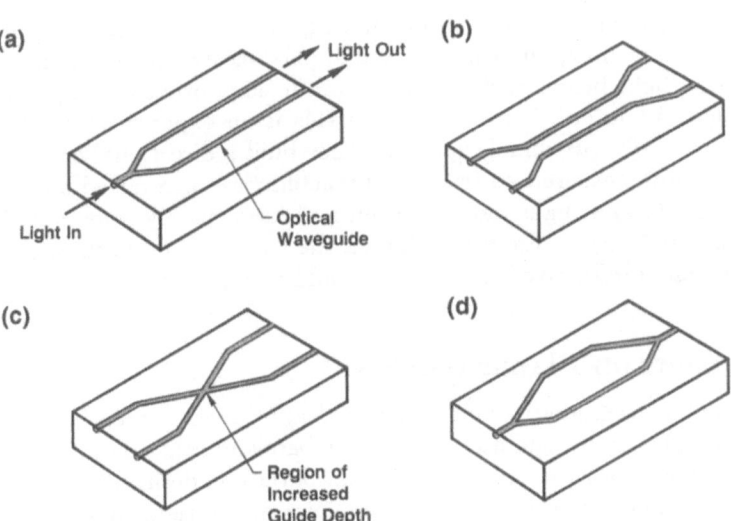

Fig. 3.3. Simple channel waveguide devices; electrodes not shown: (a) simple Y-branch; (b) proximity directional coupler; (c) cross-channel switch; (d) Mach-Zehnder interferometer (asymmetric). Branching angles exaggerated for clarity

interferometers, such as the asymmetric Mach-Zehnder shown in Fig. 3.3d, also serve well for photorefraction work, since photorefractive phase changes in either arm may be directly detected as an intensity change in a single output channel. Photorefractive changes in beam shape and effective index in a single channel do, of course, occur and have received some study [3.118, 119], see also Table 3.2 below, but structures like those in Fig. 3.3 have received most of the attention for several reasons:

— These basic components find application in devices of several types such as switching networks and multiplexers.
— Where coupling of evanescent fields between guides is significant, small photorefractive changes can manifest themselves as large changes in coupling.
— Effects of applied electric fields may be studied in a relatively straight-forward way.

On the other hand, several experimental difficulties have been encountered:

— Leakage currents, through optical isolation layers between surface electrodes and the waveguide or through high-conductivity regions on the crystal surface, are often significant.
— Dielectric relaxation in the waveguide itself may reduce the effect of applied electric field.
— Owing to the high optical power densities involved, thermal effects are difficult to avoid.
— Devices are designed to operate at a single wavelength, and are usually studied only at that wavelength.

Typical experiments simply involve monitoring the power in each output channel as a function of time, input power, and applied electric fields. In most cases the operating light beams and the beams producing optical damage are one and the same. An interesting exception to this is an experiment briefly reported [3.120] by *Hoffmann* and *Langmann*. They built a directional coupler to operate at $1.06\,\mu$m, measured its transfer characteristics, introduced a controlled amount of $0.633\,\mu$m light into one arm, and immediately remeasured the characteristics at $1.06\,\mu$m. From the shift in the switching properties, the magnitude of the photorefractive index change could be inferred.

3.4 Experiments on Planar Guides

In this section, we try to summarize what has been learned from photorefractivity experiments on planar waveguides in lithium niobate and lithium tantalate. With the benefit of hindsight, we can see that many of these experiments, albeit they represented advances at the time, were less informative than they might have been. This has occurred, as we mentioned earlier, because the importance of monitoring or controlling several experimental variables has been only gradually appreciated. We shall begin our discussion by briefly describing some of the problems that have been encountered. Much of what we say here applies also to experiments described in subsequent sections.

Time to Saturation. As we shall show, in many waveguides the buildup of optical damage is a very gradual phenomenon; and observation periods of the order of weeks, at typical input power levels, may be necessary to reach a steady-state situation. In waveguides made under strongly reducing conditions, on the other hand, it may require only a fraction of a second to reach a steady-state damage level, although (see Sect. 3.4.2d) this level may not appear to persist. Moreover, the degree of steady-state photorefractive index change cannot be predicted from the initial rate of damage. In many cases, investigators have been content with relatively short-term comparisons of effects of different waveguide preparation conditions. Statements of the kind frequently seen to the effect that waveguide treatment (or preparation condition) "A" produced a guide that showed obvious damage after exposure to $X \, \mathrm{J \, m}^{-2}$ of waveguided light, while a guide undergoing treatment "B" showed no effect after a $5X \, \mathrm{J \, m}^{-2}$ exposure, are not devoid of meaning; but they are not informative either as to the ultimate level of damage in the guide or as to the time required to attain this level. Furthermore, it should not be assumed that the saturation damage level depends only on total exposure and is independent of intensity.

Environmental Effects. Temperature changes, either as a result of ambient changes or as a result of heating by a guided beam, can significantly affect photorefractive and other waveguide experiments. Among the effects that may be present are:

1. Thermo-optic changes in guides and in coupling prisms. Materials used for prism couplers, such as $SrTiO_3$ and TiO_2, possess large temperature dependences of their refractive indices. Thermo-optic coefficients in the waveguide materials themselves are of comparable magnitudes. Ambient temperature changes of a few degrees can cause significant changes in coupling angles [3.121, 122]. Coupling efficiency may also be changed by temperature variations which alter the width or shape of the coupling gap.

2. Thermo-optic changes brought about by application of rf power. The excitation of acoustic surface waves by interdigital transducers has been shown [3.53, 123] to be accompanied in some cases by substantial heating of the waveguide region. Thermo-optic changes in effective indices of guided modes naturally result.

3. Pyroelectric effects. Transitory optically induced pyroelectric voltages of magnitudes comparable to open-circuit photorefractive voltages are readily observed [3.124, 125] in waveguides in $LiNbO_3$ and $LiTaO_3$, particularly [3.126] when some moderate optical absorption is present. Even in the absence of significant absorption, application of dc voltages can produce pyroelectric fields that are slow (take hours) to decay. These fields can particularly affect operation of channel waveguide devices on Z-cut crystals, as discussed in Sect. 3.6.1.

Such sources of temperature-dependent changes in guided-wave properties have to be understood and allowed for in photorefractivity studies, or experiments must be carried out under carefully temperature-controlled conditions.

In addition to temperature changes, relative humidity may have an effect on device stability. This has recently been demonstrated by *Beaumont* and co-workers at British Telecom [3.127], who monitored the output of an integrated Mach-Zehnder interferometer under constant temperature and electrical bias conditions at differing humidity levels. At high (80%) humidity levels, large relative changes in output were observed in devices biased near extinction. In this case the effect may be associated with changes in a buffer layer rather than with an effect intrinsic to the waveguide material; nonetheless, the importance of considering it as a possibility in device testing is evident.

Another problem, ultimately of environmental origin, that must be addressed in fabricating good quality planar as well as channel waveguide devices is that of irregular guide depth as a result of fluctuations in the concentration of titanium [3.128] or other diffusants.

Polarization Changes. It has rather recently been found [3.99, 100] that photorefractive gratings which effectively convert guided beams from modes of ordinary polarization to those of extraordinary polarization can readily be formed in LiNbO₃ waveguides. This phenomenon will be described in Sect. 3.6.2; here we simply point out that the possibility of such gratings affecting experimental results has not generally been considered. If polarization conversion has been noted in some experiments, it may have been attributed to the coupling prisms used; these prisms often produce substantial polarization changes, presumably of photoelastic origin.

3.4.1 Holographic Experiments

Much of the holographic photorefractive work that has been done in LiNbO₃ and other materials is primarily of qualitative interest because of inadequate recognition of effects such as those described above. In some cases, analysis using more appropriate theory would also have been beneficial. A selection of planar-waveguide holographic experiments, showing the wide range of conditions investigated, is summarized in Table 3.1. Only a few studies require comment beyond that given in the table.

As mentioned at the beginning of this chapter, it has been observed by many researchers that Ti-in-diffused waveguides are often markedly more susceptible to optical damage than the substrates on which they are formed. The higher optical power density in the waveguide is of course allowed for. Usual measures of the effect are the external photorefractive sensitivity

$$S_e = \Delta n / I_0 \quad , \tag{3.1}$$

where Δn is the photorefractive index change, either at saturation or after a short time, and I_0 is the total incident optical energy density required to bring this change about; and the internal photorefractive sensitivity

$$S_i = S_e/\alpha \tag{3.2}$$

where α is the absorption coefficient at the wavelength of interest. S_e and S_i are usually quoted in units of cm^2/J and cm^3/J, respectively. Authors differ on the extent to which reflection and coupling losses are allowed for in the definition of I_0. *Glass* et al. [3.124] undertook to determine whether the Ti ions in the waveguide region had a direct influence on S_e and S_i. From a combination of photorefractive measurements on bulk LiNbO$_3$ samples doped in the melt with Ti, optical absorption measurements on Ti-in-diffused channel guides, and hologram formation studies in planar Ti-in-diffused guides, they concluded that these Ti ions – presumably Ti^{4+} ions on Nb^{5+} sites – did not play a direct role in photorefractivity. They suggest that the increased sensitivity may result from stabilization of photorefractive source Fe^{2+} impurity ions by the titanium. They interpreted their results, as did all the other research groups mentioned in Table 3.1, using the simple *Kogelnik* [3.112] formula for the diffraction efficiency η of a thick phase transmission hologram

$$\eta = \sin^2 \frac{\pi \Delta n l}{\lambda \cos \theta} \tag{3.3}$$

where l is the beam interaction length and θ is the angle at which the recording beams intersect in the waveguide. Whether their LiNbO$_3$ substrates were representative of typical commercial crystals, though, is open to some question, for several reasons: (a) they were unable to record at all at 633 nm; however, they did not have much power available at this wavelength; (b) even making allowance for various experimental uncertainties, they found external and internal photorefractive sensitivities (1.3×10^{-7} cm^2/J and 1.6×10^{-6} cm^3/J, respectively) which were actually lower than those in their bulk crystals; (c) their holograms decayed rather rapidly in the dark, the $1/e$ decay time for Δn typically being around 20 minutes. None of these results are in accord with general, albeit largely undocumented, experience; they suggest that the substrates used may be particularly low in iron or other photorefractively active impurities, but may possess a high density of shallow traps. It is also possible, of course, that all other experimenters have introduced substantial amounts of iron (or whatever) during waveguide processing. At any rate, the experiments do serve to demonstrate the ineffectiveness of Ti as a photorefractive species in LiNbO$_3$. This might have been anticipated from its failure to produce any absorption bands in the visible when added to LiNbO$_3$ in the bulk [3.124, 136], although there is again the possibility that preferred ionic configurations in the waveguide may differ from those in the bulk.

An indirect influence of Ti ions on photorefractivity was demonstrated in a somewhat different way by *Nisius, Krätzig* and co-workers [3.134, 137]. Using Y-cut LiNbO$_3$ containing 10 ppm iron they prepared Ti-in-diffused waveguides supporting three TM modes at 514 nm wavelength and formed gratings in the guides using each of these modes in succession. The gratings were then erased

Table 3.1. Selected experiments on photorefractive hologram formation in planar waveguides

Substrate	Type of guide	Write			Read		
		Wave-length [nm]	Polar-iza-tion	Power level	Wave-length [nm]	Polar-iza-tion	Power level
LiNbO$_3$ LiNbO$_3$:Fe LiTaO$_3$	out-diffused out-diffused Nb-in-diffused	488	TE	~1 W/cm^2	488 633	TE TE	~0.5 W/cm^2 low
LiNbO$_3$	Fe-in-diffused	633	TM	~3 W/cm^2	633	TM	~1 W/cm^2
LiNbO$_3$	Cu-in-diffused	488			633		
LiNbO$_3$	Ti-in-diffused Ti, Cu-in-diffused	442	TM	~2 W/cm^2	633	TM TE	low
LiNbO$_3$	Ti-in-diffused	458		0.2–2.5 W/cm^2	633		low
LiNbO$_3$ LiTaO$_3$	Ti-in-diffused Proton exchanged	458 515	TE	1.8–100 W/cm^2	633	TE	low
LiNbO$_3$	Ti-in-diffused	633	TE	1.8 mW per beam	633	TE	
LiNbO$_3$:Fe	Ti-in-diffused	633	TE	1–12 W/cm^2	633	TE	half write power
LiNbO$_3$ LiNbO$_3$:Fe	Proton-exchanged TIPE Ti-in-diffused	476 531 633	TE TE	30 W/cm^2 ~10 W/cm^2	476 633 633	TE TE	half write power low 5–200 W/cm^2
LiNbO$_3$	Ti-in-diffused	514	TM		514	TM	
LiNbO$_3$	Ti-in-diffused	527 633 1055	TE TM				

Note: DE = diffraction efficiency. Except as noted, all experiments were on Y- or X-cut samples; write beams symmetrical about normal to Z-axis.

Observations	Reference
DE <1% in undoped $LiNbO_3$. DE of 3% ($\lambda = 633$ nm) obtained in shallow (5 mode) out-diffused guide in $LiNbO_3$ doped with 50 ppm Fe (in melt); DE 52% ($\lambda = 488$ nm) and 28% ($\lambda = 633$ nm) in $LiTaO_3$: Nb guide, not repoled; much above DE in bulk. All DE's approximate. Grating formation time (to saturation) 1 to 10 min; grating periods $\sim 1-2$ μm.	[3.71]
In 2 or 3 mode guides ($\Delta n \sim 0.005$ or more), DE ~ 55% obtained. Gratings written with TM_2 could be read with TM_2 or TM_1; no intermode conversion seen. Fixed grating (1% DE) made by writing at 160 °C. Grating period ~ 1.2 μm.	[3.81]
Very thick (300 μm) diffused layer. Grating formed with external beams, as in Fig. 3.1(c); grating period ~ 0.8 μm.	[3.129]
Cu diffused locally to form high sensitivity region for grating coupler, formed using external beams; grating period $0.1-0.2$ μm, record time 10 s, erase time 100 s ($\lambda = 442$ nm). Better coupling for TE than TM ($\lambda = 633$ nm); Δn_{sat}^{av} in Cu area $\sim 5 \times 10^{-4}$, elsewhere in guide $\sim 8 \times 10^{-6}$. Grating decay time >3 months.	[3.130]
Unable to record in this guide at 633 nm with ~ 3 W/cm^2 in guide. Grating spacing ~ 4.6 μm. Gratings decayed rapidly even if unilluminated; for short exposures S_e $\sim 3 \times 10^{-7}$ cm^2/J, less than in bulk or crystal. See discussion in text.	[3.124]
"Wet" vs. "dry" processing evaluated for $LiNbO_3$: Ti guides. Initial $S_e \sim 3 - 4 \times 10^{-7}$ cm^2/J for guides made at 980 °C in dry O_2 atmosphere; $S_e \sim 3 - 5 \times 10^{-9}$ cm^2/J for guides made in Ar bubbled through RT water and cooled in O_2. Nominally Ag-exchanged (actually H-exchanged) guides less sensitive, especially in $LiTaO_3$. See text.	[3.43]
Gratings of maximum DE 50% formed readily in Y-cut samples; no gratings seen in Z-cut. Grating period ~ 6 μm. Few details; see text discussion.	[3.131]
Stated Fe concentration 0.03%. Grating periods $1-8$ μm; dependence of sensitivity on period evident. Recording times to saturation DE's of 45−90% were 2−8 min. Sensitivity said to be 2 orders of magnitude higher than in undoped $LiNbO_3$. Grating decay time days or longer.	[3.132]
No grating formation evident after 1 h attempts at recording in either type of guide. Proton-exchanged guides made in benzoic acid/1% Li benzoate melt. See text discussion.	[3.96]
Grating period 9 μm. Modulation depth varied; beam interaction effects noted; phase shift and signal gain plotted for recording times up to 40 min. Erase times at high read-beam powers ($100-200$ W/cm^2) around 20 min or more; beam interaction effects still apparent at 30 min, though.	[3.133]
Waveguide supports 3 TM modes. Grating period ~ 9 μm. Low Fe (~ 10 ppm) crystal, $\Delta n_{sat}. \sim 2 \times 10^{-5}$. Emphasis on erasure of gratings using external beam; see text.	[3.134]
Using 13 ps pulses with energy in guide up to 100 μJ ($\lambda = 1055$ nm) or 10 μJ ($\lambda = 527$ nm) per pulse, no gratings generally noted in guides on X-cut or Z-cut samples, although cumulative beam-spreading effects were seen, with up to 1000 pulses (repetition rate 2 s^{-1}), and gratings easily formed in X-cut using 633 nm beams. At proper beam angles, however, a short-lived 2-photon grating could be formed in X-cut using a single 1055 nm pulse.	[3.135]

using an incoherent beam entering the crystal through the base and traversing the grating region normal to the waveguide. By monitoring the grating diffraction efficiency during erasure, it was determined that the decay of the photorefractively induced index change followed a simple exponential law

$$\Delta n(t) = \Delta n_{\text{sat}} \exp[-\kappa \alpha_1 It/\varepsilon \varepsilon_0] \quad , \tag{3.4}$$

where κ is the specific photoconductivity, α_1 is an effective optical absorption coefficient described below, I is the erase beam intensity, and ε is the relative dielectric permittivity. Dark conductivity was negligible in these samples. Although it is not explicitly indicated by the authors, the erase beam is apparently at a wavelength similar to that of the write beams. It is therefore known from experiments on bulk crystals [3.138] that the absorption of this beam is largely due to $Fe^{2+} \rightarrow Nb^{5+}$ intervalence charge transfer processes; so the absorption coefficient is closely proportional to the concentration of Fe^{2+} ions encountered. This quantity is substantially the same regardless of the guided mode used to write or read the holograms; but the effective absorption coefficient to be used in (3.4) clearly has to be weighted according to the intensity distribution of the readout beam in the grating region. That is, if a grating is read out with one of the beams with which it was formed, and if this beam has a peak of intensity in a region of high Fe^{2+} concentration, the absorption by the erase beam will be more effective in reducing the diffraction efficiency, and the photorefractive index change, than it would be otherwise. Experimentally, it was found that, for a given erase-beam intensity, gratings formed, and apparently also read out, using the TM_0 mode were much more readily erased than those using the TM_1 mode, and those formed with the latter mode were in turn more sensitive than those made using the TM_2 mode. Thus, it is inferred that lower-order modes sample regions of higher Fe^{2+} concentration. For a diffused waveguide, lower-order modes travel closer to the surface, where the diffused Ti concentration is high, than do higher-order modes. The implication is clear, namely that Ti^{4+} ions stabilize the valence of Fe^{2+} impurity ions in the near-surface region in $LiNbO_3$: Ti waveguides.

Jackel et al. [3.43] compared the rate of hologram formation in $LiNbO_3$: Ti waveguides processed in wet and dry atmospheres. They found the initial external photorefractive sensitivity was about two orders of magnitude lower in the wet-processed samples. Forming the waveguides in an atmosphere containing water vapor leads to hydrogen incorporation into the guide region. The authors suggest that this may inhibit the tendency of the titanium to stabilize the reduced valence of transition-metal impurities, since a complex consisting of a Ti^{4+} ion and a nearby OH^- has the same formal valence as the Nb^{5+}-O^{2-} that it replaces.

In a later paper [3.96] these authors and colleagues looked at photorefractive sensitivity in $LiNbO_3$ waveguides in which large amounts of hydrogen had been incorporated by proton exchange. Both guides made by exchange in diluted benzoic acid melts and TIPE guides were investigated. They were unable to record holograms in either sort of waveguide after a total exposure to visible light of around $0.1\,\mathrm{MJ/cm^2}$ in the beam intersection region. This corresponds to an external photorefractive sensitivity about 4 orders of magnitude less than that in typical Ti-in-diffused guides. A single-beam experiment at very high power, $500\,\mathrm{MW/cm^2}$ in the guide, likewise showed no evidence of photorefractive change after 24 hours. As we shall see, other workers later detected photorefractivity in proton-exchanged guides by different techniques.

Uyukin et al. [3.131] compared grating formation under similar conditions in Y- and Z-cut $\mathrm{LiNbO_3}$: Ti waveguides. Although 50 % diffraction efficiency gratings could readily be formed in the Y-cut crystals, no gratings were detected in the Z-cut. TE polarization was used in both experiments; so the effective electro-optic coefficient $n^3 r$ was smaller by a factor of almost 3 in the Z-cut samples. The primary reason why diffraction from the beam-interference region was not observed in the Z-cut guide, though, is stated to be that it is a relatively small effect compared to the large index changes occurring all along the beam regions as a result of photorefractive charge separation normal to the waveguide plane. Photorefractive mode-index decreases of around 0.001 could in fact easily be detected in single-beam experiments. Possibly when one beam is blocked to attempt to read out a grating in a Z-cut guide, any grating that may exist is very rapidly erased by the readout beam. Carriers photoexcited from originally unilluminated regions by the readout light can simply be trapped outside the beam above or below their excitation positions, reducing the modulation depth. In a Y-cut crystal, on the other hand, if these carriers are to reduce the modulation, they have to be retrapped in regions which are illuminated by the readout beam; the trapping dynamics are likely to be different.

3.4.2 Single-Beam Experiments

This section is devoted primarily to summarizing our own experiences in photorefractive beam-shape experiments and to discussing the interpretation of such experiments.

a) Waveguide Formation

Single-mode (at 633 nm and at 442 nm) crucible out-diffused lithium niobate waveguides are formed by heating polished and poled Y-cut lithium niobate single crystals in special lithium niobate reaction crucibles (made of porous $\mathrm{LiNb_3O_8}$) for 7–10 min at 1100° C in pure static oxygen with $p_{\mathrm{HOH}} > 100\,\mathrm{mm}$ Hg [3.20, 31]. A moistened oxygen atmosphere is established at room temperature by first evacuating the furnace, and then by refilling to 1 atm with pure oxygen bubbled through a heated water bath. In addition, special care is taken to eliminate chemical impurities that apparently cause competing chemical reducing conditions to develop. Contamination is eliminated from the surfaces of both the crystals and the reaction crucible by chemical precleaning. The use of platinum foil to position a crystal within a reaction crucible has been discontinued in favor of using clean lithium niobate crystal shims. The crucible out-diffusion method was developed as a way to make reproducible, low loss, single-mode out-diffused guides.

Titanium-in-diffused lithium niobate waveguides are prepared according to common practice, in the same furnace and using identical quality substrates, by depositing 25 nm titanium metal films, pre-annealing at 650° C for 4 h, and then annealing at 1000° C for 5.5 h, both in pure moistened static oxygen. Equal care is taken to avoid inadvertent chemical reducing conditions.

For comparison, both out-diffused and Ti-in-diffused guides have been made in dry oxygen atmosphere. Proton-exchanged guides discussed here were made by immersing X-cut LiNbO$_3$ crystals in pure benzoic acid at 160° C for 10–30 min. To reduce scatter and to reduce surface index so the SrTiO$_3$ coupling prisms could be used, they were annealed at 300–400° C, typically for 15 min.

b) Optical Characteristics

The apparatus for observing changes in shape of a beam transmitted through these waveguides has already been described. In most experiments, the quality of the waveguide's output beam is monitored for a minimum time of 100 h, or until a steady-state condition is evident. Continuous measurement is made of the enclosure temperature, of the incident laser power, and of the waveguide's Gaussian output beam as recorded in the far field. The spatial distribution of the output power in the waveguide plane is recorded at regular intervals.

The same experimental procedures are used for monitoring waveguide performance under the purposely stressful conditions created by higher laser powers and shorter wavelengths. In these cases, however, a photodiode and oscilloscope monitor are used in place of the optical multichannel analyzer.

c) Steady-State Laser Power-Handling Performance of Lithium Niobate Waveguides

The character of optical damage in a lithium niobate waveguide is determined by recording the spatial distribution of the waveguide's output power in a plane parallel to that of the waveguide, i.e. along an m-line [Ref. 3.2; p. 101]. Actual data for an outdiffused waveguide are given in Fig. 3.4a for a 5 % power loss, and in Fig. 3.4b for a 15 % power loss. The power lost from the central beam appears as increased scattering in the plane of the waveguide and towards the negative c-axis side of the central beam. Power loss can be seen to occur at the expense of the beam's negative c-axis portion (cited incorrectly in our early papers [3.94, 139] as the positive c-axis portion), and no line broadening is evident until the amount of power loss exceeds about 60–70 %.

This same general character was exhibited by every planar graded-index lithium niobate waveguide we studied, including those made by out-diffusion, titanium-in-diffusion, and proton exchange. Moreover, this characteristic, unlike the tendency for a waveguide to exhibit it, was found to occur completely

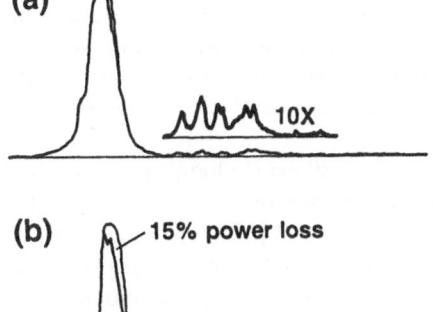

(a) 5% power loss

10X

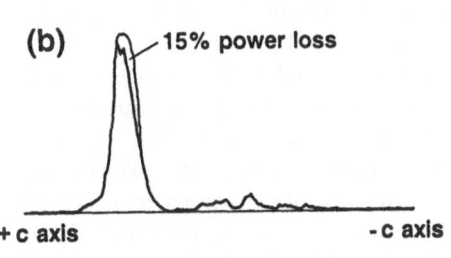

(b) 15% power loss

+c axis   -c axis

Fig. 3.4. Spatial distribution of transmitted laser power in waveguide plane after **(a)** 5 % and **(b)** 15 % central beam power loss resulting from photorefractive damage at wavelength 633 nm

Fig. 3.5. Central beam output power loss as a function of time for LiNbO$_3$ waveguides exposed to different levels of 633 nm radiation: (a) crucible out-diffused guide; (b) titanium-in-diffused guide
∇

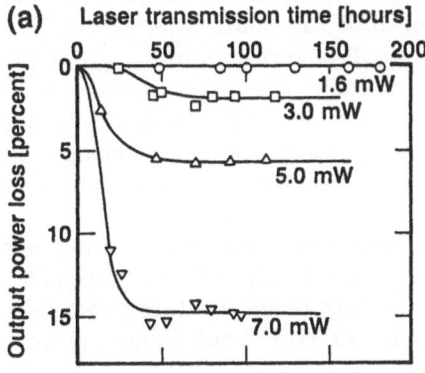

(a) Laser transmission time [hours]

1.6 mW
3.0 mW
5.0 mW
7.0 mW

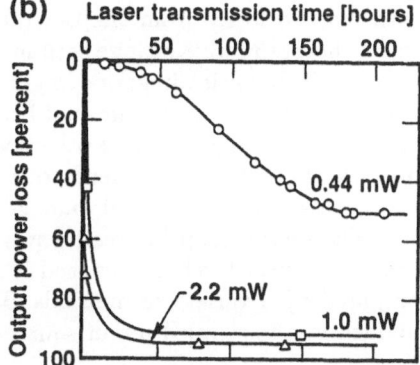

(b) Laser transmission time [hours]

0.44 mW
2.2 mW
1.0 mW

independently of a waveguide's stoichiometry, iron concentration, or degree of reduction.

The output power loss as a function of laser transmission time (100–200 h) is given in Fig. 3.5a for a representative crucible-out-diffused single TE-mode lithium niobate waveguide, prepared under oxidizing conditions, and then exposed to a HeNe laser source with a 0.9-mm beam diameter [3.94, 139]. The laser power levels reported refer to the actual total power coupled into the waveguide. In each case the power loss saturates after about 60 h of continuous exposure at a level increasing nonlinearly with the propagating power.

The behavior of an equally representative titanium-in-diffused lithium niobate waveguide, exposed to the same laser source, is described [3.94] in Fig. 3.5b. Although the power loss once again saturates after continued laser exposure, there are three notable differences in the behavior of the titanium-in-diffused and the crucible-out-diffused lithium niobate waveguides. First, the in-diffused waveguide suffers far greater induced power loss at any given laser power level than does the out-diffused waveguide. Second, while the saturation level of power loss increases nonlinearly with the laser power coupled into the in-diffused guide, it does not do so as markedly as it does in the out-diffused waveguide. Third, the time required for an in-diffused waveguide to reach saturation is roughly three times as long as the time required for a similar power loss to occur in an out-diffused waveguide.

The necessity of making observations over a sufficiently long time to ensure the establishment of steady-state conditions is demonstrated very dramatically by experiments on proton-exchanged waveguides [3.140]. In Fig. 3.6a we show the development of central-beam power loss in three such guides, processed under slightly different conditions, over 24 h of exposure to a 633 nm guided beam about 1 mm wide at a guided-wave power level of 4 mW. Guide (iii), annealed at a higher temperature, has damaged rather rapidly, but guides (i) and (ii) appear to be very damage-resistant. Extending the observations to a period of 400 h, though, we obtain quite a different picture. Guide (iii) had about reached the saturation level after 24 h, but guides (i) and (ii) do not begin to develop noticeable power loss until about 60 h exposure and do not saturate, with marked losses from the central beam, until 200–300 h have passed. The small photorefractive change seen in the first half-hour of exposure in guide (ii) is particularly misleading; one might easily conclude, after a day or two of testing, that a true steady state had been attained. These experiments also show, more emphatically that any others we have carried out, how small changes in processing conditions can lead to large changes in both the steady-state level and the dynamics of optical damage.

The relationship between waveguide output power loss at saturation and the laser power level that caused it is summarized in Fig. 3.7a for waveguides prepared by a variety of methods. The percentage of steady-state power-loss, divided by the coupled power squared, shown by a typical out-diffused lithium

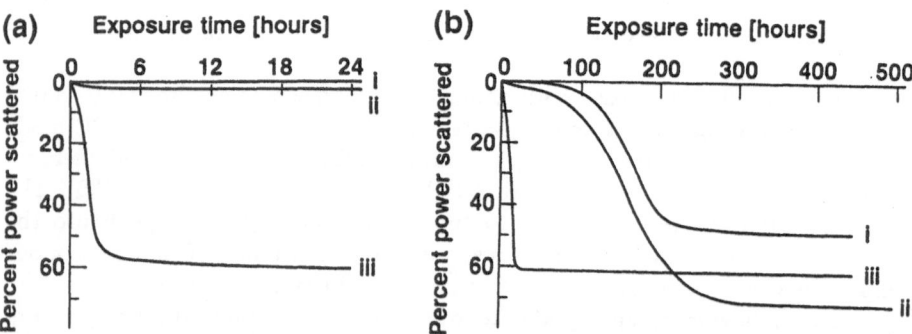

Fig. 3.6. (a) Development of output-power loss of central portion of guided beam in LiNbO$_3$ annealed, proton-exchanged planar waveguides prepared in different ways. Period of observation 24 h; (b) Same as (a), but period of observation 400 h. Preparation conditions: exchange temperature 160° C; anneal time 15 min; exchange times 10 min, 20 min and 20 min for guides (i), (ii) and (iii), respectively; anneal temperatures 300° C, 300° C and 400° C, respectively

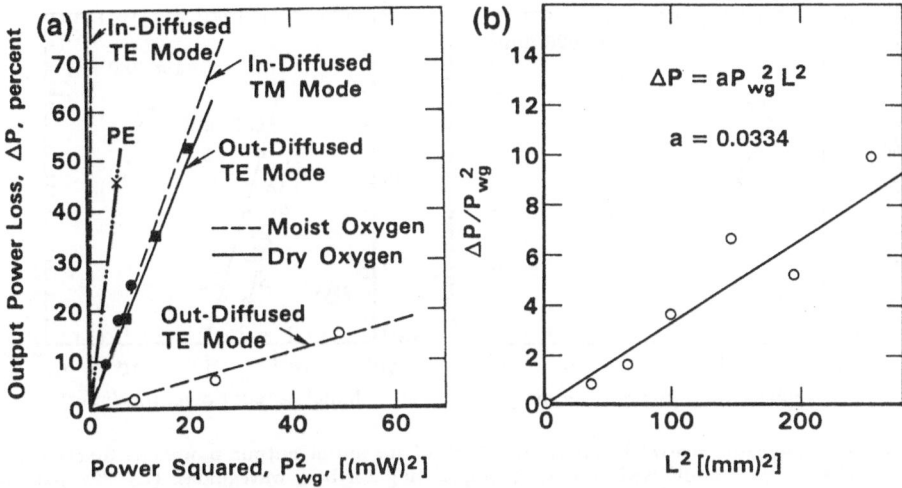

Fig. 3.7. Photorefractive central-beam saturation power loss, at wavelength 633 nm, in LiNbO$_3$ waveguides; (a) in guides prepared various ways, as function of square of input-coupled power; (b) in crucible-out-diffused guide, as function of square of beam transmission distance. Solid line is least-squares fit to equation shown. Units of coefficient **a** are percent loss/(mW-mm)^2

niobate waveguide propagating a 1 mm beam at 633 nm, is summarized in Fig. 3.7b in terms of the square of the distance in millimeters between the input and output coupling points. Each data point represents an average over 3 or 4 experiments, conducted to saturation at the indicated propagation lengths. Figs. 3.7a and b together indicate that a lithium niobate waveguide's steady-state output power loss increases as the square of both the input-coupled power and the transmission length in the guide. The results speak for themselves. Out-diffused guides reach a saturation level of optical damage lower than that in Ti-in-diffused guides. Out-diffused guides prepared in dry oxygen reach a damage level higher than those made in moistened oxygen. The proton-exchange guide shown, which supported 3 modes at 633 nm, reached the damage level indicated after 60 hours with 2.5 mW coupled into the guide.

A somewhat different way of analyzing single-beam experiments was adopted by *Handa* et al. [3.51] in an extensive study of the effects of waveguide preparation conditions on optical damage and acousto-optic properties. They conducted long-term experiments, but instead of invariably waiting for a steady-state central beam transmission level to be attained, they adopted the time for the transmitted power to drop to 90 % of its initial value as a measure of photorefractive sensitivity; they call this period the throughput decay time. The sigmoidal nature of the decay curves indicated in Fig. 3.5 indicates that a single parameter may not be entirely adequate for modeling the decay characteristics; nonetheless, their results, presented in Fig. 3.8, display very good internal consistency. The results are in good qualitative agreement with those presented in the previous paragraph. In addition, they display quite clearly the

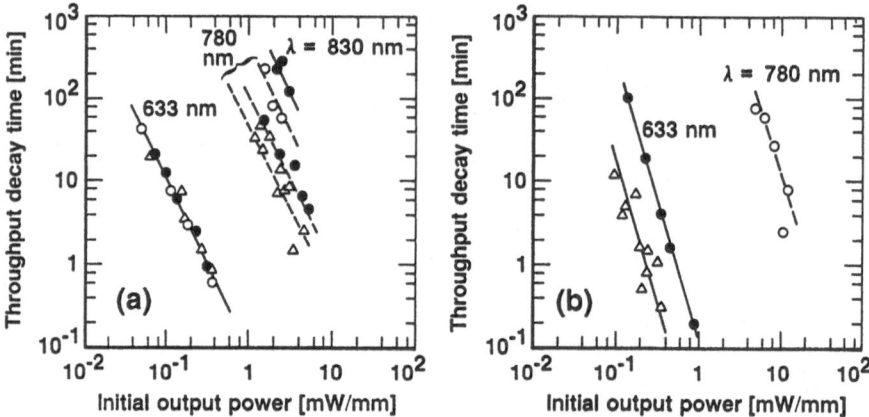

Fig. 3.8. Time for central beam to lose 10% of its initial output power, as function of that power, in Ti-in-diffused, Y-cut, X-propagating LiNbO$_3$ waveguides: (a) TE$_0$ mode; (b) TM$_0$ mode. Waveguides indicated by *solid circles* were in-diffused in atmosphere of O$_2$ bubbled through water at room temperature; *open circles*, O$_2$ bubbled through water at 80° C; *triangles*, dry O$_2$. After [3.51]

many orders of magnitude improvement in initial resistance to photorefractive beam distortion that may be obtained by working at longer wavelengths. It is interesting to note that while the throughput decay time varies inversely as the square of the initial transmitted power density for TE modes, it varies as the inverse cube of this quantity for TM modes. Reference [3.51] contains additional interesting information on waveguide properties which we have not space to summarize here.

Single-beam experiments on Ti-in-diffused guides in MgO-doped LiNbO$_3$ have recently been reported briefly by *Fejer* et al. [3.75c]. These were experiments of a somewhat different sort in that the beam studied is the second harmonic at 532 nm generated from 1.064 μm wavelength input from a Nd : YAG laser. The sample was heated to an optimum phase-matching temperature, and the intensity of the central portion of the output 532-nm beam was monitored. Under cw conditions, about 800 μW of second-harmonic could be generated, and this quantity did not decay in 1 hour. With higher-power pulsed inputs, changes in output efficiency were noted, but the relative contributions of photorefractive and thermal effects to these changes could not be readily separated. Moderate heat treatment, probably affecting the oxygen stoichiometry, strongly increased the damage susceptibility.

Proton-exchanged guides made in MgO-doped LiNbO$_3$ by immersion in pure benzoic acid do not, according to *Jackel* [3.76b], show the long-term (days to months) index fluctuations which develop [3.47, 57] in undoped guides similarly fabricated. *Digonnet* et al. [3.76a], on the other hand, reported such changes but did not give any details. Whether such guides are resistant to photorefractive damage over similar time periods has not been determined;

short-term measurements [3.76a] indicated a damage intensity threshold, at 514.5 nm, about twice that in proton-exchanged guides in undoped LiNbO$_3$.

For each type of diffused lithium niobate waveguide, the *percentage* of power loss, ΔP, is found to vary as the square of the input-coupled laser power, P_{wg}, ($\Delta P = K P_{wg}^2$), where for a fixed scattering length K is a constant depending on the substrate material and the waveguide processing conditions [3.115]. It is more useful, however, in designing practical devices, to relate the waveguide's performance to the output power delivered in steady-state operation, P_o^∞,

$$P_o^\infty = P_{wg} 10^{-\alpha L/10}(1 - K P_{wg}^2/100) \tag{3.5}$$

where α is the waveguide transmission loss (dB/cm), and L is the optical path length in the waveguide (cm). This expression conforms closely with the experimental data.

The maximum power output that can be delivered by any given waveguide is determined by calculating the maximum of (3.5),

$$P_o^\infty{}_{max} = (14.81/K)^{1/2} 10^{-\alpha L/10} \quad . \tag{3.6}$$

A typical outdiffused waveguide is able to deliver a maximum steady-state output of 6.8 mW with 10.5 mW coupled into the guide. With 75 % input coupling efficiency and a path length of 16 mm, the total insertion loss at this power level is less than 3.2 dB. The proton-exchanged and titanium-in-diffused waveguides, given equal coupling efficiency and path length, are typically able to deliver only around 1 and 0.2 mW, respectively.

The laser power-handling performance data indicated above correspond to situations where the light propagates orthogonal to the Z-axis, along either the X- or Y-axis in the lithium niobate waveguide. When the laser beam propagates along the Z-axis, however, a substantial difference results: no output power loss is observed, regardless of the power or wavelength coupled.

Z-axis propagation is allowed only in titanium-in-diffused waveguides, because only this process produces an increase in ordinary as well as extraordinary refractive index. When a 7 mW laser beam (633 nm) of 1 mm diameter propagates 10 mm along an X-cut titanium-in-diffused waveguide's Y-axis, more than 90 % of either the TE or TM input-coupled power scatters asymmetrically into the m-line. When the waveguide is remounted so that the same laser beam travels the same distance along the Z-axis, no output power changes are observed, even after more than 200 hours of surveillance. The same observation is made for Y-cut waveguides, and for similar exposures at 442 nm.

This behavior might be a consequence of the fact that photogenerated charge separates only along the crystal's Z-axis. When the relatively narrow beam travels orthogonal to the Z-axis, charges separate a short distance, and the effective fringing fields reside predominately within the path of the light. When the same beam travels along the Z-axis, the charges, which are created only in the path of the light, must also separate along this path. Interferometric experiments on bulk crystals [3.141] show that no photorefractive phase shift can be detected for beam propagation along the Z-axis.

d) Effects of Preparation Conditions; "Optical Cleanup"

A dramatic nonlinearity followed by quasi-steady-state behavior occurs when waveguides that have been strongly chemically reduced are exposed to high laser powers (1–12 mW) and, especially, short laser wavelengths (488 and 442 nm). Under such conditions, these waveguides actually develop damage quite rapidly, and then recover equally rapidly, establishing what appears to be an undamaged, or cleaned-up, steady state. This rapid self-homogenization of optical damage is shown in Fig. 3.9 for an out-diffused lithium niobate waveguide continuously propagating 10 mW at 488 nm. The waveguide appears to suffer almost complete loss of output in less than 0.25 s, recovers > 80 % of its output in 1 s, and then recovers its original output in 4 s, thereafter maintaining steady-state performance. Increasing the laser power causes this process to occur more rapidly, whereas decreasing the power causes it to occur more slowly. At a longer wavelength (633 nm) and at similar laser power levels, the same self-homogenization cycle occurs, but over a much longer period (20–40 hours). Any movement of the input laser beam, however slight, causes the entire process to replay itself.

This damage followed by recovery does not imply that optical damage occurs and then mysteriously disappears. On the contrary, the damage occurs, and then worsens in time, eventually establishing a steady-state perturbation in refractive index across the diameter of the beam. Light is scattered out of the beam only as a result of an inhomogeneous refractive index distribution. If this refractive index distribution were to become homogeneous across the beam diameter, scattering would decrease markedly. *Ohmori* et al. [3.142] demonstrated for bulk crystals of lithium niobate that the maximum amount of laser-induced refractive-index change possible for a given crystal saturates at a specific laser power density determined by the material chemistry and the laser wavelength. The same principle must apply within a waveguiding layer. When the Gaussian laser beam propagating in a waveguide possesses a power density that exceeds the saturation power density for the given material across most

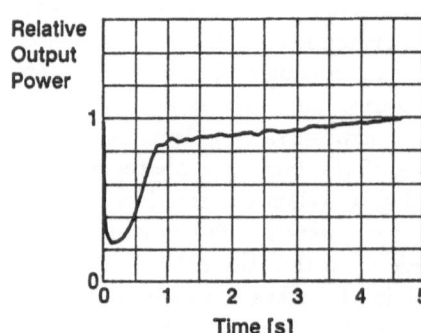

Fig. 3.9. Integrated central-beam far-field output power as function of transmission time for a chemically reduced crucible-out-diffused $LiNbO_3$ waveguide initially propagating 10 mW at wavelength 488 nm

of its beam diameter, the induced refractive-index change eventually becomes uniform across the beam. The more quickly saturation is attained, the shorter is the time required for recovery of the waveguide's original output power.

Waveguides prepared under oxidizing conditions, even those formed by titanium-in-diffusion, do not exhibit this behavior. No evidence of damage and recovery could be detected either at 633 nm or 442 nm, with 7 mW propagating. The oxidized titanium in-diffused waveguides degraded at 442 nm much as they did at 633 nm, but did so far more rapidly (2 s to a steady-state power loss of 92 %).

Waveguides become strongly reduced when they are formed in the presence of reducing agents such as low oxygen partial pressure and carbonaceous contamination, or catalysts such as platinum. Waveguides also become strongly reduced when they are exposed at lower temperatures (150–600° C) to metal surfaces or to vacuum. Such treatment of lithium niobate increases the ratio of divalent to trivalent iron ions, thereby raising the material's tendency towards optical damage [3.28a]. Recent EPR experiments [3.143] indicate that such treatment additionally produces Ti^{3+} states lying in the band gap; these states may very well also contribute to the photorefractive susceptibility.

Barnoski, Tangonan, and co-workers [3.93, 28b] first pointed out effects of this type, which they called "optical cleanup", in titanium-in-diffused lithium tantalate waveguides. In studying such guides at quite high input-coupled laser power (247 mW at 514.5 nm) they noted that, while extensive damage occurred at first, it was later followed by a relatively long period (> 45 minutes) of partial recovery. We, on the other hand, have not seen evidence of any tendency towards damage reversal in the more highly damage prone, as prepared, titanium-in-diffused lithium niobate waveguides. This suggests that during their formation the tantalate waveguides had become moderately reduced. Had we purposely subjected titanium-in-diffused lithium niobate waveguides to sufficiently nonoxidizing conditions, they might also have exhibited damage-reversing behavior.

e) Discussion

Asymmetric photorefractive scattering of light from portions of a guided beam propagating perpendicular to the optic axis is readily understood as a gradient-index lens effect. The tendency of the light to be scattered in a particular direction, as indicated in Fig. 3.4, may be attributed to dynamic hologram formation. This point of view is close to that expressed recently by *Kandidova* et al. [3.144] in interpreting their experiments on beam-shape changes in bulk crystals of Fe-doped $LiNbO_3$. As pointed out by *Chen* [3.92], the photorefractive charge redistribution leads to a reduction in the extraordinary refractive index in the center of the beam region and an increase in the wings. Direct electrostatic measurements of surface charge [3.145] confirm this model. Rays propagating in regions where there is a gradient in refractive index will tend to be deflected towards the higher-index region. Initially this tends just to broaden the beam slightly, but eventually the bulk photorefractive effect produces a field gradient which favors scattering toward the negative c-axis [3.92, 146]. (In a sufficiently reduced, iron-containing sample, this gradient may not occur [3.146]). Beams deflected in either direction intersect beams undergoing little deflection or being bent slightly back toward the center of the beam. In the intersection region, dynamic holograms will form, and for some characteristic grating vectors depending on the refractive index gradients, the propagation distance, and the beam width, these holograms will be amplified, and preferential scattering in some characteristic directions will be observed. As the beam distortion develops, though, the overall index change pattern will vary, and the grating vectors which are amplified also probably change. Detailed study of the short-term part of the process in LiNbO$_3$: Ti by *Jerominek* et al. [3.147], as well as some of our own observations and the experiments on bulk iron-doped samples [3.144, 146], show that the process can be quite complicated, with several maxima in the output optical power distribution developing, shifting, and disappearing; a quantitative theory may thus be rather difficult to devise. At present, there is not even any quantitative theory describing the dynamics of the overall long-term changes in power in the central beam region, such as illustrated in Fig. 3.5, in terms of optical scattering and waveguide preparation conditions. The corresponding problem for bulk crystals is being studied by a group at Osnabrück University [3.148]. Polarization-mixing terms in the photogalvanic tensor (see Sect. 3.6.2) may play a role in this situation [3.149].

3.5 Two-Step Photorefractive Processes

The high light intensities possible in optical waveguides should make it relatively easy to observe effects of two-photon absorption, either with or without a real intermediate state. Thus one might expect that two-photon photorefractivity, the subject of many experiments using bulk crystals, would have also been extensively studied in waveguides, particularly in view of the considerable interest in other nonlinear optical phenomena, such as second-harmonic generation [3.75c, 150] and light-intensity-dependent change in refractive index [3.151] in waveguide geometries. Surprisingly little work has in fact been done. In this section we shall quickly summarize some of the facts about two-photon photorefractivity in the bulk, since they do not seem to be covered elsewhere in these volumes; then we describe the waveguide experiments of which we are aware.

3.5.1 Two-Photon Absorption

At sufficiently high light intensities, the optical absorption coefficient below the band gap of an insulating solid can be represented in the form

$$\alpha = \alpha_0 + \alpha_2 I \quad , \tag{3.7}$$

where I is the intensity, and the nonlinear coefficient α_2 arises from terms in the third-order susceptibility tensor corresponding to simultaneous absorption of two photons through virtual intermediate states. The coefficients α_0 and α_2 naturally may depend on the polarization of the incident light relative to the crystal axes. In LiNbO$_3$, α_2 is of comparable size for ordinary and extraordinary rays [3.152], with the value for the ordinary ray being somewhat larger for two-photon total energies spanning the band gap. A typical value of α_2 at wavelength 532 nm is 5×10^{-9} cm/W [3.90]; thus high-intensity pulsed lasers are required to observe two-photon effects. If the total energy of the two absorbed photons exceeds the band gap, electron-hole pairs may be produced. For LiNbO$_3$, two 532 nm photons from the frequency-doubled output of a Nd:YAG laser easily span the energy gap, so most experiments have used such a source. 659 nm photons obtained by frequency doubling the 1318 nm Nd:YAG laser have also been used successfully. Two undoubled, 1064 nm, photons from Nd:YAG will not reach the band gap, but one 1064 nm and one 532 nm photon reach an energy of 3.5 eV, a region of measurable absorption in the band tail of most LiNbO$_3$ crystals; as we shall see, effects involving the two different wavelengths have been observed. To investigate the spectral dependence of two-photon photorefractive effects, dye-laser light sources have been used [3.153, 154], but since they have operated at somewhat lower powers than the Nd:YAG lasers, it is no longer so easy to separate linear from quadratic absorption effects.

3.5.2 Observations in Bulk Crystals

Holograms are quite readily formed in LiNbO$_3$ using high-power pulses of suitable wavelength which are beam-split and recombined at the crystal much as in cw experiments. By recording for a sufficiently long time, diffraction efficiencies of 25 % or more can sometimes be seen and easily measurable efficiencies in the tenths of a percent range can almost always be obtained. Salient features of the nonlinear hologram recording process are the following:

- The photorefractive sensitivity, often expressed in these experiments in terms of Δn/pulse, increases as the square of the incident power density.
- The effect is much less sensitive than the linear photorefractive effect to the presence and nature of impurities [3.90], although there is some evidence [3.153, 155] for increased sensitivity in iron-doped crystals.
- Recording is only possible [3.153, 154] when the total energy of the two photons reaches the range of the band edge for undoped material, around 3.8 eV. This again is in contrast to linear recording, where the sensitivity drops off only very gradually with wavelength.
- The process is *not* similar to that obtained by creating an electron-hole pair with ultraviolet excitation and then allowing the bulk photovoltaic effect to produce charge separation. For one thing, it is much more efficient. In undoped crystals, hardly any photovoltaic current at all can be obtained under 320 nm ultraviolet excitation [3.154, 156], although such a one-photon band-to-band current is not prohibited by any symmetry considerations [3.157].

- The sensitivity is enhanced [3.90] when infrared light at 1064 nm is allowed to play on the hologram region during recording.
- At low temperatures, two-photon irradiation creates color centers of two types [3.158, 159]. Similar centers can be created with x-rays [3.160, 161]. ESR experiments [3.160, 161] suggest that one type of center consists of electrons trapped at niobium sites which form Nb^{4+} ions, and the other consists of holes trapped at oxygen sites near defects. This process does not in itself seem to lead to photorefractive changes.

The most plausible picture of two-photon photorefractivity, given these observations, is along the following lines [3.156, 159]: two-photon excitation at room temperature does not simply produce mobile electron-hole pairs; the carriers produced are quickly trapped in shallow traps like those seen at low temperature. These trap states are short-lived in the dark at room temperature. In the presence of additional visible or near-ultraviolet photons, though, one or both of these kinds of traps may be emptied with a high probability. The mobile carriers thus excited from the shallow traps are retrapped preferentially along the polar axis in deeper states, and these contribute to the photorefractive effect in the usual way. The overall process is thus really one using *three* photons, in which the final step is saturated. Single-photon excitation may well also create shallow trap states, but their effect on the photorefractivity may not be noticeable because of (1) competition from photorefraction resulting from excitation from impurities, (2) lack of sufficient light intensity to empty all the traps, and particularly (3) the conventional exponential absorption characteristic, which may lead to saturation filling of traps near the front face, with few traps filled deep in the crystal.

Investigations of the dynamics and spectral dependence of photoerasure of holograms [3.162] indicate that in iron-doped $LiNbO_3$ the same final trap states are involved in holograms recorded by one-photon and by two-photon processes. In undoped samples, on the other hand, the photorefractive index

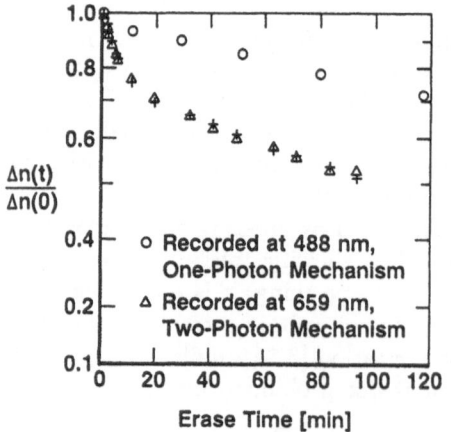

Fig. 3.10. Erasure, using 488 nm laser light, of holograms recorded in undoped bulk $LiNbO_3$ by one- and two-photon mechanisms. Plus signs represent two-exponential fit to the two-photon-hologram erasure data

change in two-photon holograms falls off with erase time in a way that is best fit by two exponentials, Fig. 3.10, indicating the persistence of some shallower trap states.

Two-photon holograms of very high sensitivity have been recorded [3.163] in $KTa_{0.65}Nb_{0.35}O_3$ crystals using 532 nm picosecond pulses. Since this material is not ferroelectric at room temperature, an applied electric field in the range of 1–10 kV/cm is necessary to produce photorefraction. In KTN crystals, dark currents generally limit the hologram storage time to 10 h or so. Chromium-doped lithium niobate contains a real impurity state near the center of the gap, so two-photon two-step excitation of photorefractivity at modest power levels is possible through this state [3.164]. Recombination through this state apparently limits the storage time to around 20 h.

3.5.3 Waveguide Experiments

With the above background, we may describe succinctly the experiments on two-photon photorefractivity in waveguides. The first observations [3.91] were on an undoped Y-cut LiNbO$_3$ crystal which had been out-diffused at 950° C in oxygen for one hour. This treatment produced a waveguide supporting a single TE mode at 532 nm. Pulses of this wavelength and of duration 140 ns were split and prism-coupled into the waveguide in a arrangement like that in Fig. 3.1a. With incident pulse powers in the 2 kW range, the refractive index change produced was found to vary as the square of the power density, indicative of the two-photon process. As a typical example, a grating of diffraction efficiency 1.2% was formed using 14 pulses of 2 kW power. The power density in the waveguide was estimated to be around 2.5 MW/cm². The external sensitivity, referred to the power in the guide, is in the range of 2 cm²/MJ. This is roughly comparable to what one might expect from extrapolating results obtained at higher powers in bulk crystals. If the sensitivity in the waveguide case is somewhat higher, it may be because the confinement of the light in one dimension leads to a large overlap of the highest-intensity parts of the recording beams.

The observation of infrared enhancement of photorefractive sensitivity in the bulk leads one to hope that it might also be possible to record holograms at infrared wavelengths in LiNbO$_3$ by "sensitizing" the beam intersection region with visible light. Just such experiments, in a bulk configuration, have been reported briefly [3.90, 159]. The holograms were formed with 1064 nm pulses from a Nd:YAG laser and sensitized with a doubled 532 nm pulse. No quantitative results were given. In a waveguide, one might attempt to form gratings with the infrared beams in the guide, while sensitizing with an external visible beam incident just on the beam intersection region from the top. This could be a very useful method of forming waveguide gratings to be used for beam splitters, for instance, or for wavelength demultiplexers. An arrangement of this type would have the following advantages:

1. many integrated devices are going to use infrared beams from laser diodes as their light sources;

2.   using the beam path and wavelength that will be used in the final device adapts the hologram to the peculiarities of the particular waveguide employed;

3.   the sensitizing visible beam needs to be incident only in the beam intersection region; thus it cannot contribute to undesirable photorefractive changes elsewhere in the waveguide;

4.   the hologram may be read out at a wavelength and power level where the material is photorefractively insensitive; so without any fixing process the grating may be addressed indefinitely without decaying.

Unfortunately we have never been able to make this external-beam sensitization procedure work.

What we have been able to do is to record holograms using 1064 nm wavelength guided beams by sensitizing the waveguide with another visible guided beam [3.165]. The basic experimental arrangement is shown in Fig. 3.11. The waveguides were made by diffusing Ti into Y-cut LiNbO$_3$ crystals at 950° C in dry oxygen to form layers supporting a few modes at both wavelengths used. TE polarization was used for both beams. A cw 1064 nm beam from a Nd : YAG laser was split and the resulting beams then crossed in the waveguide to produce an interference pattern approximately 200 μm wide with a spatial wavelength of about 4 μm. With about 1 mW of infrared power entering the waveguide, no permanently recorded photorefractive grating was produced. Focusing a 633 nm

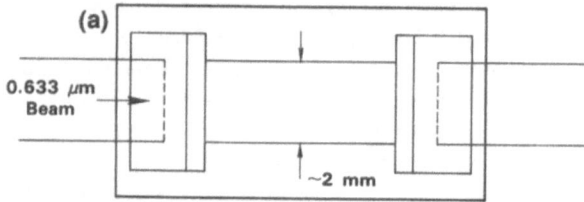

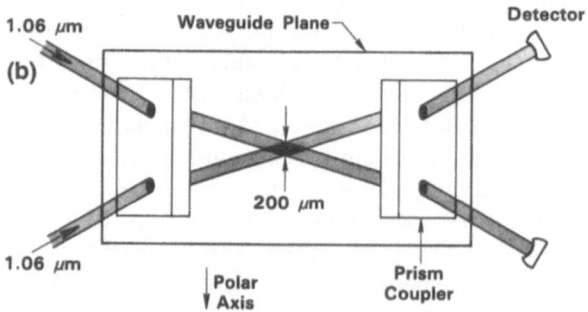

Fig. 3.11. Arrangement for recording holograms at infrared wavelengths in LiNbO$_3$: Ti optical waveguides: (a) broad sensitizing visible beam introduced immediately prior to infrared beams; (b) infrared beams intersecting in waveguide

external beam from a He-Ne laser on the guided-beam intersection region still did not yield a permanent grating. However, when a broad guided beam, 2–4 mm wide, of the red He-Ne light was passed through the intersection region, gratings were readily formed in some crystals, and their diffraction efficiency, measured by briefly blocking one infrared beam, continued to increase even when the red beam was turned off. Subsequent experiments showed that exposure of the waveguide to the guided red beam for a minute or two before the infrared beams were introduced was sufficient to sensitize the guide so that infrared holograms could be recorded. The gratings formed were stable in the dark for at least 24 h. Exposure to a total guided-wave energy of red light of as little as $12\,\mu J$ could sensitize the waveguide enough to produce a detectable grating. The gratings could be erased by exposure to another broad beam of guided visible light. Readout of a stored hologram with one infrared beam for 30 min produced little or no erasure. When the guided red beam or a guided green beam (532 nm) was focused on the infrared-beam intersection region while the infrared beams were on, no grating was recorded. Presumably in these cases the visible beam erased the holograms as fast as they were being formed. Different $LiNbO_3$: Ti waveguides displayed this phenomenon to different degrees, and not all points in a given waveguide were equally sensitive. Observed external diffraction efficiencies varied from less than 0.1 % to 25 %. This range of values results in part from variations in coupling efficiency.

Subsequent to the completion of these experiments, we came upon the remark of *Kurz* and *von der Linde* [3.159] to the effect that they formed gratings using 1064 nm beams in bulk crystals which had been sensitized by prior exposure to a green beam. Such observations indicate that in nominally undoped $LiNbO_3$ crystals there may be shallow traps which can be filled by visible-light excitation. The presence of these traps could complicate interpretation of experiments on intrinsic two-photon photorefractivity.

We are not aware of any experiments demonstrating two-photon effects in channel waveguides.

3.6 Photorefractivity in Channel Waveguides

Most of the applications for integrated-optical devices proposed in the past few years are for high-speed switching or for rotation sensing. For such devices, with several distinct input and output ports, channel waveguides are required. There has consequently been a great deal of interest in effects in channels, particularly at the infrared wavelengths being used in optical communication systems. Study of these effects is often complicated by thermal effects, by photoconductive contributions to charge transport, and by polarization conversion phenomena. In the present section we first summarize some of the experimental studies, with emphasis on recent work, and then describe the recently discovered phenomenon of polarization conversion.

Table 3.2. Selected experiments on photorefractivity in channel waveguides

Crystal cut	Propagation direction	Waveguide type	Channel width [μm]	Device	Observation wavelength [nm]	Guided wave polarization	Power level
Z		TI	3 5	Δβ reversal directional coupler switch	633 1064	TE TE	0.2 W/cm² 1–10 kW/cm²
Y	X	TI	10	straight channel	440 633	TE, TM TE, TM	>0.06 μW in >1 μW in
Z	X	TI	8	COBRA-type directional coupler switch	1150	TM	
Y	X	TI	10	straight channel	512	TE	>200 W/cm²
Z		TI	5	uniform Δβ directional coupler switch	1064	TM	1 W/cm² (at 633 nm)
X	Y	TI PE	3.8	symmetric Mach-Zehnder interferometer	850	TE	22 W/cm²
Z	Y	TI	4	passive directional coupler (no electrodes)	810 850 633	TE	0.5–290 W/cm²
Y Z		TI PE	10 3.5	straight channel; passive coupler	633	TE, TM	up to 150 μW out
Z Y X	X X Y	TI	4	asymmetric Mach-Zehnder; straight channel	830	TM TE	to 10 kW/cm²
X	Y	TI	3.8	symmetric Mach-Zehnder	458 850	TE	up to 65 W/cm²
Z	Y	TI	6	asymmetric Mach-Zehnder; directional coupler; straight channel	1300	TM TE	to 5 kW/cm²
Z	Y	TI		passive directional coupler	1300	TE	to 200 kW/cm²

Note: TI = titanium in-diffused; PE = proton-exchanged

At 633 nm, with 25 V dc applied to electrodes overlying waveguides, optical damage   [3.89]
of photoconductive origin is observed in minutes at power level indicated, as increase
in crosstalk. At 1064 nm, over range of power levels, smaller crosstalk increase seen.
See text discussion.

For TE polarization, quasiperiodic fluctuations of total output power seen for power   [3.166]
levels above threshold values listed, as result of photorefractively induced mode
hopping among transverse channel modes. Fluctuation period, 1–40 Hz at 440 nm,
proportional to input power above threshold. No effect for TM.

Short-term (time constant 5 s) drift in output intensity with 5 V dc applied shown   [3.167]
to be due to leakage through buffer layer between electrodes and guide channels.
When this is eliminated, "long-term" (time constant 1 h) drift remained. See text
discussion.

Guides made in sealed system to suppress out-diffusion. Above threshold indicated,   [3.168]
10–50% degradation of throughput power apparent within 1 min. At 400 W/cm^2,
irregular fluctuations appear.

Exposure of 1 channel to 633 nm guided light for 5 min affects the switching   [3.120]
characteristic at the 1064 nm design wavelength, as result of mode-index lowering
in exposed channel. Decay time days.

Short-term (1 min) observations reveal photovoltaic damage rate 13 x in larger in TI   [3.48, 49]
device than in PE. With applied dc field 1.5 V/μm in push-pull configuration, photoconductive damage almost 1 order of magnitude larger observed in both TI and PE
devices. See text discussion.

Threshold damage level about 32 W/cm^2. No damage observed over 24 h at lower   [3.105,
levels. At threshold, damage develops and saturates within 100 min ($\lambda = 850$ nm).   169]
Relaxation after 152 h exposure to 290 W/cm^2 non-monotonic, but relaxation time
of order of hours ($\lambda = 810$ nm). Two time constants, one associated with guide, one
with substrate suggested. Saturation mode index change $\sim 2 \times 10^{-4}$ ($\lambda = 633$ nm).
See text.

Power-handling capability $\sim$5 x larger in TI guides with out-diffusion suppression   [3.170]
(using Li$_2$CO$_3$); lower in PE guide, 10 x lower in TI coupler; similar for TE and TM
(in) in TI guides, but $\sim$70% TM to TE conversion observed. Saturation Δn in coupler
$\sim$3 x 10^{-4} in agreement with Hoffmann, Langmann.

Out-diffusion suppressed (wet O$_2$ processing). On Z-cut, power out saturates at   [3.171]
750 μW ($\sim$10 kW/cm^2 in guide); TE not saturated in 10 h at 1400 μW out. Photovoltaic damage observed above 100–500 W/cm^2 in MZs without bias. Considerable
device-to-device variations in threshold. Photoconductive damage also seen.

Damage threshold about 1.7 mW/cm^2 at 458 nm, 10 W/cm^2 at 850 nm in devices   [3.172]
made with out-diffusion suppression (wet O$_2$ processing) in short-term tests. Saturation time hours at 65 W/m^2 ($\lambda = 850$ nm). "Write-while-hot" method for damage
mitigation described; see text.

Effects much reduced compared to those in similar devices at 830 nm in short-term   [3.106]
experiments (up to 10 h). Threshold for photovoltaic damage in MZs around
5 kW/cm^2. With up to 2.5 μW in channels, no drift or saturation seen except 4%
decrease in throughput in one directional coupler. See discussion in text.

Out-diffusion suppressed (wet O$_2$ processing). Light coupled in from Nd:YAG laser   [3.173]
using polarization-preserving fiber to maintain polarization control. At 50 kW/cm^2,
photovoltaic damage develops with time constant to saturation of $\sim$100 h. On
lowering power to 5 kW/cm^2, damage decays with time constant 600 h. No damage
observed at $<$50 kW/cm^2 for up to 200 h observation. See text discussion.

3.6.1 Experiments and Interpretations

A selection of experimental observations of photorefractive effects in channel waveguides is presented in Table 3.2. To keep the table to reasonable size only representative results have been included. In addition, the reader will have to consult the original references for detailed information on waveguide preparation, beam coupling methods, etc.

Optical power densities are generally even higher in channel waveguides than they are in planar guides; so there is usually no difficulty in observing photorefractive phenomena in these guides even at infrared wavelengths; the problem is one of unambiguous interpretation of the observations. Many early investigators reported gradual changes in transfer characteristics of switches when dc voltages were applied for relatively long periods – a few seconds to a few minutes – while visible light was present in the devices. These changes were not observed with ac excitation. *Schmidt* et al. [3.89] pointed out that these changes could be explained in terms of photoconductive contributions to the overall photorefractivity, similar to those found in hologram recording in an applied electric field. This explanation is undoubtedly correct, yet these authors also noted changes in transfer characteristics in similar switches designed for use under infrared illumination at 1064 nm. The applied field was around 10^4 V/cm and the input infrared power density levels in the guides varied from 1 to 10 kW/cm^2 without much change in the maximum crosstalk levels. No explanation has been forthcoming for these observations; the power levels involved seem too small for thermal or two-photon effects to play any role. The experiments are performed by adjusting a directional coupler switch to a state of maximum power transfer and then monitoring the ratio ("crosstalk") of the output intensity in the input channel to that in the transfer channel as a function of time. The results are shown in Fig. 3.12. The experiments were done using TE polarization on Z-cut guides; so photovoltaic photorefraction is expected to be small.

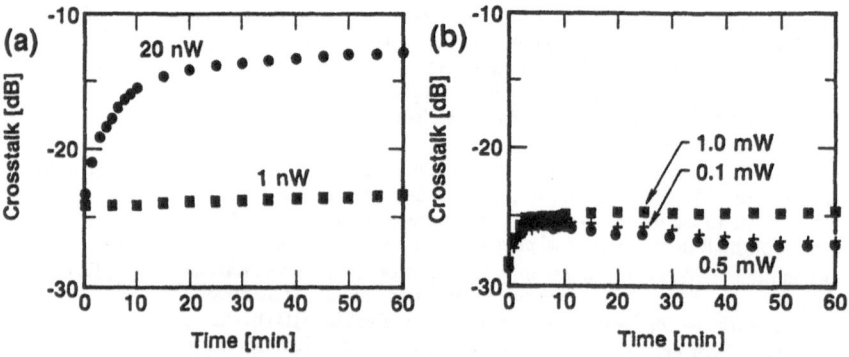

Fig. 3.12. Development of crosstalk with time in directional-coupler switches: (a) at 633 nm wavelength; (b) at 1064 nm. Output power levels indicated on graphs. After [3.89]

The realization that large photoconductive effects could make conventional photorefractivity linked to the bulk photovoltaic effect harder to observe under typical device operation conditions led to several approaches for assessing the latter effects. *Mueller* and *Garmire* [3.105, 169] and *Harvey* et al. [3.173] have used passive directional couplers. *Becker* [3.48, 49] has used symmetric, and *Bulmer* and *Burns* asymmetric [3.106, 171], integrated Mach-Zehnder interferometers with high-frequency applied fields, or no applied field. Changes in straight-channel throughput [3.106a, 166, 168, 170], and in characteristics of traveling-wave modulators [3.174] have also been used to assess channel waveguide photorefraction. The asymmetric waveguides used by Bulmer and Burns have one arm longer than the other, so there is a phase mismatch between the arms with no applied voltage. *Becker* and *Williamson* [3.175] produced an asymmetry of a slightly different sort by attenuating the beam in one arm of a symmetric device. In either case, the difference between the arms should lead to a net change in output, increasing with time, if photovoltaic optical damage is occurring. However, some such damage is found even in interferometers that are as symmetric as possible, as a result of small lithographic errors or variations in substrate properties.

An important advance in measurement technology was made by *Becker* [3.49, 75b], who showed that the electrodes used for modulation in a Mach-Zehnder arrangement could also be used as pickup electrodes for directly measuring photovoltaic and pyroelectric voltages, photoconductive and pyroelectric currents, and device resistance. The simple experimental arrangements required are shown in Fig. 3.13. A similar setup could of course be used with a single channel guide, but is not so suitable for other devices. Care is required to avoid leakage currents.

Most investigators have proceeded along similar lines. The basic experiment is to monitor the output intensity as a function of time for devices prepared various ways. A model of the dynamics of device operation is developed and used to interpret the experimental results. The principal qualitative conclusions of the various research groups are in good general agreement:

— directional couplers are more sensitive probes of optical damage than are straight channel waveguides, and suitably designed integrated interferometers are more sensitive than either;

— photovoltaic, as well as photoconductive, photorefractive effects can readily be observed in channel waveguide devices in the wavelength range around 850 nm, at power levels easily attained with butt-coupled diode lasers;

— photovoltaic effects also occur at $1.3\,\mu$m wavelength, but they are reduced;

— in directional couplers on Z-cut crystals (the only cut extensively investigated), the primary effect is overall photovoltaic lowering of the mode index in the input channel; changes in the coupling coefficient are secondary;

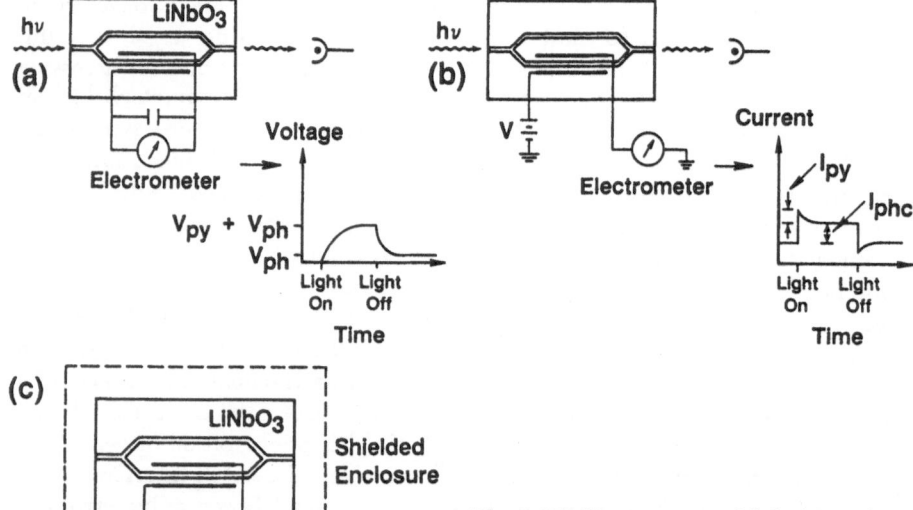

Fig. 3.13. Measurement of (a) photovoltaic and pyroelectric voltages; (b) photoconductive and pyroelectric currents; and (c) device resistance in channel-waveguide Mach-Zehnder interferometer, using modulation electrodes. After [3.49]

— the only proton-exchanged devices yet studied in detail have been made using undiluted benzoic acid; in these, photovoltaic effects are reduced, if not absent; photoconductive effects are changed a lesser amount.

A recent paper [3.75b] compares short-time, short-wavelength response of a variety of Mach-Zehnder interferometers. The devices, designed to be single-mode at 850 nm, were tested at a wavelength of 460 nm. Four different guide configurations were prepared on X-cut substrates:

1. undoped substrate, Ti-in-diffused guide, light propagation in Y-direction
2. as (1), but propagating along Z
3. as (1), but guide made by proton exchange
4. as (1), but guide made on substrate of $(LiNbO_3)_{0.95}(MgO)_{0.05}$.

These measurements indicated no measurable photovoltaic effect for light propagating along the Z-axis, but a photoconductive effect about 60 % as large as in the reference Y-propagating device. In the proton-exchanged guide, again no photovoltaic effect was measured, but the photoconductive current was the same as in the reference device. In the device on the MgO-containing substrate, both photovoltaic and photoconductive effects are reduced by about 30 % from the reference values. In bulk MgO-doped $LiNbO_3$, on the other hand, photoconductivity is increased. Evidently in $LiNbO_3$ channel waveguide devices involving long-term application of dc electric fields, the only way to avoid photoconductive optical damage will be to work at a sufficiently long wavelength.

When it is necessary to place electrodes directly over waveguides, buffer layers, often sputtered SiO_2, are usually placed between the guide and the electrode to reduce optical attenuation. These layers may be sources of current leakage and charge trapping, and thus may affect apparent photorefractive dynamics. Correct interpretation of experimental results when buffer layers are present has continued to be a problem since *Yamada* and *Minakata* [3.167] showed that current leakage through such layers (theirs were of Al_2O_3) could contribute to output drift in directional couplers. Aside from developing thoroughly reliable procedures for buffer layer deposition, several things can be done to improve the situation for experimental work. The leakage can, in principle, be eliminated by milling or etching away the buffer layer in the region between the electrodes [3.48, 167]. Charge trapping can be reduced by making the buffer layer of a transparent conductor [3.176], such as indium-tin oxide etched away between electrodes. This also leads to better long-term dynamic stability, according to the model equivalent-circuit analysis of *Gee* et al. [3.176]. Even on devices without buffer layers, transient dark currents with long time constants can sometimes appear [3.108].

Several other interesting phenomena have been observed as a result of the interest in channel guide devices. *Bulmer* and co-workers [3.177] found a long-term instability in their Mach-Zehnder interferometers operating at 1300 nm on Z-cut substrates. They traced this effect to differential pyroelectric voltages which they attribute to slight room temperature changes along with differences in the electrical conditions over the interferometer arms when modulating electrodes are connected. When one electrode is grounded, or connected to a large capacitance, pyroelectric charge cannot build up beneath it, as it can beneath the floating electrode. The differential pyroelectric charge distribution along the upper surface of the guide produces a net electric field which, in the electrode arrangement used, strongly affects propagating modes through the electro-optic effect. Even when both electrodes are contacted, as in ordinary device use, there are smaller oppositely directed Y-components of the net pyroelectric field acting on the two X-directed arms of the waveguide; these fields produce a net phase shift through the r_{22} electro-optic coefficient. Little or no instability is found in devices on X-cut substrates. For devices operating at infrared wavelengths beyond $1.1\,\mu$m, recent work of *Sawaki* et al. [3.178] shows that the susceptibility to pyroelectric effects of this type may be greatly reduced by depositing a buffer layer of poly-Si between the waveguides and the electrodes. This coating has sufficient conductivity to keep any differential fields of pyroelectric origin from building up, yet it is not photoconductive at the operating wavelengths.

Susceptibility to photorefractive damage, whether photovoltaic or photoconductive in origin, can be reduced by lowering the number of photoionizable impurity centers. In $LiNbO_3$ channel guides this can be done, as demonstrated by *Becker* [3.172], by a "write-while-hot" technique similar to that used in fixing bulk holograms [3.179]. The channel is exposed to a high intensity of visible light (458 nm in Becker's experiments) and heated to 150° C for 15 minutes, then cooled while still illuminated. The details of the process are not understood, but apparently electrons which are ionized from Fe^{2+} and similar impurity centers and trapped outside the beam are neutralized as photorefraction sources by hydrogen ions, which are quite mobile at 150° C. Thus there is no net photorefractive field, yet most of the impurities in the channel region have been oxidized. Short-term (1 hour) tests showed that a guide treated this way was about ten times less sensitive at 458 nm than a similar guide not treated. The improvement in the infrared damage resistance is even larger. This "fixing" treatment appears permanent; no increase in damageability was noted in intermittent testing over a period of at least ten months.

Among these experimental studies, that of *Harvey* et al. [3.173] deserves mention because it is one of the few in which long-term observations were carried out. To assess the possible effect of photorefraction in the 1300 nm wavelength regime on the operation of Ti-in-diffused $LiNbO_3$ directional couplers, they monitored the coupling efficiency for periods of up to 200 hours, with optical power densities in the input arm in the range of $0.1\,MW/cm^2$. Under such conditions, they were able to establish unambiguously that photorefractive index changes were developing with a time constant of around 100 hours. The effect under these conditions is quite small, representing an index change in the 10^{-4} range. *Mueller* and *Garmire* [3.105, 169], working on similar couplers at 850 nm, found photorefractive changes occurring much more rapidly (within an hour) and at lower power densities ($32\,W/cm^2$). Their saturation index changes were larger, closer to 10^{-3}. Differences in waveguide processing technique may have as much as the shorter wavelength to do with the differences between these observations and those of Harvey and his colleagues. Mueller and Garmire's waveguides were made by heating in dry argon, following by cooling in oxygen to reoxidize the sample. The guides of Harvey et al. were made in flowing O_2, bubbled through warm water to compensate for out-diffusion. At these long wavelengths, where direct excitation from transition-metal impurity states is unlikely, the photoionizable species and the available trap states may depend markedly on waveguide processing.

3.6.2 Polarization Conversion

The most interesting development in waveguide photorefractivity in recent years is the discovery of polarization conversion in channel guides [3.99]. This phenomenon promises to provide further insight into the nature of photorefractive processes, and may be particularly helpful in elucidating the sources of the effect in waveguides. From the limited amount of information that has appeared so far [3.99–102, 174, 180], we can piece together the following summary of the essential experimental characteristics of the phenomenon:

- The effect involves the conversion of an incident ordinary-mode wave to the extraordinary mode. Thus in a guide on a Y-cut crystal, incident TM polarization is converted to TE, while on a Z-cut sample, TE is changed to TM.

- The effect is not reciprocal; incident light of the wrong (extraordinary) polarization is not converted at all unless the material has first been "sensitized" with an ordinary-polarization beam [3.102]; then extraordinary to ordinary conversion can be seen on initial exposure to the extraordinary beam, but the effect soon disappears.
- The effect is seen only in channel guides, and is absent in planar guides. There is one report [3.181] of planar-guided to unguided wave conversion with change of polarization, which may be related; however we will not attempt to discuss it here.
- The fraction of light converted to extraordinary polarization increases gradually, typically requiring 10 s to 10 minutes to approach saturation. At low light levels, the time development of the process may not be monotonic, as indicated in Fig. 3.14.
- In all experiments reported so far, the incident light has been at wavelength 633 nm and the waveguides have been made by diffusion of Ti into LiNbO$_3$. Typical channel widths are in the 3 to 5 μm range, and lengths from 0.5 to 2 cm. Generally endfire coupling to these channels has been used.

To try to make some sense of these observations, it will be helpful to focus on a particular case. We will consider a Z-cut crystal with a channel running

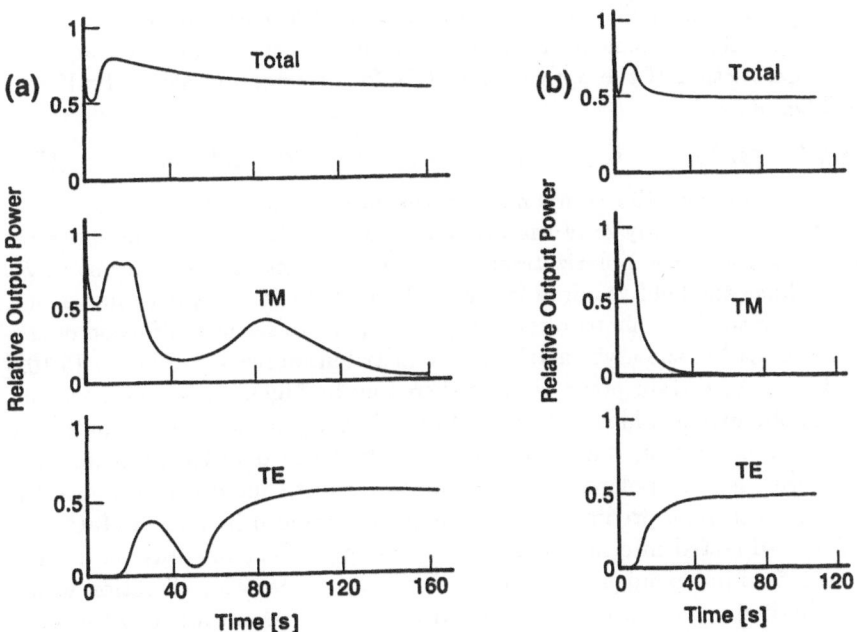

Fig. 3.14. Development with time of TM $\rightarrow$ TE mode conversion in an X-propagating channel waveguide on Y-cut LiNbO$_3$: (a) initial power level in waveguide 25 μW; (b) initial power level 50 μW. After [3.99]

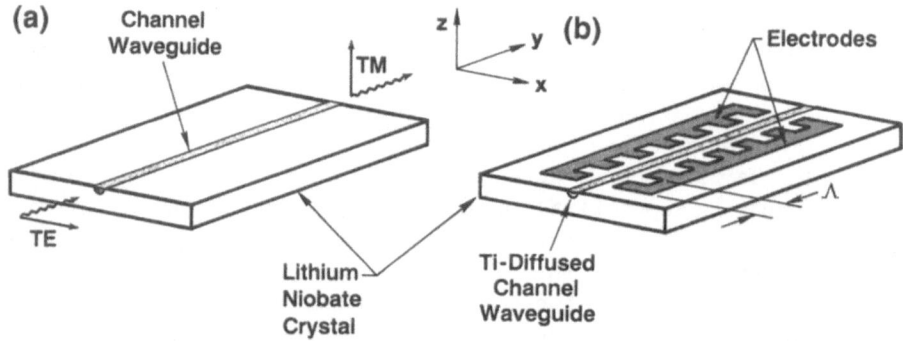

Fig. 3.15. TE → TM mode conversion in Z-cut channel waveguides: (a) arrangement for observation of photorefractive conversion; (b) electro-optic modulator constructed by *Alferness* [3.187]

in the Y-direction, Fig. 3.15a, as this turns out to be the simplest situation to analyze. The time development of the polarization change and the nature of the non-reciprocity make it clear that we are dealing with some sort of photorefractive phenomenon. Unpublished experiments [3.182] apparently indicate that the non-monotonic changes sometimes observed may be due to nothing more than multiple reflections in the resonant cavity formed between the polished waveguide ends. The mode coupling may be considered to result from an electro-optic effect arising from an internal field generated photorefractively. The normal section of the optical indicatrix for propagation along the Y-axis in LiNbO$_3$ is

$$(n_o^{-2} - r_{22}E_y + r_{13}E_z)x^2 + (n_e^{-2} + r_{33}E_z)z^2 + 2xzr_{51}E_x = 1 \quad , \qquad (3.8)$$

where E_i represents the component of the internal quasistatic electric field along the i-axis. Clearly only the last term on the left of (3.8) contributes to mixing of TE and TM polarizations. For simplicity, let us ignore terms in E_y and E_z. Since the field required is directed along the X-axis, it cannot result directly from the bulk photovoltaic effect. It is hard to see how diffusion of carriers would lead to the asymmetric charge distribution needed. *Phillips* [3.102] suggests that in certain polar crystals there can be higher-order contributions to the photocurrent which are proportional in magnitude to the Glass bulk photovoltaic coefficient, but which are directed orthogonal to the polar axis. If such terms lead to polarization rotation, the effect should vanish above the Curie temperature, a prediction which might be tested using, say, LiTaO$_3$.

The X-directed internal field will not alone produce one-way mode conversion; a field independent of distance along the propagation direction would lead to periodic changes in the mode structure, so the output TE/TM ratio would depend on the crystal length. To obtain one-way conversion, we need a periodic variation of the internal field along the direction of propagation such that the propagating orthogonal modes are phase-matched; i.e.,

$$E_x = C \cos (y/\Lambda) \quad \text{and} \tag{3.9a}$$

$$k_0 N_{\mathrm{TM}} = k_0 N_{\mathrm{TE}} + 2\pi/\Lambda \quad , \tag{3.9b}$$

where N_{TM} and N_{TE} are the effective refractive indices of the coupled modes. The dynamics of the mode conversion process must be such that where the TM mode is enhanced, it tends to reinforce the electric field, but where TM is small, the TE light tends to erase the field already built up. This observation is helpful in identifying the terms in the photogalvanic tensor primarily responsible for the effect. Once the photorefractive grating is formed, it will also couple an incident TM wave to TE, but the TE light thus produced gradually destroys the grating.

The X-directed field required is generated by the corresponding components j_x of the photogalvanic current. In the general phenomenological theory of photogalvanic effects as expounded by *Belinicher* and *Sturman* [3.183], this current is related to the optical electric fields by

$$j_x = \sum_{m,n} \beta_{xmn} \mathcal{E}_m \mathcal{E}_n^* \quad . \tag{3.10}$$

Since $\mathcal{E}_m$ and $\mathcal{E}_n$ may have different propagation constants, their product is in general complex; thus β_{xmn} must also be complex to produce a real quasistationary current. We may write it as

$$\beta_{xmn} = \beta'_{xmn} + i\beta''_{xmn} \quad \text{where} \tag{3.11a}$$

$$\beta'_{xnm} = \beta'_{xmn} \quad , \quad \text{and} \tag{3.11b}$$

$$\beta''_{xnm} = -\beta''_{xmn} \quad . \tag{3.11c}$$

The Glass-type photogalvanic currents, dependent only on the intensity and polarization direction of a light beam, appear only along the polar axis; thus only terms with $m \neq n$ appear in (3.10). The parts of β_{xmn} symmetric in mn transform like the electro-optic tensor; for LiNbO$_3$, we may write the terms in j_x arising from this, in the usual condensed notation, as

$$j_x^{\mathrm{S}} = \beta'_{15}(\mathcal{E}_x \mathcal{E}_z^* + \mathcal{E}_x^* \mathcal{E}_z) - 2\beta'_{22}(\mathcal{E}_x \mathcal{E}_y^* + \mathcal{E}_x^* \mathcal{E}_y) \quad . \tag{3.12}$$

The remaining terms can be written, using (3.11c), in the form

$$j_x^{\mathrm{A}} = i \sum_{m,n} \beta''_{xmn}(\mathcal{E}_m \mathcal{E}_n^* - \mathcal{E}_m^* \mathcal{E}_n) \quad . \tag{3.13}$$

We see from (3.13) that the part of the β tensor antisymmetric in m and n relates the polar vector j^{A} to the axial vector $\vec{\mathcal{E}} \times \vec{\mathcal{E}}^*$. It can thus be considered as an axial tensor of second rank, and consultation of standard tables [3.184] shows that for LiNbO$_3$ it has only one component, β''_{15}; thus we can write

$$j_x^{\mathrm{A}} = i\beta''_{15}(\mathcal{E}_x \mathcal{E}_z^* - \mathcal{E}_x^* \mathcal{E}_z) \quad . \tag{3.14}$$

This is the term that turns out to be responsible for the coupling leading to polarization rotation. Since it couples the TE and TM modes, some initial TM-polarized light is necessary to get the process started. Scattering is suggested in the published discussions as the likely way of obtaining some TM light in the guide. With both polarizations present, the quasistatic field can build up,

$$E_x(y) = j_x(y)/\sigma_{\mathrm{ph}} \quad , \tag{3.15}$$

where σ_{ph} is the x-component of the photoconductivity tensor; this field thus further modifies the optical electric field components as indicated by (3.8). Several coupled-mode theories describing the spatial dynamics of this feedback process have been presented [3.101, 102, 185, 186]; they all predict the one-way conversion process that is observed. *Lam* and *Yen* [3.186] included the symmetric part of the β tensor – at least the β'_{15} term – in their calculation. They showed that it led to a change in the phase of the output beam, but not to any difference in polarization.

The process is one in which a phase-matched longitudinal grating is built up automatically through the photogalvanic effect. One may inquire whether it would not be possible to create a similar phase grating using real electrodes. The answer, of course, is yes; just such a grating was built by *Alferness* [3.187]. It is sketched in Fig. 3.15b. By trial and error he found that for a $2\,\mu$m wide channel and an operating wavelength of 599.5 nm, a grating period Λ of $7\,\mu$m gave high conversion efficiency. He did not report any conversion without a field applied.

The principal outstanding question in waveguide polarization conversion is why it has been observed only in channel guides. An explanation of this, or an observation of a similar effect in a planar guide, would be very welcome.

3.7 Other Materials

Our discussion of waveguide photorefractivity in materials other than LiNbO$_3$ and LiTaO$_3$ is necessarily very brief as very little information has been published.

Bykovskii and co-workers [3.188] formed gratings in a GaAs$_{0.15}$P$_{0.85}$ layer on a GaP substrate using two external beams, as in Fig. 3.1c. The gratings were formed using beams of 532 nm wavelength and read out using a guided beam at 1064 nm. These are dynamic gratings resulting from creation of free carriers, and there is thus nothing specifically photorefractive about them. However, the authors demonstrate that gratings formed this way are primarily phase, rather than amplitude, gratings and that high diffraction efficiencies (45 %) may be obtained in this way.

Antsigin et al. [3.189] prepared oriented polycrystalline thin films of Sr$_{0.5}$Ba$_{0.5}$Nb$_2$O$_6$ by rf-sputtering in oxygen. They noted a strong dependence of the apparent electro-optic coefficient on the magnitude of an applied dc field, but did not specifically attribute any part of this effect to photorefraction.

Single-crystal films of PLZT have been under development by a group at Matsushita [3.190, 191]. Films are grown epitaxially on sapphire by rf planar magnetron sputtering. Ridge waveguides 20 μm wide and 50 nm high were formed in a sample of composition $Pb_{0.72}La_{0.28}TiO_3$ by ion-beam etching. A beam from a 3 mW He-Ne laser was coupled into one of these channels for 6 months without any form of optical damage being observed. The material shows good electro-optic properties at frequencies up to 1 GHz.

3.8 Summary and Conclusions

Many researchers have devoted much time and ingenuity to the measurement and understanding of photorefractive effects in waveguides in $LiNbO_3$ and $LiTaO_3$. As a result of their efforts, quite a lot has been learned about the waveguides themselves and about photorefractive phenomena in them.

— The high photorefractive susceptibility in $LiNbO_3$ Ti-in-diffused guides, compared to bulk samples, has been studied in several ways. It has been interpreted in terms of the effect of Ti in reducing the valence of impurity ions such as iron. The effects of waveguide processing conditions on the photorefractivity have been determined qualitatively, and several damage-mitigation procedures have been devised.

— The reduced level of damage in proton-exchanged, MgO-doped, and Z-propagating waveguides has been measured and related to other waveguide properties. The suitability of various types of guides for specific applications has been elucidated.

— The wavelength dependence of the effect has been studied from 0.44 μm to 1.3 μm in several experimental arrangements.

— The unusual temporal and spatial development of photorefractive scattering in planar guides has been investigated in guides made by several processes.

— New phenomena, such as polarization conversion, have been discovered.

When one considers that for many of the workers in this field photorefractivity has not been a primary topic of their research, but rather a side issue to be dealt with in the course of developing some prototype integrated-optical device or component, the progress in the past five years has been remarkable. Nonetheless, the reader of the foregoing pages can scarcely have failed to notice that in many cases interpretations of experiments were based on limited observations on samples of uncertain origin, and thus contain large measures of uncertainty and speculation. Reducing the uncertainty will not be an easy task, although interest in the subject remains strong and we expect considerable progress in the near future.

Some of the specific topics that should receive attention are:

— **Adequately Characterized Waveguides.** Comparison of experimental results, particularly from different laboratories, is almost impossible unless the experiments are done on guides fabricated by a tested and reproducible method. Knowledge of substrate stoichiometry and principal impurity concentrations is also important. Progress is being made in this area as a result of cooperation of commercial crystal suppliers in providing information about their products, and through the development of microprocessor-controlled furnaces for waveguide fabrication.

– **Long-Term Experiments.** More long-term studies of development and saturation level of photorefractive optical damage need to be carried out, particularly at longer wavelengths. The data we have summarized show clearly that observation periods of *many hundreds of hours* are often necessary to determine saturation damage levels. A better theory describing the time dependence of the development of such damage is also needed, as is a better theory of the spatial distribution of the scattered light in planar-waveguide single-beam experiments.

– **Other Processes.** Most work reported to date on photorefraction in proton-exchanged guides has been on guides made in undiluted benzoic acid. Since guides made by proton exchange in melts diluted with a small amount of lithium benzoate have several advantages, the photorefractive sensitivity as a function of lithium concentration needs to be determined. There is also renewed interest in forming thin films of $LiNbO_3$ and other ferroelectrics by vapor deposition processes. When the optical quality of these films is adequate, photorefractive effects will no doubt be looked for.

– **Other Phenomena.** From the device designer's standpoint, photorefraction is one of the more annoying properties to deal with because of its cumulative nature. This does not mean, of course, that other photoinduced phenomena can be ignored, particularly at the high optical power densities and high switching rates likely to be encountered in the next generation of integrated-optics components. We are presently seeing the beginnings of research aimed at providing all the information needed by the device designer.

– **Other Coefficients.** Polarization conversion measurements provide in principle a way of getting information about the β_{15} coefficient of the photogalvanic tensor, which is not easy to measure otherwise [3.18]. No attempt has been made to find ways of determining in waveguides all the remaining components of the photogalvanic tensor, much less to compare such values with those obtained by similar measurements on bulk samples. Such measurements should contribute to understanding the charge transport mechanisms involved in photorefraction.

– **Other Imperfections.** In common with other authors, we have had occasion to refer to shallow electron traps of unknown origin – in our case, to try to explain the sensitization of waveguides for formation of holograms with infrared beams. These traps deserve careful investigation. Can their number or their energy levels be influenced by changes in stoichiometry or doping? Are they different in waveguides and in bulk? Are they associated with native defects, or with impurities, or both? Is there more than one kind?

— **Other Materials.** If one were going to select a crystal for studying photorefractivity, whether in bulk or in waveguides, one might not choose $LiNbO_3$. The complicated phase diagram, the tendency to phase separation, the importance of impurities at the parts-per-million level, the remoteness of the Curie temperature, all make it difficult to repeat experiments or to interpret them unambiguously. On the other hand, this complexity, together with the practical importance of the material, has brought perforce a level of understanding which might not otherwise have been attained. Moreover, lithium niobate does have its advantages, some of which we mentioned earlier. It is a well-behaved ferroelectric with a sharp phase transition, it is not biaxial, it doesn't show optical activity, and it is relatively free of strain birefringence, striations, and optical absorption. Few other crystals, and even fewer waveguides, possess such advantages. As preparation and analysis techniques improve, however, other interesting and useful materials will doubtless appear. Effects occurring in layers of III-V and II-VI compounds will certainly be investigated. Strained epilayers may display photorefractive effects not present in bulk samples of the same material. More work in $LiTaO_3$ waveguides would be desirable. Scattered reports indicate that waveguides in this crystal develop damage more slowly than comparable $LiNbO_3$ guides, but the long-term steady-state power-loss levels do not appear to have been determined. Experiments on new materials are likely to improve our general understanding of photorefraction, and thus to tell us something about this effect in the materials presently being studied as well.

Acknowledgements. We wish to acknowledge the contributions of our present and former colleagues J.R. Busch, C.M. Chapman, N.F. Hartman, D. Hicks, R.P.Kenan, M. Parmenter, J.F. Revelli, R.C. Sherman, and D.W. Vahey to the research at Battelle and at Xerox Corporation described in this chapter. Work at Battelle benefited from financial support at various times from the United States Office of Naval Research, the Air Force Office of Scientific Research, the Army Research Office, and the NASA Langley Research Center, and from Battelle Corporate Technical Development funds. We thank H.A. Haus for copies of references [3.101, 102], and A.M. Glass for information on unpublished results. We also thank the editors and publishers for their patience.

References

3.1 R.P. Kenan, D.W.Vahey, N.F. Hartman, V.E. Wood, C.M. Verber: Opt. Eng. **15**, 12 (1976); W.S. Goruk, P.J. Vella, R. Normandin, G.I. Stegeman: Appl. Opt. **20**, 4024 (1981)

3.2 T. Tamir (ed.): *Integrated Optics*, 2nd ed., Topics Appl. Phys., Vol. 7 (Springer, Berlin, Heidelberg 1979)

3.3 R.G. Hunsperger: *Integrated Optics: Theory and Technology*, 2nd ed., Springer Ser. Opt. Sci., Vol. 33 (Springer, Berlin, Heidelberg 1984)

3.4 D. Marcuse (ed.): *Integrated Optics* (IEEE Press, New York 1973)

3.5 E.A.S. Marcatili: Bell. Syst. Tech. J. **48**, 2071 (1969)

3.6 M. Papuchon, Y. Combemale, X. Mathieu, D.B. Ostrowsky, L. Reiber, A.M. Roy, B. Sejourne, M. Werner: Appl. Phys. Lett. **27**, 289 (1975)

3.7 O. Wada, T.Sakurai, T. Nakagami: IEEE J. QE-**22**, 805 (1986)

3.8 S. Sakano, H. Inoue, K. Nakamura, T. Katsuyama, H. Matsumura: Electron. Lett. **22**, 594 (1986)

3.9 H. Inoue, K. Hiruma, K. Ishida, T. Asai, H. Matsumura: IEEE J. LT-**3**, 1270 (1985)

3.10 A.M. Glass, J. Strait: In *Photorefractive Materials and Applications I*, ed. by P. Günter, J.-P. Huignard, Topics Appl. Phys., Vol. 61 (Springer, Berlin, Heidelberg 1988) Chap. 7

3.11 G.H. Haertling, C.E. Land: J.Amer. Ceram. Soc. **54**, 1 (1971)
3.12 T. Kawaguchi, H. Adachi, K. Setsune, O. Yamazaki, K. Wasa: Appl. Opt. **23**, 2187 (1984)
3.13 A.A. Ballman, H.Brown, P.K. Tien, R.J. Martin: J. Cryst. Growth **20**, 251 (1973); K. Tada, Y. Kuhara, M. Tatsumi, T. Yamaguchi: Appl. Opt. **21**, 2953 (1982)
3.14 T. Mitsuyu, K. Wasa, S. Hayakawa: J. Crystal Growth **41**, 151 (1977)
3.15 P. Günter, J.-P. Huignard: In *Photorefractive Materials and Applications I*, ed. by P. Günter, J.-P. Huignard, Topics Appl. Phys., Vol. 61 (Springer, Berlin, Heidelberg 1988) Chap. 2
3.16 G.A.Alphonse, W. Phillips; RCA Rev. **37**, 184 (1976), and references therein
3.17 D. von der Linde, A.M. Glass: Appl. Phys. **8**, 85 (1975)
3.18 R.S. Weis, T.K. Gaylord: Appl. Phys. A**37**, 191 (1985)
3.19 P. Lerner, C. Legras, J.P. Dumas: J. Cryst. Growth **3,4**, 231 (1968)
3.20 R.L. Holman: "Novel Uses of Gravimetry in the Processing of Crystalline Ceramics", in *Processing of Crystalline Ceramics*, ed. by H. Palmour III, R.F. Davis, T.M. Hare (Plenum, New York 1979), pp. 343–358
3.21 R.J. Holmes, D.M. Smyth: J. Appl. Phys. **55**, 3531 (1984)
3.22 L. Castelli: Laser Focus/Electro-Optics **21**, No. 12, 120 (1985)
3.23 G.J. Sellers, S. Sriram: Laser Focus/Electro-Optics **22**, No. 9, 74 (1986)
3.24 W.A. Stallard, T.G. Hodgkinson, K.R. Preston, R.C. Booth: Electron. Lett. **21**, 1077 (1985)
3.25 R.A. Linke, B.L. Kasper, N.A. Olsson, R.C. Alferness: Electron. Lett. **22**, 30 (1986)
3.26 H. Shimizu, R. Ishikawa, K. Kaede: Electron. Lett. **22**, 334 (1986)
3.27 S.K. Korotky, E.A.J. Marcatili, G. Eisenstein, J.J. Veselka, F. Heismann, R.C. Alferness: Appl. Phys. Lett. **49**, 10 (1986)
3.28 a) A.M. Glass, G.E. Peterson, T.J. Negran: "Optical Index Damage in Electrooptic Crystals", in *Laser Induced Damage in Optical Materials*, ed. by A.J. Glass and A.H. Guenther, U.S.Nat. Bur. Stand. Spec. Publ. 372 (1972), pp. 15–26; b) G.L. Tangonan, M.K. Barnoski, J.F. Lotspeich, A. Lee: Appl. Phys. Lett. **30**, 238 (1977)
3.29 J.R. Carruthers, I.P. Kaminow, L.W. Stulz: Appl. Opt. **13**, 2333 (1974)
3.30 I.P. Kaminow, J.R. Carruthers: Appl. Phys. Lett. **22**, 326 (1973)
3.31 R.L. Holman, P.J. Cressman, J.F. Revelli: Appl. Phys. Lett. **32**, 280 (1978)
3.32 R.J. Esdaile: Appl. Phys. Lett. **33**, 733 (1978)
3.33 R.V. Schmidt, I.P. Kaminow: Appl. Phys. Lett. **25**, 458 (1974)
3.34 J.L. Jackel, V. Ramaswamy, S.P. Lyman: Appl. Phys. Lett. **38**, 509 (1981)
3.35 J.L. Jackel: J. Opt. Commun. **3**, 82 (1982)
3.36 A. Rasch, M. Rottschalk, W. Karthe: J. Opt. Commun. **6**, 14 (1985); C. Canali, M.N. Armenise, A. Carnera, M. De Sario, P. Mazzoldi, G. Celotti: Proc. SPIE **46**, 34 (1984)
3.37 L.O. Svaasand, M. Eriksrud, G. Nakken, A.P. Grande: J. Cryst. Growth **22**, 232 (1974)
3.38 R.L. Holman, P.J. Cressman, J.A. Anderson: Ferroelectrics **27**, 77 (1980); M.N. Armenise, C. Canali, M. De Sario, A. Carnera, P. Mazzoldi, G. Celotti: J. Appl. Phys. **54**, 62, 6223 (1983)
3.39 J. Noda, N. Uchida, S. Saito, T. Saku, M. Minakata: Appl. Phys. Lett. **27**, 19 (1975)
3.40 B.-U. Chen: Digest Tech. Papers, Topical Mtg. Integr. Guided Wave Opt. (Opt. Soc. Amer., Washington 1976), p. WA5-1; Thin Solid Films **64**, 173 (1979)
3.41 S. Forouhar, G.E. Betts, W.S.C. Chang: Appl. Phys. Lett. **45**, 207 (1984); O. Eknoyan, A.S. Greenblatt, W.K. Burns, C.H. Bulmer: Appl. Opt. **25**, 737 (1986)
3.42 T.R. Ranganath, S. Wang: Appl. Phys. Lett. **30**, 376 (1977); B.-U. Chen, A.C. Pastor: Appl. Phys. Lett. **30**, 570 (1977); S. Miyazawa, R. Guglielmi, A. Carenco: Appl. Phys. Lett. **31**, 742 (1977); W.K. Burns, C.H. Bulmer, E.J. West: Appl. Phys. Lett. **33**, 70 (1978)
3.43 J.L. Jackel, D.H. Olson, A.M. Glass: J. Appl. Phys. **52**, 4855 (1981)
3.44 J.L. Jackel, C.E. Rice, J.J. Veselka: Appl. Phys. Lett. **41**, 607 (1982); Ferroelectrics **50**, 165 (1983)
3.45 J.L. Jackel, C.E. Rice: Ferroelectrics **38**, 801 (1981); C.E. Rice, J.L. Jackel: J. Solid State Chem. **41**, 308 (1982)
3.46 a) C. Canali, A. Carnera, P. Mazzoldi, R.M. De La Rue: Proc. SPIE **517**, 119 (1984); b) A. Campari, C. Ferrari, G. Mazzi, C. Summonte, S.M. Al-Shukri, A. Dawar, R.M. De La Rue, A.C.G. Nutt: J. Appl. Phys. **58**, 4521 (1985); c) W.E. Lee, N.A. Sanford, A.H. Heuer: J. Appl. Phys. **59**, 2629 (1986)

3.47 A. Yi-Yan: Appl. Phys. Lett. **42**, 633 (1983)
3.48 R.A. Becker: Appl. Phys. Lett. **43**, 131 (1983)
3.49 R.A. Becker: Proc. SPIE **460**, 95 (1984)
3.50 R.L. Davis: Proc. SPIE **517**, 75 (1984)
3.51 Y. Handa, M. Miyawaki, S. Ogura: Proc. SPIE **460**, 101 (1984)
3.52 A.L. Dawar, S.M. Al-Shukri, R.M. De La Rue: Appl. Phys. Lett. **48**, 1579 (1986), and references therein
3.53 A. Dawar, R.M. De La Rue, G.F. Doughty, N. Findlayson, S.M. Al-Shukri, J. Singh: Proc. SPIE **517**, 64 (1984)
3.54 S.M. Al-Shukri, J. Duffy, R.M. De La Rue, M.N. Armenise, C. Canali, A. Carnera: Proc. SPIE **578**, 2 (1985)
3.55 K.K. Wong, R.M. De La Rue, S. Wright: Opt. Lett. **7**, 546 (1982)
3.56 M. De Micheli, J. Botineau, S. Neveu, P. Sibillot, D.B. Ostrowsky, M. Papuchon: Opt. Lett. **8**, 114 (1983)
3.57 J.L. Jackel, C.E. Rice: Proc. SPIE **460**, 43 (1984)
3.58 J.L. Jackel, C.E. Rice, J.J. Veselka: Electron. Lett. **19**, 387 (1983)
3.59 M. De Micheli, J. Botineau, P. Sibillot, D.B. Ostrowsky, M. Papuchon: Opt. Commun. **42**, 101 (1982); M. De Micheli: J. Opt. Commun. **4**, 1 (1983)
3.60 M.L. Shah: Appl. Phys. Lett. **26**, 652 (1975); J. Jackel: Appl. Phys. Lett. **37**, 739 (1980)
3.61 Y.-X. Chen, W.S.C. Chang, S.S. Lau, L. Wielunski, R.L. Holman: Appl. Phys. Lett. **40**, 10 (1982); J.L. Jackel, C.E. Rice: Appl. Phys. Lett. **41**, 508 (1982)
3.62 P.R. Meek, L. Holland, P.D. Townsend: Thin Solid Films **141**, 251 (1986), and references therein; A. Okada: Ferroelectrics **14**, 739 (1976)
3.63 S. Miyazawa, K. Sugii, N. Uchida: J. Appl. Phys. **46**, 2223 (1975), and references therein
3.64 S. Kondo, S. Miyazawa, H. Iwasaki: Mat. Res. Bull. **15**, 243 (1980), and references therein; A. Baudrant, H. Vial, J. Daval: Mat. Res. Bull. **10**, 1373 (1975)
3.65 S. Fushimi, K. Sugii: Jpn. J. Appl. Phys. **13**, 1895 (1974)
3.66 R.A. Betts, C.W. Pitt: Electron. Lett. **21**, 960 (1985)
3.67 B.J. Curtis, H.R. Brunner: Mat. Res. Bull. **10**, 515 (1975)
3.68 D.P. Gia Russo, C.S. Kumar: Appl. Phys. Lett. **23**, 229 (1973)
3.69 J.-C. Baumert, C. Walther, P. Buchmann, H. Kaufmann, H. Melchior, P. Günter: Appl. Phys. Lett. **46**, 1018 (1985)
3.70 W. Phillips, J.M. Hammer: J. Electron. Mater. **4**, 549 (1975)
3.71 V.E. Wood, N.F. Hartman, C.M. Verber, R.P. Kenan: J. Appl. Phys. **46**, 1214 (1975)
3.72 D.A. Bryan, R. Gerson, H.E. Tomaschke: Appl. Phys. Lett. **44**, 847 (1984)
3.73 H. Wang, G. Shi, Z. Wu: Phys. Status Solidi A**89**, K211 (1985); D.A. Bryan, R.R. Rice, R. Gerson, H.E. Tomaschke, K.L. Sweeney, L.E. Halliburton: Opt. Engineering **24**, 138 (1985)
3.74 R.J. Holmes, Y.S. Kim, C.D. Brandle, D.M. Smyth: Ferroelectrics **51**, 41 (1983)
3.75 a) C.H. Bulmer: Electron. Lett. **20**, 902 (1984); b) R.A. Becker: Proc. SPIE **578**, 12 (1985); c) M.M. Fejer, M.J.F. Digonnet, R.L. Byer: Opt. Lett. **11**, 230 (1986)
3.76 a) M. Digonnet, M. Fejer, R. Byer: Opt. Lett. **10**, 235 (1985); b) J.L. Jackel: Electron. Lett. **21**, 509 (1985)
3.77 D.J. Channin: Appl. Phys. Lett. **19**, 128 (1971)
3.78 J.M. Hammer, W. Phillips: Appl. Phys. Lett. **24**, 545 (1974)
3.79 J. Noda, T. Saku, N. Uchida: Appl. Phys. Lett. **25**, 308 (1974)
3.80 R.V. Schmidt, I.P. Kaminow: IEEE J. QE-11, 57 (1975)
3.81 C.M. Verber, V.E. Wood, R.P. Kenan, N.F. Hartman: Ferroelectrics **10**, 253 (1976)
3.82 H.F. Taylor, A. Yariv: Proc. IEEE **62**, 1044 (1974)
3.83 I.P. Kaminow, V. Ramaswamy, R.V. Schmidt, E.H. Turner: Appl. Phys. Lett. **24**, 622 (1974)
3.84 R.D. Standley, V. Ramaswamy: Appl. Phys. Lett. **25**, 711 (1974)
3.85 J. Noda, N. Uchida, M. Minakata, T. Saku, S. Saito, Y. Ohmachi: Appl. Phys. Lett. **26**, 298 (1975)
3.86 W.K. Burns, A.B. Lee, A.F. Milton: Appl. Phys. Lett. **29**, 790 (1976)
3.87 R.V. Schmidt, L.L. Buhl: Electron. Lett. **12**, 575 (1976)
3.88 H. Sasaki, I. Anderson: IEEE J. QE-14, 883 (1978)

3.89   R.V. Schmidt, P.S. Cross, A.M. Glass: J. Appl. Phys. **51**, 90 (1980)
3.90   D. von der Linde, A.M. Glass, K.F. Rodgers: Appl. Phys. Lett. **25**, 155 (1974)
3.91   C.M. Verber, N.F. Hartman, A.M. Glass: Appl. Phys. Lett. **30**, 272 (1977)
3.92   F.S. Chen: J. Appl. Phys. **40**, 3389 (1969)
3.93   M.K. Barnoski, J.F. Lotspeich: Digest Tech. Papers, Topical Mtg. Integr. Guided Wave Opt. (Opt. Soc. Amer., Washington 1976), p. Tuc3-1
3.94   R.L. Holman, P.J. Cressman: IEEE Trans. CHMT-4, 332 (1981)
3.95   R.G. Smith, D.B. Fraser, R.T. Denton, T.C. Rich: J. Appl. Phys. **39**, 4600 (1968)
3.96   J. Jackel, A.M. Glass, G.E. Peterson, C.E. Rice, D.H. Olson, J.J. Veselka: J. Appl. Phys. **55**, 269 (1984)
3.97   Yu.L. Kopylov, V.B. Kravchenko, E.N. Mirgorodskaya, A.V. Bobylev: Pis'ma Zh. Tekh. Fiz. **9**, 601 (1983) [English transl.: Sov. Tech. Phys. Lett. **9**, 259 (1983)]
3.98   R.L. Holman, J. Busch, M. Parmenter, P.J. Cressman: Ferroelectrics **50**, 171 (1983)
3.99   E.M. Zolotov, P.G. Kazanskii, V.A. Chernykh: Pis'ma Zh. Tekh. Fiz. **7**, 924 (1981) [English transl.: Sov. Tech. Phys. Lett. **7**, 397 (1981)]
3.100 A. Lattes, C. Gabriel, H. Haus: Digest Tech. Papers, Topical Mtg. Integr. Guided Wave Opt. (Opt. Soc. Amer., Washington 1982), p. ThA4-1; H.A. Haus, E.P. Ippen, A. Lattes, C. Gabriel, F.J. Leonberger: Appl. Phys. B28, 161 (1982)
3.101 A.L. Lattes: "Ultrafast Nonlinear Effects in Optical Waveguides"; Ph.D. Thesis, Mass. Inst. Technol. (1983)
3.102 M.R. Phillips: "Photorefractive Processes in Lithium Niobate"; M.S. Thesis, Mass. Inst. Technol. (1983)
3.103 J.H. Hammer, C.C. Neil: IEEE J. QE-18, 1751 (1982)
3.104 E.P. Harris, M.L. Dakss: IBM J. Res. Develop. **13**, 722 (1969)
3.105 C.T. Mueller, E. Garmire: Proc. SPIE **412**, 37 (1983)
3.106 a) C.H. Bulmer, W.K. Burns, S.C. Hiser: Digest Tech. Papers, 7th Topical Mtg. Integr. Guided Wave Opt. (Opt. Soc. Amer., Washington 1984), p. WC1-1; b) C.H. Bulmer, S.C. Hiser: Proc. SPIE **517**, 177 (1984)
3.107 P.H. Smakula, P.C. Claspy: Trans. Metall. Soc. AIME **239**, 421 (1967)
3.108 R.A. Becker: Digest Tech. Papers, 7th Topical Mtg. Integr. Guided Wave Opt. (Opt. Soc. Amer., Washington 1984), p. WC3-1
3.109 S. Thaniyavarn: Appl. Phys. Lett. **47**, 674 (1985); Opt. Lett. **11**, 39 (1986)
3.110 N.A. Sanford, W.C. Robinson: Proc. 6th IEEE Internatl. Symp. Appl. Ferroelectrics (IEEE, New York, 1987), p. 4
3.111 R.P. Kenan, IEEE J. QE-14, 924 (1978)
3.112 H. Kogelnik: Bell Syst. Tech. J. **48**, 2909 (1969)
3.113 C.M. Verber, D.W. Vahey, V.E. Wood, R.P. Kenan, N.F. Hartman: "Feasibility Investigation of Integrated Optics Fourier Transform Devices", NASA Contractor Report 2869 (NASA, Washington 1977)
3.114 R.P. Kenan, J. Appl. Phys. **46**, 4545 (1975)
3.115 R.L. Holman, P.J. Cressman: Opt. Eng. **21**, 1025 (1982)
3.116 R.L. Holman, P.J. Cressman: Ferroelectrics **27**, 85 (1980)
3.117 G.E. Betts, W.S.C. Chang: IEEE J. QE-22, 1027 (1986), and references therein
3.118 P.W. Smith, I.P. Kaminow, P.J. Maloney, L.W. Stulz: Appl. Phys. Lett. **33**, 24 (1978)
3.119 O.N. Bikeev, L.N. Deryugin, A.I. Reutov: Pis'ma Zh. Tekh. Fiz. **5**, 1496 (1979) [English transl.: Sov. Tech. Phys. Lett. **5**, 632 (1979)]
3.120 D. Hoffmann, U. Langmann: Digest Tech. Papers, Topical Mtg. Integr. Guided Wave Opt. (Opt. Soc. Amer., Washington 1982), p. WD4-1
3.121 R.L. Holman, P.J. Cressman: Appl. Phys. Lett. **38**, 409 (1981)
3.122 E. Schneider, P.J. Cressman, R.L. Holman: J. Appl. Phys. **53**, 4054 (1982)
3.123 J. Singh, J.R.Tobin, R.M. De La Rue: J. Appl. Phys. **58**, 1990 (1985)
3.124 A.M. Glass, I.P. Kaminow, A.A. Ballman, D.H. Olson: Appl. Opt. **19**, 276 (1980)
3.125 K.H. Haegele, R. Ulrich: Opt. Lett. **4**, 60 (1979)
3.126 V.V. Lemanov, B.V. Sukharev: Pis'ma Zh. Tekh. Fiz. **9**, 505 (1983) [English transl.: Sov. Tech. Phys. Lett. **9**, 218 (1983)]
3.127 A.R. Beaumont, B.E. Daymond-John, R.C. Booth: Electron. Lett. **22**, 263 (1986)
3.128 L. McCaughan, K.D. Choquette: IEEE J. QE-22, 947 (1986)
3.129 O. Mikami: Opt. Commun. **19**, 42 (1976)

3.130 S.I. Bozhevol'nyi, E.M. Zolotov, V.A. Kiselev, E.A. Shcherbakov: Kvantovaya Elektron. **6**, 367 (1979) [English transl.: Sov. J. Quantum Electron. **9**, 216 (1979)]

3.131 E.M. Uyukin, J. Ctyroky, J. Janta: Kvantovaya Electron. **10**, 2358 (1983) [English transl.: Sov. J. Quantum Electron. **13**, 1536 (1983)]

3.132 O.V. Kandidova, V.V. Lemanov, B.V. Sukharev: Pis'ma Zh. Tekh. Fiz. **9**, 777 (1983) [English transl.: Sov. Tech. Phys. Lett. **9**, 335 (1983)]

3.133 O.V. Kandidova, V.V. Lemanov, B.V. Sukharev: Zh. Tekh. Fiz. **54**, 1748 (1984) [English transl.: Sov. Phys. Tech. Phys. **29**, 1019 (1984)]

3.134 J.P. Nisius, E. Krätzig: Solid State Commun. **53**, 743 (1985)

3.135 O.B. Mavritskii, A.N. Petrovskii: Kvantovaya Elektron. **13**, 197 (1986) [English transl.: Sov. J. Quantum Electron. **16**, 130 (1986)]

3.136 P.A. Arsenev, B.A. Baranov: Phys. Status Solidi **A9**, 673 (1972)

3.137 J.P. Nisius, P. Hertel, E. Krätzig, H. Pape: In *Integrated Optics*, Proc. Third Europ. Conf. (Springer, Berlin, Heidelberg 1985), p. 62

3.138 H. Kurz, E. Krätzig, W. Keune, H. Engelmann, U. Gonser, B. Dischler, A. Räuber: Appl. Phys. **12**, 355 (1977)

3.139 R.L. Holman, P.J. Cressman: Tech. Digest, Third Int. Conf. Integr. Opt. and Opt. Fiber Commun. (IEEE, New York 1981), p. 90

3.140 R.L. Holman, Proc. SPIE **460**, 90 (1984)

3.141 R.L. Holman, J.R. Busch, C.M. Verber, V.E. Wood, N.F.Hartman, P.J. Cressman: Digest Tech. Papers, 7th Topical Mtg. Integr. Guided Wave Opt. (Opt. Soc. Amer., Washington 1984), p. WC2-1

3.142 Y. Ohmori, M. Yamaguchi, K. Yoshino, Y. Inuishi: Jpn. J. Appl. Phys. **18**, 79 (1979)

3.143 S. Juppe, O.F. Schirmer: Phys. Lett. **A117**, 150 (1986)

3.144 O.V. Kandidova, V.V. Lemanov, B.V. Sukharev: Fiz. Tverd. Tela **28**, 762 (1986) [English transl.: Sov. Phys. Solid State **28**, 424 (1986)]

3.145 L.B. Schein, P.J. Cressman, F.M. Tesche: J. Appl. Phys. **48**, 4844 (1977)

3.146 L.F. Kanaev, V.K. Malinovsky, B.I. Sturman: Optics Commun. **34**, 95 (1980)

3.147 H. Jerominek, R. Tremblay, C. Delisle: IEEE J. LT-3, 1105 (1985); H. Jerominek, C. Delisle, R. Tremblay: Appl. Opt. **25**, 732 (1986)

3.148 R.A. Rupp, F.W. Drees: Appl. Phys. **B39**, 223 (1986); J. Marotz, K.H. Ringhofer, R.A. Rupp, S. Treichel: IEEE J. QE-22, 1376 (1986)

3.149 R.A. Rupp, K.H. Ringhofer, F.W. Drees, J. Marotz, S. Treichel, E. Krätzig: Proc. 6th IEEE Internat. Symp. Appl. Ferroelectrics (IEEE, New York, 1987), p. 72

3.150 W. Sohler, H. Suche: Appl. Phys. Lett. **33**, 518 (1978)

3.151 A. Lattes, H.A. Haus, F.J. Leonberger, E.P. Ippen: IEEE J. QE-19, 1718 (1983); H.A.Haus, N.A. Whitaker Jr.: Phil. Trans. R. Soc. London **313**, 311 (1984); Proc. SPIE **517**, 226 (1984)

3.152 N.M. Bityurin, V.I. Bredikhin, V.N. Genkin: Kvantovaya Elektron. **5**, 2453 (1978) [English transl.: Sov. J. Quantum Electron. **8**, 1377 (1978)]; V. Dimitriev, V. Konovalov, E. Shvom: Summaries, 8th Nat. Conf. Laser and Non-linear Opt., Tbilisi, 1976 (Metsniereba, USSR 1976), p. 79

3.153 V.E. Wood, N.F. Hartman, C.M. Verber: Ferroelectrics **27/28**, 237 (1980)

3.154 A.M. Glass: "Efficient Multiphoton Photorefractive Processes in LiNbO$_3$ Crystals and Waveguides", paper presented at Int. Mtg. Ferroelectricity, Leningrad 1977 (unpublished)

3.155 C.M. Verber, D.W. Vahey, R.P. Kenan, V.E. Wood, N.F. Hartman, C.F. Chapman: "An Investigation for the Development of an Integrated Optical Data Preprocessor", NASA Contractor Report 3151 (NASA, Washington 1979)

3.156 A.M. Glass: Opt. Eng. **17**, 470 (1978)

3.157 R. von Baltz, W. Kraut: Phys. Rev. **B23**, 5590 (1981)

3.158 D. von der Linde, O.F. Schirmer, H. Kurz: Appl. Phys. **15**, 153 (1978)

3.159 H. Kurz, D. von der Linde: Ferroelectrics **21**, 621 (1978)

3.160 O.F. Schirmer, D. von der Linde: Appl. Phys. Lett. **33**, 35 (1978)

3.161 O.F. Schirmer: J. Appl. Phys. **50**, 3404 (1979)

3.162 V.E. Wood, R.C. Sherman, N.F. Hartman, C.M. Verber: Ferroelectrics **34**, 175 (1981); "Optical Erasure of One- and Two-Photon Holograms in Lithium Niobate", paper presented at Int. Mtg. Ferroelectricity, University Park, PA 1981 (unpublished)

3.163 D. von der Linde, A.M. Glass, K.F. Rodgers: Appl. Phys. Lett. **26**, 22 (1975)

3.164 D. von der Linde, A.M. Glass, K.F. Rodgers: J. Appl. Phys. **47**, 217 (1976)
3.165 V.E. Wood, R.C. Sherman: Ferroelectrics **50**, 155 (1983)
3.166 S.I. Bozhevol'nyi, E.M. Zolotov, V.A. Chernykh: Pis'ma Zh. Tekh. Fiz. **6**, 852 (1980) [English transl.: Sov. Tech. Phys. Lett. **6**, 366 (1980)]
3.167 S. Yamada, M. Minakata: Jpn. J. Appl. Phys. **20**, 733 (1981)
3.168 A.D. McLachlan, R.M. De La Rue, J.A.H. Wilkinson: Proceedings, First Europ. Conf. Integr. Opt. (1981), p. 4
3.169 C.T. Mueller, E. Garmire: Proc. SPIE **460**, 109 (1984); Appl. Opt. **23**, 4348 (1984); Proc. SPIE **517**, 57 (1984)
3.170 K.S. Buritskii, E.M. Zolotov, R.F. Tarlykaev, V.A. Chernykh: Zh. Tekh. Fiz. **54**, 1839 (1984) [English transl.: Sov. Phys. Tech. Phys. **29**, 1078 (1984)]
3.171 C.H. Bulmer, R.P. Moeller, W.K. Burns: Proceedings, 2nd Europ. Conf. Integt. Opt. (1983), p. 140; C.H. Bulmer, W.K. Burns: IEEE J. LT-**2**, 512 (1984)
3.172 R.A. Becker: Appl. Phys. Lett. **45**, 121 (1984); Proc. SPIE **517**, 194 (1984)
3.173 G.T. Harvey, G. Astfalk, A.Y. Feldblum, B. Kassahun: IEEE J. QE-**22**, 939 (1986)
3.174 C.M. Gee, G.D. Thurmond: Proc. SPIE **477**, 17 (1984)
3.175 R.A. Becker, R.C. Williamson: Appl. Phys. Lett. **47**, 1024 (1985)
3.176 C.M. Gee, G.D. Thurmond, H. Blauvelt, H.W. Yen: Appl. Phys. Lett. **47**, 211 (1985)
3.177 C.H. Bulmer, W.K. Burns, S.C. Hiser: Appl. Phys. Lett. **48**, 1036 (1986); P. Skeath, C.H. Bulmer, S.C. Hiser, W.K. Burns: Appl. Phys. Lett. **49**, 1221 (1986)
3.178 I. Sawaki, H. Nakajima, M. Seino, K. Asama: "Thermally Stabilized Z-Cut Ti : LiNbO$_3$ Waveguide Switch", paper J-61, Tech. Digest, CLEO '86 (IEEE, New York, 1986)
3.179 J.J. Amodei, D.L. Staebler: Appl. Phys. Lett. **18**, 540 (1971)
3.180 E.M. Zolotov, P.G. Kazanskii, V.A.Chernykh: Pis'ma Zh. Tekh. Fiz. **9**, 360 (1983) [English transl.: Sov. Tech. Phys. Lett. **9**, 155 (1983)]
3.181 E.M. Zolotov, A.G. Kazanskii, P.G. Kazanskii, V.A. Chernykh: Pis'ma Zh. Tekh. Fiz. **9**, 1409 (1983) [English transl.: Sov. Tech. Phys. Lett. **9**, 605 (1983)]
3.182 T. Schurr: unpublished work (1983), quoted in [3.101]
3.183 V.I. Belinicher, B.I. Sturman: Usp. Fiz. Nauk **130**, 415 (1980) [English transl.: Sov. Phys. Usp. **23**, 199 (1980)]
3.184 R.R. Birss: *Symmetry and Magnetism* (North-Holland, Amsterdam 1964), Ch. 2
3.185 S.I. Bozhevol'nyi, E.M. Zolotov, P.G. Kazanskii, A.M. Prokhorov, V.A. Chernykh: Pis'ma Zh. Tekh. Fiz. **9**, 690 (1983) [English transl.: Sov. Tech. Phys. Lett. **9**, 297 (1983)]
3.186 J.F. Lam, H.W. Yen: Appl. Phys. Lett. **45**, 1172 (1984)
3.187 R.C. Alferness: Appl. Phys. Lett. **36**, 513 (1980)
3.188 Yu.A. Bykovskii, Yu.Yu. Vaitkus, E.P. Gaubas, Yu.N. Kul'chin, V.L. Smirnov, K.Yu. Yarashyunas: Kvantovaya Elektron. **9**, 676 (1982) [English transl.: Sov. J. Quantum Electron. **12**, 418 (1982)]
3.189 V.D. Antsigin, E.G. Kostsov, V.K. Malinovsky, L.N. Sterelyukhina: Ferroelectrics **38**, 761 (1981)
3.190 T. Kawaguchi, H. Adachi, K. Setsune, O. Yamazaki, K. Wasa: Appl. Opt. **23**, 2187 (1984); K. Wasa, O. Yamazaki, H. Adachi, T. Kawaguchi, K. Setsune: IEEE J. LT-**2**, 710 (1984)
3.191 O. Yamazaki, K. Wasa, T. Kawaguchi, Y. Manabe, H. Adachi, H. Higashino, K. Setsune: Digest Tech. Papers, 7th Topical Mtg. Integr. Guided Wave Opt. (Opt. Soc. Amer., Washington 1984), p. TuA6-1

Addendum (April, 1988)

This review was originally completed in December, 1986. The ensuing period has seen the introduction of several more packaged LiNbO$_3$ waveguide devices, an increased interest in ion-implanted waveguides, and the publication of a number of further studies of waveguide photorefraction. We have not deemed it necessary, though, to modify our text; rather, we have simply appended a list of additional references. In addition to those papers dealing with waveguide photorefractivity per se, a few papers dealing with other topics touched on in our review have been included in the list of additional references on page 353.

4. Wave Propagation in Photorefractive Media

Jeffrey O. White, Sze-Keung Kwong, Mark Cronin-Golomb,
Baruch Fischer, and Amnon Yariv

With 34 Figures

In the twenty years since the discovery of the photorefractive effect [4.1], it has proven to be both a nuisance and an asset to the optics community. Originally, it was labelled "optical damage", and even today it is a concern in the field of integrated optics. Soon after this discovery, photorefractive media were recognized as a possible replacement for photographic film, making dynamic holography and real-time van der Lugt filters accessible. Now photorefractive crystals have the status of excellent nonlinear media for implementing phase conjugate mirrors. They are also now viewed as parametric gain media with which it is possible to construct a new type of optical oscillator. An understanding of the wave propagation in photorefractive media is crucial to solving the problems and exploiting the opportunities which they represent.

Photorefractive crystals, e.g. $LiNbO_3$, $BaTiO_3$, and $Bi_{12}SiO_{20}$, have unique optical properties which distinguish them from all other nonlinear materials, e.g. Kerr media, atomic vapors, liquid crystals, and saturable absorbers. For example, intensity thresholds in photorefractive crystals are extremely low or nonexistent. Also, the refractive index is modified only when the incident wave has spatial variations in intensity.

The nonlinearities are very large, of a magnitude more commonly associated with resonant phenomena. The photorefractive effect is resonant in the sense that if the spatial intensity variations are created by the interference of two coherent laser beams, their difference frequency is very important. Because of the slow response time, difference frequencies typically have to be on the order of Hertz, a very narrow resonance indeed. The photorefractive effect is broadband in the sense that the incident light does not have to be of any particular wavelength, but not in the sense of responding to rapid temporal variations of intensity.

The optimum distance scale for spatial variations is quite small, and is most easily accessed by the focusing, or interference of two or more laser beams. Not surprisingly, holographic storage was the first application envisioned for the photorefractive effect. The advantages over photographic film are several: (1) no developing is required; (2) the process is reversible; (3) three-dimensional holograms are easily realizable; and (4) the refractive index changes rather than the absorption, hence larger diffraction efficiencies are possible. There are some similarities, for example, the sensitivities can be comparable. Also, in both cases, during an initial transient regime, the optical properties change in response to the total energy flux absorbed.

The simplest hologram is a perfectly periodic index variation, or grating, In previous chapters, the grating formation in photorefractive crystals was explained in terms of constitutive equations which describe the photo-induced excitation, migration, and retrapping of the mobile charges within the crystal. A spatially varying intensity was found to cause a redistribution of charge which modulates the index of refraction through the dc linear electro-optic effect. In this chapter, using Maxwell's equations, we describe how the various plane wave components of the incident light are subsequently coupled by this self-induced index change.

The coupled wave equations used throughout this chapter are nonlinear because the index of refraction is a function of the incident light intensity. The change in index is essentially proportional to the product of two optical fields, and it is a stationary change when their frequencies cancel. The nonlinearity is considered third order because the product of the optical field and the induced index change is a polarization that is cubic in the field. In addition to this nonlinearity, a dc electric field is sometimes applied to the crystal to enhance the interaction.

4.1 Two-Wave Interactions Via a Third Order Nonlinearity

The index of a photorefractive crystal is perturbed by a spatially varying, time averaged intensity. The simplest intensity variation has a single Fourier component, e.g., that created by the interference of two plane waves. Accordingly, the subject of this section is the nonlinear coupling of two plane waves propagating in a photorefractive crystal, two-wave mixing (2WM). Both the transmission and reflection geometries are discussed. Intensity coupling is shown to represent a form of coherent optical amplification without stimulated emission. An understanding of a photorefractive amplifier is crucial to the oscillator described in the next section. Before discussing nonlinear optics, this section begins with a review of fixed gratings, and an introduction to coupled wave equations.

4.1.1 Diffraction from Fixed Gratings, Coupled Wave Theory

Fixed, i.e., non-erasable gratings can be produced in photographic emulsions, dichromated gelatin [4.2], photopolymer materials [4.3], and photorefractive crystals that have been recorded at high temperature and then cooled [4.4]. When the thickness is less than or comparable to a fringe spacing, the transmitted field is given simply by the incident field times a multiplicative amplitude to be transmitted, t. The analysis is simple because the diffraction occurs outside the medium, in a region of uniform index.

If the transmission varies in a sinusoidal fashion, e.g., $t = \frac{1}{2}(1 + m \sin Kx)$, then two diffracted components appear (Fig. 4.1a). If either the exponential absorption constant or the index of refraction is given a sinusoidal modulation, e.g., $t = \exp(i2\pi m \sin Kx)$, as many diffracted orders appear as there are

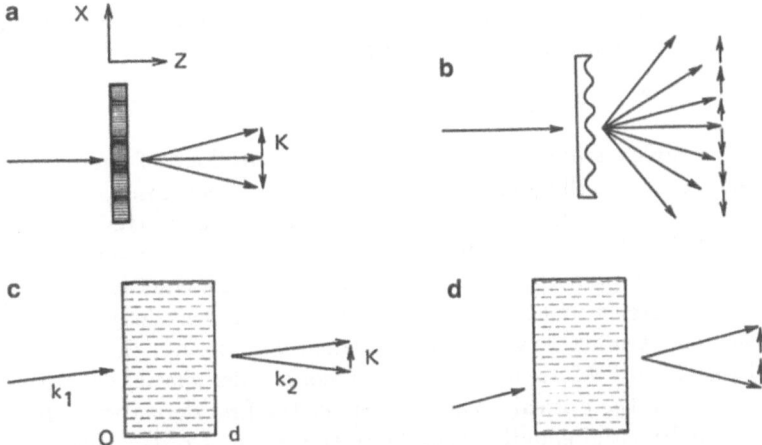

Fig. 4.1a–d. Diffraction from thin and thick gratings, (a) thin amplitude, (b) thin phase or absorption grating, (c) zero and first order Bragg diffraction within a thick grating, (d) second order Bragg diffraction

Fourier components in t (Fig. 4.1b). The angular spectrum of diffracted light is limtied to wavevectors whose x components are multiples of the grating vector. This is simply constructive interference or one-dimensional phase matching. When the medium is thin compared to a fringe spacing, so that multiple reflections do not occur, there is no analogous constraint on phase matching in the z direction. Consequently, t is independent of the incident angle, within the paraxial approximation. The diffraction efficiency of most thin gratings is limited, however, hence the interest in thick gratings.

In a sample of thickness much greater than fringe spacing, the interaction and the diffraction occur simultaneously, within the same volume. The wave equation must be solved in a region of periodic index variation. In contrast to thin gratings, only a single diffracted beam is radiated, and only when the incident beam is close to Bragg incidence (Fig. 4.1c). The transfer of energy into a diffracted component is cumulative in z only if the diffracted wavevector equals the incident minus the grating vector, $k_2 = k_1 - K$. Put another way, multiple reflections only interfere constructively when there is phase matching in both x and z. Higher order diffraction can occur, given the proper angle of incidence, i.e., $k_2 = k_1 - 2K$ (Fig. 4.1d).

Propagation inside volume gratings, including the effects of phase matching, is well described by coupled wave equations derived from the scalar wave equation as follows [4.5]: For a grating with only one sinusoidal component, n_1, the surfaces of constant index are planes, normal to the grating vector, K, with phase ϕ at the origin:

$$n = n_0 + \frac{n_1}{2}e^{-i(K \cdot r + \phi)} + \frac{n_1}{2}e^{i(K \cdot r + \phi)} \quad .$$

The optical electric field consists of two components, incident at the Bragg angle, whose wavevectors are given by $k_1 - k_2 = K$:

$$E(r) = A_1(z)e^{-ik_1 \cdot r} + A_2(z)e^{-ik_2 \cdot r} \quad .$$

The A_i are the complex amplitudes representing the envelope of the optical field. The z dependence of the amplitudes is due to the linear absorption and the coupling caused by the grating. The above expressions are inserted into the scalar wave equation. Terms with different, rapidly varying exponentials are assumed to go to zero separately, so are collected in separate equations.

The amplitudes are assumed to obey the slowly varying envelope approximation, i.e., they are assumed to vary by only a small fraction over the distance of one optical wavelength. Mathematically, the second order spatial derivative is neglected relative to the optical wavevector times the first derivative. Then, the second order differential equation reduces to the two first order equations:

$$\cos \theta_1 \frac{dA_1}{dz} = -\frac{\alpha}{2} A_1 - i \frac{\pi n_1}{\lambda} e^{-i\phi} A_2 \quad ,$$

$$\cos \theta_2 \frac{dA_2}{dz} = -\frac{\alpha}{2} A_2 - i \frac{\pi n_1}{\lambda} e^{i\phi} A_1 \quad ,$$

where θ_i is the angle between the z axis and k_i. Separating the amplitudes into a magnitude and phase with the definition $A_j \equiv \sqrt{I_j} \exp(-i\phi_j)$, the equations for the intensities can be found:

$$\cos \theta_1 \frac{dI_1}{dz} = -\alpha I_1 + \frac{2\pi n_1}{\lambda} \sqrt{I_1 I_2} \sin(\phi_1 - \phi_2 - \phi) \quad ,$$

$$\cos \theta_2 \frac{dI_2}{dz} = -\alpha I_2 + \frac{2\pi n_1}{\lambda} \sqrt{I_1 I_2} \sin(\phi_1 - \phi_2 - \phi) \quad . \tag{4.1}$$

Note that the phases of A_1 and A_2 appear in the equations, so that at any point in space, if the interference pattern and the grating are out of phase so that $\phi_1 - \phi_2 - \phi \neq 0$, the intensities are coupled. When $\theta_1 = -\theta_2$, the diffraction efficiency is given by

$$\eta \equiv \frac{I_2(d)}{I_1(0)} = \exp\left(-\frac{\alpha d}{\cos \theta_1}\right) \sin^2\left(\frac{\pi n_1 d}{\lambda \cos \theta_1}\right) \quad .$$

Such formulae are commonly used to interpret the experimental data described in other chapters of this book.

4.1.2 The Coupled Wave Theory of Dynamic Gratings

To extend the analysis to the dynamic case where the grating is written by the very waves that are coupled by it [4.6], we will derive equations coupling the complex amplitudes of two linearly polarized plane waves

$$E = \hat{e}_1 A_1(z)e^{-ik_1 \cdot r} + \hat{e}_2 A_2(z)e^{-ik_2 \cdot r} \quad ,$$

which interfere inside the medium to produce an intensity distribution

$$I \equiv \boldsymbol{E} \cdot \boldsymbol{E} = |A_1|^2 + |A_2|^2 + \hat{e}_1 \cdot \hat{e}_2 A_1 A_2^* e^{-\mathrm{i}(\boldsymbol{k}_1 - \boldsymbol{k}_2) \cdot \boldsymbol{r}}$$
$$+ \hat{e}_1 \cdot \hat{e}_2 A_1^* A_2 e^{\mathrm{i}(\boldsymbol{k}_1 - \boldsymbol{k}_2) \cdot \boldsymbol{r}}$$
$$= I_0 \left(1 + \hat{e}_1 \cdot \hat{e}_2 \frac{A_1 A_2^*}{I_0} e^{-\mathrm{i}(\boldsymbol{k}_1 - \boldsymbol{k}_2) \cdot \boldsymbol{r}} \right.$$
$$\left. + \hat{e}_1 \cdot \hat{e}_2 \frac{A_1^* A_2}{I_0} e^{\mathrm{i}(\boldsymbol{k}_1 - \boldsymbol{k}_2) \cdot \boldsymbol{r}} \right)$$

where $I_0 \equiv I_1 + I_2$. We restrict the discussion to waves whose polarization vectors do not change in space, i.e., they are individually eigenpolarizations of the medium, and they are either perpendicular or parallel to the plane of incidence so that the coupling does not alter the polarization of either wave. See Ref. [4.7] and TAP61 [1], Chap. 4 for treatments of the general case. The form for the index follows from the expression for the time averaged intensity:

$$n = n_0 + n_1 e^{-\mathrm{i}\phi} \frac{A_1 A_2^*}{I_0} e^{-\mathrm{i}(\boldsymbol{k}_1 - \boldsymbol{k}_2) \cdot \boldsymbol{r}} + \mathrm{c.c.}$$

n_0 is the unperturbed index of the material accompanied by any uniform electro-optic effect. Experiments have shown the amplitude of the grating to be proportional to the fringe visbility, or modulation index of the intensity distribution, $2A_1 A_2^*/I_0$ [4.8]. The real quantity n_1 is the material response to a unit modulation index, and is given by the amplitude of the space-charge field, E_{sc}, and an effective electro-optic coefficient. E_{sc} can be calculated from the constitutive equations of the photorefractive effect, (TAP61, Chap. 3), for example

$$n_1 e^{-\mathrm{i}\phi} = r_{\mathrm{eff}} \frac{n_0^3}{4} E_{\mathrm{sc}} = -\mathrm{i} r_{\mathrm{eff}} \frac{n_0^3}{4} \frac{E_q (E_0 + \mathrm{i} E_D)}{E_0 + \mathrm{i}(E_D + E_q)} \hat{e}_1 \cdot \hat{e}_2 \quad . \qquad (4.2)$$

E_0 is an externally applied dc field. E_q is the limiting space-charge field, i.e., the field that would exist if all of the available charge carriers, within one fringe period, were separated by exactly one fringe spacing. E_D is a pseudo-field, actually the conduction band (or valence band) current, due to carrier diffusion, divided by the conductivity. ϕ represents a spatial phase shift between the grating and the interference pattern. A nonzero phase shift represents a "nonlocal" response of the medium.

In general, ϕ depends on intrinsic crystal properties, e.g., the polarity, density, and mean free path of the mobile charges, and extrinsic factors, e.g., an externally applied dc field, grating period, temperature, and light intensity. Most of the experimental work described in this chapter concerns $BaTiO_3$, with no applied field, in which the phase shift is 90°. This can be visualized as follows: after successive cycles of photoexcitation, diffusion, and recombination, the holes migrate away from the intensity maxima, leaving a net negative charge.

---

[1] TAP is the Springer Series "Topics in Applied Physics".

They collect in the intensity minima, creating a positive charge. The space-charge field is a maximum half way between the intensity maxima and minima, 90° out of phase.

Substituting these expressions into the scalar wave equation and using the slowly varying envelope approximation yields the following equations:

$$\cos\theta_1 \frac{dA_1}{dz} = -\frac{\alpha}{2}A_1 - i\frac{\pi n_1}{\lambda}e^{-i\phi}\frac{A_1 A_2^* A_2}{I_0}$$

$$\cos\theta_2 \frac{dA_2}{dz} = -\frac{\alpha}{2}A_2 - i\frac{\pi n_1}{\lambda}e^{i\phi}\frac{A_1^* A_2 A_1}{I_0} \quad .$$

The complex coupling constant, $\gamma = i\pi n_1 \exp(-i\phi)/\lambda$, will appear throughout the rest of this chapter. The complex number i in the coupling constant represents the 90° phase shift which occurs upon distributed reflection from a periodic index variation. The intensities and phases of the two beams are coupled according to the grating phase, ϕ:

$$\cos\theta_1 \frac{dI_1}{dz} = -\alpha I_1 - \frac{2\pi n_1}{\lambda}\sin\phi\frac{I_1 I_2}{I_0} \quad ; \quad \cos\theta_1 \frac{d\phi_1}{dz} = \frac{\pi n_1}{\lambda}\cos\phi\frac{I_2}{I_0} \quad ;$$

$$\cos\theta_2 \frac{dI_2}{dz} = -\alpha I_2 + \frac{2\pi n_1}{\lambda}\sin\phi\frac{I_1 I_2}{I_0} \quad ; \quad \cos\theta_2 \frac{d\phi_2}{dz} = \frac{\pi n_1}{\lambda}\cos\phi\frac{I_1}{I_0} \quad .$$

$$(4.3)$$

Note that the intensities do not depend on the phase of the optical field, in contrast to (4.1). For $\phi = \pi/2$, there is no phase coupling, which makes it possible for a diffraction limited signal wave to be amplified with impunity by an aberrated pump wave (Sects. 4.2.4 and 4.4.1). The energy balance for the dynamic system is the same as that for a fixed grating:

$$\frac{d}{dz}(I_1 \cos\theta_1 + I_2 \cos\theta_2) + \alpha(I_1 + I_2) = 0 \tag{4.4}$$

which shows that, for zero α, energy flow in the z direction is conserved.

4.1.3 Dynamic Gratings in the Transmission Geometry

In the case of a transmission hologram (Fig. 4.2a) with symmetric angles of incidence, i.e., $\theta_1 = -\theta_2$, the intensity equations can be decoupled using the relation $I_1 + I_2 = [I_1(0) + I_2(0)]\exp(-\alpha r)$. The path length is denoted by $r = z/\cos\theta_1$. The variable z ranges from 0 to d, and r ranges from 0 to l throughout this chapter. Changing to new variables $I_j \exp(\alpha r)$ is helpful in integrating the decoupled equations

$$I_1 = I_1(0)e^{-\alpha r}\frac{I_1(0) + I_2(0)}{I_1(0) + I_2(0)e^{\Gamma r}} \quad ;$$

$$I_2 = I_2(0)e^{-\alpha r}\frac{I_1(0) + I_2(0)}{I_1(0)e^{-\Gamma r} + I_2(0)} \quad .$$

$$(4.5)$$

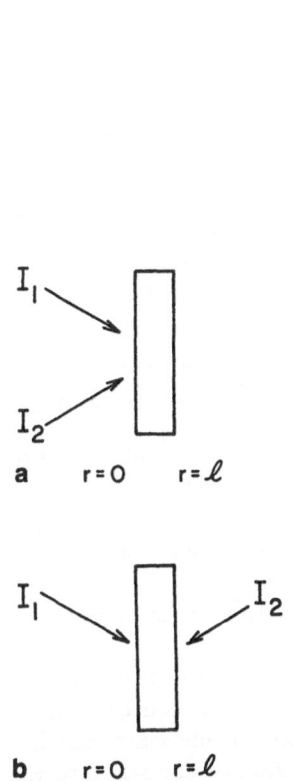

a   r = 0   r = ℓ

b   r = 0   r = ℓ

Fig. 4.2a,b. Two-wave mixing, **(a)** in the transmission geometry, **(b)** in the reflection geometry

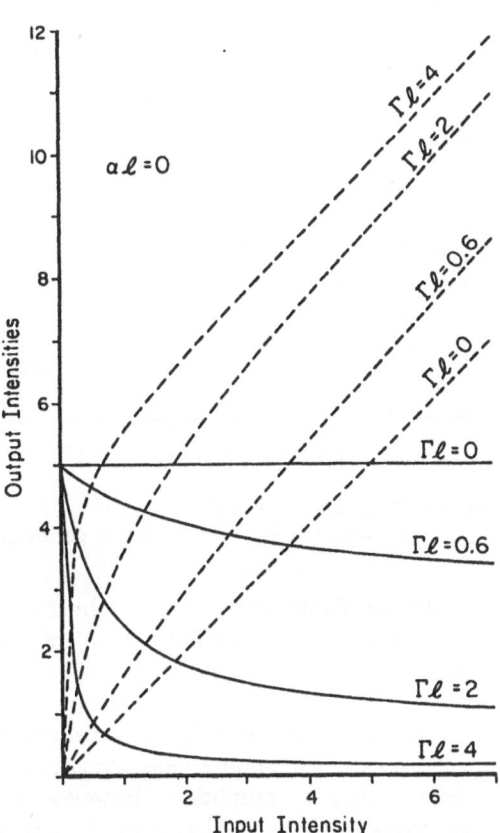

Fig. 4.3. Diagram showing the operation of a photorefractive amplifier based on two-wave mixing

The intensity coupling constant is given by $\Gamma \equiv 2\pi n_1 (\sin\phi)/\lambda$. Aside from the linear absorption, for large, positive Γ, the sum of the intensities of both beams appears in I_2. This represents coherent optical gain for beam 2 and loss for beam 1. Input/output curves for such an amplifier are shown in Fig. 4.3. Both output intensities are plotted as a function of one of the input intensities. The donor, or pump, beam intensity is fixed at five arbitrary units and the signal beam is varied from zero to seven to show gain saturation. The dashed line is the amplified beam and the solid line is the attenuated beam.

In the small signal limit, $I_2(0) \ll I_1(0)\exp(-\Gamma l)$, Γ is the exponential gain coefficient for I_2, i.e., $I_2 = I_2(0)\exp[(\Gamma - \alpha)r]$. In this regime, there is no transfer of intensity variations from a pump to a signal, which is another desirable characteristic for an amplifier (see Sect. 4.4.1).

Experimental data taken with a 5 mm crystal of BaTiO$_3$ is shown in Fig. 4.4a [4.9]. This type of energy transfer has also been used to coherently

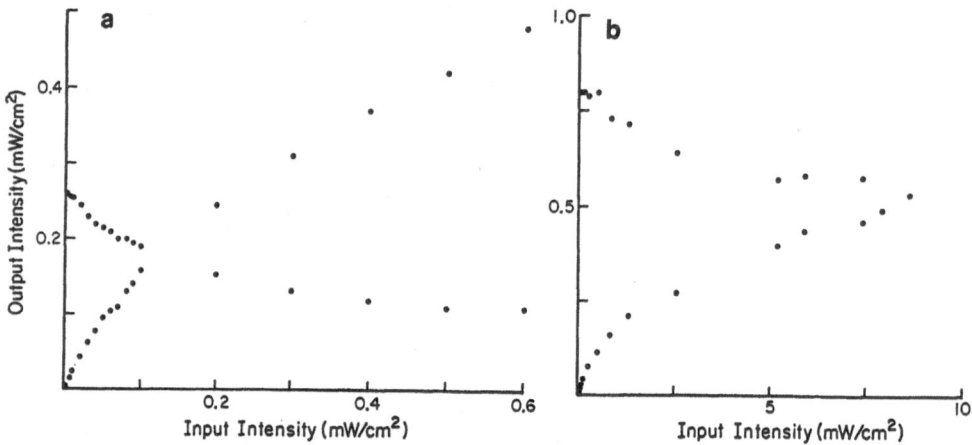

Fig. 4.4. Experimental results showing the operation of a photorefractive amplifier based on two-wave mixing, (a) in the transmission geometry, (b) in the reflection geometry

amplify nonplanar object waves which carry information. A gain of 4000 has been measured for a $15\,\mathrm{nW/cm^2}$ object beam passing through a 2.4 mm thick crystal of $BaTiO_3$ pumped by 15 mW of 514.5 nm light from an argon ion laser [4.10].

Unlike the situation in amplifiers that store energy in the form of population inversions, spontaneous emission is not an important source of noise in photorefractive amplifiers. However, noise does arise from scattering imperfections in the crystal and random thermal excitation of carriers into the conduction band (as opposed to deliberate photoexcitation). At present, the slow response time of photorefractive crystals limits their use to coupling two beams supplied by the same laser. Using a portion of the laser output to pump an amplifier of this type would appear to be most useful when the object could be damaged by directing all the laser power into the object wave.

4.1.4 Dynamic Gratings in the Reflection Geometry

In the reflection geometry, the two waves are incident at opposite faces (Fig. 4.2b). In practice, reflection gratings are written by beams that are more nearly counterpropagating, and transmission gratings are written by beams that are more nearly copropagating. Very short period transmission gratings, and very long period reflection gratings can be written, but as the beams approach grazing incidence, a hologram of finite extent will intercept less and less of the beam. A special case of the reflection geometry is where the beams are counterpropagating. It is important to understand the coupling between counterpropagating beams because they are one constituent of the four-wave mixing implementation of phase conjugate mirrors (Sect. 4.3).

In the case of symmetric angles of incidence, i.e., $\theta_2 = \pi \pm \theta_1$, the key to solving the two-point boundary value problem can be derived from (4.3):

$$\frac{d}{dr}(I_1 I_2) = \Gamma I_1 I_2 \quad \text{or} \quad I_1(l)I_2(l) = e^{\Gamma l} I_1(0)I_2(0) \tag{4.6}$$

independent of the absorption constant α. All the intensity dependence here is explicit; Γ should only depend on the grating vector and the crystal orientation.

Before solving the reflection grating problem, an experimental test of (4.6) will be described. The experimental apparatus is shown in Fig. 4.5. The half-wave plate and polarizing beamsplitter combination serves as a no-loss beamsplitter with a transmission/reflection ratio that can be varied by simply rotating the half-wave plate. The sum of the two input intensities was held constant and the coupling constant Γ was calculated as the ratio was varied. The results are shown in Fig. 4.6 in which Γ changes by only a factor of two while the ratio of the two input intensities changes by six orders of magnitude. A similar relationship has been confirmed in the transmission geometry [4.11].

Equation (4.6) can be used to uncouple the intensity equations. A change of variables to $I_i \exp(-\Gamma r)$ is helpful in integrating the equations and solving for the output intensities [4.9, 12, 13]

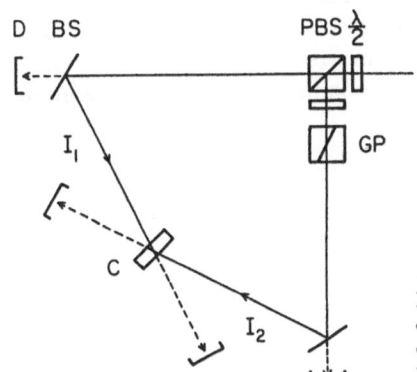

Fig. 4.5. Experimental configuration for two-beam coupling in the reflection geometry (C: crystal; D: detector; GP: glan prism; BS: beamsplitter; PBS: polarizing beamsplitter; $\lambda/2$: half-wave plate)

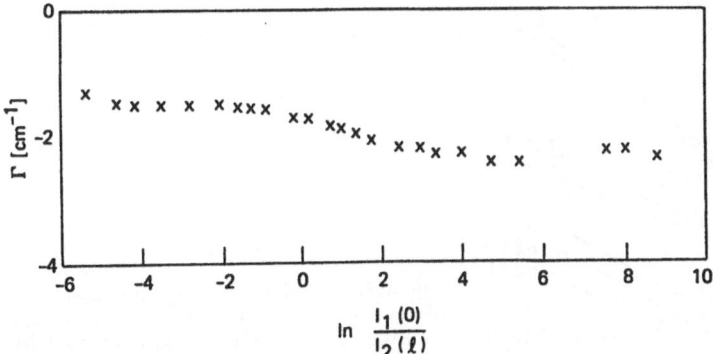

Fig. 4.6. The two-beam coupling constant vs the ratio of the intensities of the two input beams

$$e^{-(\Gamma+2\alpha)r} = \left[\frac{(\Gamma+2\alpha)I_1^2(0) - (\Gamma+2\alpha)I_1(0)I_2(0)}{(\Gamma+2\alpha)I_1^2 e^{\Gamma r} - (\Gamma+2\alpha)I_1(0)I_2(0)}\right]^{2\Gamma/(\Gamma+2\alpha)}$$

$$\times \left(\frac{I_1}{I_1(0)}\right)^2 e^{\Gamma r} \quad ; \tag{4.7a}$$

$$e^{-(\Gamma-2\alpha)r} = \left[\frac{(\Gamma+2\alpha)I_2^2(0) - (\Gamma-2\alpha)I_1(0)I_2(0)}{(\Gamma+2\alpha)I_2^2 e^{\Gamma r} - (\Gamma-2\alpha)I_1(0)I_2(0)}\right]^{2\Gamma/(\Gamma+2\alpha)}$$

$$\times \left(\frac{I_2}{I_2(0)}\right)^2 e^{\Gamma r} \quad . \tag{4.7b}$$

$I_2(0)$ can be found by evaluating (4.7b) at $r = l$.

In the limit $\alpha \to 0$, one can show, from (4.7), that

$$I_1(l) = I_1(0)\frac{I_1(0) + I_2(l)}{I_1(0) + I_2(l)e^{\Gamma l}} \quad , \qquad I_2(0) = I_2(l)\frac{I_1(0) + I_2(l)}{I_1(0)e^{-\Gamma l} + I_2(0)} \quad ,$$

which are identical to (4.5) under the interchange $I_2(0) \to I_2(l)$. Therefore an amplifier with $\alpha = 0$ and a reflection geometry will have the same operating curves as the transmission geometry (Fig. 4.3). The design of an amplifier should be concerned with maximizing Γl and need not be prejudiced in favor of either geometry. Figure 4.4b shows data taken with the apparatus of Fig. 4.5.

A difference between the two amplifier geometries appears for a nonzero α, and this is shown in Fig. 4.7. The pump intensity is fixed at five, and the input

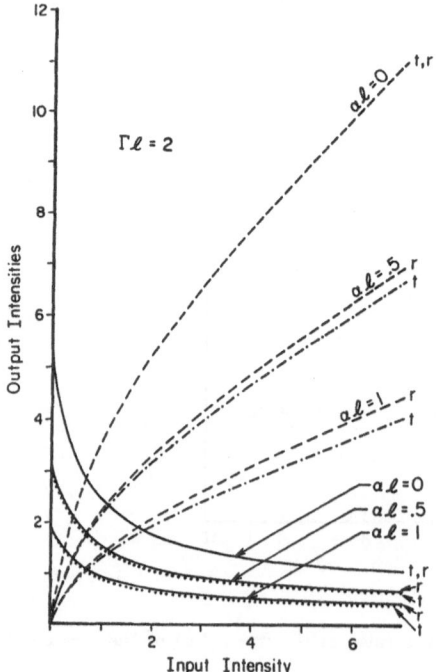

Fig. 4.7. Diagram showing the operation of two-beam coupling amplifiers, in the transmission and reflection geometries, with absorption present

signal intensity is varied, as before. Note that if two beams are sent through a photorefractive crystal in the reflection geometry, the combined intensity of the two beams which emerge is larger than if the same two beams are sent through in the transmission geometry. (Same coupling strength and absorption constant in both cases.)

This analysis has neglected Fresnel reflections at the crystal surfaces, which give rise to multiple gratings. The influence of Fresnel reflections has recently been analyzed theoretically [4.14].

4.1.5 Coupling Between Counterpropagating Waves in a Ring Resonator

The power flow from one wave to another in two-wave mixing suggests the use of a photorefractive crystal as a unidirectional element inside a ring laser cavity (Fig. 4.8). The preferred one-directional power flow could enhance the clockwise oscillation at the expense of the counter-clockwise oscillation if the two counterpropagating waves were coupled with a reflection grating.

A uni-directional ring laser is of interest because it has no standing wave nodes which prevent complete utilization of the inversion inside the gain medium [4.15, 16]. Increased mode stability and a four-fold increase in power output result from eliminating the oscillation in one direction with an optical diode. Currently, Faraday rotators are combined with Brewster windows to provide a 1 % difference in round trip loss between the competing directions. This difference in loss, combined with the competition for the gain, is sufficient to keep the oscillation in one direction below threshold.

The holographic coupling only exists when both beams are present, so the best one could hope for would be a situation where one direction was dominant and only a small amount of power remained in the other direction to maintain the coupling. Unfortunately, for gain media that do not possess any directionality in their own right, such a steady state solution does not exist. For example, for gain media characterized by $\gamma = \gamma_0/(1 + I/I_{\text{sat}})$ where I_{sat} is the saturation intensity, the intensities within the gain medium obey the relation

$$\frac{d}{dz}(I_1 I_2) = 0 \quad \text{or} \quad I_1(0)I_2(0) = I_1(l)I_2(l) \quad ,$$

which is incompatible with (4.6), except when one of the intensities is zero. The solution would be stable in the sense that if I_1 starts to increase, the system

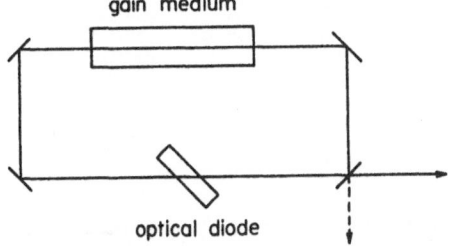

gain medium

optical diode

Fig. 4.8. A ring laser containing a gain medium and an optical diode, or unidirectional element

tends to counteract it. But if the directionality develops slowly compared to a cavity roundtrip time, the mean square fluctuations may be large.

An analysis of the directional properties of a lossless photorefractive crystal in a ring resonator with gain has been made assuming that the intensities are known [4.17]. The clockwise and counter-clockwise beams experience different phase shifts as well as different gains, giving rise to different resonant frequencies. Such a frequency bias is one method of avoiding the frequency locking problem in ring gyroscopes.

Another way to observe the asymmetric interaction is by allowing a single incident wave to interfere with its Fresnel reflection from the exit surface [4.18]. If the crystal orientation is correct, power flow in the bulk will be from the incident to the reflected wave, increasing the net loss. A wave incident in the opposite direction will recoup some of its surface reflection loss because power flow will be from the reflected wave to the incident wave.

4.2 Oscillation in a Resonator with Photorefractive Gain

The gain through two beam coupling, described in the first section, was first observed in 1972 [4.19]. In 1982, it was combined with feedback to produce oscillation in a unidirectional ring resonator [4.20]. In the configuration of Fig. 4.9, the pump I_1 is supplied externally, but I_2 is not. Scattered light passing through the crystal in many directions is amplified at the expense of I_1. Scattered light heading in the proper direction is fed back, by the mirrors, into the crystal to be amplified again. In this way, an infinitesimally weak initial beam (noise) can build up to an intensity I_2, comparable to I_1.

In the previous section, the two interacting beams had the same frequency, and both were plane waves. To describe the unidirectional ring resonator, the analysis is generalized to nondegenerate two-wave mixing, which implies a moving grating (Sect. 4.2.1). Expressions for the oscillation frequency and the threshold coupling constant will be obtained. Experimental results are presented in Sect. 4.2.2. In Sect. 4.2.3, the theoretical treatment is extended beyond the plane wave case, to an interference pattern which produces a complicated hologram, rather than a simple grating. This analysis explains how a photorefractive ring resonator could be used to "clean up" a beam with a distorted wavefront (Sect. 4.2.4).

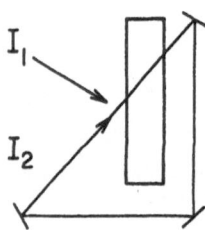

Fig. 4.9. A unidirectional ring resonator with photorefractive gain

4.2.1 Nondegenerate Two-Wave Mixing in a Ring Resonator

To allow for an oscillation beam, A_2, whose frequency is different from that of the pump beam, A_1, the field within the photorefractive crystal can be written as

$$E = \hat{e}_1 A_1(z) \exp i(\omega_1 t - k_1 \cdot r) + \hat{e}_2 A_2(z) \exp i(\omega_2 t - k_2 \cdot r) \quad .$$

The induced index variation will take the form of a travelling wave.

$$n = n_0 + n_1 e^{-i\phi} \frac{A_1 A_2^*}{I_0} \exp i(\Omega t - K \cdot r) + \text{c.c.}$$

where $\Omega = \omega_1 - \omega_2$, and $K = k_1 - k_2$. If $\Omega \ll \omega_1, \omega_2$, then most of the results from the prvious section can be carried over, provided that the amplitude of the index variation is changed [4.21] from (4.2) to

$$n_1 e^{-i\phi} = -i r_{\text{eff}} \frac{n_0^3}{4} \frac{E_q(E_0 + iE_D)\hat{e}_1 \cdot \hat{e}_2}{E_0 + i(E_D + E_q) + (E_D + E_\mu - iE_0)\Omega\tau} \tag{4.8}$$

where $1/\tau$ is the sum of the carrier photoexcitation rate and recombination rate. It is treated as a property of the crystal, although, in reality, it is approximately proportional to the total intensity, $1/\tau \sim I_0$. An additional pseudo-field, $E_\mu \equiv \gamma_R N_A / \mu |K|$, appears in the case of a moving grating. γ_R is the coefficient that characterizes the recombination of mobile changes (mobility μ) with traps (density N_A). When $E_0 = 0$, the coupling constant assumes a Lorentzian form

$$\gamma = \frac{\gamma_0}{1 + i\Omega\tau'} \quad \text{where} \quad \gamma_0 = \frac{\pi r_{\text{eff}} n_0^3}{4\lambda} \frac{E_q E_D}{E_q + E_D} \quad \text{and} \quad \tau' = \tau \frac{E_D + E_\mu}{E_D + E_q} \quad .$$

As before, in the transmission geometry, the intensities and phases are coupled according to

$$\frac{dI_1}{dr} = -\alpha I_1 - \Gamma \frac{I_1 I_2}{I_0} \quad ; \quad \frac{d\phi_1}{dr} = \frac{\Gamma' I_2}{2I_0} \quad ;$$

$$\frac{dI_2}{dr} = -\alpha I_2 + \Gamma \frac{I_1 I_2}{I_0} \quad ; \quad \frac{d\phi_2}{dr} = \frac{\Gamma' I_1}{2I_0} \quad ;$$

where $\Gamma + i\Gamma' = 2\gamma$.

The boundary condition appropriate for a ring resonator is $I_2(0) = M I_2(l)$, where M is the product of the mirror reflectivities, including the output coupler. The oscillating intensity, I_2, and the threshold coupling strength are then given by

$$\frac{I_2(0)}{I_1(0)} = \frac{M - e^{(\alpha-\Gamma)l}}{e^{\alpha l} - M} \quad \text{and} \quad (\Gamma l)_t = \alpha l - \ln M \quad . \tag{4.9}$$

In general, the ring cavity will not be resonant at the pump frequency. The oscillation frequency is determined by the round trip phase condition for the loaded cavity

$$\frac{\omega_2 L}{c} + \phi_2(l) - \phi_2(0) = \frac{\omega_a L}{c}$$

where ω_a is the resonant frequency of the empty cavity that is closest to ω_1. The differential equations for the phases can be solved to yield

$$\phi_2(l) - \phi_2(0) = -(\alpha l - \ln M)\cot\phi \quad .$$

Substituting into the round trip phase condition and solving for the oscillation frequency, we see that there is a shift away from ω_a, toward ω_1 [4.22]:

$$\omega_2 - \omega_1 = \frac{\omega_a - \omega_1}{1 + \tau'/\tau_a} \quad , \tag{4.10}$$

where the cavity decay time is $\tau_a = L/c(\alpha l - \ln M)$. Typically, photorefractive response times are much longer than cavity decay times, so the oscillation frequency is much closer to the pump frequency than to the cavity resonance. This frequency pulling is also characteristic of lasers with atomic gain media, although there the lasing frequency is typically closer to the cavity resonance than to the atomic resonance.

4.2.2 Experimental Results

The photorefractive ring laser has exhibited an intrinsic tendency toward unidirectional oscillation, in fact a bidirectional oscillation has not been observed to date. It has always been the case that the beam which oscillates is that which interacts with the pump in the transmission geometry, i.e., the clockwise beam in Fig. 4.9. Evidently, the Γl coupling the pump to the counterclockwise beam has always been below threshold. The unidirectional behavior is partly a result of the crystal's spatial frequency response, which typically prefers grating periods of approximately $2\,\mu m$. Transmission gratings with this period can be easily formed, whereas reflection gratings typically have shorter periods. Whether it is possible for two counterpropagating beams to be pumped by the same third beam also depends on the relative orientation of the crystal through the electro-optic coefficient appearing in (4.8).

The large electro-optic coefficients in $BaTiO_3$ ($r_{42} = 820\,pm/V$) and SBN ($r_{33} \cong 1\,nm/V$) are accessed by beams of extraordinary polarization, for which the effective electro-optic coefficient is given by

$$r_{\rm eff} = \left(n_e^4 r_{33} \sin\alpha \sin\beta + 2n_e^2 n_0^2 r_{42} \cos^2\frac{\alpha+\beta}{2} + n_0^4 r_{13} \cos\alpha \cos\beta \right)$$
$$\times \frac{1}{n_e n_0^3} \sin\frac{\alpha+\beta}{2} \quad ,$$

where $\alpha(\beta)$ is the angle of the pump (oscillation) beam with respect to the optic axis of the crystal.

If we only consider the r_{42} term, when two waves interact in $BaTiO_3$, the one whose wavevector lies closest to the c axis gets amplified at the expense of the other, via the grating formed between them. Given an acute angle between the pump beam and the c axis, the pump will amplify one member of a pair of counterpropagating beams and attenuate the other (Fig. 4.10a). Given an obtuse angle between the pump beam and the c axis, the pump will amplify each of two counterpropagating beams *individually* (Fig. 4.10b). However, a two-beam coupling analysis cannot predict the gain when all three beams are present because each beam will interact with the superposition of two or three different gratings. We have not been able to observe in our laboratory a bidirectional ring oscillator pumped by a single beam.

The lack of a pumping threshold is another unique property, which can be traced back to the photorefractive effect being driven by a spatial modulation of intensity, rather than by absolute intensity. The pump intensity dependence of the coupling constant enters implicitly through τ (4.8). In the degenerate case, (4.10) indicates that $\Omega\tau$ is independent of intensity, if $\tau' \gg \tau_a$. This conclusion was checked experimentally with a multi-longitudinal mode He-Ne laser operating at 632.8 nm. Oscillation was observed to build up even for a pump intensity as low as $15\,\mathrm{mW/cm^2}$, setting an upper limit for the threshold. The time to reach steady state was approximately 8 min.

An experimental test of the frequency pulling (4.10) was performed with a $BaTiO_3$ crystal in a 38 cm ring resonator by pumping with a single longitudinal mode argon ion laser operating at 514.5 nm. Part of the oscillation beam and part of the pump beam were directed to a beamsplitter outside the resonator, where they were combined at a shallow angle to generate observable interference fringes. The frequency shift was determined by measuring the speed of the moving fringes. The resonant frequency, ω_a, was changed by moving one of the cavity mirrors on a piezoelectric mount. Figure 4.11a shows the frequency detuning, $\Omega = \omega_2 - \omega_1$, as a function of mirror displacement, for an oscillating TEM_{00} mode and a TEM_{01} mode. The frequency detuning had a period of $\lambda/2$, and was nearly linear within each period. The gain linewidth is $1/\tau \cong 1\,\mathrm{Hz}$ in $BaTiO_3$, so the maximum detuning observed is a few Hz.

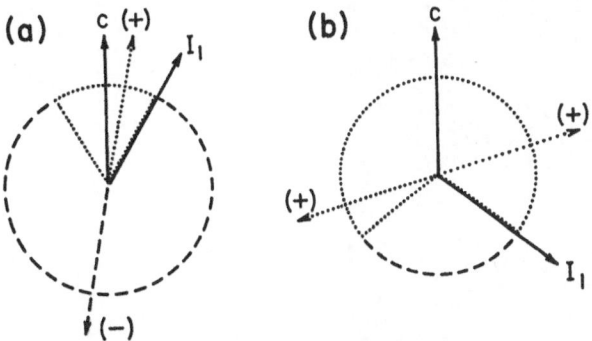

Fig. 4.10. Diagram showing amplification and attenuation through two beam coupling. The pump wave is I_1 and the optic axis is denoted by c. The angle between the pump wave and the c axis is (**a**) acute and (**b**) obtuse. Signal waves heading in directions spanned by the dotted region of the circle will be amplified. Signal waves heading in directions spanned by the dashed region of the circle will be attenuated

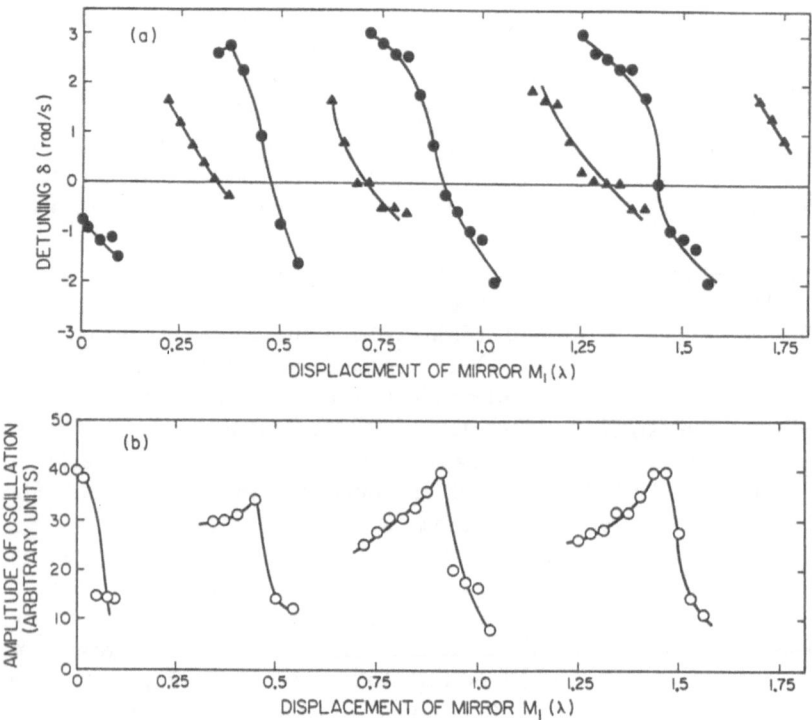

Fig. 4.11. (a) Measurement of frequency detuning δ vs displacement of mirror M_1. Results for both the TEM_{00} mode (•) and the TEM_{01} mode ($\triangle$). (b) The oscillating beam power of TEM_{00} mode vs displacement of mirror M_1

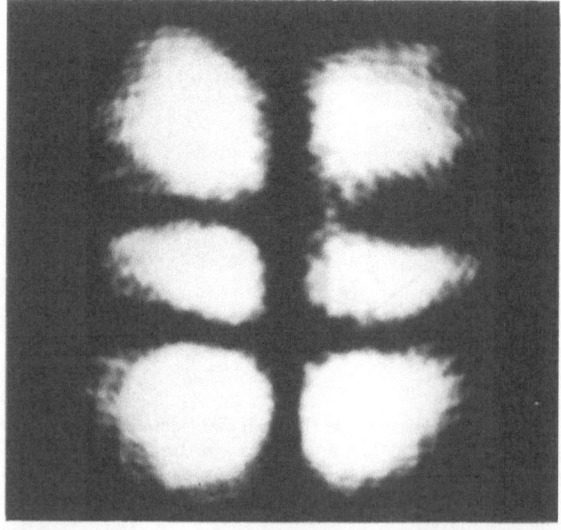

Fig. 4.12. A pure high order mode of a ring resonator with photorefractive gain [4.10]

The variation of intracavity power with cavity length is shown in Fig. 4.11b. The maximum power occurred at zero frequency detuning, where γ is a maximum.

By placing the crystal inside a stable ring cavity formed with curved mirrors, and using an intracavity aperture, it has been possible to observe the oscillation of pure high order modes (Fig. 4.12).

4.2.3 Relaxing the Plane Wave Constant

The theoretical treatment in this section is more general than that in the first section which was restricted to plane waves. Now the (time dependent) oscillation field, $\vec{\mathcal{E}}_2$ is taken to be one of a complete set of Slater modes, and the pump field is completely arbitrary. The simultaneous presence of both fields induces a time dependent nonlinear polarization in the medium, $\vec{\mathcal{P}}_{\mathrm{NL}}$. Following Lamb's self-consistent analysis of an inhomogeneous laser [4.23], we require that the field radiated by $\vec{\mathcal{P}}_{\mathrm{NL}}$ be identical to $\vec{\mathcal{E}}_2$.

The known pump beam is taken to be

$$\vec{\mathcal{E}}_1(\boldsymbol{r},t) = \boldsymbol{E}_1(\boldsymbol{r})e^{i\omega_1 t} \quad ,$$

where $\boldsymbol{E}_1$ contains the propagation factor and any other spatial modulation. The oscillating field inside the resonator can be expanded in the complete set of Slater modes of the empty cavity:

$$\vec{\mathcal{E}}_2(\boldsymbol{r},t) = \sum_a -\frac{1}{\sqrt{\varepsilon}}p_a(t)\boldsymbol{E}_a(\boldsymbol{r}) \quad , \quad \vec{\mathcal{H}}_2(\boldsymbol{r},t) = \sum_a \frac{1}{\sqrt{\mu}}\omega_a q_a(t)\boldsymbol{H}_a(\boldsymbol{r}) \quad ,$$

where p_a and q_a contain both the fast and slow temporal behavior of mode a. ε is the spatially averaged dielectric constant, and μ is the magnetic permittivity. The frequency and wavenumber of the ath mode of the empty, unperturbed resonator are related by $\omega_a = k_a/\sqrt{\mu\varepsilon}$. $\boldsymbol{E}_a$ and $\boldsymbol{H}_a$ satisfy $k_a\boldsymbol{E}_a = \vec{\nabla} \times \boldsymbol{H}_a$, $k_a\boldsymbol{H}_a = \vec{\nabla} \times \boldsymbol{E}_a$, as well as the resonator boundary conditions for electric and magnetic fields. The property which makes them particularly useful is that they are orthonormal when integrated over the resonator volume:

$$\int_{\mathrm{res}} \boldsymbol{E}_a \cdot \boldsymbol{E}_b dV = \delta_{ab} \quad \text{and} \quad \int_{\mathrm{res}} \boldsymbol{H}_a \cdot \boldsymbol{H}_b dV = \delta_{ab} \quad .$$

The distributed polarization field in the resonator contains a linear component, $\vec{\mathcal{P}}_{\mathrm{L}}$, which is incorporated in the dielectric constant, and a nonlinear component due to the wave interaction. The total fields, $\vec{\mathcal{E}}$ and $\vec{\mathcal{H}}$, obey Maxwell's equations:

$$\vec{\nabla} \times \vec{\mathcal{H}} = \vec{\imath} + \frac{\partial}{\partial t}(\varepsilon_0\vec{\mathcal{E}} + \vec{\mathcal{P}}_{\mathrm{L}} + \vec{\mathcal{P}}_{\mathrm{NL}}) = \sigma\vec{\mathcal{E}} + \varepsilon\frac{\partial\vec{\mathcal{E}}}{\partial t} + \frac{\partial}{\partial t}\vec{\mathcal{P}}_{\mathrm{NL}} \quad ,$$

$$\vec{\nabla} \times \vec{\mathcal{E}} = -\mu\frac{\partial\vec{\mathcal{H}}}{\partial t} \quad ,$$

where σ is the effective conductivity, introduced to account for the losses. Assuming that the resonator field consists of only a single Slater mode, a, using

117

the orthonormality condition, and eliminating q_a, we arrive at an equation for p_a

$$\ddot{p}_a + \frac{\omega_a}{Q_a}\dot{p}_a + \omega_a^2 p_a = \frac{1}{\sqrt{\varepsilon}}\frac{\partial^2}{\partial t^2}\int_{\text{res}}\vec{\mathcal{P}}_{\text{NL}}(\boldsymbol{r},t)\cdot\boldsymbol{E}_a(\boldsymbol{r})dV$$

where $Q_a = \omega_a\varepsilon/\sigma$ is the quality factor of the resonator for mode a. The distributed nonlinear polarization term driving the oscillation of the resonator field is that produced by the incidence of the pumping field $\vec{\mathcal{E}}_1$ on the index variation, Δn, created by the interaction of the pumping field with the resonator field $\vec{\mathcal{E}}_2$, i.e., $\vec{\mathcal{P}}_{\text{NL}}(\boldsymbol{r},t) = \text{Re}[\varepsilon_0\Delta n\vec{\mathcal{E}}_1]$. If we assume that only mode a oscillates, the index variation is given by

$$\Delta n = -\frac{2\mathrm{i}c\gamma}{\omega}\frac{\boldsymbol{E}_1^*\cdot\boldsymbol{E}_a(p_{a0}/\sqrt{\varepsilon})\mathrm{e}^{\mathrm{i}(\omega_a-\omega_1)t}}{|\boldsymbol{E}_1|^2+|p_{a0}\boldsymbol{E}_a|^2/\varepsilon}+\text{c.c.}$$

where p_a was taken to be the product of a slowly varying amplitude and an optical frequency oscillation, $p_a = p_{a0}\exp(\mathrm{i}\omega_2 t)$. ω_2 is the unknown oscillation frequency, which becomes ω_a in the limit of an unperturbed cavity. In steady state $\dot{p}_{a0}$ and $\ddot{p}_{a0}$ vanish, $\partial/\partial t \to \mathrm{i}\omega_2$ and $p_{a0}(t)\to p_{a0}(\infty)$. The oscillation condition becomes

$$(\omega_a^2-\omega_2^2)+\mathrm{i}\frac{\omega_a\omega_2}{Q_a}=\frac{\varepsilon_0}{\varepsilon}2\mathrm{i}c\omega_2 f\gamma = \frac{\varepsilon_0}{\varepsilon}2\mathrm{i}\omega_2 f\frac{\gamma_0}{1+\mathrm{i}(\omega_2-\omega_1)\tau'} \tag{4.11}$$

where in the second equality we used the zero external field form of γ. The real, dimensionless quantity f is given by

$$f = \int_{\text{crystal}}\frac{|\boldsymbol{E}_1^*(\boldsymbol{r})\cdot\boldsymbol{E}_a(\boldsymbol{r})|^2 dV}{|\boldsymbol{E}_1|^2+|p_{a0}(\infty)\boldsymbol{E}_a|^2/\varepsilon}\quad.$$

The real part of the oscillation condition can be solved for the oscillation frequency, which, in the limit $\tau_a \ll \tau$, reduces to the result from the coupled wave theory (4.10). The overlap integral, f, can be rewritten as

$$f = f_0\Big/\left(1+\frac{\langle|\boldsymbol{E}_a|^2\rangle}{\langle|\boldsymbol{E}_1|^2\rangle}\frac{p_{a0}^2(\infty)}{\varepsilon}\right)\quad\text{where}\quad f_0 = \int_{\text{crystal}}\frac{|\boldsymbol{E}_1^*\cdot\boldsymbol{E}_a|^2 dV}{|\boldsymbol{E}_1|^2}\quad.$$

The brackets indicate averaging over the crystal volume. The real part of (4.11) then becomes

$$\frac{1}{\omega_a\tau_a}=\frac{2\varepsilon_0 c\gamma_0 f_0/\omega_a\varepsilon}{1+4(\omega_1-\omega_a)^2\tau_a^2}\left(1+\frac{\langle|\boldsymbol{E}_a|^2\rangle}{\langle|\boldsymbol{E}_1|^2\rangle}\frac{p_{a0}^2(\infty)}{\varepsilon}\right)^{-1}\quad.$$

At the start of oscillation, $\boldsymbol{E}_a = 0$, and the right-hand side must be greater than or equal to the left-hand side. Once oscillation starts, $\boldsymbol{E}_a$ will grow until both sides are equal. The threshold coupling constant is thus

$$\gamma_0 \geq \frac{1+4(\omega_1-\omega_a)^2\tau_a^2}{2\varepsilon_0 cf_0\tau_a/\varepsilon}$$

118

in agreement with 2WM result, (4.9). One can also solve for the oscillation intensity

$$\langle|E_a|^2\rangle = \langle|E_1|^2\rangle\left[\frac{2\gamma_0\varepsilon_0 c f_0\tau_a/\varepsilon}{1+4(\omega_1-\omega_a)^2\tau_a^2}-1\right]$$

which is reminiscent of the expression for the power output of homogeneously broadened lasers [4.23].

4.2.4 One-Way, Real-Time Wavefront Conversion

The holographic nature and low threshold of photorefractive gain indicates potential in the area of beam cleanup, i.e., transforming a beam with a distorted wavefront into one with a planar or spherical wavefront with minimal loss of power. One way to accomplish this is to spatially filter a small part of the distorted beam, and amplify this seed using the bulk of the distorted beam as a pump. Stimulated Raman scattering (SRS) and stimulated Brillouin scattering (SBS) have already proven useful in this capacity because power can be transferred from a distorted pump beam to a Stokes beam without transferring the phase [4.24]. The distortion is transferred instead to the locally excited molecular excitation or acoustic wave.

In the photorefractive recording of a hologram between a clean and distorted beam, the distortion is transferred to the spatial variation of the refractive index. Diffraction of the distorted beam by the hologram reconstructs (and amplifies) the clean beam. This is just the opposite of the conventional way in which a hologram is used to reconstruct an object beam by illuminating with a spherical or planar reference beam. An advantage of photorefractive media is that they can be operated at low intensity, while SRS and SBS require high intensities to reach threshold.

An alternative to injecting a seed beam would be to induce a diffraction limited oscillation inside a resonator with photorefractive gain. This concept has been tested using an argon ion laser at 514.5 nm to pump a BaTiO$_3$ crystal inside a ring resonator composed of two high reflectors, an intracavity aperture, and an output coupler. One figure of merit for a wavefront converter is the ratio of the photometric brightness of the output beam to that of the input. In this experiment, the input beam had a power of 50 mW, a diameter of 1.5 mm, and a divergence of 50 mrad. The output beam had a power of 7.5 mW, a diameter of 0.4 mm, and a divergence of 1.15 mrad. The spectral bandwidth of the radiation remained unchanged, so these figures represent an increase in brightness by a factor of 4000 [4.25]. This compares favorably with a figure of 4300 recently achieved by amplifying a Stokes seed with SRS [4.26].

The power conversion efficiency can be obtained from (4.9), assuming that the output coupler has reflectivity M, and the other mirrors have reflectivity one.

$$\eta = \frac{I_{\text{out}}}{I_1(0)} = \frac{1-M}{M}\frac{I_2(0)}{I_1(0)} = \frac{1-M}{M}\frac{M-e^{(\alpha-\Gamma)l}}{e^{\alpha l}-M} \quad .$$

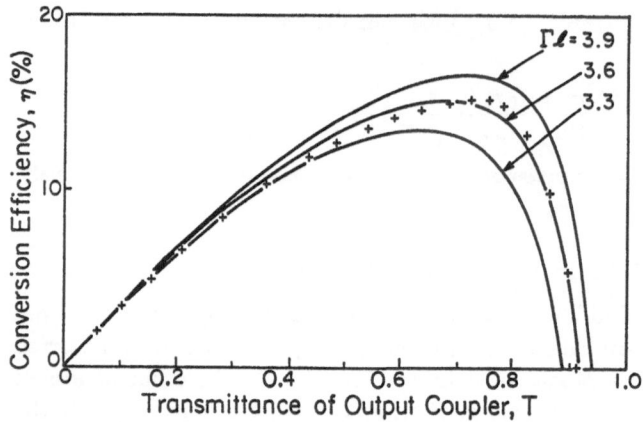

Fig. 4.13. Power conversion efficiency of wavefront converter based on a photorefractive ring resonator

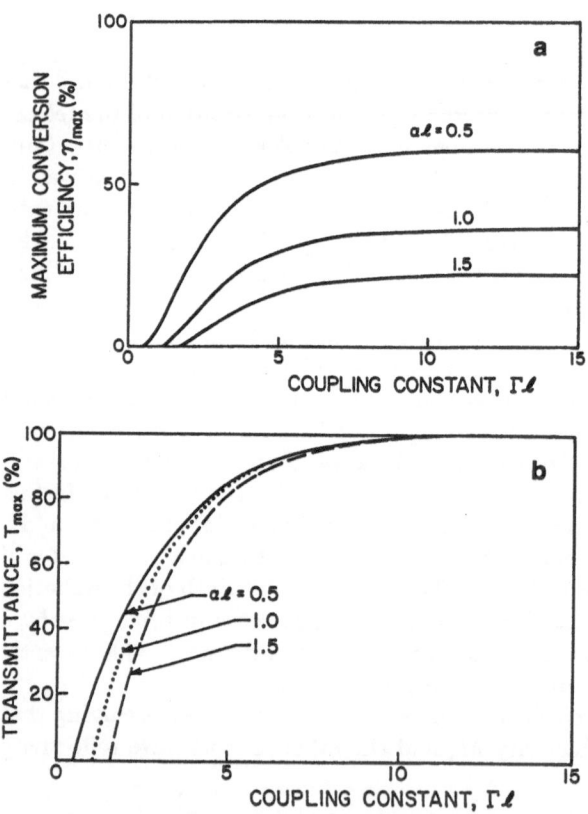

Fig. 4.14. (a) Maximum conversion efficiency and (b) optimum output coupling of a photorefractive wavefront converter based on a ring resonator

The efficiency is plotted in Fig. 4.13 vs the transmittance of the output coupler, $T = 1 - M$. The experimental data points were taken with the apparatus described above. The optimum output transmittance, T_{opt}, gives the largest efficiency, η_{max}:

$$1 - T_{opt} = \frac{1 - e^{\alpha l}\sqrt{(e^{-\Gamma l} - 1)(1 - e^{\alpha l})}}{1 + e^{\Gamma l}(e^{-\alpha l} - 1)} \quad .$$

T_{opt} and η_{max} are plotted vs Γl in Fig. 4.14.

4.3 Four-Wave Interactions Via a Third Order Nonlinearity

The holographic formulation of four-wave mixing (4WM) is an extension of the two-wave mixing formulation presented in the first section. We continue to consider the same third order nonlinearity, but now four waves are present in the medium. Four-wave mixing has a commensurate increase in versatility over two-wave mixing, and, in particular, it is the most important nonlinear technique for generating phase conjugate wavefronts.

The phase conjugate replica of a monochromatic optical field is a field, of the same frequency, whose wavefronts take the same shape throughout space, but propagate in the opposite direction at every point. A wave travelling essentially in the positive z direction is denoted by

$$E(\mathbf{r}, t) = A(\mathbf{r}) \exp[i(\omega t - kz)] + A^*(\mathbf{r}) \exp[-i(\omega t - kz)] \quad ,$$

where the complex amplitude $A(\mathbf{r})$ can describe any spatial amplitude or phase information impressed upon E. Mathematically, wavefront reversal is obtained by complex conjugating the spatial part

$$E_{PC}(\mathbf{r}, t) = A^*(\mathbf{r}) \exp[i(\omega t + kz)] + A(\mathbf{r}) \exp[-i(\omega t + kz)]$$

or by changing $t \to -t$, hence the identification of phase conjugation with "time reversal". A phase conjugate mirror (PCM) is a device which generates the phase conjugate of an incident wavefront. In an ideal two mirror laser cavity, the counterpropagating fields are phase conjugates of each other, so both mirrors are performing wavefront reversal. However, a true PCM will generate the phase conjugate replica of an incident wavefront possessing *arbitrary* spatial variation of amplitude, phase, and polarization.

Much of the interest in PCM's is due to their distortion correcting capability. Figure 4.15 illustrates a scenario in which an undistorted wave (1) passes through a lossless region of nonuniform index of refraction, the distorted wave (2) is incident upon a PCM, the phase conjugate wave (3) returns through the distortion and emerges aberration free (4). The distortion in the figure could represent modal dispersion in a fiber, atmospheric turbulence, thermal blooming, imperfect optics, etc.

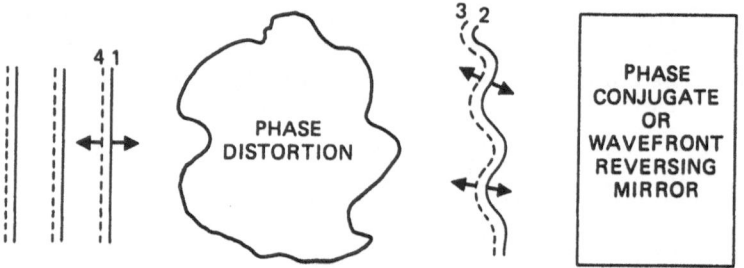

Fig. 4.15. Distortion correction via phase conjugation

The first realization of phase conjugate mirrors were via coherent optical adaptive techniques [4.27]. These systems have achieved compensation for aberrated wavefronts with the use of deformable "rubber" mirrors which are thin metallic reflectors supported by arrays of piezoelectric actuators and driven by wavefront error sensors.

Nonlinear optical implementations of PCM's began with stimulated Brillouin backscattering, but the study of all-optical parametric processes led directly to the current activity in the field. Four-wave mixing has become the most important nonlinear optical technique for generating phase conjugate waves because the interaction is automatically phase matched for all angular components of the incident wave, the requisite third order susceptibility is not forbidden in any material on symmetry grounds, and an amplified reflection is possible [4.28].

4.3.1 Holographic Formulation of Four-Wave Mixing

The four-wave mixing implementation of a phase conjugate mirror consists of a medium possessing a third order susceptibility pumped by two counterpropagating beams. Let the electric field amplitude associated with the jth beam be

$$E_j(r,t) = A_j(r) \exp\left[i(k_j \cdot r - \omega t)\right] + \text{c.c.}$$

By convention, A_1 and A_2 are taken to be the counterpropagating pump beams. A_4 is the incident object, or probe beam and A_3 is the phase conjugate wave (Fig. 4.16a). $A_1(0)$, $A_2(l)$, and $A_4(0)$ are the known, inputs. A_3 is not input, i.e., $A_3(l) = 0$, but is generated within the medium. The unknown outputs are $A_1(l)$, $A_2(0)$, $A_3(0)$, and $A_4(l)$.

Each plane wave component, k_4, of the object, or probe beam induces a corresponding component, k_3, of the phase conjugate wave travelling in exactly the opposite direction (Fig. 4.16b). In the grating picture, this comes about because A_4 and A_1 write a grating, k_I, which creates A_3 by diffracting A_2 (Fig. 4.16c). Simultaneously, A_4 and A_2 may write a grating, k_{II}, which also contributes to A_3 by diffracting A_1.

The four beams may write a total of four gratings, whose vectors are given by

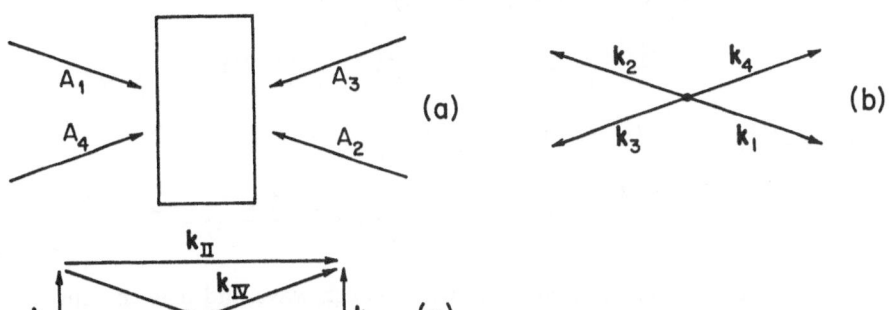

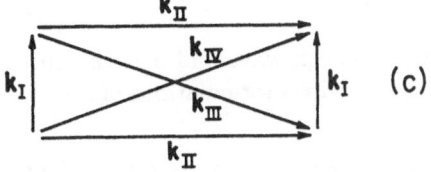

Fig. 4.16. Geometry of four-wave mixing in a nonlinear medium

$$k_I \equiv k_4 - k_1 = k_2 - k_3 \quad , \quad k_{II} \equiv k_1 - k_3 = k_4 - k_2 \quad ,$$
$$k_{III} \equiv k_1 - k_2 \quad , \qquad k_{IV} \equiv k_4 - k_3 \quad .$$

The index of refraction may have four spatial Fourier components.

$$n = n_0 + \frac{n_I e^{i\phi_I}}{2} \frac{A_1^* A_4 + A_2 A_3^*}{I_0} e^{ik_I \cdot r} + \text{c.c.}$$
$$+ \frac{n_{II} e^{i\phi_{II}}}{2} \frac{A_1 A_3^* + A_2^* A_4}{I_0} e^{ik_{II} \cdot r} + \text{c.c.}$$
$$+ \frac{n_{III} e^{i\phi_{III}}}{2} \frac{A_1 A_2^*}{I_0} e^{ik_{III} \cdot r} + \text{c.c.} + \frac{n_{IV} e^{i\phi_{IV}}}{2} \frac{A_3^* A_4}{I_0} e^{ik_{IV} \cdot r} + \text{c.c.}$$

where $I_0 \equiv \sum_{j=1}^{4} I_j$.

The complex amplitudes, $n_I \exp(i\phi_I)$, etc., are in general different, but are all given by equations of the form (4.2). Note that grating n_I has contributions from two pairs of interfering waves, as does n_{II}.

Substituting this form for the index into the Helmholtz wave equation, $(\nabla^2 + \omega^2 \mu \varepsilon)E = 0$, and using the slowly varying envelope approximation (Sect. 4.1.1), we arrive at the following coupled wave equations [4.29]:

$$\cos \theta \frac{dA_1}{dz} = -i \frac{\pi n_I}{\lambda} e^{-i\phi_I} \frac{A_1 A_4^* + A_2^* A_3}{I_0} A_4 - i \frac{\pi n_{II}}{\lambda} e^{i\phi_{II}}$$
$$\times \frac{A_1 A_3^* + A_2^* A_4}{I_0} A_3 - i \frac{\pi n_{III}}{\lambda} e^{-i\phi_{III}} \frac{A_1 A_2^*}{I_0} A_2 - \frac{\alpha}{2} A_1$$

$$(4.12a)$$

$$\cos \theta \frac{dA_2}{dz} = i \frac{\pi n_I}{\lambda} e^{i\phi_I} \frac{A_1^* A_4 + A_2 A_3^*}{I_0} A_3 + i \frac{\pi n_{II}}{\lambda} e^{-i\phi_{II}}$$
$$\times \frac{A_1^* A_3 + A_2 A_4^*}{I_0} A_4 + i \frac{\pi n_{III}}{\lambda} e^{-i\phi_{III}} \frac{A_1^* A_2}{I_0} A_1 + \frac{\alpha}{2} A_2$$

$$(4.12b)$$

$$\cos\theta \frac{dA_3}{dz} = i\frac{\pi n_I}{\lambda}e^{-i\phi_I}\frac{A_1 A_4^* + A_2^* A_3}{I_0}A_2 + i\frac{\pi n_{II}}{\lambda}e^{-i\phi_{II}}$$

$$\times \frac{A_1^* A_3 + A_2 A_4^*}{I_0}A_1 + i\frac{\pi n_{IV}}{\lambda}e^{-i\phi_{IV}}\frac{A_3 A_4^*}{I_0}A_4 + \frac{\alpha}{2}A_3$$

$$\cos\theta \frac{dA_4}{dz} = -i\frac{\pi n_I}{\lambda}e^{i\phi_I}\frac{A_1^* A_4 + A_2 A_3^*}{I_0}A_1 - i\frac{\pi n_{II}}{\lambda}e^{i\phi_{II}} \quad\quad (4.12c)$$

$$\times \frac{A_1 A_3^* + A_2^* A_4}{I_0}A_2 - i\frac{\pi n_{IV}}{\lambda}e^{i\phi_{IV}}\frac{A_3^* A_4}{I_0}A_3 - \frac{\alpha}{2}A_4 \quad .$$

$$(4.12d)$$

Here $\pm\theta$ is the angle of incidence common to each wave and α is the absorption constant. This formulation differs from previous formulations of four-wave mixing [4.28] in several respects:

1) The complex part of the coupling coefficients represents a spatial phase shift (between a grating and the interference pattern that generates it), rather than the more common temporal phase shift (between an optical frequency electric field and the response of an atomic dipole moment, for example). The four beams propagate in different directions in space, but the same direction in time. Accordingly, any temporal phase shift will have the same sign for all beams, while any spatial phase shift will be positive with respect to two beams, and negative with respect the other two.

2) The normalizing factor in the denominator (I_0) is peculiar to the photorefractive effect. This additional intensity dependence represents nonlinearities higher than third order.

3) Missing from here are terms such as $A_1 A_1^* A_1$, which give rise to self-focusing in some media, but are not holographic in origin. For a plane wave A_1, the quantity $A_1^* A_1$ is constant in the plane perpendicular to k_1, and contains no grating fringes. Also, in some media possessing a true third order susceptibility, waves with perpendicular polarizations can interact, whereas in the holographic formulation, only terms representing waves that interfere are included in the expression for the index.

4.3.2 Single Grating, Undepleted Pumps Approximation

Having generated all the terms in (4.12), it was found that only a small number of terms are needed to describe BaTiO$_3$ in many situations. In the single grating approximation, we assume that, out of a possible four index gratings, only one is present. There are several reasons why the other gratings may be absent:

1) If beams two and four are incoherent with respect to each other, or have orthogonal polarizations, then the time averaged $A_2^* A_4$ terms vanish, and do not represent a static interference pattern.

2) If two waves do interfere, but the period of the interference pattern is shorter than the mean distance between traps, then it will not produce a space-charge field. The spatial frequency response of the medium can thus limit grating formation.

3)   Even if two waves do produce a space-charge field, if the field is in a direction for which the effective electro-optic coefficient vanishes, then it will not produce an index grating.

In the undepleted pumps approximation, we assume that the two pump beams are unaffected by the nonlinear interaction. This is typically true when the pump beams are much more intense than the other beams. This is a safe assumption in media with local response, but in photorefractive media one has to be aware of the possible energy exchange between the pump beams themselves, as discussed in Sect. 4.1.4.

4.3.3 Transmission Grating, Undepleted Pumps

If the transmission grating is dominant, we have from (4.12)

$$\frac{dA_1}{dr} = -\frac{\alpha}{2}A_1 \quad , \quad \frac{dA_2}{dr} = \frac{\alpha}{2}A_2 \quad ,$$

$$\frac{dA_3}{dr} = \frac{\alpha}{2}A_3 + \gamma\frac{A_1A_4^* + A_2^*A_3}{I_0}A_2 \quad ,$$

$$\frac{dA_4^*}{dr} = -\frac{\alpha}{2}A_4^* + \gamma\frac{A_1A_4^* + A_2^*A_3}{I_0}A_1^* \quad , \tag{4.13}$$

where $\gamma \equiv i\pi n_I \exp(-i\phi_I)/\lambda$, and $r = z/\cos\theta$. The first two equations are immediately integrable to give

$$A_1 = A_1(0)e^{-\alpha r/2} \quad A_2 = A_2(l)e^{\alpha(r-l)/2}$$

$$I_0(r) = I_1(0)e^{-\alpha r} + I_2(l)e^{\alpha(r-l)} \quad .$$

To uncouple the remaining two equations we make use of the relation

$$\frac{d}{dr}(A_1^*A_3 - A_2A_4^*) = 0 \quad \text{or} \quad A_1^*A_3 - A_2A_4^* \equiv c \quad . \tag{4.14}$$

The equation for A_3 can then be written

$$\frac{dA_3}{dr} = \left(\gamma + \frac{\alpha}{2}\right)A_3 - \frac{c\gamma A_1}{I_0} \quad \text{or}$$

$$A_3 = c\gamma A_1(0)\int_r^l \frac{e^{(\gamma+\alpha/2)(r-r')}dr'}{I_0(r')} \equiv c\gamma A_1(0)/J(r)$$

using the boundary condition $A_3(l) = 0$. To solve for c, the expression for A_3 can be inserted into (4.14) and evaluated at $r = 0$:

$$c = \frac{A_2(l)A_4^*(0)e^{-\alpha l/2}}{\gamma I_1(0)/J(0) - 1} \quad .$$

The phase conjugate amplitude reflectivity is given by [4.30]

$$\varrho = \frac{A_3(0)}{A_4^*(0)} = \frac{\gamma A_1(0)A_2(l)}{e^{\alpha l/2}(\gamma I_1(0) - J(0))} \quad .$$

In Fig. 4.17 the phase conjugate intensity reflectivity, $R = |\varrho|^2$, is plotted versus the pump ratio, p. The pump ratio is defined according to the *function* of the two pump beams with respect to the probe wave, A_4, i.e., p is the ratio of the intensity of the pump which reads out the grating written by A_4, to the intensity of the pump which writes the grating with A_4. From (4.13), we can see that for the transmission grating, $p \equiv I_2(l)/I_1(0)$. The reflectivity is plotted for $\gamma l = \pm 3$ and for several values of α. We see that the effect of increasing linear absorption is primarily to decrease the reflectivity, with the greater decrease being for negative γl.

The dependence of phase conjugate reflectivity on the pump ratio is a manifestation of the same phase shift which produced the directionality in two-wave mixing. Figure 4.18 shows how the optimum pump ratio varies with

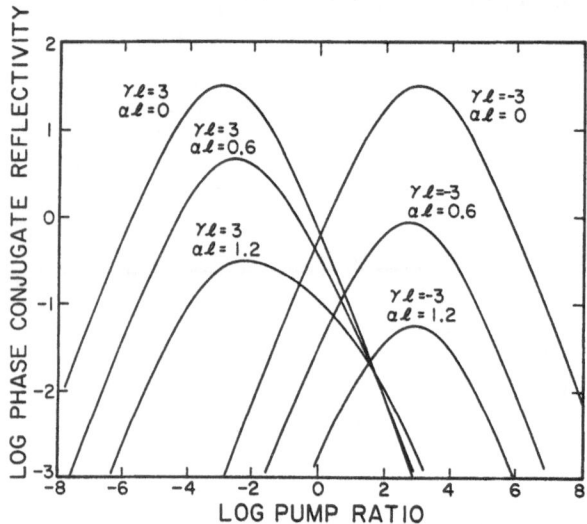

Fig. 4.17. Phase conjugate reflectivity of the transmission grating in the undepleted pumps approximation vs the pump ratio $I_2(l)/I_1(0)$. The coupling strength is $\gamma l = \pm 3$ and the reflectivity is shown for various values of the linear absorption α

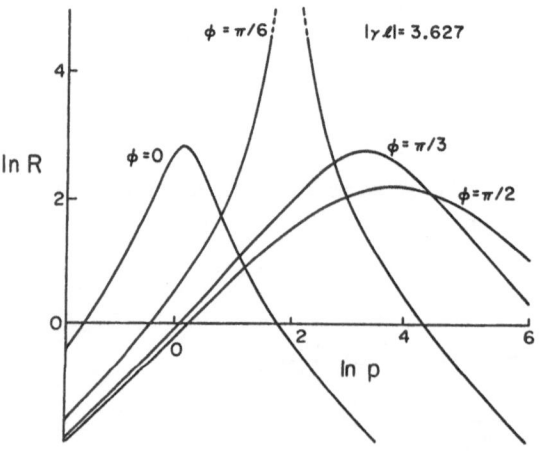

Fig. 4.18. Phase conjugate reflectivity in the undepleted pumps approximation for coupling strength $\gamma l = -3.627$ and various phase angles, ϕ. Mirrorless self-oscillation occurs here for $\phi = \pi/6$ and $p = 6.13$

ϕ_I. Media with local response ($\phi_I = 0$) display an optimum pump ratio of unity. Figure 4.19 shows the apparatus used to measure the phase conjugate reflectivity vs pump ratio data shown in Fig. 4.20 [4.31]. The nonlinear medium is a poled, $4 \times 4 \times 7\,\mathrm{mm}^3$ single crystal of $BaTiO_3$. All three input beams at 514.5 nm are supplied by the same argon ion laser. Their total intensity is $0.3\,\mathrm{W/cm}^2$.

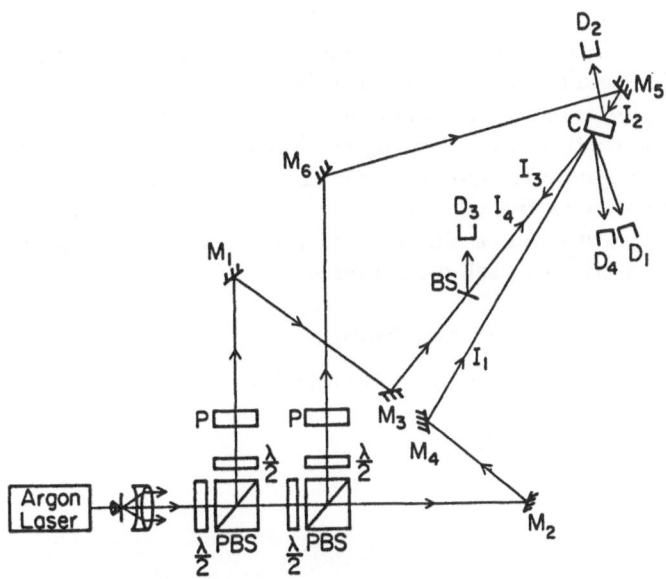

Fig. 4.19. Experimental apparatus for measuring phase conjugate reflectivity (C: crystal; D: detector; P: polarizer; BS: beamsplitter; PBS: polarizing beamsplitter; $\lambda/2$: half-wave plate)

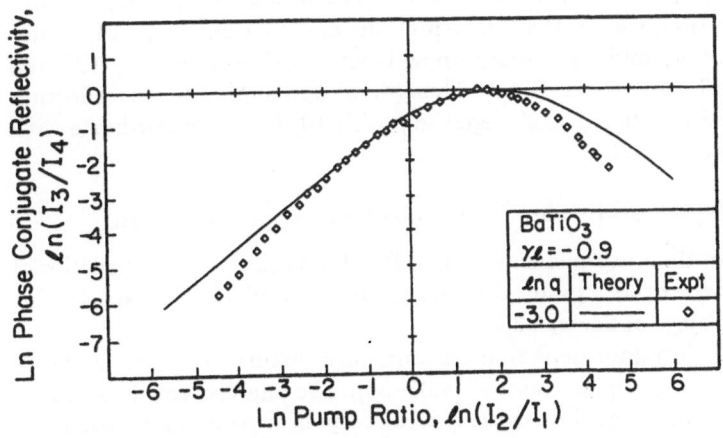

Fig. 4.20. Experimental and theoretical curve of phase conjugate reflectivity vs the pump ratio, in the undepleted pumps regime, for $BaTiO_3$

The expression for the reflectivity can be simplified when $\alpha = 0$ to

$$R = p \left[\frac{e^{-\gamma l} - 1}{e^{-\gamma l} + p} \right]^2 . \tag{4.15}$$

R is invariant under the interchange $p \to 1/p$ and $\gamma l \to -\gamma l$, meaning that probe beams travelling in opposite directions to each other, incident upon opposite faces of a photorefractive PCM, will experience the same reflectivity.

The conditions for an infinite reflectivity are $\text{Im}\{-\gamma l\} = \pm\pi, \pm 3\pi, \ldots$ and $\text{Re}\{-\gamma l\} = \ln p$. Of course, pump depletion occurs long before infinite reflectivities. While this condition cannot be satisfied for $\phi_I = \pi/2$, reflectivities exceeding 100 % are within reach of finite coupling strengths at all phase angles. The phase angle can be controlled by applying an electric field to the crystal, and also by detuning the pump and probe beams. In the latter case the finite response time of the photorefractive effect causes the grating to lag behind the moving fringes. This departure from a ninety degree phase shift can restore the possibility of infinite reflectivity or self-oscillation [4.32].

4.3.4 Reflection Grating, Undepleted Pumps

If the reflection grating is dominant, we have from (4.12)

$$\frac{dA_1}{dr} = -\frac{\alpha}{2} A_1 \quad , \qquad \frac{dA_2}{dr} = \frac{\alpha}{2} A_2 \quad ,$$

$$\frac{dA_3}{dr} = \frac{\alpha}{2} A_3 + \gamma \frac{A_1^* A_3 + A_2 A_4^*}{I_0} A_1 \quad ,$$

$$\frac{dA_4^*}{dr} = -\frac{\alpha}{2} A_4^* + \gamma \frac{A_1^* A_3 + A_2 A_4^*}{I_0} A_2^* \quad ,$$

where the definition of the coupling constant for the reflection grating, $\gamma = i\pi n_{II} \exp(-i\phi_{II})/\lambda$, is analogous to that for the transmission grating. In the absence of absorption, we see that the equations are the same as (4.13) under the interchange of A_1 and A_2, because now beam 2 writes a grating with the object beam, and beam 1 reads it out. Thus (4.15) also applies to the reflection grating case, when $\alpha \to 0$, if p is changed to $p = I_1(0)/I_2(l)$ in accordance with its definition in Sect. 4.3.3.

4.3.5 Pump Depletion in the Single Grating Approximation

As the intensity of the wave incident upon a PCM approaches the intensities of the pumps, we must expect pump depletion. The theoretical approach to this problem is the subject of Sects. 4.3.5–7.

The influence of pump depletion on four-wave mixing has been analyzed previously using a Lagrangian method [4.33–35]. The analysis considered four *colinear* waves inside a medium with a *local* response. Another method has been used to solve a very general class of nonlinear parametric processes including 4WM [4.36]. However, it has not been applied to nonlocal nonlinear susceptibilities, such as are present in photorefractive media. The method for treating *nonlocal* media presented in this section is not restricted to a colinear geometry, but is restricted to a symmetrical geometry where all waves have the same angle of incidence [4.37].

As a starting point, we allow for the r dependence of beams 1 and 2 but do not consider the effect of more than a single grating on the interactions of the four beams, and restrict ourselves to the case where $\alpha l \ll 1$. The equations now under examination are intermediate in complexity between (4.12) and (4.13).

4.3.6 Transmission Grating with Pump Depletion

When only n_I is present, and $\alpha = 0$, the equations (4.12) take the form

$$\frac{dA_1}{dr} = -\gamma \frac{A_1 A_4^* + A_2^* A_3}{I_0} A_4 \quad ,$$

$$\frac{dA_2^*}{dr} = -\gamma \frac{A_1 A_4^* + A_2^* A_3}{I_0} A_3^* \quad ,$$

$$\frac{dA_3}{dr} = \gamma \frac{A_1 A_4^* + A_2^* A_3}{I_0} A_2 \quad ,$$

$$\frac{dA_4^*}{dr} = \gamma \frac{A_1 A_4^* + A_2^* A_3}{I_0} A_1^* \quad .$$

We are chiefly interested in calculating phase conjugate reflectivities so it is enough to solve for the ratios A_3/A_4^* and A_1/A_2^*. The differential equations for the ratios can be decoupled with the use of the relations

$$A_1 A_2 + A_3 A_4 \equiv \frac{c}{2} \quad \text{and} \quad A_1 A_3^* - A_2^* A_4 \equiv \frac{d}{2} \quad . \tag{4.16}$$

Several fluxes are constant with respect to z, as in the two-wave mixing case (4.4):

$$f_+ \equiv I_1 + I_4 \quad f_- \equiv I_2 + I_3 \quad f \equiv f_+ - f_- \quad .$$

The fluxes are known, i.e., they are given by the boundary conditions. Substituting the above constants into the differential equations for $A_{34} \equiv A_3/A_4^*$, and $A_{12} \equiv A_1/A_2^*$, we obtain

$$\frac{dA_{34}}{dr} = \frac{\gamma}{I_0} \left[\frac{c}{2} - f(A_{34}) - \frac{c^*}{2}(A_{34})^2 \right]$$

$$\frac{dA_{12}}{dr} = \frac{-\gamma}{I_0} \left[\frac{c}{2} - f(A_{12}) - \frac{c^*}{2}(A_{12})^2 \right]$$

which can be integrated directly, because I_0 is a constant with respect to r, to yield

$$\frac{-c^* A_{34})_l - f - \sqrt{f^2 + |c|^2}}{-c^* A_{34})_l - f + \sqrt{f^2 + |c|^2}} \times \frac{-c^* A_{34})_0 - f + \sqrt{f^2 + |c|^2}}{-c^* A_{34})_0 - f - \sqrt{f^2 + |c|^2}}$$

$$= \exp\left(\sqrt{f^2 + |c|^2}\,\frac{\gamma l}{I_0}\right) \quad .$$

The corresponding equation for A_{12} is obtained by replacing $f \to -f$, and $\gamma \to -\gamma$. Making the following substitutions [the latter two from (4.16)]

$$A_{34})_0 = \varrho \quad A_{34})_l = 0 \quad A_{12})_0 = \frac{I_1(0)}{(c^*/2) - \varrho^* I_4(0)} \quad A_{12})_l = \frac{c}{2 I_2(l)}$$

we obtain a pair of transcendental equations which can be solved for $|c|^2$, and $R = |\varrho|^2$. The three input intensities appear only in the two combinations $p = I_2(l)/I_1(0)$ and $q = I_4(0)/[I_1(0) + I_2(l)]$. As the probe ratio, q, increases, we expect to see more pump depletion. A contour plot of R vs p and q, for $\gamma l = -3$, is shown in Fig. 4.21. The figure suggests that R remains finite as $p \to \infty$. Experimental data, obtained using the configuration of Fig. 4.19, is shown in Fig. 4.22.

A plot of R vs $|\gamma l|$ is shown in Fig. 4.23. The phase shift between the grating and the interference fringes is 5°. The intensities of the two pumping beams are equal ($p = 1$) and the probe intensity is 20 % of the total pumping intensity ($q = 0.2$). The top of the graph corresponds to the reflectivity that

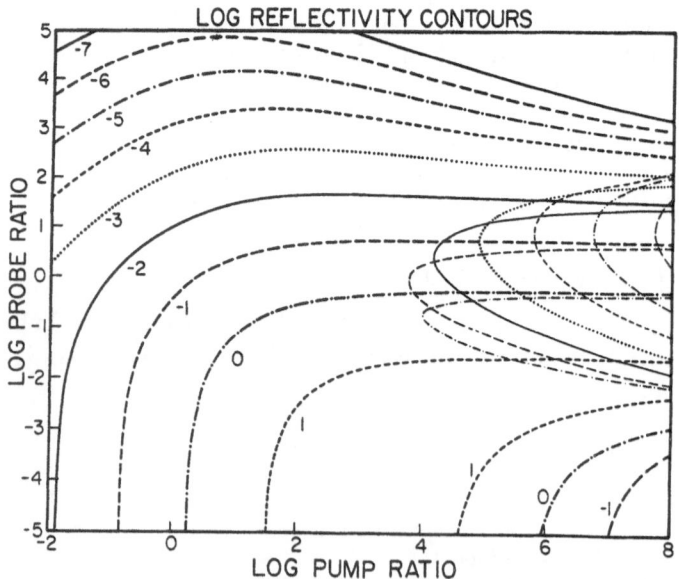

Fig. 4.21. Contour plot of phase conjugate reflectivity for $\gamma l = -3$ as a function of the pump ratio $I_2(l)/I_1(0)$ and the probe ratio $I_4(0)/[I_1(0) + I_2(l)]$. The transmission grating is operative

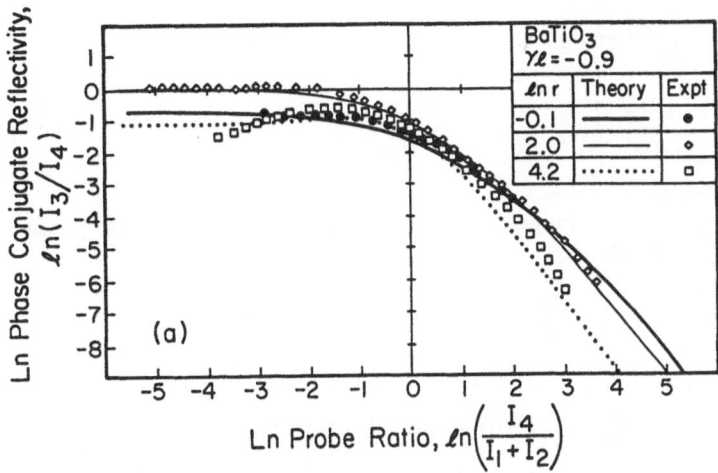

Fig. 4.22. Experimental and theoretical curve of phase conjugate reflectivity vs the probe ratio, for BaTiO$_3$

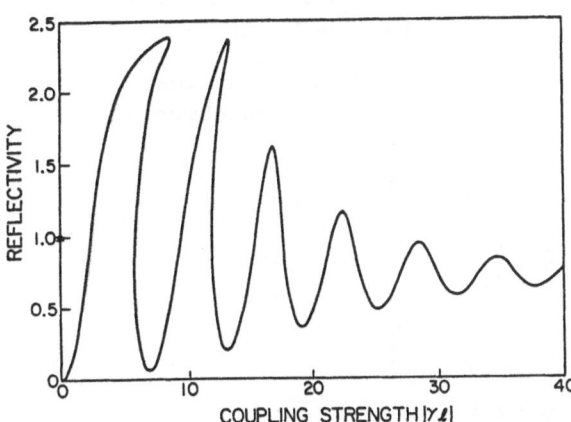

Fig. 4.23. Reflectivity of a photorefractive PCM vs coupling strength magnitude $|\gamma l|$. The transmission grating is operative. The phase shift between the grating and the interference fringes is 5°

would result if all the power of beam 2 were transferred to beam 3. This is the maximum reflectivity consistent with the conservation laws (4.16). The peaks in the curve correspond to the poles in the reflectivity of a PCM with no pump depletion and no phase shift between the grating and the interference fringes.

4.3.7 Reflection Grating with Pump Depletion

When the reflection grating is dominant, equations (4.12) take the form

$$\frac{dA_1^*}{dr} = \gamma \frac{A_1^* A_3 + A_2 A_4^*}{I_0} A_3^* \quad ,$$

$$\frac{dA_2}{dr} = \gamma \frac{A_1^* A_3 + A_2 A_4^*}{I_0} A_4 \quad ,$$

$$\frac{dA_3}{dr} = \gamma\frac{A_1^*A_3 + A_2A_4^*}{I_0}A_1 \quad ,$$

$$\frac{dA_4^*}{dr} = \gamma\frac{A_1^*A_3 + A_2A_4^*}{I_0}A_2^* \quad , \tag{4.17}$$

where $\gamma \equiv i\pi n_{II}\exp(-i\phi_{II})/\lambda$.

The intensity, I_0, is no longer a constant, but is coupled to the grating term $g/2 \equiv A_1^*A_3 + A_2A_4^*$:

$$\frac{dI_0}{dr} = \frac{|g|^2}{2I_0}(\gamma + \gamma^*) \quad ; \quad \frac{dg}{dr} = \gamma g \quad .$$

This system is easy to solve, since g is not coupled to I_0:

$$g(r) = g(0)e^{\gamma r} \quad ; \quad I_0^2(r) - I_0^2(0) = |g(r)|^2 - |g(0)|^2 \quad . \tag{4.18}$$

When $\gamma \in \mathbb{R}$ we can make use of the change of variables

$$du = \frac{\gamma}{I_0}e^{\gamma r}dr \quad \text{or} \quad e^{|g_0|u(r)} = \frac{|g_0|e^{\gamma r} + \sqrt{|g_0|^2e^{2\gamma r} + I_0^2(-\infty)}}{|g_0| + \sqrt{|g_0|^2 + I_0^2(-\infty)}} \quad ,$$

to transform the system of equations (4.17). Note that u has the same sign as γ. g_0 is an abbreviation for $g(0)$, i.e., $g_r \equiv g(r)$. One then obtains

$$\frac{d}{dr}\begin{pmatrix}A_1\\A_3\end{pmatrix} = \frac{\gamma}{I_0(r)}\begin{pmatrix}0 & g^*(r)\\g(r) & 0\end{pmatrix}\begin{pmatrix}A_1\\A_3\end{pmatrix}$$

$$\rightarrow \frac{d}{du}\begin{pmatrix}A_1\\A_3\end{pmatrix} = \begin{pmatrix}0 & g_0^*\\g_0 & 0\end{pmatrix}\begin{pmatrix}A_1\\A_3\end{pmatrix} \quad ,$$

$$\frac{d}{dr}\begin{pmatrix}A_4\\A_2\end{pmatrix} = \frac{\gamma}{I_0(r)}\begin{pmatrix}0 & g^*(r)\\g(r) & 0\end{pmatrix}\begin{pmatrix}A_4\\A_2\end{pmatrix}$$

$$\rightarrow \frac{d}{du}\begin{pmatrix}A_4\\A_2\end{pmatrix} = \begin{pmatrix}0 & g_0^*\\g_0 & 0\end{pmatrix}\begin{pmatrix}A_4\\A_2\end{pmatrix} \quad . \tag{4.19}$$

The solutions, in terms of the unknown g_0 and $L \equiv u(l)$, are given by:

$$A_1(u) = A_1(0)\frac{\cosh|g_0|(u-L)}{\cosh|g_0|L} + A_3(l)\frac{g_0^*\sinh|g_0|u}{|g_0|\cosh|g_0|L}$$

$$A_3(u) = A_1(0)\frac{g_0\sinh|g_0|(u-L)}{|g_0|\cosh|g_0|L} + A_3(l)\frac{\cosh|g_0|u}{\cosh|g_0|L} \tag{4.20}$$

and a corresponding pair for A_4 and A_2. When $A_3(l) = 0$, the phase conjugate reflectivity is given by

$$R = \left|\frac{A_3(0)}{A_4^*(0)}\right|^2 = \frac{I_1(0)}{I_4(0)}\tanh^2|g_0|L \quad . \tag{4.21}$$

To solve for $|g_0|L$, we return to (4.18), used to separate the problem, and evaluate g at $r = 0$ and $r = l$ using (4.20):

$$\frac{g_0}{2} = -[I_1(0) + I_4(0)]\frac{g_0}{|g_0|}\tanh|g_0|L + \frac{A_2(L)A_4^*(0)}{\cosh|g_0|L} \quad ,$$

$$\frac{g_0}{2}e^{\gamma l} = I_2(L)\frac{g_0}{|g_0|}\tanh|g_0|L + \frac{A_2(L)A_4^*(0)}{\cosh|g_0|L} \quad .$$

Solving the above equations for g_0 and taking the magnitude yields two equations in $|g_0|$ and $|g_0|L$. Eliminating $|g_0|$, we obtain:

$$\sinh|g_0|L = \frac{\sqrt{I_2(L)I_4(0)}(1 \pm e^{-\gamma l})}{I_1(0) + I_2(L)e^{-\gamma l} + I_4(0)} = \frac{\sqrt{(p+1)q}(1 \pm e^{-\gamma l})}{p + e^{-\gamma l} + (p+1)q} \quad . \quad (4.22)$$

Combining (4.21) and (4.22), we have an explicit expression for the reflectivity of the reflection grating as a function of the pump and probe ratios. As the probe ratio $q \to 0$, we expect the reflectivity to reduce to the solution obtained, in Sect. 4.3.4, with the undepleted pumps approximation. Of the two solutions above, only the one with the minus sign reduces to the proper answer. A contour plot of this solution, for $\gamma l = -3$, is shown in Fig. 4.24. Associated with each proper solution is an improper solution, obtained with the minus sign, and shown by the fine solid lines. In contrast to the transmission grating, Fig. 4.21, here $R \to 0$ in the limit $p \to 0$. Expressions have been obtained for the complete

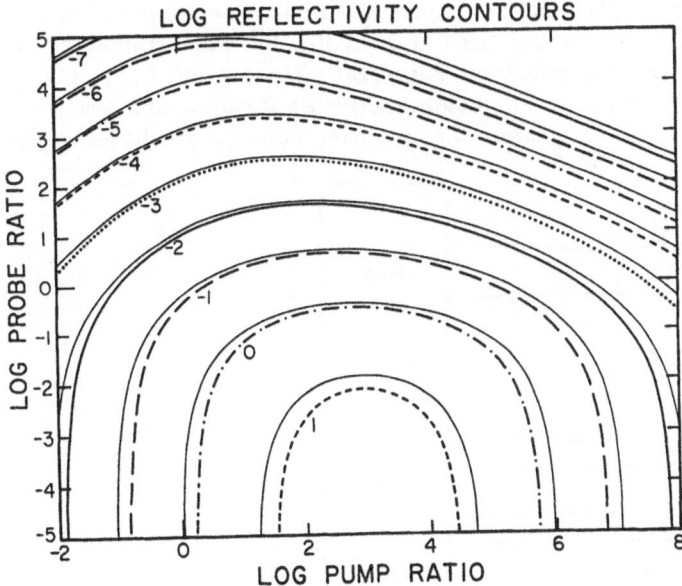

Fig. 4.24. Contour plot of phase conjugate reflectivity for $\gamma l = -3$ as a function of the pump ratio $I_1(0)/I_2(l)$ and the probe ratio $I_4(0)/[I_1(0) + I_2(l)]$. The reflection grating is operative

z dependence of all four beams and also numerical solutions which include the effects of absorption [4.38].

4.3.8 Oscillation in Four-Wave Mixing

Even if no signal is supplied, a PCM with gain can produce oscillation when provided with positive feedback such as in a mirror resonator. In this case, pump depletion determines the steady state oscillation strength. The simplest phase conjugate resonator (PCR) would consist of a PCM and a normal mirror (Fig. 4.25). A PCM with infinite reflectivity can oscillate even with no auxiliary mirror.

Oscillation with a PCM based on four-wave mixing was predicted in 1977 [4.28] and was first observed in 1979, using CS_2 as the nonlinear medium [4.39]. Subsequently, continuous wave oscillators have been constructed with a single domain crystal of $BaTiO_3$ pumped with 514.5 nm light from an argon ion laser [4.40], strontium barium niobate pumped with 488.0 nm light [4.41], and with a Na vapor cell pumped with a resonant wavelength from a tunable dye laser [4.42].

Much of the interest in PCR's is due to their unique stability and frequency spectrum. The standard stability analysis uses matrices to trace the evolution of the characteristics (spot size and wavefront curvature) of a Gaussian beam as it traverses a resonator. If the stability criterion is that the characteristics exactly repeat after one round trip, then the matrix analysis indicates that, at the normal mirror, the radii of curvature of the Gaussian beam and the mirror must be equal. Thus, both concave and convex normal mirrors are stable, independent of cavity length. Another unique feature is the existence of modes that repeat only after two round trips (Fig. 4.25b), for which there are no criteria for stability, i.e., *any wavefront will repeat after two round trips* [4.39].

The oscillation strength in the photorefractive PCR can be obtained from Fig. 4.21 or Fig. 4.24. In steady state, the operating point (p, q) will lie on the

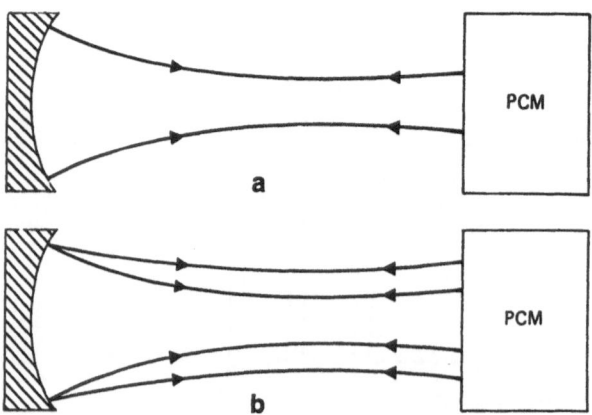

Fig. 4.25. Sketch of a typical allowed PCR Gaussian mode for the degenerate case and demanding self-consistent field solutions for one (**a**) and two (**b**) round trips

contour $R = 1/M$, where M is the round-trip loss, including the feedback mirror. The specific location on the contour is determined by the intersection with a vertical line representing the appropriate pump ratio. The oscillation strength is given by the q coordinate.

4.4 Self-Pumped Phase Conjugate Mirrors and Lasers

One of the limitations of the 4WM implementations of phase conjugate mirrors discussed in the last section is the need to supply pump waves that are coherent with the wavefront to be reversed. Stimulated Brillouin backscattering is free of this limitation, but it is subject to an intensity threshold, and a frequency downshift is imposed on the reflected wave. A big step toward the development of practical wavefront reversing devices was the invention of the self-pumped or passive phase conjugate mirror (PPCM) [4.20]. A PPCM combines photorefractive gain and feedback to generate pump beams which are automatically phase conjugates of each other, and are coherent with the probe wave.

The most successful applications of PPCM's have been as distortion correcting and frequency tuning end mirrors in laser resonators. Other potential applications include one-way field transmission through distorting media [4.43, 44], ring gyroscopes [4.45], and interferometers [4.46].

4.4.1 Self-Pumped Mirror Based on a Fabry-Perot Resonator with Photorefractive Gain

One of the most interesting configurations in which oscillation has been observed, and the first to display self-pumped photorefractive phase conjugation, is shown in Fig. 4.26a. The gain is provided by the single probe beam, A_4, via the two-beam coupling discussed in Sect. 4.1. Feedback is provided by the Fabry-Perot resonator, which allows the buildup of an oscillation shown by the dashed line. To the extent that the oscillating beams are phase conjugates of each other, they will act as the pump waves A_1 and A_2 of Fig. 4.16, and generate the phase conjugate of A_4. Indeed, a wave has been observed propagating backward relative to A_4. The phase conjugate character of the backward wave has been confirmed by a double pass distortion correction experiment using a field containing pictorial information [4.47].

For the counterpropagating waves in the cavity to be phase conjugates of each other, their wavefronts must match the spherical mirror surfaces, i.e., they must be Gaussian beams. Because of the holographic nature of the gain, a probe beam with an arbitrary wavefront can generate a cavity mode with spherical wavefronts. When the phase shift is 90°, no phase information is transferred from the probe beam to the cavity mode (Sect. 4.1.2). When the oscillation intensities are sufficiently weak, no amplitude information is transferred, either.

Reflectivities greater than one are impossible in steady state, but reflectivities approaching one are theoretically possible and 30 % has been observed. The alignment of the Fabry-Perot cavity is critical to the quality of phase con-

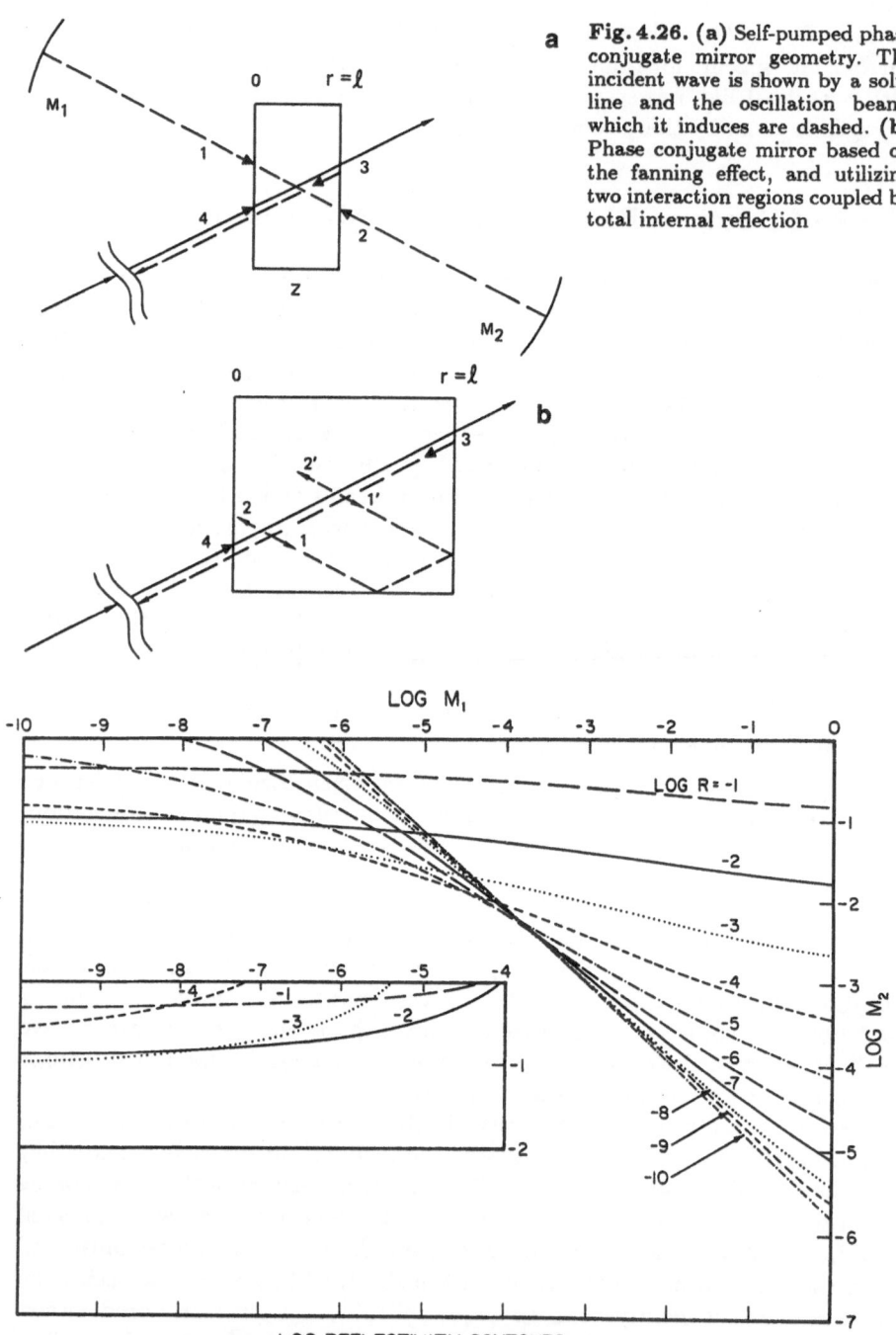

a **Fig. 4.26.** (a) Self-pumped phase conjugate mirror geometry. The incident wave is shown by a solid line and the oscillation beams which it induces are dashed. (b) Phase conjugate mirror based on the fanning effect, and utilizing two interaction regions coupled by total internal reflection

Fig. 4.27. Contour plots of the reflectivity of the self-pumped PCM. The coupling strength is $\gamma l = -3$. Some of the contours at low M_1 and high M_2 have been redrawn as an insert

jugation, but the alignment of the cavity relative to the incident wave is not. The acceptance angle has been measured to be 20° on either side of the crystal.

A similar experiment has been performed with $LiNbO_3 : Fe$ [4.48]. In this material the writing of a shifted grating is driven by the action of circular (oscillating in space) photogalvanic currents. A multiple longitudinal mode laser operating at $0.44\,\mu m$ provided gain. The parallel sample faces provided the feedback rather than external mirrors. Lasing even occurred when the interface reflectivity was lowered to 3.1 % by immersing the crystal in bromoform. Measurements of the lasing threshold implied a net gain of $69\,cm^{-1}$.

If one grating is dominant, the single grating analysis of Sect. 4.3.6 can be applied to this PPCM [4.47]. If the cavity is assumed to be in resonance, then the boundary conditions are the mirror reflectivities $M_1 = I_1(0)/I_2(0)$ and $M_2 = I_2(l)/I_1(l)$. The coupling strength threshold is given by $M_1 M_2 = \exp(\gamma + \gamma^*)l$, so, for ideal feedback, $(\gamma l)_t = 0$. A contour plot of R vs M_1 and M_2, for $\gamma l = -3$, is shown in Fig. 4.27. It is apparent that R can be multivalued and is insensitive to M_1 in the region of small M_1 and large M_2. This corresponds to the region of Fig. 4.21 where R is insensitive to p, since as $M_1 \to 0$, $p \to \infty$. The next two sections describe experiments which prove that R remains finite and large even when $M_1 = 0$.

4.4.2 Self-Pumped Mirrors Based on the Fanning Effect

For buildup of an oscillation from zero intensity in the Fabry-Perot cavity of the last section, the single grating analysis indicates a threshold coupling strength given by $M_1 M_2 = \exp(\gamma + \gamma^*)l$. Clearly an oscillation cannot begin from zero in the absence of M_1 or M_2. However, self-starting behavior has been obtained, with $M_1 = 0$, when the photorefractive crystal is placed in the center of curvature of M_2 (Fig. 4.26a) [4.30].

Buildup of oscillation in the crystal$-M_2$ cavity is dependent upon the crystal manifesting the fanning effect [4.49], which can diffract more than 99 % of an incident collimated laser beam into a semi-continuous, two-dimensional angular spectrum of tens of degress. This process has been attributed to asymmetrical self-defocusing [4.50], and also an energy transfer, via 2WM, into a cascade of diffracted beams that originates at a scattering imperfection in the crystal [4.51].

The mirror redirects the fanning rays back into the crystal forming a crude approximation to counterpropagating pump beams. The diffuse scattering then collapses slightly into a well-defined beam, whose wavefront matches the curvature of M_2. The extent to which the pumps are phase conjugates of each other can be inferred from the fidelity of the backward wave A_3. If the single grating analysis is applied to this device, and ideal feedback assumed, the threshold should be $(\gamma l)_t = -2.49$.

The self-starting PPCM has exhibited two provocative capabilities, multicolor phase conjugation and bistability. Simultaneous reflection has been achieved for up to six different lines from a multi-mode argon ion laser: 457, 476, 488, 496, 501, and 514 nm [4.52]. Since the pump beams are stimulated, no *a priori* knowledge of the spectrum of the incident wave is required. Multiline phase conjugation has also been demonstrated with photorefractive crystals when pump beams of the proper wavelengths are supplied externally [4.53].

A configuration based on the self-starting PPCM has exhibited a large hysteresis [4.54]. The incident intensity is divided into two channels: the transmitted beam and the diffuse fan. Mirrors provide feedback by retroreflection of the light in each channel. Feedback into either channel tends to quench the other. Such behavior could be exploited in a device and also in investigations into the nature of the fanning effect itself.

Another passive phase conjugate mirror based on the fanning effect makes use of total internal reflection from the faces of the crystal (Fig. 4.26b) [4.55]. Pump beams develop from that part of the diffuse fan which intersects the original beam after retroreflection from a crystal edge. Two interaction regions appear, each involving four waves, which may be the only fundamental difference between this configuration and that described above. Feedback can be obtained with internal reflection or external mirrors in either case. A single grating analysis indicates a threshold of $(\gamma l)_t = -4.68$, assuming ideal feedback.

4.4.3 Laser with Dynamic Holographic Intracavity Distortion Correction Capability

Distortions in laser resonators with normal mirrors commonly arise because of defects in laser rods, turbulence in gaseous gain media, imperfect optics, thermal effects, and nonlinear effects. The possibility of using real-time holography to compensate for intracavity distortions had been proposed in 1975 [4.56]. The double pass distortion correction scenario of Fig. 4.15 can also work equally well inside a resonator, shown schematically in Fig. 4.28.

A laser with this self-correcting capability has been demonstrated [4.57]. For comparison, both the standard and the phase conjugate resonator experimental configurations are shown in Fig. 4.29a,c. The distortion correcting component is the self-pumped photorefractive PCM described in Sect. 4.4.1. The distortion is simulated with a piece of etched glass. The gain medium is that of an argon ion laser tuned with a prism to the high gain line at 488.0 nm. Lasing in the PCR is initiated with the aid of a mirror situated between the plasma tube and the distortion, since the coherence of the fluorescence is insufficient to allow the formation of the required refractive index grating in the crystal. Once the grating is established, the auxiliary mirror can be removed. The configuration of Fig. 4.29c corresponds to an equilibrium state, and the grating in the crystal is continuously maintained by the very beams which it couples together.

The distortion correcting capability of the laser is apparent from pictures taken 1 m from the output mirror. Inserting the etched glass into the normal laser cavity drastically lowered the spatial mode quality and the output power decreased from 2 W to 1 mW (Fig. 4.29b). Replacing the normal mirror with a PPCM restored the mode shape and boosted the power output back to 500 mW

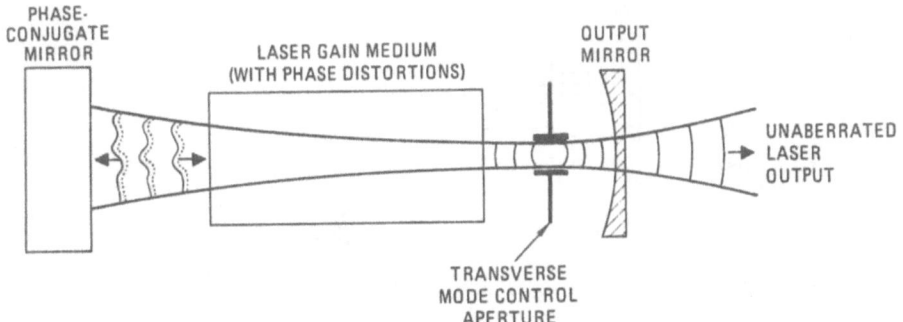

Fig. 4.28. Laser based on a phase conjugate resonator

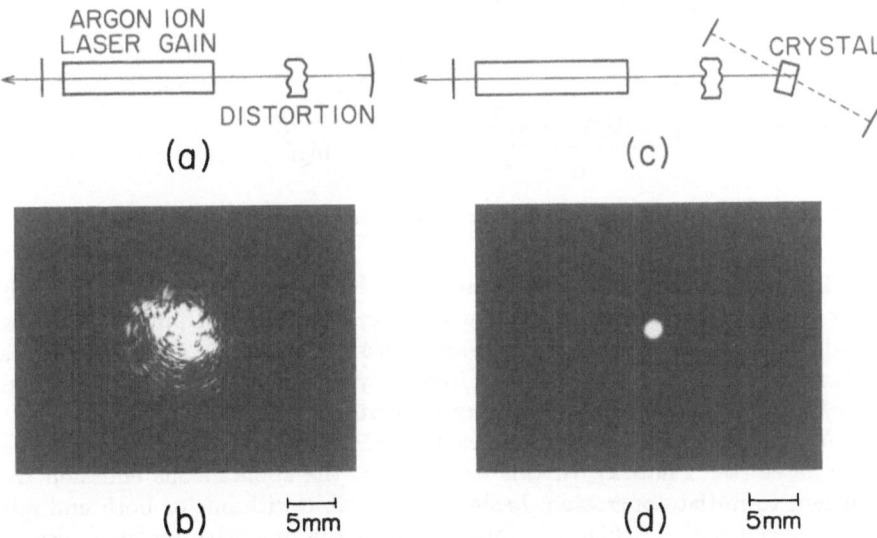

Fig. 4.29. Laser performance with and without a PCM

(Fig. 4.29d). This mode of oscillation may not be the only allowed stable configuration but in the presence of spatial filters such as the plasma bore tube, it is the minimum diffraction loss configuration and thus the one surviving in a laser oscillator.

According to the single grating analysis, above a coupling strength of $\gamma l = -2.49$, it should be possible to maintain the oscillation with only one of the two mirrors forming a cavity about the crystal (Sect. 4.4.2). Such an oscillation has been demonstrated, although it is not self-starting. In this configuration, the crystal looks much like a distortion correction element to be inserted into an existing laser cavity without replacing either mirror (Fig. 4.30a). Such a correction element may also find use in a ring laser (Fig. 4.30b).

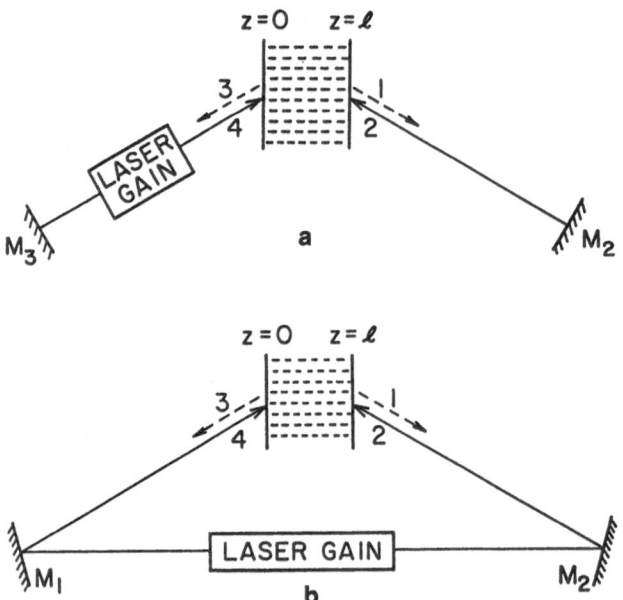

Fig. 4.30. Lasers with intracavity distortion correction elements based on 4WM

The pump beams in a PPCM are not independent of each other or of the input beam, in contrast to a PCM with externally supplied pumps. This loss of independence means that the half-axial, $c/4L$ longitudinal modes of an ideal PCR are not expected. The observed frequency spectrum was multi-mode, with the $c/2L$ mode spacing of an ordinary resonator.

A somewhat different photorefractive PPCM has been used in a copper vapor laser (see Chap. 2). In this experiment, the spontaneous emission was sufficient to initiate operation. Lasing was possible with one or both end mirrors replaced with $LiNbO_3$ crystals, in which the operative gratings were of the reflection type. The bandwidth was $0.3\,cm^{-1}$, practically identical to the superradiant spectrum [4.58].

4.4.4 Self-Pumped Mirrors as Tuning Elements

A self-pumped photorefractive phase conjugate mirror has also been used as the end mirror on a pulsed dye laser [4.59]. When the normal mirror was replaced with the photorefractive mirror, the laser linewidth narrowed from 160 GHz to 2 GHz. The system was shown to be insensitive to slowly changing intracavity aberrations, and also capable of adapting to gross changes in the position of the normal mirror. The cavity shown in Fig. 4.31 is marginally stable (concentric) when mirrors M_1 and M_2 are planar. If M_1 is moved into an unstable region, a sharp decrease in output power is observed (Fig. 4.32). If it is replaced with the $BaTiO_3$ crystal, which is then translated, the output power remains essentially constant.

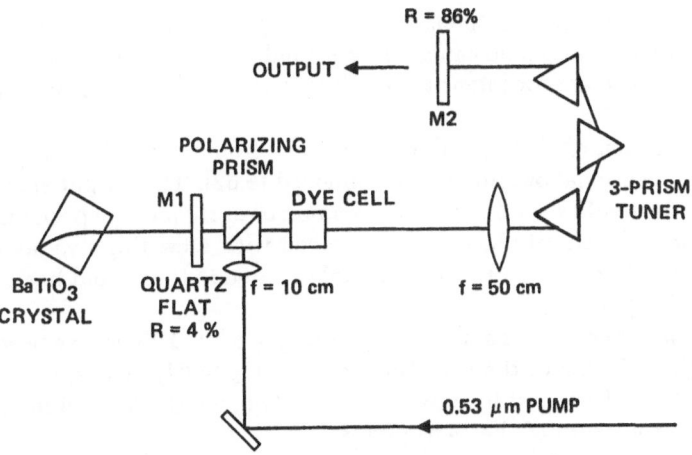

Fig. 4.31. Dye laser resonator using self-pumped PCM. M_1 is removed after gratings are written in BaTiO$_3$

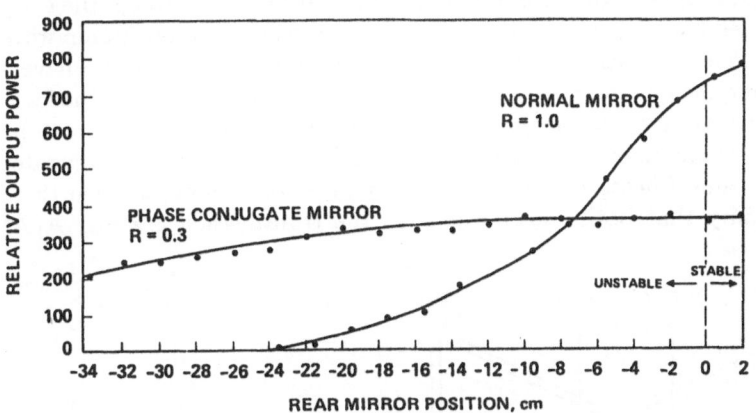

Fig. 4.32. Output power of laser in Fig. 4.31 as a function of the rear mirror position

Frequency self-scanning as well as narrowing of the instantaneous spectrum was observed in a cw dye laser with external feedback from a PR-PPCM placed outside the conventional output coupler [4.60]. The multi-cavity system had a linewidth of 4 GHz, which was a factor of 400 less than the single cavity untuned linewidth. The quasi-linear recurrent wavelength sweep occurred over tens of nanometers.

Speculation about the cause of the nondegenerate reflection centered on factors which influence the phase shift between the index grating and interference patterns inside the crystal. A stationary, uniform electric field, possibly arising from the bulk photovoltaic effect [4.61], can alter the charge migration and cause the phase shift to depart from $\frac{\pi}{2}$ (TAP61, Chap. 5). If the two interfering beams differ in frequency, the moving interference pattern can also alter the phase shift, due to the finite response time of the photorefractive effect. A frequency-induced phase shift could cancel the field-induced phase shift, restore the $\frac{\pi}{2}$ shift which is optimum for intensity coupling, and yield a nondegenerate reflection.

Further confirmation of this hypothesis has come from study of the same system with the conventional output coupler removed [4.62]. The output spectrum became a doublet split by 1.5 GHz. An even narrower frequency spectrum (1.3 GHz) and a slower scan rate have been observed with a cw ring dye laser modified to accept feedback from a BaTiO$_3$ crystal situated in the output beam, external to the cavity [4.63].

Picosecond pulse operation of a dye laser containing a PR-PPCM has been achieved in a further display of the versatility of BaTiO$_3$ [4.64]. The spectral bandwidth of this laser decreased by only a factor of two, which was still large enough to span the spectrum of the 14 ps pulses.

4.4.5 Self-Pumped Mirror Based on a Ring Cavity

In a third type of PR-PPCM [4.65], the incident wave passes through the crystal, is redirected back in, crosses its original path, and serves as one pump wave (Fig. 4.33a). The counterpropagating pump grows from noise. Again we have a situation where the possibility of gain through wave mixing exists, and the feedback paths are strongest when the four waves in the interaction region can be grouped into two pairs of conjugate beams. If fields A_1 and A_2 are conjugates of each other, and if fields A_3 and A_4 are conjugates of each other, then there is the possibility that both contributions to the transmission grating, $A_1 A_4^*$

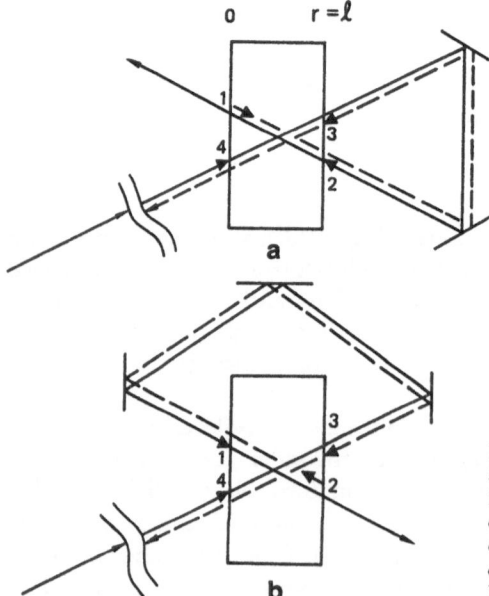

Fig. 4.33. (a) Geometry of the ring self-pumped PCM. The incident wave is shown by the solid line and the oscillation beam which it induces is dashed. (b) Geometry of a PCM based on parametric feedback and stimulated Brillouin scattering

and $A_2^* A_3$, can constructively interfere throughout the crystal. Alternatively, one could say that noise at the face $z = r = 0$ that is conjugate to A_2 can pass through the interaction region and be amplified twice, in analogy with the serpentine path argument (see Appendix 4.A).

If the analysis of Sect. 4.3.6 is applied to this configuration with the boundary conditions $I_4(0)/I_2(0) = M$ and $I_1(0)/I_3(0) = M$, the threshold is found to be $(\gamma l)_t = \ln(1/2 + 1/2M)[(M+1)/(M-1)]$. With ideal feedback, $M = 1$, and $(\gamma l)_t = -1$. This is a self-starting threshold: when it is exceeded, oscillation beams of infinitesimal intensity experience gain.

The characteristics of the four PPCM's discussed in this section can be compared in Fig. 4.34, which illustrates reflectivity vs coupling strength. The calculations assumed a single grating, no absorption, and ideal feedback [4.30]. The two mirrors with the highest threshold also have non-zero reflectivities at turn-on, which shows that they cannot be self-starting, since infinitesimal oscillation beams cannot yield finite reflectivities.

The low threshold of the ring PPCM has made passive photorefractive phase conjugation possible in the infrared, where large coupling strengths are difficult to obtain because of a reduction in the ionization cross section for trapped charge carriers. The longest wavelength sources used to date are an argon ion laser operating at 1090 nm and GaAlAs lasers operating at 815–865 nm [4.66]. At the latter wavelength, operation of the ring mirror required a threshold intensity of 50 mW/cm^2. Below that point, it is possible that the (nonproductive) thermal excitation of charges dominated the photon excitation. At 126 mW/cm^2, the reflectivity reached 16 % and the risetime was 40 s.

In spite of the long writing times, this device can be surprisingly vibration resistant [4.67]. Beam 1 is formed by the diffraction of beam 4, and both propagate a distance L before intersecting again, so they are automatically coherent on both the first and second passes through the crystal. Accordingly, the formation of the transmission grating should be independent of (1) changes in L, (2) laser coherence length relative to L, and (3) laser pulse length relative to

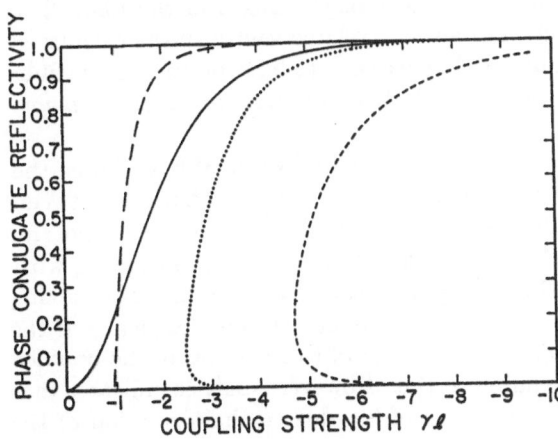

Fig. 4.34. Reflectivity of a self-pumped PCM based on: *(solid line)* a Fabry-Perot; *(dots)* a Fabry-Perot with one mirror removed; *(long dashes)* a ring cavity; *(dashed)* two interaction regions

L. The first contention was verified by vibrating one of the feedback mirrors. The phase conjugate reflectivity actually increased with frequency, possibly because a counterproductive reflection grating was being washed out. The lack of a coherence requirement was verified by conjugating a beam from a multi-longitudinal-mode argon ion laser which had a coherence length less than the 20 cm ring path length. The lack of a pulse length requirement was verified with 5 ps (1.7 mm) pulses from a mode-locked dye laser operating at 591 nm. The beams were present simultaneously only in pairs, never all four at once. Nevertheless, oscillation built up almost as in the cw case. Successive reinforcement of the grating by the forward and backward travelling pairs, together with the interpulse memory inherent in the photorefractive holograms, was sufficient to generate the required grating.

As noted in the last paragraph, if beams 1 and 4 travel the same distance L before intersecting a second time, then they will automatically be coherent on the second pass, and, furthermore, the two interference patterns generated on the two passes will overlap in phase. If, however, a nonreciprocal path difference exists, then the two will not constructively interfere, an unfavorable circumstance for the buildup of a grating. A grating can build up, however, if there is a slight frequency difference between the beams going in opposite directions around the ring cavity.

This frequency difference is the basis for a new type of ring gyroscope [4.45]. Like a conventional laser gyro, rotating it produces a frequency shift rather than a phase shift. A long fiber can be used in the ring to increase the sensitivity, as in conventional fiber gyros. The latter detect phase shifts between counterpropagating beams in a Sagnac interferometer. The accuracy of a conventional fiber gyro is limited by nonreciprocal phase differences due to modal dispersion, unless single-mode fibers are used. However, in the new ring gyro, to the extent that the counterpropagating beams in Fig. 4.33a are phase conjugates of each other, such a restriction is unnecessary.

This concept has been tested with a $BaTiO_3$ crystal and a 1 m multi-mode fiber ring. A nonreciprocal phase shift of ~ 0.1 radian was introduced via the Faraday effect by applying a magnetic field along a length of the fiber. The observed fringe movement of ~ 0.1 radian/s correlated well with the calculated frequency shift. Nonreciprocal phase differences due to birefringence in the fiber are automatically minimized because the electro-optic tensor of $BaTiO_3$ favors the coupling of extraordinary beams.

A very similar geometry has been used in backward SBS to lower the threshold for a phase conjugate reflection, illustrating the similarities between the two types of media [4.68]. The gratings which build up from noise are of the reflection type. The feedback, A_1, in this case is close to co-propagating with the probe, A_4, (Fig. 4.33b) so that the waves scattered from all three moving gratings ($k_1 - k_3$, $k_1 - k_2$, and $k_3 - k_4$) will be at the same frequency. If the threshold input intensity went down by a factor of two because of the doubled path length, the result would be uninteresting. However, analysis indicates that the threshold should decrease by a factor of 20, due to the formation of the additional gratings. An experiment has been performed using a waveguide filled with acetone as the nonlinear medium. The measured threshold of $1.8\,\mathrm{MW/cm^2}$ was only slightly above the calculated threshold of $1.2\,\mathrm{MW/cm^2}$.

4.A Appendix

There is a close mathematical and conceptual analogy between the propagation of light in photorefractive, Brillouin, Raman, and Rayleigh active media. As the understanding in each field progresses, the overlap appears greater and the differences appear to be more quantitative than qualitative. The underlying physics, which determines the coupling coefficients and time constants, is entirely different, but the coupled wave treatments are almost identical. Therefore, the contents of this chapter are germane to the study of stimulated Brillouin (SBS), Raman (SRS), and Rayleigh (SRLS) scattering. The term stimulated photorefractive scattering (SPS) can be used to emphasize the similarity. The field of phase conjugation has helped in the unification because photorefractive and Brillouin active media have emerged as the media of choice for many applications. This appendix is intended to give a brief overview of SBS and SRS, and some references to the extensive literature.

In a Brillouin scattering event, an acoustic phonon is created in the medium when an incident (laser) photon is absorbed and a Stokes, or lower energy, photon is emitted. The density of the medium and the optical field are coupled through the electrostrictive force. In the stimulated process (SBS), two optical and one hypersonic wave are coherently coupled. The interference between two optical plane waves is a periodic intensity distribution, characterized by the difference wavevector and the difference frequency. If the frequency and wavevector lie on the acoustic phonon dispersion curve, then electrostrictive forces can generate a phonon. In turn, the periodic change in dielectric constant, which propagates in unison with the acoustic wave, can scatter one optical wave into the other. This moving grating imposes a Doppler frequency shift on the reflected wave. In the classical treatment of SBS, the medium is taken to be an elastic continuum driven by the electrostrictive force and damped by viscosity and by thermal conductivity.

In a vibrational Raman scattering event, a molecule makes a transition from a $v = 0$ state to a $v = 1$ state when an incident (laser) photon is absorbed, and a Stokes, or lower energy, photon is emitted. The nuclear motion and the optical field are coupled because the molecule's polarizability is a function of the vibrational coordinate. In the stimulated process (SRS), the molecular excitation and optical waves are coherently coupled. The molecular excitation is a quadrupole excitation rather than a change in vibrational level populations. It can be thought of as an optical phonon in contrast to the acoustic phonon of SBS. Two optical waves can generate an optical phonon if their difference frequency is close to the vibrational resonance. At the same time, the periodic change in polarizability will scatter each optical wave into the other, through the creation of sidebands separated by the vibrational frequency. In the classical treatment, the molecules are taken to be independent harmonic oscillators. They are damped by collisions and driven by a force derived from the dependence of the electrostatic stored energy density on the vibrational normal coordinate.

In backward SBS, the laser wave at ω_L and the Stokes wave at ω_S are counterpropagating. On resonance and in steady state, the coupling is described by the following equations [4.23, 69]:

$$\frac{dI_S}{dz} = -g_B I_L I_S + \alpha I_S \quad , \quad \frac{dI_L}{dz} = -g_B I_L I_S - \alpha I_L \quad ,$$

$$g_B = \frac{\omega_S^2 \varrho_0 \tau_B}{c^3 n v} \left(\frac{\partial \varepsilon}{\partial \varrho} \right)_T^2 \quad .$$

Here $I = cn|E|^2/8\pi$, ϱ_0 is the average density of the medium, τ_B the lifetime of the acoustic phonon, v the sound velocity, n the index of refraction, ε the dielectric constant, and T the temperature. These equations are identical in form to (4.3), except for the absence of the denominator I_0.

The analogous equations for SRS are [4.23, 69]:

$$\frac{dI_S}{dz} = -g_R I_S I_L + \alpha I_S \quad , \quad \frac{dI_L}{dz} = -g_L I_S I_L - \alpha I_L \quad ;$$

$$g_R = \frac{8\pi^2 \omega_S N}{n^2 c^2 m \Omega} \left(\frac{\partial \alpha}{\partial q} \right)^2 \frac{1}{1 + \beta} \quad , \quad g_L = \frac{\omega_L}{\omega_S} g_R \quad .$$

Here N is the molecular number density, m is the reduced mass of the vibration, Ω is the vibrational frequency, q is the vibrational normal coordinate, $\partial \alpha / \partial q$ is the differential polarizability, and β is the depolarization ratio. If photon fluxes were used instead of intensities, the two equations would have equal coupling constants.

The equations for other forms of stimulated scattering, e.g., Rayleigh and polariton, are also similar [4.69]. We have neglected them because they are less well known than Raman and Brillouin scattering, not because they are less similar. In fact, the magnitude of the frequency shift in Rayleigh scattering is closer to the SPS shift than to any other.

Stimulated photorefractive scattering has several unique properties and several characteristics in common with the more conventional mechanisms (SBS, SRS, and SRLS). Insight can be obtained from considering just the steady state small signal regime. For example, a signal wave can be amplified by any of the four mechanisms. The gain is due to a coupling of pump and signal amplitudes, not to the release of any energy stored in the medium. The magnitude and direction of energy flow between two optical waves is determined by the sine of a phase shift between their mutual interference pattern and the associated index variation. Phase transfer is proportional to the cosine of the phase shift. If only one wave is present initially, the existence of gain is crucial for the buildup of a second wave from noise.

In conventional stimulated scattering, a nonzero phase shift is due to the nonzero response time, which causes the index variation to lag behind a moving interference pattern. The resulting energy flow favors the lower frequency wave, i.e., the Stokes wave experiences gain[1]. Waves that are degenerate in frequency have a stationary interference pattern and will not exchange energy in steady state because the phase shift goes to zero. The conventional mechanisms only support travelling waves, so the amplitude also goes to zero, and the waves will not exchange phase either.

The photorefractive effect is unique in that a large phase shift and amplitude exist in the degenerate case, due to the mechanism of charge diffusion. The electro-optic effect can arise from a static distortion of the lattice by the space-charge field, which corresponds to a zero frequency phonon. In the non-degenerate case, fringe motion imparts an additional component to the phase shift, which can change the direction and magnitude of the energy transfer.

In conventional stimulated scattering media, the decay time is so short that the induced excitations do not typically travel to a region of pump intensity different from that of their origin. Thus the gain experienced by a signal wave at a given point in space is a function of the pump intensity at that point only. A local relationship also holds for the photorefractive case[3]. However, the gain is independent of the pump intensity in the small signal limit. Mathematically, this is because of the normalizing denominators in (4.3). Physically, it is because spatial variations in intensity cause a redistribution of trapped charge, not absolute intensity.

Under certain conditions, SBS and SRS backscattering give rise to the wavefront reversed replica of an incident wave. Wavefront reversal in SBS was discovered first [4.70] and has been studied more thoroughly. The incident and reflected waves are nearly degenerate, so the latter is closer to a time reversed replica. The wavefront reversing characteristics of SRS were also identified long after the backscattering was first observed [4.71]. The large frequency shift imposed upon the backward wave limits the use of SRS in, for example, double pass distortion correction scenarios.

A serpentine path argument has been offered to explain why SBS backscattering generates a wavefront reversed replica of an incident pump wave [4.72]. Of all the optical noise components at the exit face which are heading back into the medium, those whose amplitude and phase variations mimic those of the pump will experience the highest gain. This is because such a wave will retrace the path of the pump wave and therefore will favor the regions of high intensity (high gain), and avoid the regions of low intensity (low gain).

In the past, most experimental realizations and theoretical treatments of wavefront reversal by SBS consisted of only one incident wave, focused into either a bulk medium or a waveguide. In contrast, wavefront reversal by SPS has always been implemented with 4WM, which involves two additional incident

---

[2] An exception to this occurs when absorptive heating is responsible for the interaction.

[3] The "nonlocal response" referred to in the literature indicates instead a nonzero phase shift.

147

Table 4.1

	Stimulated Raman scattering	Stimulated Brillouin scattering	Stimulated photorefractive scattering
Type of media	Solid, liquid, gas		Electro-optic insulators[a]
	Bulk and waveguide		Bulk[b]
Nonlinear gain	$g \sim 5 \times 10^{-3}$ cm MW^{-1} [c]	$g \sim 10^{-2}$ cm MW^{-1} [c]	$10^{-1} < \gamma < 10^2$ cm^{-1}
Linear loss	Negligible[d]		$10^{-1} < \alpha < 10^1$ cm^{-1}
Frequency shift	10^{13} Hz 10^3 cm^{-1} [c]	10^{10} Hz 10^0 cm^{-1} [c]	10^0 Hz 10^{-10} cm^{-1} [e]
Response time/ bandwidth	10^{-12} s 5 cm^{-1} [c]	10^{-9} s 5×10^{-3} cm^{-1} [c]	$10^{-2} - 10^2$ s $10^{-10} - 10^{-14}$ cm^{-1} [f]
Steady-state on-resonance grating phase shift	$90°$		$90°$ [g]
Type of source	Nd:YAG, ruby, excimer		Argon ion, HeCd, dye, semiconductor diode
Intensity	$10^6 < I < 10^9$ W cm^{-2}		$10^{-1} < I < 10^2$ W cm^{-2} [f]
Pulse length	$10^{-13} - 10^{-7}$ s [h]	$10^{-9} - 10^{-7}$ s [h]	cw[f]
Intensity gain threshold for self-pumped phase conjugation	$10 < gll < 30$ [i]		$0 < \Gamma l < 10$ [i]
Most common grating type	Reflection		Transmission

[a] Semiconductors have also recently been used (TAP 61, Chap. 8).
[b] SPS is a problem called "optical damage" in LiNbO$_3$ for example, but waveguides have not been used to study the effect.
[c] These values have been taken from Table 1 of Ref. [4.69], with permission.
[d] Exceptions are resonant Raman scattering and stimulated thermal Brillouin scattering.
[e] This is not a property of the medium alone, it depends on the light intensity.
[f] A few pulsed experiments have been done. See, e.g., [4.78]. The intensities have been larger than the norm, and the response times are approximately inversely proportional to the intensities.
[g] This can vary with an applied dc electric field or an internal bulk photovoltaic field.
[h] A few cw experiments have been done using wave guides.
[i] For a discussion of these values, see [4.30, 72]. The SPS values apply to 4WM rather than simple backscattering.

waves, the counterpropagating pump waves of Sect. 4.3. As an illustration of the overlap between the two disciplines, phase conjugation via 4WM in Brillouin activa media is now being studied [4.73, 74]. Similarly, photorefractive backscattering in the conventional SBS geometry has been recently observed [4.75]. No frequency shift was detected on the reflected wave, but the expected shift was very small, on the order of Hz. In the case of SBS, the backscattered wave originates in scattering of the pump wave from random density variations.

The origin of SPS is not well understood, but it is also believed to originate from optical inhomogeneities due to random space-charge fields or defects.

The backscattered Stokes wave in SRS is usually accompanied by a forward traveling Stokes wave and a forward traveling anti-Stokes (higher frequency) wave [4.76]. The photorefractive analog has been recently identified as the competition between backscattering (wavefront reversal) and forward scattering (the fanning effect) [4.77].

Table 4.1 illustrates the similarities and differences between SRS, SBS, and SPS. The stated values, for example the intensities, are meant only to be representative. Where the numbers vary more than an order of magnitude on either side of a given value, a range of values is given, at the risk of appearing to be all inclusive. The entries say perhaps as much about what specific materials and regimes people have chosen to study as they say about the nature of the three phenomena.

Acknowledgements. One of us (J.W.) would like to thank G.C. Valley and J.F. Lam for valuable discussions concerning the relationship between SPS, SBS, and SRS.

References

4.1   A. Ashkin, G.D. Boyd, J.M. Dziedzic, R.G. Smith, A.A. Ballman, J.J. Levinstein, K. Nassau: Appl. Phys. Lett. **9**, 72 (1966)
4.2   T. Shankoff: Appl. Opt. **7**, 2101 (1968)
4.3   D.H. Close, A.D. Jacobson, J.D. Margerum, R.G. Brault, F.J. McClung: Appl. Phys. Lett. **14**, 159 (1969)
4.4   D.L. Staebler, W.J. Burke, W. Phillips, J.J. Amodei: Appl. Phys. Lett. **26**, 182 (1975)
4.5   H. Kogelnik: Bell Syst. Tech. J. **48**, 2909 (1969)
4.6   D.W. Vahey: J. Appl. Phys. **46**, 3510 (1975)
4.7   N.V. Kukhtarev, G.E. Dovgalenko, V.N. Starkov: Appl. Phys. **A33**, 227 (1984)
4.8   J.P. Huignard, J.P. Herriau, P. Auborg, E. Spitz: Opt. Lett. **4**, 21 (1979)
4.9   J.O. White: "Four-Wave Mixing and Phase Conjugation in Photorefractive Crystals"; Ph.D. Thesis, California Institute of Technology (1984)
4.10 F. Laeri, T. Tschudi, J. Albers: Opt. Commun. **47**, 387 (1983)
4.11 N.V. Kukhtarev, V.B. Markov, S.G. Odulov, M.S. Soskin, V.L. Vinetskii: Ferroelectrics **22**, 961 (1979)
4.12 Y.H. Ja: Opt. Quant. Electron. **14**, 547 (1982)
4.13 P. Yeh: Opt. Commun. **13**, 323 (1983)
4.14 Y.H. Ja: Opt. Commun. **53**, 153 (1985)
4.15 M. Sargent III: Appl. Phys. **9**, 127 (1976)
4.16 G. Marowsky, K. Kaufman: IEEE J. QE-12, 207 (1976)
4.17 P. Yeh: Appl. Opt. **23**, 2974 (1984)
4.18 M.Z. Zha, P. Günter: Opt. Lett. **10**, 184 (1985)
4.19 D.L. Staebler, J.J. Amodei: J. Appl. Phys. **43**, 1042 (1972)
4.20 J.O. White, M. Cronin-Golomb, B. Fischer, A. Yariv: Appl. Phys. Lett. **40**, 450 (1982)
4.21 G.C. Valley: JOSA B1, 868 (1984)
4.22 A. Yariv, S.-K. Kwong: Opt. Lett. **10**, 454 (1985)
4.23 A. Yariv: *Quantum Electronics*, (Wiley, New York, 1975)
4.24 N.G. Basov, A.Z. Grasyuk, Yu.I. Karev, L.L. Losev, V.G. Smirnov: Sov. J. Quantum Electron. **9**, 780 (1979)
4.25 S.-K. Kwong, A. Yariv: Appl. Phys. Lett. **48**, 564 (1986)
4.26 R.S.F. Chang, R.H. Lehmberg, M.T. Duignan, N. Djeu: IEEE J. QE-21 , 477 (1985)
4.27 D.L. Fried (ed.): Special issue on adaptive optics JOSA **67**, No. 3 (1977)
4.28 A. Yariv, D.M. Pepper: Opt. Lett. **1**, 16 (1977)
4.29 B. Fischer, M. Cronin-Golomb, J.O. White, A. Yariv: Opt. Lett. **6**, 519 (1981)

4.30   M. Cronin-Golomb, B. Fischer, J.O. White, A. Yariv: IEEE J. QE-20, 12 (1984)
4.31   S.-K. Kwong, Y.-H. Chung, M. Cronin-Golomb, A. Yariv: Opt. Lett. 10, 359 (1985)
4.32   J.F. Lam: Appl. Phys. Lett. 42, 155 (1983)
4.33   J.H. Marburger, J.F. Lam: Appl. Phys. Lett. 34, 389 (1979)
4.34   J.H. Marburger, J.F. Lam: Appl. Phys. Lett. 35, 249 (1979)
4.35   H.G. Winful, J.H. Marburger: Appl. Phys. Lett. 36, 613 (1980)
4.36   A.R. Kessel, V.M. Musin: Opt. Commun. 44, 133 (1982)
4.37   M. Cronin-Golomb, J.O. White, B. Fischer, A. Yariv: Opt. Lett. 7, 313 (1982)
4.38   M.R. Belić: Phys. Rev. A31, 3169 (1985)
4.39   J. Au Yeung, D. Fekete, D.M. Pepper, A. Yariv: IEEE J. QE-15, 1180 (1979)
4.40   J. Feinberg, R.W. Hellwarth: Opt. Lett. 5, 519 (1980); 6, 257 (1981)
4.41   B. Fischer, M. Cronin-Golomb, J.O. White, A. Yariv, R. Neurgaonkar: Appl. Phys. Lett. 40, 863 (1982)
4.42   R.C. Lind, D.G. Steel: Opt. Lett. 6, 554 (1981)
4.43   A. Yariv, T.L. Koch: Opt. Lett. 7, 113 (1982)
4.44   B. Fischer, M. Cronin-Golomb, J.O. White, A. Yariv: Appl. Phys. Lett. 41, 141 (1982)
4.45   B. Fischer, S. Sternklar: Appl. Phys. Lett. 47, 1 (1985)
4.46   M.D. Ewbank, P. Yeh, M. Khoshnevisan, J. Feinberg: Opt. Lett. 10, 282 (1985)
4.47   M. Cronin-Golomb, B. Fischer, J.O. White, A. Yariv: Appl. Phys. Lett. 41, 689 (1982)
4.48   S.G. Odulov: Sov. J. Quantum Electron. 14, 360 (1984)
4.49   V.V. Voronov, I.R. Dorosh, Yu.S. Kus'minov, N.V. Tkachenko: Sov. J. Quantum Electron. 10, 1346 (1980)
4.50   J. Feinberg: JOSA 72, 46 (1983)
4.51   S. Odoulov, K. Belabaev, I. Kiseleva: Opt. Lett. 10, 31 (1985)
4.52   M. Cronin-Golomb, S.K. Kwong, A. Yariv: Appl. Phys. Lett. 44, 727 (1984)
4.53   T.Y. Chang, D.L. Naylor, R.W. Hellwarth: Appl. Phys. B28, 156 (1983)
4.54   S.K. Kwong, M. Cronin-Golomb, A. Yariv: Appl. Phys. Lett. 45, 1016 (1984)
4.55   J. Feinberg: Opt. Lett. 7, 486 (1982)
4.56   Yu.A. Anan'ev: Sov. J. Quantum Electron. 4, 7 (1975)
4.57   M. Cronin-Golomb, B. Fischer, J. Nilsen, J.O. White, A. Yariv: Appl. Phys. Lett. 41, 219 (1982)
4.58   A. Litvinenko, S. Odoulov: Opt. Lett. 9, 68 (1984)
4.59   R.A. McFarlane, D.G. Steel: Opt. Lett. 8, 208 (1983)
4.60   W.B. Whitten, J.M. Ramsey: Opt. Lett. 9, 44 (1984)
4.61   A.M. Glass, D. von der Linde, T.J. Negran: Appl. Phys. Lett. 25, 233 (1974)
4.62   J. Feinberg, G.D. Bacher: Opt. Lett. 9, 420 (1984)
4.63   J.M. Ramsey, W.B. Whitten: Opt. Lett. 10, 362 (1985)
4.64   R.K. Jain, K. Stenersen: Opt. Lett. 9, 546 (1984)
4.65   M. Cronin-Golomb, B. Fischer, J.O. White, A. Yariv: Appl. Phys. Lett. 42, 919 (1983)
4.66   M. Cronin-Golomb, K.Y. Lau, A. Yariv: Appl. Phys. Lett. 47, 567 (1985)
4.67   M. Cronin-Golomb, J. Paslaski, A. Yariv: Appl. Phys. Lett. 47, 1131 (1985)
4.68   V.I. Odintsov, L.F. Rogacheva: JETP Lett. 36, 344 (1982)
4.69   W. Kaiser, M. Maier: "Stimulated Rayleigh, Brillouin and Raman Spectroscopy", in Laser Handbook, ed. by F.T. Arecchi and E.O. Schulz-Dubois, (North-Holland, Amsterdam 1972) pp. 1077–1150
4.70   B.Ya. Zel'dovich, V.I. Popovichev, V.V. Ragulsky, F.S. Faizullov: JETP Lett. 15, 109 (1972)
4.71   B.Ya. Zel'dovich, N.A. Mel'nikov, N.F. Pilipetsky, V.V. Ragulsky: JETP Lett. 25, 36 (1977)
4.72   B.Ya. Zel'dovich, N.F. Pilipetsky, V.V. Shkunov: Principles of Phase Conjugation, Springer Ser. Opt. Sci. Vol. 42 (Springer, Berlin, Heidelberg 1985)
4.73   R.L. Abrams, C.R. Giuliano, J.F. Lam: Opt. Lett. 6, 131 (1981)
4.74   A.M. Scott: Opt. Commun. 45, 127 (1983)
4.75   T.Y. Chang, R.W. Hellwarth: Opt. Lett. 10, 408 (1985)
4.76   Yu.E. D'Yakov, S.Yu. Nikitin: Sov. J. Quantum Electron. 12, 796 (1982)
4.77   G.C. Valley: JOSA B4, 14 (1987)
4.78   L.K. Lam, T.Y. Chang, J. Feinberg, R.W. Hellwarth: Opt. Lett. 10, 475 (1981)

5. Phase-Conjugate Mirrors and Resonators with Photorefractive Materials

Jack Feinberg and Kenneth R. MacDonald

With 35 Figures

The concept of time-reversing an optical wave is an intriguing one. The time-reversed (or phase-conjugate) version of an optical wave can be generated by nonlinear mixing in an appropriate material [5.1–3]. Photorefractive materials have been ideal for demonstrating the possibilities of phase conjugation, such as correcting a distorted picture, tracking a moving mirror (or a shiny hamburger spatula), scanning the frequency of a dye laser, and creating images having sub-micron resolution. Because photorefractive materials have a large nonlinearity to even weak light beams, new ideas can be tested quickly and easily with low-power lasers. In addition, the very large optical nonlinearity has produced unpredicted and desirable new effects, such as self-pumped phase conjugation.

Although photorefractive materials are ideal for testing ideas and demonstrating devices, they are often less than ideal for implementing a device outside of the laboratory. No one has yet used phase conjugation in a photorefractive crystal (or in anything else!) to send an undistorted picture through the atmosphere to an orbiting satellite or through the ocean to a submerged submarine. Nor have any commercial applications yet been adopted, such as high-resolution lithography, or tracking a fast-moving satellite, or improving the spatial mode profile of a high-power laser. Photorefractive crystals can be fragile, expensive, commercially unavailable, temperature sensitive, difficult to grow, exceedingly sensitive to impurities, or distressingly slow (response times of a second) when used with low-powered (milliwatt) lasers. Some photorefractive crystals are optically active requiring careful manipulation of beam polarizations for efficient coupling; others are too absorptive or have nonlinearities that are too small. At present there is no crystal that embodies all of the desirable properties of a photorefractive material without any of the drawbacks.

Nevertheless, the recent demonstrations of self-pumped phase conjugation, optical processing, coherent beam amplification, and associative memories using photorefractive crystals have focused academic and commercial interest on these materials. New effects that were not predicted have been observed, and have led to new research areas. Crystal growers are producing new and higher-quality photorefractive crystals, and quite a few research laboratories are studying the origins and properties of the charge carriers in these crystals, with a view to understanding and improving the photorefractive effect itself.

This chapter will review some of the applications of photorefractive materials, and will highlight phase conjugation, self-pumped phase conjugation, and the interesting properties of phase-conjugating optical resonators. Section 5.1 presents a brief and biased history of the important discoveries in this field [5.1–3]. Section 5.2 describes the photorefractive effect. Section 5.3 describes two-wave and four-wave mixing of optical beams in photorefractive materials. Section 5.4 discusses ring resonators. Section 5.5 discusses self-pumped phase conjugation. Section 5.6 describes several types of "stimulated scattering" in photorefractive materials. Section 5.7 discusses the origins of the frequency shifts observed in a number of self-pumped phase conjugators. Finally, Sect. 5.8 describes some of the many applications of phase conjugation in photorefractive materials.

5.1 A Brief History of Phase Conjugation in Photorefractive Materials

5.1.1 Holography

The holographic "time-reversal" of an image-bearing beam was first demonstrated [5.4] by *Gabor* in 1949. The interference pattern of an image-bearing beam and a plane reference beam was recorded on a photographic plate, as shown in Fig. 5.1. The photographic plate was then developed, fixed, and dried. When the resulting hologram was illuminated by a plane wave traveling antiparallel to the original reference wave, the original image was reconstructed [5.5, 6].

Kogelnik [5.7] demonstrated the distortion-correcting property of phase conjugation in 1965. He repeated Gabor's experiment with a distorting plate between the original object and the photographic plate as shown in Fig. 5.2. The developed hologram was then read out by a plane wave as before. The reconstructed beam − the phase-conjugate replica of the original beam − was allowed to pass back through the same distorter, from which it emerged with the original image intact and undistorted. The phase-conjugate beam traversed

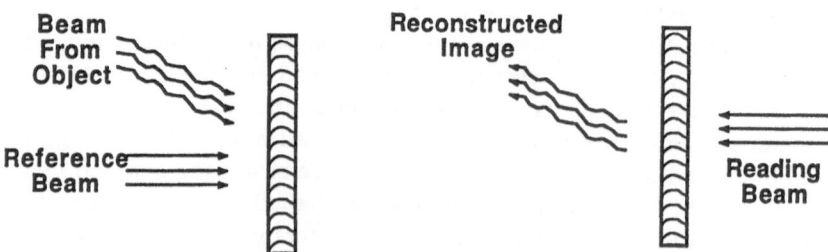

Fig. 5.1. Holography. The interference pattern of the two writing beams is recorded on photographic film. The film is developed and then illuminated by a reading beam. The diffracted light forms a real image of the original object

WRITING THE HOLOGRAM:

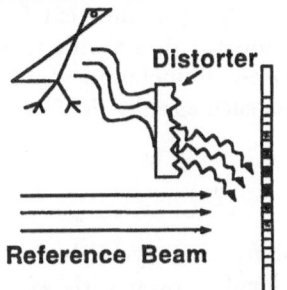

Distorter

Reference Beam

READING THE HOLOGRAM:

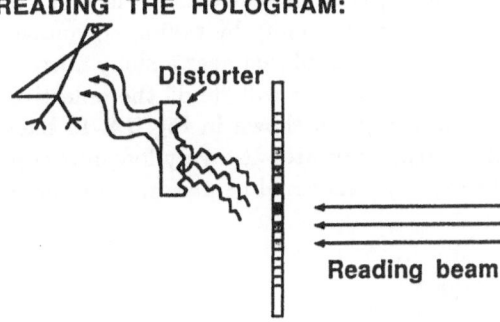

Distorter

Reading beam

Fig. 5.2. Holographic correction of phase distortion. The distorted image is recorded on a hologram, and the reconstructed beam is made to pass back through the same distorter. The final image is free of distortion. The incident beam and the reconstructed image-bearing beam are phase conjugates of each other

exactly the same path as the incident beam, but in the opposite direction, as though it had been reversed in time.

For most applications of phase conjugation, photographic film is an impractical medium: it can only be used once and there is a long delay between exposing and reading the hologram. *Real-time* holography only became feasible with the recent development of nonlinear optical materials that respond rapidly and repeatedly to an incoming optical wavefront.

5.1.2 Real-Time Holography

Yariv [5.8, 9] proposed generating a phase-conjugate wave by three-wave mixing in a noncentrosymmetric material in 1976. This scheme requires careful adjustment of the frequencies and incident angles of the optical beams in order to satisfy a phase-matching condition. In 1977, *Hellwarth* [5.10] proposed four-wave mixing to create the phase-conjugate replica of an optical wave, as shown in Fig. 5.3. If all three input waves have the same frequency, the phase-matching condition is automatically satisfied in this geometry. Four-wave mixing exper-

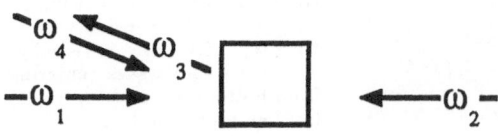

Fig. 5.3. Four-wave mixing in a nonlinear medium. The input beam 4 mixes with the pumping beams 1 and 2 to create the signal beam 3. Beams 3 and 4 are phase-conjugates of each other

iments were first performed using megawatt-power laser beams, with liquid carbon disulfide (CS_2) as the nonlinear medium [5.11, 12]. Over the next few years phase conjugation by four-wave mixing was demonstrated in a wide variety of materials (involving many different physical processes), including atomic vapors, [5.13, 14] absorbing dyes [5.15] and laser media such as ruby [5.16] and neodymium-doped glass [5.17].

5.1.3 Photorefractive Phase Conjugators

Feinberg et al. [5.18] showed in 1980 that the photorefractive crystal barium titanate ($BaTiO_3$) can act as an efficient phase conjugator, even with weak optical beams. *Feinberg* and *Hellwarth* [5.19] then demonstrated the first cw phase conjugator having a reflectivity greater than unity by taking advantage of barium titanate's large Pockels coefficient r_{42}. If placed near a shiny mirror, or even a shiny kitchen spatula, this phase conjugator will "find" the reflective surface and direct a beam of light normal to it, as shown in Fig. 5.4. In barium titanate, the photorefractive nonlinearity saturates at a very low intensity (less than 100 milliwatts/cm^2), so the experiments could be performed using a milliwatt-power laser.

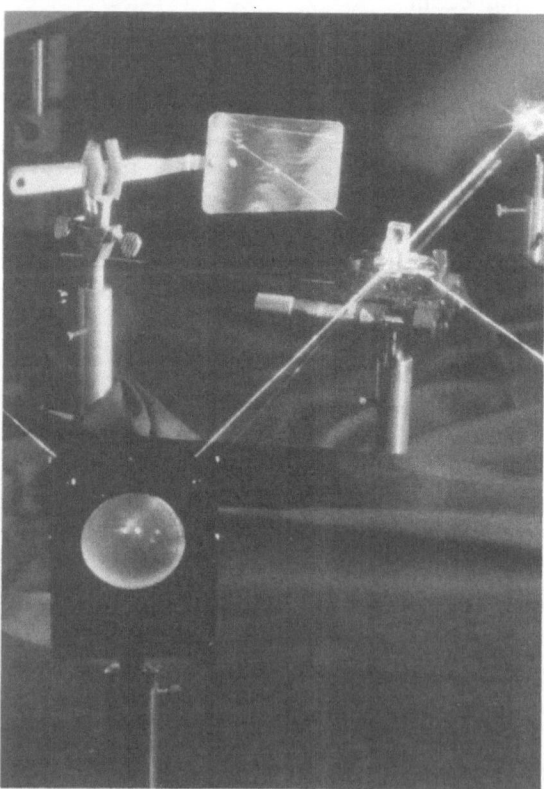

Fig. 5.4. Spatula laser: a laser resonator formed by a kitchen spatula and a phase-conjugating mirror. The phase-conjugating mirror has a reflectivity greater than unity, and provides the gain and the aberration correction needed to sustain lasing. The phase-conjugator is a crystal of photorefractive barium titanate, seen here inside a glass cuvette, with two externally-supplied pump waves (entering from bottom left and top right) from an argon-ion laser

5.1.4 Self-Pumped Phase Conjugators

In 1982, *White* and coworkers [5.20] demonstrated a remarkable new device: a "self-pumped" phase conjugator. It consists of a photorefractive crystal placed between two mirrors that have been carefully aligned to form a resonator cavity, as shown in Fig. 5.5a. The input wave causes a pair of counterpropagating beams to form between the two mirrors. These self-generated beams serve as "pumping" beams to generate the phase-conjugate signal by four-wave mixing. Once the device is working, one of the resonator mirrors can be removed (Fig. 5.5b), although the device needs both mirrors in order to start [5.21].

In 1982, *Feinberg* demonstrated a self-pumped phase conjugator consisting of a photorefractive crystal alone [5.22]. Like White's device, this device also generates its own pumping beams, but the self-generated pumping beams are contained completely inside the crystal by internal reflection at the crystal faces, as shown in Fig. 5.5c. In addition, the pumping beams automatically self-align to maximize the phase-conjugate gain. The phase-conjugate reflectivity of the device is as high as 30 % (even with Fresnel reflection losses) [5.23].

In 1983, *Cronin-Golomb* and co-workers [5.24] demonstrated a self-pumped phase conjugator that uses mirrors to deflect the incident beam into a ring that loops back into the photorefractive crystal, as shown in Fig. 5.5d. Because this conjugator has a low nonlinear coupling threshold, it works in crystals and at wavelengths where the optical nonlinearity is small.

All of the above phase conjugators rely on four-wave mixing to produce a phase-conjugate wave, even though all four waves are derived from a single incident wave. Recently *Chang* and *Hellwarth* [5.25] demonstrated a phase conjugator that uses *two*-wave mixing between the incident and phase-conjugate beams. Although the geometry, shown in Fig. 5.5e, resembles that of stimulated

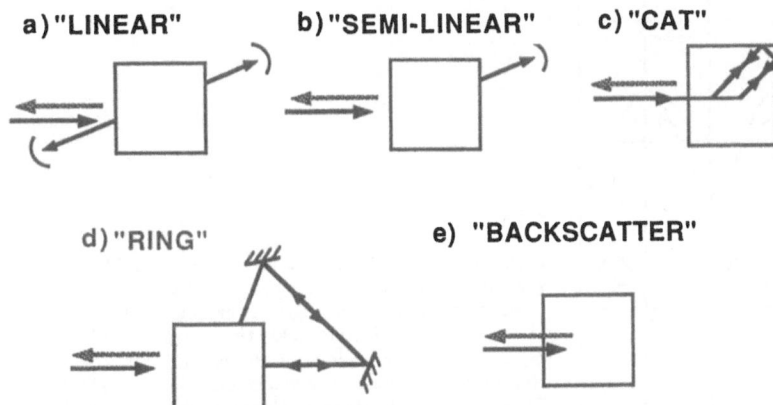

Fig. 5.5a–e. Various self-pumped phase conjugators (see also Sect. 5.5) using a photorefractive crystal. **(a)** The "LINEAR" requires a resonator cavity around the crystal; **(b)** the "SEMILINEAR" requires a single mirror and a crystal; **(c)** the "CAT" requires only the crystal itself; **(d)** the "RING" requires the input beam to be redirected back into the crystal; **(e)** the "BACKSCATTER" creates its own reflection grating inside the crystal

Brillouin scattering [5.26, 27], this conjugator has the decided advantage that a high-intensity input beam is not required.

5.2 The Photorefractive Effect

Detailed descriptions of the photorefractive effect are found in other chapters of this volume. Here we emphasize two important features of the photorefractive effect: (1) the coupling strength is *independent of the intensity* of the optical beams in most practical situations; (2) there is generally a *spatial shift* between the intensity pattern and the resulting refractive-index variation.

Photorefractive materials are noncentrosymmetric photoconductors: they contain charges that are trapped in the dark but become free to move through the crystal in the presence of light. These charges originate from impurities or defect sites in the crystal lattice. The photorefractive effect consists of the sequence of events illustrated in Fig. 5.6: light having a spatially-varying pattern $I(x)$ redistributes charges to form a charge-density pattern $\varrho(x)$. The static space-charge electric field $E_{sc}(x)$ corresponding to the charge density $\varrho(x)$ causes a first-order change $\Delta n(x) \propto E_{sc}(x)$ in the refractive index of the material via the Pockels effect (which exists only in noncentrosymmetric crystals). Note that the refractive-index pattern $\Delta n(x)$ is spatially shifted with respect to the incident light pattern $I(x)$ as a consequence of the derivative in Poisson's equation, as we show below.

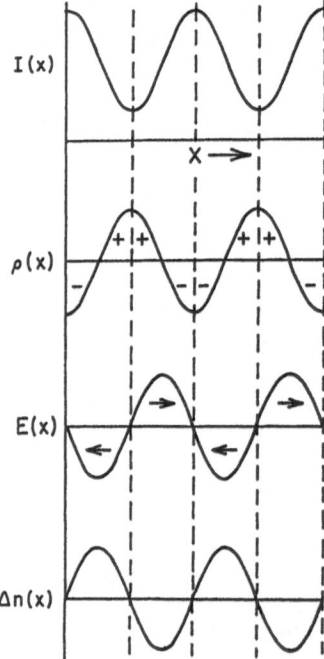

Fig. 5.6. The photorefractive effect. Light with a spatially periodic intensity pattern $I(x)$ causes charges to drift and diffuse through the crystal. In the absence of any uniform electric field, the charges tend to pile up in the darker regions of the crystal. The resulting charge density $\varrho(x)$ creates a strong electric field $E(x)$. This electric field distorts the crystal lattice as shown by the arrows, to create a periodic refractive-index variation $\Delta n(x)$ by the linear electro-optic (Pockels) effect. Note the spatial shift between the intensity pattern $I(x)$ and the refractive-index variation $\Delta n(x)$ [5.3]

Most optical nonlinearities increase as the intensity of the optical beams increases. In contrast, the photorefractive nonlinearity saturates at a large, steady-state value at a very low light intensity (typically less than $0.1\,\mathrm{W/cm^2}$). Since the photorefractive effect is caused by the light-induced displacement of charge inside the material, a weak light displaces charge just as well as a bright light, albeit more slowly. (The photorefractive nonlinearity will decrease if the light intensity is so weak that the photoconductivity is less than or comparable to the dark conductivity.) Some photorefractive materials, when illuminated by very intense $(\mathrm{MW/cm^2})$ optical pulses, exhibit an enhancement [5.28] or a reduction [5.29] of the nonlinear coupling strength.

The *speed* of the photorefractive effect, on the other hand, increases as the intensity of the optical beams increases. While a few hundred milliwatts/cm^2 of light produces a response time on the order of seconds [5.18] in barium titanate, a few megawatts/cm^2 of light produces a sub-nanosecond response [5.29] in the same crystal, an increase in speed of nine orders of magnitude. The speed of the photorefractive effect has been observed to increase linearly or sublinearly with optical intensity, depending on the material, the particular crystal, and the character of the traps and charge carriers participating in the photorefractive effect. (The physical origin of this peculiar intensity-dependence of the speed is not understood at present). For example, in those barium titanate crystals where the majority photorefractive charge carriers are electrons, the speed increases linearly with optical intensity; in hole-dominated samples, on the other hand, the speed increases sublinearly with intensity, I^x, with x varying between 0.6 and 0.9 depending on the sample [5.30, 31]. The speed of the photorefractive effect also depends on the geometry of the beams.

Assume for the moment that there is no uniform electric field (either applied or photovoltaic) in the crystal. The stationary light pattern due to the interference of two optical plane waves $E_1 \exp[\mathrm{i}(k_1 \cdot x - \omega t)]$ and $E_2 \exp[\mathrm{i}(k_2 \cdot x - \omega t)]$ varies periodically in space according to

$$I(x) = I_0[1 + \mathrm{Re}\{m \exp(\mathrm{i}K \cdot x)\}] \ , \tag{5.1}$$

where $m \equiv 2E_1 \cdot E_2^*/I_0$ is the modulation ratio, $I_0 \equiv I_1 + I_2$ is the total intensity, and $K = k_1 - k_2$ is the grating wavevector. Free charges generated in the crystal by absorption of light diffuse from regions of higher intensity into regions of lower intensity, resulting in a steady-state charge-density pattern $\varrho(x)$. This final charge pattern $\varrho(x)$ represents a balance between the diffusion of charges due to the nonuniform light intensity and the tendency of the resulting space-charge field $E_{sc}(x)$ to restore the charge to its initial uniform distribution. The magnitude of the resulting space-charge field $E_{sc}(x)$ can be found from the transport equations [5.32, 18] and Poisson's equation

$$\varrho(x) = \frac{\varepsilon}{4\pi} \nabla \cdot E_{sc}(x) \ . \tag{5.2}$$

For the case of no applied or intrinsic uniform electric field, and if the charge transport length is small compared to the grating spacing $2\pi/|K|$, the space charge field is [5.18]:

$$E_{sc}(\boldsymbol{x}) = \mathrm{Re}\left\{ im\xi(K)\frac{k_\mathrm{B}T}{q}\frac{K}{1+(K/k_0)^2}\exp\left(i\boldsymbol{K}\cdot\boldsymbol{x}\right)\right\} \quad , \tag{5.3}$$

where $k_\mathrm{B}T/q$ is the thermal energy per unit photorefractive charge and $k_0 \equiv (4\pi Nq^2/\varepsilon k_\mathrm{B}T)^{1/2}$ is the space-charge screening wavevector. This screening wavevector depends on the effective density of photorefractive charge carriers N, and on $\varepsilon = (\boldsymbol{K}\cdot\underset{\sim}{\varepsilon}\cdot\boldsymbol{K})/K^2$, which is the dimensionless dc dielectric constant along the direction of the grating wavevector $\boldsymbol{K}$ ($\underset{\sim}{\varepsilon}$ is the relative dc dielectric tensor). The scalar $K = |\boldsymbol{K}|$ is the magnitude of the grating wavevector $\boldsymbol{K}$. The factor $\xi(K)$ takes into account the possibility of competition between electrons and holes as the photorefractive charge carriers ($|\xi(K)| \leq 1$) [5.33, 33a]. $\xi(K)$ is defined as:

$$\xi(K) = (1-C)/(1+C) \quad , \tag{5.4}$$

where

$$C \equiv \frac{s_\mathrm{h}N_0(K^2+K_\mathrm{e}^2)}{s_\mathrm{e}(N_\mathrm{D}-N_0)(K^2+K_\mathrm{h}^2)} \quad . \tag{5.5}$$

In (5.5), s_h and s_e are the hole and electron photoexcitation cross-sections, N_0 is the average density of ionized dopants, N_D is the total density of dopants, and K_h^{-1} and K_e^{-1} are the average transport lengths for holes and electrons.

The space-charge field $\boldsymbol{E}_{sc}(\boldsymbol{x})$ in the crystal generates a refractive-index variation $\Delta n(\boldsymbol{x})$ by the linear electro-optic (Pockels) effect:

$$\Delta n(\boldsymbol{x}) = -r_{\mathrm{eff}}(n^3/2)\boldsymbol{E}_{sc}(\boldsymbol{x}) \quad . \tag{5.6}$$

Note that $\boldsymbol{E}_{sc}(\boldsymbol{x})$ and the refractive-index pattern $\Delta n(\boldsymbol{x})$ are phase-shifted by 90° from the original light pattern given in (5.1) above. This is due to the factor of i in (5.3), which comes from the derivative in (5.2).

In (5.6), the effective Pockels coefficient is given by

$$r_{\mathrm{eff}} = e_1^* \cdot \{\underset{\sim}{\varepsilon}\cdot[\underset{\sim}{R}\cdot(\boldsymbol{K}/|\boldsymbol{K}|)]\cdot\underset{\sim}{\varepsilon}\}\cdot e_2/n^4,$$

where e_1 and e_2 are the polarization vectors of the beams writing the grating and $\underset{\sim}{R}$ is the third-rank tensor of electro-optic coefficients. For crystals with point symmetry 4 mm (barium titanate, for example) and ordinary rays:

$$r_{\mathrm{eff}} = r_{13}\,\sin\left[\frac{\alpha_1+\alpha_2}{2}\right] \quad ; \tag{5.7}$$

and for extraordinary rays:

$$r_{\mathrm{eff}} = \left\{ n_\mathrm{o}^4 r_{13}\cos(\alpha_1)\cos(\alpha_2) + 2n_\mathrm{e}^2 n_\mathrm{o}^2 r_{42}\cos\left[\frac{\alpha_1+\alpha_2}{2}\right]\right.$$
$$\left. + n_\mathrm{e}^4 r_{33}\sin(\alpha_1)\sin(\alpha_2)\right\}\sin\left[\frac{\alpha_1+\alpha_2}{2}\right]\bigg/ n^4 \quad . \tag{5.8}$$

The directions of propagation of beams 1 and 2 make angles α_1 and α_2, respectively, with the positive c-axis of the crystal.

In Sect. 5.4, where we discuss the coupled-wave equations, a convenient parameter will be γ, the coupling strength per unit length. If there is no frequency shift between the writing beams 1 and 2, and if there is no dc electric field E_0 in the crystal, then the coupling strength is independent of the total optical intensity (ignoring dark conductivity), and is given by

$$\gamma_0 \equiv \frac{(2\omega/c)n^3 r_{\text{eff}}}{\cos\left[(\alpha_1 + \alpha_2)/2\right]} \frac{k_B T}{q} \frac{K}{1 + (K/k_0)^2} e_1 \cdot e_2^* \tag{5.9}$$

where c is the speed of light in a vacuum. Note that here the coupling strength is real, which implies a 90° phase shift between the optical intensity fringes and the refractive-index grating.

In the most general case [5.18, 34], there may be an applied (or photovoltaic) uniform electric field E_0 in the crystal, and there may be a frequency shift $\delta\omega = \omega_1 - \omega_2$ between the optical "writing" beams 1 and 2. Now the photorefractive charge no longer accumulates in the intensity troughs of the optical interference pattern, as depicted in Fig. 5.6, but tends to accumulate on the shoulders of the intensity pattern, and the phase shift between the light-intensity pattern and the resulting refractive-index pattern is no longer 90°. The coupling strength γ becomes complex [5.18]:

$$\gamma \equiv \frac{(2\omega/c)n^3 r_{\text{eff}}}{\cos\left[(\alpha_1 + \alpha_2)/2\right]}$$
$$\times \frac{(k_B T/q)K + iE_0}{1 + i[(K/k_0)(E_0/f_0) - \delta\omega \cdot \tau] + (K/k_0)^2} e_1 \cdot e_2^* \quad , \tag{5.10}$$

where $E_0 = E_0 \cdot K/|K|$ is the component of the electric field along the grating direction K, and $f_0 \equiv k_0 k_B T/q$ is a characteristic field ($f_0 = 2300\,\text{V/cm}$, for a Debye screening length of $2\pi/k_0 = 0.7\,\text{micron}$, at room temperature). The response time τ is determined by the total intensity in the crystal.

If there is no electric field E_0, the coupling strength has a somewhat simpler form:

$$\gamma = \gamma_0 \frac{1 + (K/k_0)^2}{1 - i\tau \cdot \delta\omega + (K/k_0)^2} \tag{5.11}$$

which shows that for $E_0 = 0$, the coupling strength decreases monotonically with the frequency shift $\delta\omega$ as the grating is forced to translate in the crystal. In both (5.10) and (5.11), the coupling strength has become intensity dependent due to the intensity-dependent time constant τ in the denominator.

If the spatial phase shift between $I(x)$ and $\Delta n(x)$ is nonzero, there is a steady-state transfer of energy between the two beams that are writing the grating, as described in Sect. 5.3. This energy coupling is important in many applications of photorefractive materials, and is useful for studying the photorefractive effect itself. For example, by observing which of the two optical beams is amplified in relation to the direction of the crystal's +c-axis, it is possible to determine the sign of the dominant charge carrier in the material. Energy coupling can be quite strong in materials with large Pockels coefficients such as barium titanate and strontium barium niobate; one beam can transfer nearly all of its energy to another beam in only a few millimeters of interaction length. Energy coupling causes the ring to appear in ring resonators and contributes to the phase-conjugate beam in self-pumped phase conjugators. This transfer of energy from one beam to another is at the heart of the various stimulated effects discussed below.

5.3 Two-Wave and Four-Wave Mixing in Photorefractive Materials

The nonlinear mixing of light beams consists of two (usually inseparable) parts: the action of the light on the material and, further, the action of the material back on the light. The first part is covered in the theory of the photorefractive effect, for example, in which light redistributes charges and so perturbs the refractive index of the material. The second part is described by Maxwell's equations of electrodynamics, which detail how these variations in the refractive index scatter the light passing through the material. In the steady-state coupled-wave equations, the intermediary role of the material is hidden in the "coupling strength", leaving what appears to be a direct interaction of the light beams, with the coupling strength parametrizing the strength of the interaction. (The coupling strength can itself become dependent on the optical intensities through a photovoltaic field or through the response time of the material.)

In this section we give the steady-state coupled-wave equations for the mixing of optical waves in photorefractive materials. Because the coupled-wave equations are highly nonlinear, analytic solutions are possible in only a few, somewhat idealized, cases. Nevertheless, these solutions have been applied with reasonable success to a variety of experiments.

The coupled-wave equations are derived from Maxwell's equations for the total optical electric field $E \exp(-i\omega t)$:

$$\nabla^2 E + (n\omega/c)^2 E = \frac{4\pi\omega^2}{c^2} \chi' \cdot E \qquad (5.12)$$

where $E = \sum E'_j$, $j = 1, 2, \ldots$. The nonlinear susceptibility tensor $\chi' = -\underset{\sim}{\varepsilon} \cdot (\underset{\sim}{R} \cdot E_{\mathrm{sc}}) \cdot \underset{\sim}{\varepsilon}$ depends on the space-charge field $E_{\mathrm{sc}}(x)$, which is in turn a function of the optical field amplitudes, resulting in nonlinear coupling between the optical fields. In (5.12) n is the index of refraction, c is the speed of light, $\underset{\sim}{\varepsilon}$ is the optical dielectric tensor, and $\underset{\sim}{R}$ is the third-rank tensor of nonlinear (Pockels) coefficients.

If the individual optical fields $E'_j(x)$ are expressed as a product of a plane wave with wavevector k_j and a slowly varying amplitude $E_j(x)$:

$$E'_j(x) = E_j(x) \exp(i k_j \cdot x) \quad , \qquad (5.13)$$

then the coupled-wave equations are derived by substituting (5.13) into (5.12), equating terms having the same wavevector, and neglecting the second space derivatives of the amplitudes $E_j(x)$. The latter "slowly-varying-envelope approximation" [5.35] assumes that the field amplitudes do not change significantly over a distance of a few wavelengths.

In the following, we will consider only plane waves. Extending the coupled-wave equations to account for transverse variation in the incident waves is straightforward; *solving* the resulting equations is not.

5.3.1 Two-Wave Mixing

Consider the mixing of two optical waves in a photorefractive crystal. If the waves are either nearly copropagating or nearly counterpropagating ("nearly" meaning to within about 10 degrees of angle), the coupling causes the waves to vary only along their directions of propagation, so that we can take $E_j(x) = E_j(z)$, and the coupled-wave equations for the beam amplitudes and phases can be solved exactly.

We consider only copropagating waves here. (For counterpropagating waves, see [5.36, 37].) The coupled-wave equations are then [5.35, 38–41]:

$$\frac{dE_1}{dz} = \gamma \frac{E_1 E_2^* E_2}{I_1 + I_2} - \frac{\alpha E_1}{2} \quad ; \quad \frac{dE_2}{dz} = -\gamma^* \frac{E_2 E_1^* E_1}{I_1 + I_2} - \frac{\alpha E_2}{2} \qquad (5.14)$$

where E_j are the slowly varying amplitudes of the optical electric fields; $I_j = |E_j|^2$ are the respective intensities; α is the intensity absorption per unit length, and γ is the (generally complex) coupling strength per unit length (see (5.9–11)). As long as the beams have the same frequency and the crystal's dark conductivity can be neglected, γ is *independent* of the total intensity in the steady state.

If the field amplitudes are written $E_j = \sqrt{I_j} \exp(i\phi_j)$ and the real and imaginary parts collected, (5.14) becomes:

$$\frac{dI_1}{dz} = 2 \operatorname{Re}\{\gamma\} \frac{I_1 I_2}{I_1 + I_2} - \alpha I_1 \quad ; \quad \frac{dI_2}{dz} = -2 \operatorname{Re}\{\gamma\} \frac{I_1 I_2}{I_1 + I_2} - \alpha I_2 \quad (5.15)$$

$$\frac{d\phi_1}{dz} = \operatorname{Im}\{\gamma\} \frac{I_2}{I_1 + I_2} \quad ; \quad \frac{d\phi_2}{dz} = \operatorname{Im}\{\gamma\} \frac{I_1}{I_1 + I_2} \quad . \qquad (5.16)$$

By noting that the total intensity $I_0 = I_1 + I_2$ varies with z only by absorption, (5.15) for the intensities of the individual beams can be uncoupled and solved. Once the z-dependence of the intensities is known, (5.16) for the phases can be integrated. The results are [5.39–41]:

$$I_1(z) = I_{10} e^{-\alpha z} \frac{I_0}{I_{10} + I_{20} \exp[-2 \operatorname{Re}\{\gamma\} z]} \equiv G_1(z) I_{10} e^{-\alpha z} \qquad (5.17)$$

$$I_2(z) = I_{20} e^{-\alpha z} \frac{I_0}{I_{20} + I_{10} \exp[2 \operatorname{Re}\{\gamma\} z]} \equiv G_2(z) I_{20} e^{-\alpha z} \qquad (5.18)$$

$$\phi_1(z) = \frac{1}{2}\frac{\operatorname{Im}\{\gamma\}}{\operatorname{Re}\{\gamma\}}\ln[G_1(z)] \qquad\qquad (5.19)$$

$$\phi_2(z) = -\frac{1}{2}\frac{\operatorname{Im}\{\gamma\}}{\operatorname{Re}\{\gamma\}}\ln[G_2(z)] \qquad\qquad (5.20)$$

where I_{10} and I_{20} are the incident intensities. If $\operatorname{Re}\{\gamma\} > 0$, beam 1 is amplified at the expense of beam 2.

The amplification of a weak probe beam by a strong pump beam bears a resemblance to stimulated scattering: the interaction causes the weak beam to increase in intensity, which strengthens the interaction, which strengthens the weak beam even more, etc. The analogy is especially striking when the weak probe beam arises from stray scattered light (as in "beam fanning"), which we discuss further in Sect. 5.5.

Two-wave mixing is a convenient method for measuring the coupling gain $2\operatorname{Re}\{\gamma\}L$ for a particular beam geometry. By measuring the transmitted intensities of beams 1 and 2 with and without coupling, the gain is obtained generally from

$$2\operatorname{Re}\{\gamma\}L = \ln\left[\frac{I_{20}I_1(L)}{I_{10}I_2(L)}\right] \;, \qquad\qquad (5.21)$$

where L is the interaction length. If I_{10} is made small enough that the depletion of beam 2 is negligible (that is, $I_{10} \ll I_{20}\exp[-2\operatorname{Re}\{\gamma\}L]$), then the amplification of beam 1 is *independent* of the intensity of beam 2: $I_1(L) = I_{10}e^{-\alpha z}\exp[2\operatorname{Re}\{\gamma\}L]$, and (5.21) reduces to $2\operatorname{Re}\{\gamma\}L = \ln[I_1(L)/I_{10}e^{-\alpha L}]$. The gain can then be determined by measuring the intensity of beam 1 transmitted through the crystal with and without beam 2.

There is a simple physical explanation for two-wave mixing when there is no applied or photovoltaic electric field ($E_0 = 0$) in the photorefractive material (Fig. 5.7a). Consider the case where the two optical beams have the *same* frequency, so that there is a spatial phase shift of 90° between the intensity interference pattern and the refractive-index variation, as shown in Sect. 5.2. [This 90° phase shift corresponds to a real coupling strength γ_0 defined in (5.9).] Scattering from the grating introduces another 90° phase shift [5.35]. However, when beam 2 is scattered by the grating into beam 1, the two 90° shifts cancel and the scattered light adds to beam 1 exactly in phase, causing beam 1 to grow with distance. On the other hand, when beam 1 is scattered by the grating into beam 2, the two 90° shifts add. The scattered beam and beam 2 add 180° out of phase, causing beam 2 to be depleted with distance. The net result is a transfer of light from beam 1 to beam 2 (5.17, 18). Note that when there is no frequency shift between the two beams, the nonlinear interaction affects only their *amplitudes*, and *not their phases* [$\phi_1(z) = \phi_2(z) = 0$ in (5.19, 20)].

Now assume that the two waves have *different* frequencies so that $\delta\omega \equiv \omega_1 - \omega_2 \neq 0$. The intensity interference pattern now translates and drags the refractive-index variation along with it. Because the material cannot respond instanta-

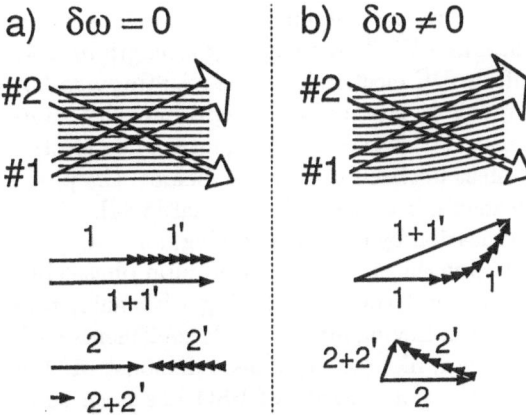

a) $\delta\omega = 0$ b) $\delta\omega \neq 0$

Fig. 5.7a,b. Two-wave mixing in a photorefractive crystal. (a) The writing beams have the same frequency ($\delta\omega = 0$), and create a stationary refractive-index grating in the crystal. In the absence of any uniform electric field, the diffracted beam 1′ adds to the transmitted beam 1 exactly in phase, producing a transmitted beam with an increased amplitude $1 + 1'$ and unchanged phase. The diffracted beam 2′ adds to the transmitted beam 2 exactly 180° out of phase, producing a transmitted beam with decreased amplitude $2 + 2'$ and unchanged phase. (b) The writing beams have a slight frequency difference ($\delta\omega \neq 0$), and the photorefractive grating bends as shown. The two beams now change both their amplitude *and* their phase. Note that the phases of the transmitted beams both change in the same direction, but by different amounts

neously to changes in the intensity pattern ($\tau \neq 0$), the refractive-index variation cannot keep up with the motion of the interference fringes, so that the spatial phase relation between them is no longer 90° [and, accordingly, γ is complex, as seen in (5.11)]. Consider, for example, the phase $\phi_1(z)$ of beam 1 when $\delta\omega > 0$ (the interference pattern moves in the direction of beam 1; see Fig. 5.7b). Since the grating lags, the light scattered from beam 2 into beam 1 is slightly "delayed" with respect to beam 1. Therefore the phase of beam 1 as it exits the crystal lags behind what it would be if $\delta\omega$ were zero; this corresponds to $\phi_1(L) > 0$ in (5.19). At the same time, partly because of the changed phase relationship and partly because of smearing of the charge-density pattern, the amount of power transferred between the beams decreases. Therefore, two-wave mixing with a frequency shift changes both the *amplitudes and phases* of the waves as they interact. Note also that since $\phi_1 - \phi_2 \neq 0$, a frequency shift causes the grating planes to bend.

5.3.2 Example: Obtaining a Large Coupling Strength in BSO

The photorefractive crystal BSO (bismuth silicon oxide: $Bi_{12}SiO_{20}$) has a response time about 40 times faster than barium titanate [5.42], making it a potentially attractive material for photorefractive applications. On the other hand, BSO has the drawback that without an applied electric field its optical nonlinearity is about 100 times smaller than barium titanate due to its small Pockels coefficients. (BSO is also optically active $- \sim 40°$ per mm at $0.5\,\mu m$

[5.43] – which causes a further decrease in the coupling strength unless care is taken with the input beam polarizations.) While the coupling strength of BSO can be increased by applying a uniform DC electric field, the resulting non-90° phase shift between the refractive-index variation and the light intensity interference fringes is not optimum in general. However, by using an applied electric field *and* introducing a frequency offset between the writing beams, the phase shift can be set at the value which maximizes the desired signal [5.44]. (For a particular application – image amplification by two-wave mixing, for example – the optimum phase shift is generally *not* the same as the optimum phase shift for four-wave mixing.) Both four-wave and two-wave mixing with gain, processes normally associated with materials having large Pockels coefficients such as barium titanate and strontium barium niobate, have been demonstrated in BSO [5.45]. Using these techniques, the nonlinearity of BSO has been sufficiently increased to enable its use as a self-pumped phase conjugator [5.46].

5.3.3 Four-Wave Mixing

The coupled-wave equations for four-wave mixing are [5.47]:

$$\frac{dE_1}{dz} = +\gamma\frac{(E_1E_2^* + E_3^*E_4)E_2}{I_0} \quad , \tag{5.22}$$

$$\frac{dE_2^*}{dz} = -\gamma\frac{(E_1E_2^* + E_3^*E_4)E_1^*}{I_0} \quad , \tag{5.23}$$

$$\frac{dE_3^*}{dz} = +\gamma\frac{(E_1E_2^* + E_3^*E_4)E_4^*}{I_0} \quad , \tag{5.24}$$

$$\frac{dE_4}{dz} = -\gamma\frac{(E_1E_2^* + E_3^*E_4)E_3}{I_0} \quad , \tag{5.25}$$

where now $I_0 = I_1 + I_2 + I_3 + I_4$.

The form (5.22–25) of the coupled-wave equations assumes that absorption is negligible and that one photorefractive grating predominates out of as many as four. In general the latter can be a reasonably good approximation because of the strong dependence of the coupling strength on the orientation of the grating wavevector. If the reading beam is incoherent with the two writing beams, then the assumption is exact.) Note that the four-wave mixing equations "contain" the two-wave mixing equations [for example, the first terms on the right-hand sides of (5.22, 23) appear in (5.14)].

Equations (5.22–25) were first solved by making the linearizing assumption that the probe beam 2 is so much weaker than the other beams that the intensity of the pumping beams are unchanged by the interaction [5.47] (the "undepleted-pumps approximation"). *Cronin-Golomb* et al. [5.48] uncoupled (5.22–25) by noting that these equations conserve the following quantities:

$$d_1 = I_1 + I_2 \quad ,$$
$$d_2 = I_3 + I_4 \quad ,$$
$$c = E_1 E_3 + E_2 E_4 \quad . \tag{5.26}$$

They obtained:

$$\frac{E_1(z)}{E_3^*(z)} = -\frac{(\Delta - r)D\exp(\mu z) - (\Delta + r)\exp(-\mu z)}{2c^*[D\exp(\mu z) - \exp(\mu z)]} \quad , \tag{5.27}$$

$$\frac{E_4(z)}{E_2^*(z)} = \frac{(\Delta - r)E\exp(\mu z) - (\Delta + r)\exp(-\mu z)}{2c^*[E\exp(\mu z) - \exp(\mu z)]} \quad , \tag{5.28}$$

where $\Delta = d_2 - d_1$, $r = (\Delta^2 + 4|c|^2)^{1/2}$, $\mu = \gamma r/2I_0$, and D and E are integration constants. This solution has proved useful in analyzing self-pumped phase conjugation, where pump depletion cannot be neglected.

Both two-wave and four-wave mixing play prominent roles in the devices and applications described in the following sections.

5.4 Ring Resonators

In this section we discuss examples of ring resonator cavities containing a photorefractive crystal. These resonators have unique properties which are potentially useful for ring laser gyroscopes, ring lasers, position sensors, and holographic memories.

In the unidirectional ring resonator, the photorefractive crystal and one pump beam provide gain for the resonator beam by two-wave mixing. We consider the unidirectional ring resonator in some detail because the theoretical equations are exactly solvable and experimentally well-supported.

In the bidirectional ring resonator, the two counterpropagating waves in a ring laser interact by two-wave mixing in the photorefractive material.

We also describe an example of ring resonator which uses four-wave mixing in a photorefractive crystal to sustain the resonator beam.

5.4.1 Unidirectional Ring Resonator

The unidirectional ring resonator, shown in Fig. 5.8, consists of a photorefractive crystal placed in a ring cavity and pumped by one external beam [5.20]. The orientation of the crystal in the cavity and the angle of incidence and polarization of the pumping beam are chosen so that light circulating in the cavity in one direction experiences two-wave-mixing gain. If this gain is large enough to support oscillation, the resonating beam builds up from the amplification of stray light.

The unidirectional ring resonator has two interesting properties: (1) the oscillating beam generally has a slightly different frequency (on the order of a few Hz) than the pumping beam [5.49]; (2) oscillation occurs almost regardless

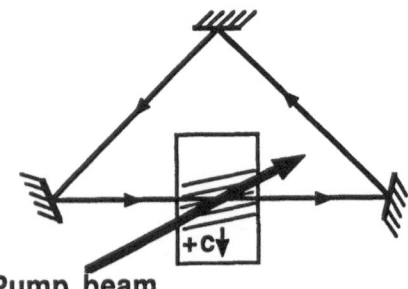

Fig. 5.8. Unidirectional ring resonator. A single pumping wave in a photorefractive crystal creates gain by two-wave mixing for a beam that travels around the ring resonator in one direction

Pump beam

of the optical path length of the cavity, even when only one spatial mode is allowed to oscillate. The latter observation appears to violate the usual resonator condition that oscillation can only take place if the beam's round-trip optical path length through the resonator equals an integral number of wavelengths.

The above two properties are related, as we describe in more detail below. We show that for (almost) any resonator length, the two-wave mixing interaction, through the spontaneous appearance of a $\sim 1\,\mathrm{Hz}$ frequency shift between the resonator beam and the pumping beam, supplies the correct amount of phase to satisfy the resonator condition [5.50, 51]. (Note that the corresponding change in the *wavelength* of the resonator beam − a few parts in 10^{15} − does not substantially contribute to satisfying the resonator condition.) In the following, we assume that the coupling strength is small enough so that the two-wave-mixing interaction does not change the phase of the resonator beam by more than half a wave, although in principle a larger phase change is possible [see (5.32)].

Consider a ring resonator containing a photorefractive crystal that is externally pumped by one beam, as shown in Fig. 5.8. Let M be the total reflective and absorptive loss per round trip in the cavity, and $\delta\phi$ be the phase detuning of the resonator in the absence of the two-wave mixing interaction (i.e., the round-trip optical path length of the resonator beam differs from a whole number of wavelengths by $\lambda \cdot \delta\phi/2\pi$). The resonator condition requires that the intensity and phase (modulo 2π) of the oscillator beam repeat after one round trip in the cavity:

$$MG_1(L) = 1 \tag{5.29}$$

$$\phi_1(L) - \phi_1(0) = -\delta\phi \tag{5.30}$$

where $G_1(z)$ and $\phi_1(z)$ are the two-wave-mixing gain (5.17) and phase shift (5.19) for beam 1, respectively, and L is the two-wave-mixing interaction length. Combining (5.29), (5.17), and (5.30), and using $\mathrm{Im}\{\gamma\}/\mathrm{Re}\{\gamma\} = \omega\tau/[1 + (K/k_0)^2]$ [from (5.11)], gives a linear relation between the frequency shift of the resonator beam and the cavity detuning:

$$\delta\omega = 2\delta\phi[1 + (K/k_0)^2]/[\tau \ln(M)] \quad . \tag{5.31}$$

Experimental results [5.52a] confirming that the resonator frequency shift $\delta\omega$ varies linearly with cavity detuning are shown in the top of Fig. 5.9. Furthermore, if the cavity detuning is too large, the two-wave-mixing gain is insufficient to overcome the cavity losses, and the resonator beam is extinguished (bottom of Fig. 5.9).

The physical situation can be summarized as follows: When the phase detuning of the resonator $\delta\phi$ is zero, the frequency shift $\delta\omega$ is also zero and the cavity intensity I_{10} is high enough to satisfy (5.29) by saturating the gain $G_1(L)$ (assuming, of course, that the coupling strength $\gamma_0 L > -\ln(M)/2$, so that oscillation is possible). On the other hand, when the detuning $\delta\phi$ is nonzero, the circulating beam has the frequency shift $\delta\omega$ (5.17) which causes the two-wave mixing interaction to change the beam's round-trip phase by $-\delta\phi$ and thereby satisfy the resonator condition. At the same time, the unsaturated gain decreases, causing the circulating intensity to decrease. At the limit of allowable detuning, the circulating intensity has dropped to zero.

Using photorefractive barium titanate, *Ewbank* and *Yeh* [5.52a] verified this theory in detail, including the dependence of the maximum frequency shift

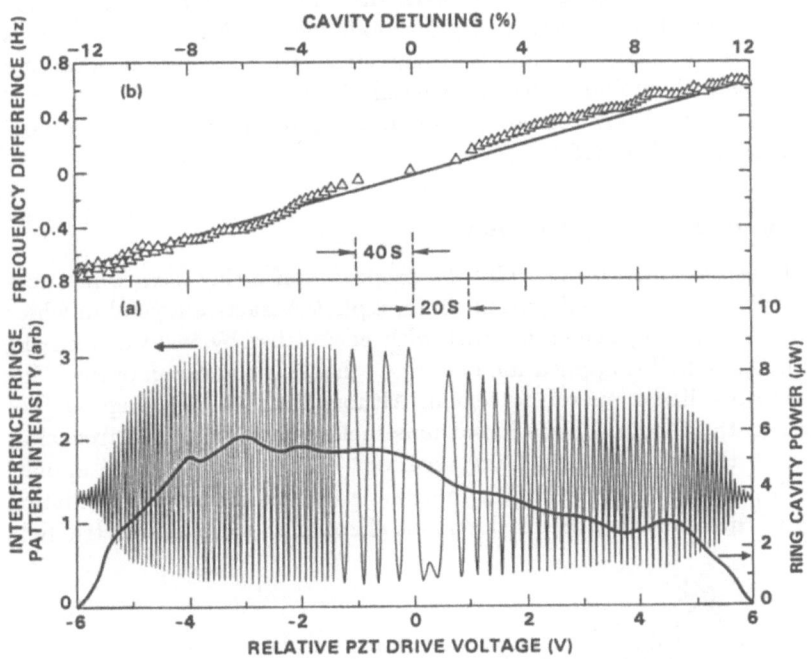

Fig. 5.9. Frequency of the unidirectional ring resonator beam vs. the length of the ring resonator cavity. *Top:* Measured instantaneous frequency shift between the pumping beam and the resonator beam as the resonator length is changed. *Bottom:* Beat note between the resonator and pumping beams as the resonator length is changed. The solid line is the power of the optical beam in the ring cavity. Note that if the cavity is tuned too far from resonance, the power in the ring resonator drops to zero [5.52a]

and cavity detuning on the incident pump beam power and nonlinear coupling strength. They proposed that the unidirectional ring resonator may be a model for understanding the frequency shifts that occur in self-pumped phase conjugation. We will consider this issue below in Sect. 5.6.

Rajbenbach et al. [5.46] obtained oscillation of a unidirectional ring resonator using photorefractive BSO with an applied electric field. As described in Sect. 5.3.2, in order to obtain a large two-wave mixing gain in BSO it is necessary to simultaneously apply an electric field and to use writing beams of different frequencies. The analysis of the frequency shifts observed in this resonator is the same as above, except that the applied field must be accounted for by using the form (5.10) of the coupling strength. (Note that the effects of a frequency shift scale inversely with the response time of the medium; therefore, since the response time per unit intensity for BSO is ~ 40 times that of barium titanate, typical frequency shifts observed in experiments with BSO are tens of Hz rather than Hz.)

A more detailed analysis of the unidirectional ring oscillator having photorefractive gain is found in *Anderson* and *Saxena* [5.52b], in which coupling between various transverse modes is considered. They show that, in the weak coupling limit, the gratings formed between the various modes themselves are quite weak, so that there is no tendency for mode-locking. However, the modes do compete for the available gain, with the strongest competition occurring between modes with similar transverse spatial distributions.

An associative memory based on a photorefractive unidirectional ring oscillator is described in Sect. 5.8.7.

5.4.2 Bidirectional Ring Resonator

Yeh [5.53] has analyzed theoretically the properties of a ring laser which contains, in addition to the usual gain medium, a photorefractive crystal in which the counterpropagating beams interact with each other by two-wave mixing (Fig. 5.10). One possible application of such a scheme is to bias an optical ring gyroscope into a linear operating region. Although cavity detuning was not considered in the theory, it was found nonetheless that there is a frequency shift between the counterpropagating oscillator beams. If there is no applied electric field, the frequency split is symmetric about the "natural" oscillation frequency of the resonator; the predominant effect of applying an electric field

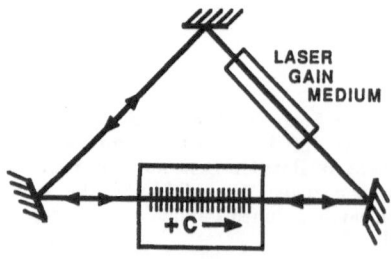

LASER
GAIN
MEDIUM

Fig. 5.10. Bidirectional ring resonator. A conventional laser gain medium provides gain for beams traveling in either direction around the ring resonator. These two beams interact with each other via two-wave mixing in a photorefractive crystal

to the crystal is that one of the beams is shifted from the natural frequency more than the other.

To our knowledge, this active resonator has not been realized experimentally. Taking the resonator condition into consideration as in Sect. 5.4.1 above suggests that both beams cannot oscillate simultaneously: although a frequency shift between the beams changes the effective number of wavelengths in the cavity, the change is *different* for each of the beams [5.53]. Therefore, the only situation for which oscillation of both beams might occur is when the cavity is *not* detuned and the beams have the same frequency. However, the above work [5.53] concluded that this latter situation is unstable to a frequency split which would drive the beams out of resonance with the cavity. The likely experimental result is that oscillation will occur only when the cavity is in resonance and then in only one direction, since the two-wave mixing in the crystal makes the loss for one oscillation direction larger than for the other. In this way a photorefractive crystal can serve as a unidirectional transmission device [5.54, 54a].

5.4.3 Four-Wave-Mixing Ring Resonator

Consider a photorefractive crystal in a ring cavity that is pumped by two counterpropagating beams (Fig. 5.11): counterpropagating oscillating waves are generated in the cavity if the coupling strength is large enough. This four-wave mixing ring resonator, first demonstrated by *White* et al. [5.20], is of interest because the two oscillation beams are a phase-conjugate pair. In addition, the oscillating beams should be expected to have frequency shifts relative to the pumping beams and which should depend on the cavity length, the relative intensities of the pump beams, cavity losses, etc. *Yeh* [5.55] showed theoretically that in the undepleted pump limit there is a minimum round trip gain required for oscillation which depends strongly on the cavity length detuning. To our knowledge, this resonator has not been investigated further either theoretically or experimentally (in a photorefractive material). (Oscillation of such a resonator using sodium vapor as a nonlinear medium has been demonstrated [5.56]).

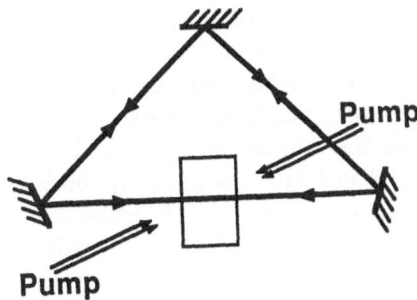

Pump

Pump

Fig. 5.11. Four-wave mixing resonator. Two pumping waves in a photorefractive crystal create a phase-conjugate pair of resonator beams

5.5 Self-Pumped Phase Conjugation

The phase-conjugate replica of a complicated, image-bearing beam can be generated by four-wave mixing [5.10] in a photorefractive material. The quality of the phase-conjugate beam is critically dependent on the alignment, intensities, and phase profiles of the pump waves; any "defect" in the pump waves degrades the fidelity of the phase-conjugate wave. In general, in order to produce an accurate phase-conjugate replica of a complicated wave, it is necessary to generate pump waves that are every bit as complicated as the image wave itself and, moreover, might depend on the *particular* input wave (which would certainly limit the usefulness of ordinary four-wave mixing for making phase-conjugate waves!). Moreover, to our knowledge, the problem of which pump waves are "best" for conjugating a particular complicated input wave is unsolved except in special cases (the weak-probe or weak-coupling limits).

Fortunately, in "self-pumped" phase conjugators the pump waves are generated from the incident beam itself. By a process that is not entirely understood, the self-generated pump waves form in a configuration that tends to optimize the phase-conjugate wave. Self-pumped phase conjugators have been demonstrated in photorefractive crystals using a variety of geometries. The experimenter need only send a single wave into the crystal and wait for the phase-conjugate wave to appear. (The waiting time ranges from nanoseconds to minutes, depending on the intensity and wavelength of the incident beam.) Nearly all of these devices self-start through the amplification of stray light. As we discuss further below, the fidelity of phase conjugation and ease of alignment vary considerably from one self-pumped phase conjugator to another.

A large optical gain is a prerequisite for self-pumped phase conjugation. *Feinberg* and *Hellwarth* [5.19, 19a] demonstrated a continuous-wave phase conjugator with gain (~ 100) using a barium titanate crystal. They also showed that an oscillation beam would spring up between an external mirror and the barium titanate crystal, and thus made the first phase-conjugating resonator. By closing an aperture on the beam waist of a distorted mode of this resonator, the mode was cleaned up without significantly decreasing the intracavity power. The crystal was pumped by two externally-supplied pumping waves.

White et al. [5.20] demonstrated a number of resonators in which mirrors in various configurations give rise to oscillating beams through "self-induced" refractive-index gratings in a crystal of barium titanate. Among these were the unidirectional ring resonator (Sect. 5.4.1) and the "linear mirror", which was the first self-pumped phase conjugator. The latter (shown in Fig. 5.5) consists of a crystal of barium titanate placed in a linear resonator cavity. A probe beam incident at the proper angle on the crystal leads to the build-up of a pair of counterpropagating pumping waves in the resonator cavity, thereby generating the phase-conjugate replica of the probe beam. Subsequently, *Cronin-Golomb* et al. [5.21] showed that phase conjugation still takes place when one of the mirrors in the linear mirror resonator cavity is removed (the "semi-linear mirror").

The same group [5.24] demonstrated a self-pumped phase conjugator having a ring configuration. In this device, mirrors direct the transmitted probe beam back into the crystal. This phase conjugator has a low coupling-strength threshold of $\gamma L = 1$, and it is the only configuration to date for which self-pumped phase conjugation in barium titanate has been achieved in the near infrared [5.57, 58]. However, the response time of barium titanate is very long when using optical beams in the near infrared: the rise time (per unit incident intensity) of the output signal from the ring mirror at the 1090 nm argon-ion line is ~8 minutes per watt/cm².

As practical phase conjugators, the devices described above share the common disadvantage of relying on external mirrors. Three self-pumped phase conjugators have been demonstrated that do not need external mirrors.

Feinberg [5.22, 59] demonstrated a self-pumped phase conjugator in which all of the interacting beams are contained within a barium titanate crystal by total internal reflection from the crystal faces themselves, as shown in Fig. 5.12.

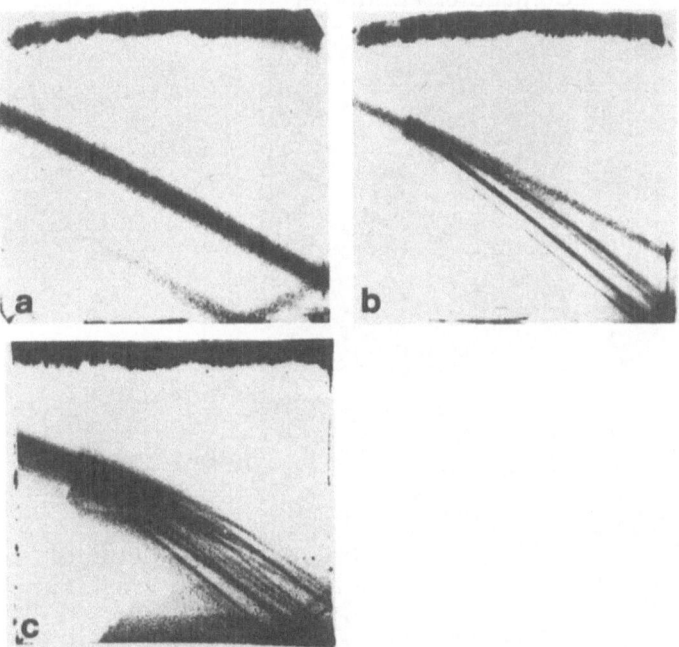

Fig. 5.12a–c. Three photomicrographs of barium titanate crystal acting as a "CAT" self-pumped phase conjugator. The pumping waves are self-generated from the single incident beam, which enters near the top of the crystal's left face. The dark horizontal line across the top of the crystal is light scattered from a crystal face that was damaged during crystal poling. The c-axis of the crystal is directed from top to bottom. (a) The incident beam is an ordinary ray, and the stimulated gain is below threshold. No pumping waves are formed and no phase-conjugate beam is generated. (b) The incident beam is an extraordinary ray, and the stimulated gain is above threshold. Loops of light containing the pumping waves are seen between the incident beam and the lower right-hand corner of the crystal. (c) The incident beam contains the image of the cat shown in Fig. 5.13, and many loops have formed [5.22]

Because the self-generated pump waves have the freedom to choose the spatial pattern and orientation that optimize the phase-conjugate signal, the quality of the reproduction is excellent, as shown in Fig. 5.13. A phase-conjugate reflectivity exceeding 30 % has been obtained with the "CAT" conjugator in spite of Fresnel reflection losses [5.23] and a phase-conjugate reflectivity of 60 % has been reported using a 45° cut crystal of barium titanate [5.23a].

One outstanding question about the CAT conjugator is the mode structure of the self-generated pump waves. An interesting clue is provided by an experiment in which the pump waves self-generated during phase conjugation of one image in a CAT conjugator were used as the pump waves to phase conjugate a different image by ordinary four-wave mixing [5.60]. Figure 5.14 shows that the phase conjugates of both images, a resolution chart and a cat, were formed with high fidelity: there was no crosstalk between the two images. This result implies that the self-generated waves in the crystal form a nearly perfect

Conjugator   Conjugator + Distorter

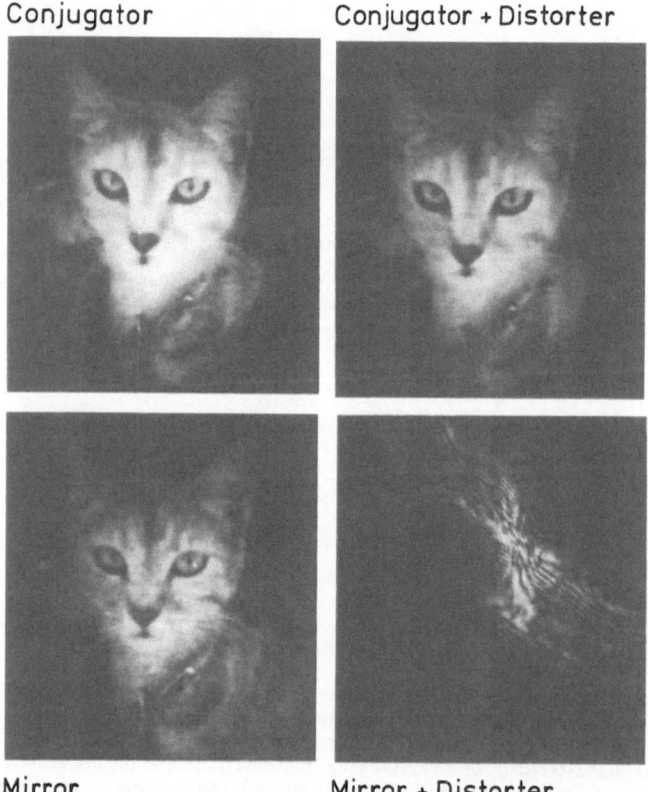

Mirror   Mirror + Distorter

Fig. 5.13. Phase-conjugate images using the "CAT" self-pumped phase conjugator shown in Figs. 5.5 and 5.12. Top row: The phase conjugator can restore an image that has passed through a distorter (clear glue smeared on a glass slide). Bottom row: An ordinary mirror cannot correct for the distorter [5.22]

phase-conjugate pair of pump waves. Apparently the pump waves are not specific to the particular images (resolution chart or cat), but could form the phase conjugate of *any* image.

A second phase conjugator that needs no external mirrors was demonstrated by *Chang* et al. [5.25]. In this device, phase conjugation takes place by "stimulated backscattering" in barium titanate due to two-wave mixing between the counterpropagating incident and phase-conjugate waves, as shown in Fig. 5.15. This geometry resembles phase conjugation by stimulated Brillouin scattering (SBS). In SBS, if the plane-wave gain is large enough ($\sim e^{20}$), the backscattered mode that is the phase conjugate of a complicated input wave has an (exponential) gain coefficient that is approximately twice as large as any other mode, so that the phase-conjugate beam dominates [5.26, 61]. Undoubtedly, a similar competition between modes occurs during self-pumped phase conjugation in photorefractive materials [5.27]. However, because of the different physical properties of Kerr and photorefractive materials, the properties of these phase conjugators are quite different. For example, stimulated backscattering in a Kerr medium such as liquid carbon disulfide (CS_2) requires an incident laser intensity of at least $\sim 10^6$ watts/cm^2. In barium titanate, on the other hand, the intensity need only be large enough to overcome the dark conductivity of the material, making the effective intensity "threshold" about 10^{-1} watts/cm^2 [5.31] which is seven orders of magnitude smaller than the typical stimulated Brillouin threshold.

Double-beam phase conjugators were recently demonstrated by *Sternklar* et al. [5.62], *Smout* and *Eason* [5.63], and *Ewbank* [5.63a]. These devices require that two separate beams be incident on a photorefractive crystal; the phase-conjugate replica of each of the beams emerges. The two input beams need not be mutually coherent, although their wavelengths must be within a few tens of nanometers of each other.

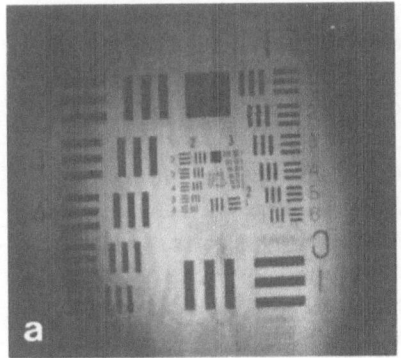

Fig. 5.14. Simultaneous phase conjugation of two different images. A beam that has passed through a resolution chart and its phase-conjugate return beam are used as the pumping waves to phase-conjugate a different image (a wistful cat) by traditional four-wave mixing. Note the lack of any crosstalk between the two images [5.60]

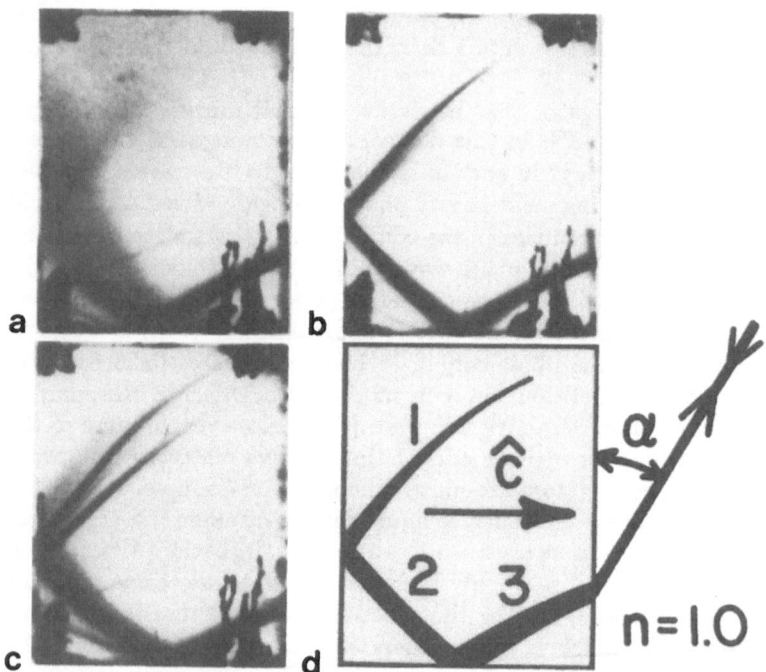

Fig. 5.15a–c. Phase conjugation by two-wave stimulated backscattering. (a) The beam enters the photorefractive crystal of barium titanate near the bottom of the right face, bends, reflects twice, and (b) eventually finds a corner of the crystal. (c) A loop sometimes forms. The corner reflection probably helps "seed" the backscattered beam [5.25]

Ewbank et al. [5.64] performed a qualitative test of a number of different phase conjugators by visually comparing the quality of their phase-conjugate images using a weakly converging incident beam. As seen in Fig. 5.16, the "linear mirror" produced multiple interfering images due to multiple spatial modes between the resonator mirrors. The "ring" mirror was unable to reproduce the beam divergence in the vertical direction due to walk-off of the incident beam after traversing the ring. The CAT conjugator and the "semi-linear mirror" (not shown in Fig. 5.16) produced images of comparable quality, although only the CAT mirror would self-start.

With the exception of the linear resonator configuration, all of the self-pumped phase-conjugators described above share the desirable feature of being relatively insensitive to vibration. Because these devices generate their own pump waves from the incident beam itself, any phase shift imparted to the incident beam (for example through vibration of a steering mirror) is imparted to all of the beams, so that the relative phase shift between any two of the beams is zero. In addition, the coherence length of the incident beam can be fairly short ($\sim 1\,\mathrm{mm}$) [5.65], so that the incident laser need not be especially narrowband. These properties of self-pumped phase conjugators have made them ideal for a host of novel device applications, as will be described in Sect. 5.8.

174

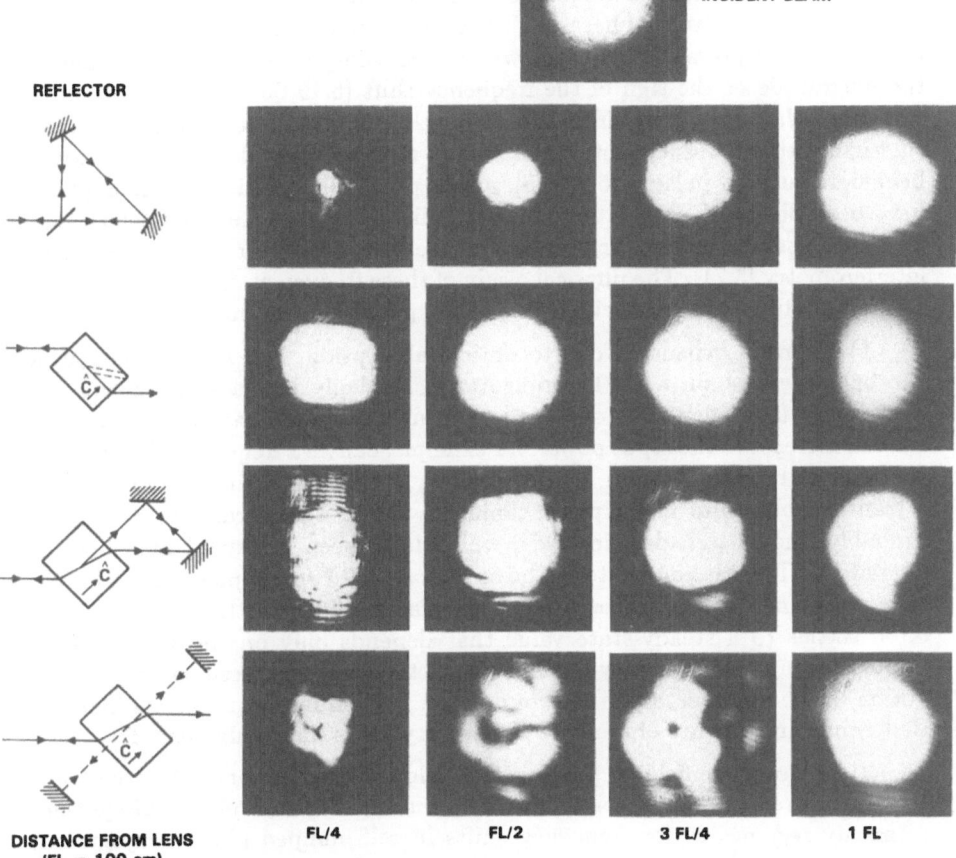

Fig. 5.16. Comparison of the image quality from the output of a Sagnac interferometer and from three different self-pumped phase conjugators. The incident Gaussian beam is focused by a lens into each device which is located at the distance shown at the bottom of the figure. The return beam passes back through the lens and is photographed. Note that the diameter of the return beam from the Sagnac interferometer (*top*) varies with distance, while the diameter of return beam from each of the phase conjugators is the same for any distance between the focusing lens and the conjugator, thereby illustrating the difference between a retroreflector and a true phase conjugator. The CAT conjugator (*next to top*) shows little distortion for any distance between it and the focusing lens. The distortion seen with the "RING" conjugator (*next to bottom*) is caused by poor beam overlap between the incident beam and the beam that has traversed the ring, and can be minimized by a judicious choice of focusing lens and ring length. The distortion evident with the "LINEAR" phase conjugator (bottom) is caused by multimode oscillations between the two mirrors; it can be minimized by tightly focusing the incident beam into the crystal as in the far right photograph, so that only one transverse mode of the linear resonator can oscillate [5.64]

5.6 Frequency Shifts in Self-Pumped Phase Conjugation

Most of the self-pumped conjugators described in the previous section produce a phase-conjugate wave that has a slightly different frequency than the input wave (by $\sim 1\,\text{Hz}$). At present it is not known with certainty which parameters control the magnitude or the sign of the frequency shift [5.49, 66, 67]. Observation of *time-dependent* frequency shifts that regularly change sign [5.66] indicates that the sign of the frequency shift is not simply random. Complex time-dependent behaviors such as pulsing [5.66–68], and a possible transition to chaos [5.69] have been observed. Depending on the application, these frequency shifts can be either useful, such as for scanning a dye laser [5.70], or a nuisance, as in interferometry [5.71]. The physical origin of these frequency shifts is the subject of some controversy; we discuss some of the possible mechanisms below.

1) Thermal expansion due to optical absorption. When the intensity of the light incident on a CAT conjugator is suddenly increased or decreased, a transient frequency shift of the phase-conjugate beam can be observed. A sudden increase in the input power, for example, causes a frequency shift toward the red end of the spectrum, and, conversely, a decrease in intensity results in a transient blue shift in the phase-conjugate beam. These transient effects are probably due to pyroelectric fields, which have been measured in a barium titanate CAT phase conjugator to be as high as $200\,\text{V/cm}$ (open circuit) [5.68]. However, once the conjugator reaches thermal equilibrium, the frequency shift often settles to a steady-state value that depends only on the steady input power (through the intensity-dependence of the material's response time) and not on the thermal history of the conjugator. Preliminary experiments indicate that similar results are obtained when the crystal is uniformly heated.

2) Photovoltaic field. A photovoltaic field of $\sim 10\,\text{V/cm}$ at $\sim 1\,\text{W/cm}^2$ has been measured in a CAT conjugator by *Gower* [5.68]. If a photovoltage were primarily responsible for frequency shifts in self-pumped phase conjugation, the frequency shifts would have only one sign. (See, for example, the scheme described in Sect. 5.3.2 to obtain large coupling in BSO [5.44, 45, 72], where the electric field was externally applied). In fact, not only do frequency shifts of both signs occur during self-pumped phase conjugation, but time-varying shifts that alternate sign have also been observed [5.66]. Although photovoltaic fields may play a crucial role in producing frequency shifts, a piece of the puzzle is clearly missing.

Chang et al. [5.25] did not observe a frequency shift in the output of a self-pumped phase conjugator in barium titanate which works by "stimulated backscattering". In such a scheme, the backscattered field tries to maximize its own amplification (which is why the phase-conjugate wave is produced – see Sect. 5.5). Therefore, in analogy with the experiments in BSO [5.44, 45, 72], the observation of a frequency shift would have indicated the presence of a uniform electric field in the crystal. However, it should be noted that the internal electric field needed to produce a given frequency shift for gratings formed by *counterpropagating* beams is an order of magnitude larger than for gratings formed by nearly *copropagating* beams, so a backscattering experiment is not a particularly sensitive indicator of a uniform internal electric field.

3) Resonator condition. If the self-generated pumping beams form a *closed* loop inside the CAT conjugator, then the circulating light must satisfy a resonator condition. If the circulating light is amplified by two-wave mixing, the nonlinear interaction can allow an otherwise nonresonant wavelength to satisfy a resonator condition (Sect. 5.4.1). It has been proposed [5.52a] that the frequency shifts seen in self-pumped phase conjugation in barium titanate result from the crystal's attempt to satisfy a resonator condition. There is no doubt that a resonator condition plays a role in determining the frequency shift in phase conjugators in which the resonator is well-defined by ordinary mirrors, such as in the "linear mirror" phase conjugator [5.52a, 73]. However, in the unidirectional ring resonator, unless the beam path is severely restricted, many transverse modes oscillate, each having a different frequency shift [5.49, 73]. Similarly, in both the "semilinear mirror" and the CAT phase conjugators, the length of the resonator cavity is not well defined. Finally, if a closed ring resonator were formed by the crystal faces in the CAT conjugator, then heating the whole crystal would change the length of the resonator and cause a steady-state change in the frequency shift. However, as noted in (1) above, this has not been observed.

4) Enhancement of four-wave-mixing efficiency. *MacDonald* and *Feinberg* have suggested [5.74] that the photorefractive gratings in self-pumped conjugators translate (and therefore shift the frequency of the diffracted beams) in order to increase their four-wave-mixing efficiency. The enhancement is a fundamental property of the four-wave mixing process and occurs under fairly general conditions when the nonlinear interaction is strong, as is certainly the case in self-pumped conjugators. To see how a frequency shift can increase the four-wave mixing efficiency, assume first that the two beams writing the photorefractive grating have the same frequency. The two-wave mixing interaction causes a net transfer of energy from one beam to the other (Sect. 5.3). In materials such as barium titanate, this interaction can be so strong that the energy transfer is nearly total in a distance that is short compared to the width of the amplified beam. The ability of a grating to scatter a third (reading) beam to form a phase-conjugate wave by four-wave mixing (the grating's "diffraction efficiency") depends, at each point, on the *relative* intensities of the writing beams (through the modulation of their interference fringes (Sect. 5.3)). The rapid depletion with distance of one of the writing beams by two-wave mixing results in a small average fringe modulation and a correspondingly small diffraction efficiency.

Introducing a small frequency shift into one of the writing beams changes the phase relationship between the interference fringes and the refractive-index variation (5.11) and reduces two-beam coupling. By thus reducing the depletion of the writing beam, the average modulation of the fringes, and hence the diffraction efficiency of the grating, *increases*. (Note, however, that too large a frequency shift simply washes out the grating.)

A plane-wave calculation of the diffraction efficiency as a function of frequency shift shows that under general conditions, the maximum four-wave diffraction efficiency occurs at two nonzero values of the frequency shift symmetrically displaced from zero frequency shift, as seen in Fig. 5.17. The enhancement occurs whenever the average amplitude of the grating (with no frequency shift) is small due to depletion of the probe beam by energy coupling.

The experimental dependence of the four-wave-mixing efficiency on the frequency shift between two (Gaussian) writing beams (Fig. 5.18) is in qualitative agreement with the theory: two maxima of the diffraction efficiency occur at nonzero values of the frequency shift.

It has been noted [5.73] that the plane-wave model of the CAT conjugator as presented in [5.59] does not allow frequency shifts. We expect that a less restrictive and more realistic model which, for example, includes reflection gratings, will allow frequency shifts through the mechanism described above, and may explain some of the more complex behavior of the CAT conjugator [5.66—69].

5) *Fischer* et al. [5.76] showed theoretically using the plane-wave approximation that any phase nonreciprocity in the "ring" of the ring self-pumped phase conjugator causes a frequency shift between the incident and phase-conjugate beams. Experimentally, it has been shown that a nonreciprocal phase shift deliberately introduced into the ring (using a Faraday rotator) *does* induce a frequency shift [5.76, 77]. Additionally, if the nonreciprocity is large enough, the device no longer generates the phase-conjugate replica of the input beam, but instead generates a new mode that compensates for the nonreciprocity [5.77]. Nonetheless, even without an externally impressed nonreciprocal phase shift, frequency shifts of both signs are observed in a ring phase conjugator,

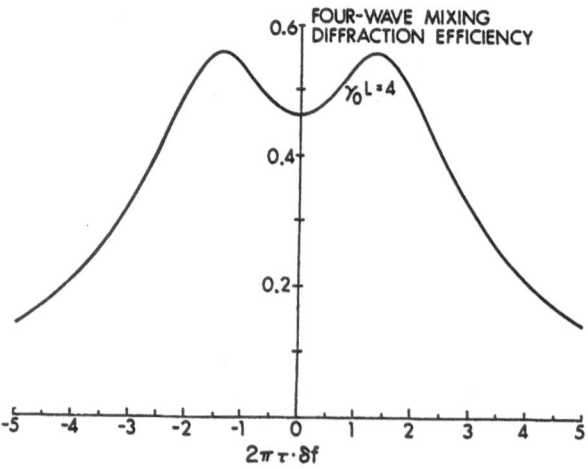

Fig. 5.17.
Predicted enhancement of the four-wave mixing efficiency with a moving grating in a photorefractive crystal. The frequency shift δf between the two writing beams causes the grating to move in the crystal, which has a response time τ. Note the two pronounced peaks for nonzero frequency shifts [5.74]

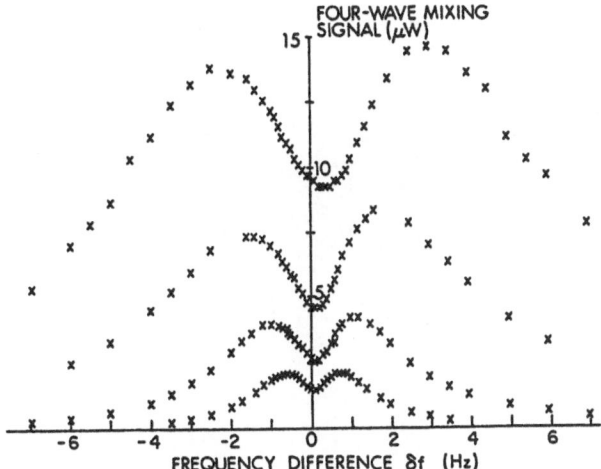

FOUR-WAVE MIXING SIGNAL (μW)

FREQUENCY DIFFERENCE δf (Hz)

Fig. 5.18. Measured four-wave mixing signal in a barium titanate crystal vs the frequency difference between the two writing beams. Note the enhanced efficiency for a nonzero frequency shift. The four data sets correspond to different optical intensities [5.74]

depending on the precise alignment of the beams in the crystal [5.75, 76, 77]. This seems to rule out a photovoltaic field as a primary cause of the frequency shift, although frequency shifts caused by applied electric fields have also been demonstrated [5.75]. Also, frequency shifts are still observed even when reflection gratings have been eliminated (by reducing to coherence length of the laser to be less than the length of the ring).

5.7 Stimulated Scattering That Does Not Produce a Phase-Conjugate Wave

Our discussion of self-pumped phase conjugators in photorefractive crystals might leave the impression that self-generated beams always conspire to produce a phase-conjugate wave. For completeness we illustrate a few examples in which a single beam incident on a photorefractive crystal generates not a phase-conjugate beam, but beams in the shape of broad fans, rings, crosses, or even "dancing" Gaussian modes.

5.7.1 Dancing Modes

If the ring phase conjugator shown in Fig. 5.5d is altered by inserting a nonreciprocal phase element in the ring (such as Faraday glass in a strong magnetic field), then the device no longer generates the phase-conjugate replica of the incident wave. For small nonreciprocity $\Delta\phi$, the output wave is simply frequency shifted a few Hz due to translation of the hologram inside the photorefractive crystal [5.75]. However, when $\Delta\phi \sim \frac{\pi}{2}$, the spatial mode abruptly changes to a new one, as shown in Fig. 5.19. The modes resemble Gaussian resonator modes, even though there is no closed resonator. *Jiang* and *Feinberg* [5.77] showed that this new output mode enhances its gain over the phase-conjugate

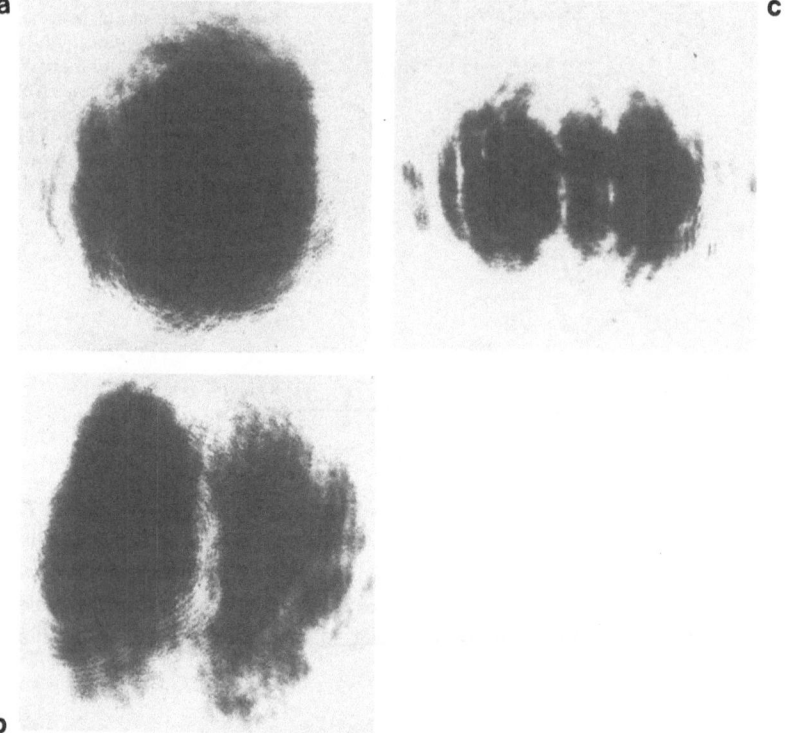

a

c

b

Fig. 5.19. Dancing modes. The effect of inserting a phase nonreciprocity $\Delta\phi$ in the ring of the "ring conjugator" shown in Fig. 5.5d. From left to right, the output mode changes from the phase conjugate of the incident beam (TEM$_{00}$ mode) to higher order Gaussian modes as $\Delta\phi$ is increased [5.77]

mode by partially cancelling the effective nonreciprocity of the ring, thereby allowing the hologram to translate more slowly. During the transition from one mode to the other, the modes beat and cause a lively dancing output beam. As the nonreciprocity is further increased, the output mode pattern becomes more complex, until $\Delta\phi \sim 2\pi$, after which the entire sequence repeats.

5.7.2 Stimulated Scattering

Stimulated scattering occurs in photorefractive materials when a single input beam with a well-defined wavevector k_1 creates gain for other beams with wavevectors k_2 which start from noise (random scattering of the incident beam from defects and inclusions in the crystal) and are amplified by two-wave energy coupling with the incident beam (Sect. 5.3.1). The dominant Fourier components of the gratings formed by the incident and scattered beams have spatial frequencies:

$$K^+ = k_1 - k_2 \quad \text{and} \quad K^- = k_2 - k_1 = -K^+ \quad . \tag{5.32}$$

These gratings scatter the incident beam into the scattered beams, and the pattern of light that emerges depends on the direction of incidence and polarization of the input beam, and the existence of a phase-matching channel for the scattered light. Three different examples are examined below:

a) In the case of "beam fanning" in cerium-doped strontium barium niobate ($Sr_{1-x}Ba_xNb_2O_6$) and in barium titanate, the secondary beams emerge in a wide "fan" of light which has the same polarization as the incident beam [5.78, 79]. The fanned beams, which start from noise, have a wide array of wavevectors k_2 and write gratings having a corresponding array of grating wavevectors $K^- = k_2 - k_1$. Because the coupling strength switches sign if K^- is reversed, the noise is amplified on one side of the c-axis, and extinguished on the other. Consequently, the observed "fan" lies on one side of the incident beam, as determined by the direction of the positive c-axis and by the sign of the dominant photorefractive charge carrier. (Observing the direction of beam fanning in a poled photorefractive crystal is, in practice, the simplest method for determining the sign of the dominant charge carrier.) There is no "phase-matching" restriction on the directions of the scattered beams, since the incident beam is automatically Bragg-matched to all of the gratings. Closer inspection of the fan often reveals distinct stripes of high intensity, perhaps because those particular "beamlets" originated by scattering from particularly large defects in the crystal.

b) A single incident beam of light can also produce a distinct "ring" of forward scattered light in barium titanate [5.80, 81], iron-doped lithium niobate ($LiNbO_3$) [5.80], and copper-doped lithium tantalate ($LiTaO_3$) [5.82]. The polarization of the scattered light is *orthogonal* to that of the incident beam. The pattern of the scattered light changes from circular to elliptical to teardrop-shaped as the incident beam angle is varied, as shown in Fig. 5.20. The intensity of the ring is largest in the direction perpendicular to the plane containing the incident beam in the crystal's c-axis, and decreases to zero in this plane.

In barium titanate, scattering of a beam of one polarization to produce a beam of the orthogonal polarization takes place via the tensor element R_{yzy} ($\equiv r_{42}$) of the third-rank electro-optic tensor $\underline{R}$. Note that the initial and the scattered beams, being orthogonally polarized, cannot interfere to write a grating. Therefore, the scattered beam cannot reinforce the grating that is scattering it, in contrast to the case of "beam-fanning" described above.

The rings can be explained as follows:

i) Two-beam coupling between the incident beam k_1 and noisy beams with an array of wavevectors k_2 causes amplification of the noisy beam via the grating component K^-, as described in (a) above;

ii) The grating component $K^+ (\equiv -K^-)$ is unavoidably also present. Of all the gratings present, Bragg-scattering of the incident beam to form a beam with orthogonal polarization can occur only from those gratings for which the phase-matching condition:

$$|k_{\text{scattered}}| = n_{\text{eff}}\omega/c \tag{5.33}$$

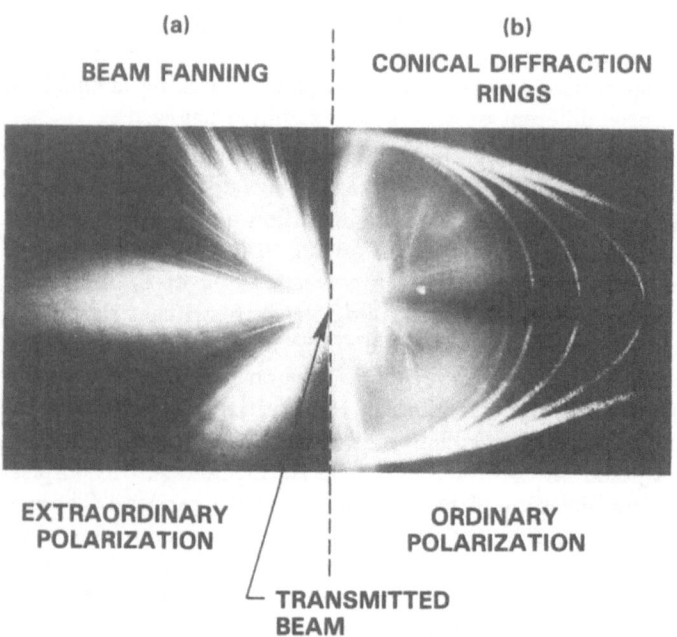

(a)

BEAM FANNING

(b)

**CONICAL DIFFRACTION
RINGS**

**EXTRAORDINARY
POLARIZATION**

**ORDINARY
POLARIZATION**

**TRANSMITTED
BEAM**

Fig. 5.20. Stimulated scattering in barium titanate. A single incident beam causes broad fans of light to appear in the $+c$-axis direction (*left side of photograph*) and a ring of light in the $-c$-axis direction (*right side of photograph*). Each angle of incidence produces a single ring. The different rings seen here result from a multiple exposure using five different angles of incidence [5.81]

is satisfied, where

$$k_{\text{scattered}} = K^+ + k_1 = (k_1 - k_2) + k_1 = 2k_1 - k_2 \tag{5.34}$$

is the wavevector of the scattered light, and n_{eff} is the effective index of refraction. The condition (5.33) determines the pattern formed by the scattered light, and is shown in Fig. 5.21. Note that phase-matching is possible only if the refractive index n_{eff} of the scattered beam is greater than the refractive index of the writing beams, so the polarization of the scattered beam must differ from that of the incident beam. If the ordinary index exceeds the extraordinary index ($n_o > n_e$), as in barium titanate, the incident ray must be extraordinary and the scattered ray ordinary. No rings are observed if the incident ray is polarized to be an ordinary ray. (In lithium tantalate, the situation is reversed because $n_o < n_e$ [5.82].) Because the cone angle of the generated ring of light is a sensitive function of the birefringence $n_o - n_e$ for the crystal, a careful measurement of the scattering angle yields accurate values for the refractive indices of the crystal [5.81].

c) Anisotropic beam fanning, in which fans of light appear on opposite sides of the transmitted incident beam to make a bowtie-shaped pattern, has been

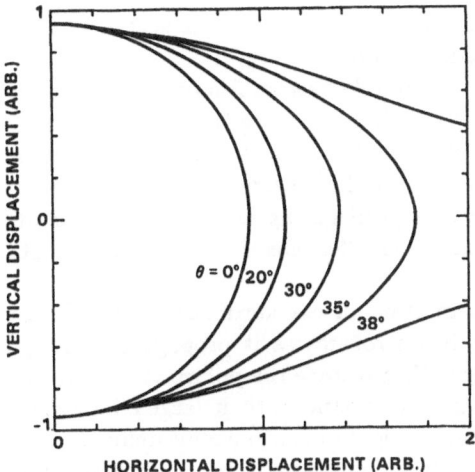

Fig. 5.21. Stimulated scattering in barium titanate. Calculated positions for the conical diffraction light rings with the five different angles of incidence used in Fig. 5.20 [5.81]

(figure axis labels: VERTICAL DISPLACEMENT (ARB.), HORIZONTAL DISPLACEMENT (ARB.), $\theta = 0°$, $20°$, $30°$, $35°$, $38°$)

observed in iron-doped lithium niobate [5.83, 84] and copper-doped lithium tantalate [5.84]. The scattered light has polarization orthogonal to that of the incident beam. The mechanism proposed [5.84] for this effect is a rather unusual form of energy coupling [5.85–87]: through the imaginary part β_{24}^i of the β_{24} component of the photovoltaic tensor, two orthogonally polarized optical beams generate localized "circular" photovoltaic currents which write a photorefractive grating. The incident beam then scatters from this grating with a change of polarization via the nonlinear coefficient r_{42}, as described in (b). However, in contrast to (b), the interaction between incident and scattered light reinforces the original grating (if the incident polarization is correct) and the scattered light is amplified. Since each scattered beam is automatically Bragg-matched to the grating that scattered it (as in beam fanning in (a)), there is no angular restriction on the scattered light, which therefore appears as broad fans. In ordinary energy coupling, the direction of coupling is determined by the direction of the positive c-axis of the crystal; here the direction of coupling is determined by the *polarizations* of the beams, with the sign of β_{24}^i determining which polarization is amplified at the expense of the orthogonal polarization. It is for this reason that fans form on *both* sides of the incident beam. In lithium niobate, for example, an incident ordinary wave produces fans that have extraordinary polarization implying that $\beta_{24}^i < 0$. (The situation is reversed in lithium tantalate, so that $\beta_{24}^i > 0$.)

Experimental evidence for circular photovoltaic currents in tellurium [5.88] and lithium niobate [5.89] has also been reported.

d) A single incident beam of light can produce lively backscattered patterns of rings and crosses in iron-doped LiNbO$_3$ [5.90]. These scattered beams originate deep in the crystal, possibly on the far face, and are amplified as they propagate back toward the entrance face by two-beam coupling with the incident beam. The observed pattern and polarization of the scattered light depend on the polarization of the incident beam.

If the incident beam enters as either a purely extraordinary ray or a purely ordinary ray, its polarization is preserved as it propagates through the crystal. The scattered light is more strongly amplified if it keeps the same polarization as the incident beam. The locus of all such polarization-preserving paths for the scattered beams forms a cross. (Beams propagating in other directions are in a mixed state of polarization and have a smaller effective interaction length with the incident beam.)

If the incident beam enters as a mixture of extraordinary and ordinary rays, the polarization of the incident beam rotates as it propagates through the birefringent crystal. Scattered beams that retrace the polarization-rotation of the incident beam have the largest gain by virtue of their large effective interaction length with the incident beam. The locus of such beams defines a cone centered about the c-axis, and gives rise to a conical pattern of backscattering.

Note that because the scattered beams propagate in directions approximately opposite to the direction of k_1, the photorefractive grating wavevector K^- has a high spatial frequency ($\sim 2k_1$). Therefore, the gain is significant only if there is a high density of charge carriers in the crystal. Additionally, in order to observe the backscattered beam, the crystal must be oriented so that the backscattered beam experiences two-wave-mixing gain.

5.8 Applications of Phase Conjugation

Phase conjugation has a host of potential applications, including correcting spatial phase distortion due to turbulence, spatial- and temporal-frequency filtering of optical beams, ring gyroscope deadband removal and biasing, and forming active laser "mirrors" for beam clean-up and cavity stabilization [5.3]. Just as lasers remained a laboratory curiosity with "no real application" for almost 15 years, phase conjugators do not yet enjoy the respectability that comes with commercial success. The drawbacks of current phase conjugators – they are too slow, too insensitive to light, or too expensive – will be overcome only by breakthroughs in materials research. Nevertheless, a number of promising applications of phase conjugation have already been demonstrated, some of which we briefly describe below. The first three applications rely on the *phase* of the phase-conjugated wave; the remaining applications use the *image content* of the generated wave.

5.8.1 Phase-Conjugating Laser Cavity

Consider a dye laser having a linear resonator in which one of the end mirrors is a phase conjugator [5.19], as shown in Fig. 5.22; such a phase-conjugating dye laser was demonstrated by *Feinberg* and *Bacher* using a barium titanate crystal as a CAT self-pumped phase conjugator [5.49]. This laser produces a spatially

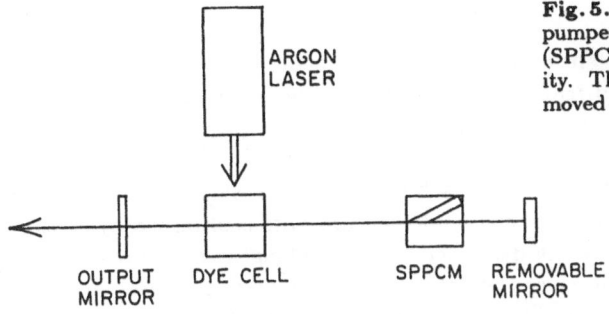

Fig. 5.22. Dye laser having a self-pumped phase-conjugating mirror (SPPCM) inside the resonator cavity. The removable mirror is removed once lasing has been initiated

uniform output beam despite the astigmatism of the resonator cavity and the thermal lensing of the dye cell. In addition to compensating for these optical distortions, the phase conjugator narrows the spectral linewidth of the dye laser, and was unexpectedly observed to scan its own wavelength with time. The direction of the wavelength scan is towards the blue or the red end of the spectrum, depending on the precise location and orientation of the crystal [5.49, 91, 92], as shown in Fig. 5.23. If the phase conjugator *replaces* one of the dye laser mirrors, the dye laser inexorably self-scans its wavelength to the end of the dye tuning range, where lasing stops. However, if the phase conjugator is placed *outside* of the intact dye laser, then at the end of the wavelength scan the dye laser jumps back to its original color as it resumes lasing with its ordinary mirror, and the wavelength scan repeats [5.91]. Inserting a neutral density-2 (1 % transmission) filter between the dye laser and the phase conjugator does not prevent self-scanning [5.93].

Further experiments revealed that the wavelength scanning is caused by the self-pumped phase conjugator itself, which imparts a $\sim 1\,\mathrm{Hz}$ frequency shift

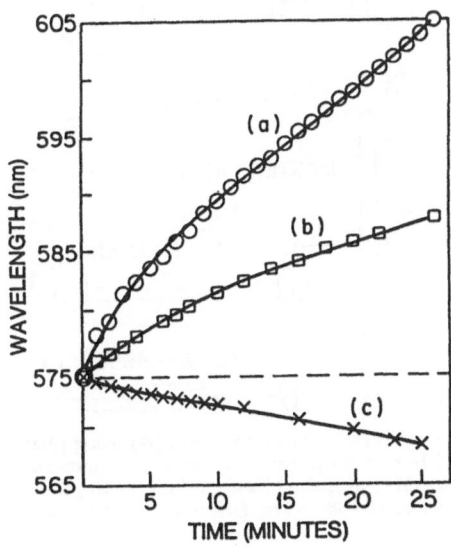

Fig. 5.23. Self-scanning of the wavelength of a dye laser having a self-pumped phase-conjugating mirror inside the dye laser cavity, as shown in Fig. 5.22. Scans towards the red (a) and (b), and towards the blue (c) are observed. The fast red scan (a) results from tapping on the optical table [5.49]

to the phase-conjugate beam [5.94, 49]. This miniscule frequency shift is compounded with every round trip of the laser cavity, about 10^9 times per second, to produce an easily observable wavelength scan. (The physical origin of the frequency shift is a matter of some debate, as discussed in Sect. 5.6.)

Ramsey and *Whitten* used such a self-scanning dye laser to perform spectroscopic measurements [5.70]. A constant scanning rate is obtained by controlling the optical intensity of the dye laser. The wavelength scans are obtained with no moving parts, and cover the entire tuning range of the laser dye.

Recent experiments involving a semiconductor diode laser coupled to a self-pumped phase conjugator [5.96] also revealed a rapid self-scanning of the diode laser, although the magnitude of the scanning rate is greater than that expected from frequency shifts alone.

5.8.2 Phase-Locking Lasers

There are many applications that require that two or more lasers be locked together in frequency. For example, the output beams of individual semiconductor diode lasers can be coherently combined into a single intense beam if their optical frequencies and phases are locked together. As another example,

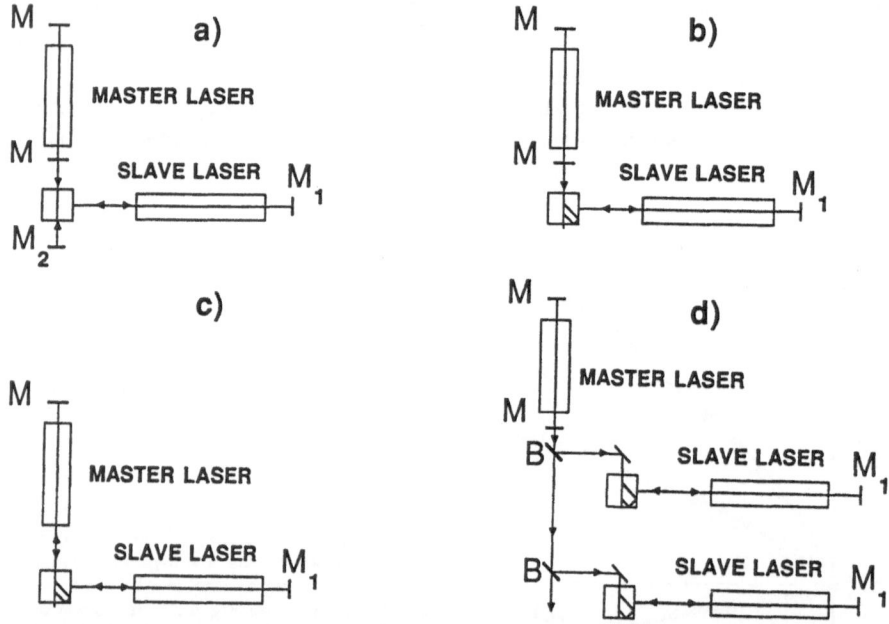

Fig. 5.24. Various schemes for phase-locking two lasers (a–c) or many lasers (d) using phase conjugation. The master laser provides pumping beams for phase conjugation by four-wave mixing. The slave laser (which has only one mirror M_1) uses the phase conjugator to provide optical feedback. The frequency of the slave laser is locked to the frequency of the master laser [5.95]

the image received by an optical fiber can be coherently processed if a local laser is phase-locked to the frequency of the incoming beam.

Feinberg and *Bacher* [5.95] demonstrated phase-locking of two separate lasers using phase conjugation. The output of the "master" argon-ion laser was incident on a crystal of photorefractive barium titanate, which generated a self-pumped phase conjugate beam. A "slave" argon-ion laser, with its output mirror removed, pointed at the barium titanate crystal. Lasing took place between the phase conjugator and the remaining mirror in the slave laser cavity, as shown in Fig. 5.24b. The frequencies of the two lasers stayed locked indefinitely to within less than 1 Hz unless disturbed. Transient phase-locking of two argon lasers was also demonstrated by *Sternklar* et al. using a barium titanate crystal in a similar geometry [5.62].

Cronin-Golomb and *Yariv* recently showed [5.96] that strong coupling could be obtained between beams in the near infrared (850 nm) using a barium titanate crystal cooled to near its tetragonal-orthorhombic transition temperature of $\sim 10°$ C. This suggests the possibility of using a cooled barium titanate phase conjugator to phase-lock arrays of semiconductor lasers.

5.8.3 Interferometry with a Self-Pumped Phase-Conjugating Mirror

Interferometry with visible light is capable of measuring submicron distance changes. In principle, the minute motion of a distant mirror can be measured by clearing an optical path to the mirror, either through the atmosphere or through an optical fiber, and interferometrically comparing any change in this round-trip optical path length to a fixed reference path length. One problem with this scheme is that the wave to and from the distant mirror may be

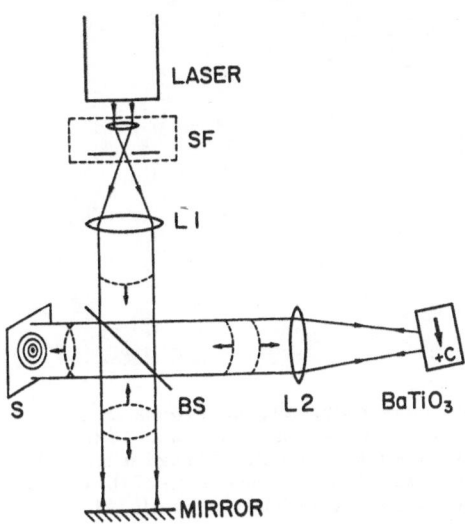

Fig. 5.25. Interferometry with a phase-conjugating mirror. The spatial filter (SF) and the lens (L1) provide an approximately collimated laser beam (here shown slightly diverging) to the beamsplitter (BS), which splits the beam into two directions. The beams returning from the phase conjugator and from the ordinary mirror are recombined by the beamsplitter to make an interference pattern on the screen (S). The BaTiO₃ crystal acts as CAT self-pumped phase conjugator: it corrects for any distortion between the crystal and the beamsplitter, yet preserves information about the distance between them [5.97]

distorted by atmospheric turbulence, or by modal dispersion if a multimode fiber is used.

These distortions can be eliminated by replacing the remote mirror with a phase conjugator, as shown in Fig. 5.25. For example, if a plane wave becomes scrambled by modal dispersion as it propagates through a multimode fiber, after being phase conjugated at the far end of the fiber it will be unscrambled as it retraces its path back through the fiber. This interferometer, however, calls for a rather peculiar phase conjugator: it must phase conjugate only the high spatial-frequency phase information (the noise) caused by modal dispersion and not the low spatial frequency phase change (the signal) caused by propagating to and from the mirror. This is precisely the behavior of a *self-pumped* phase conjugator [5.97]. Since a self-pumped phase conjugator has no external phase

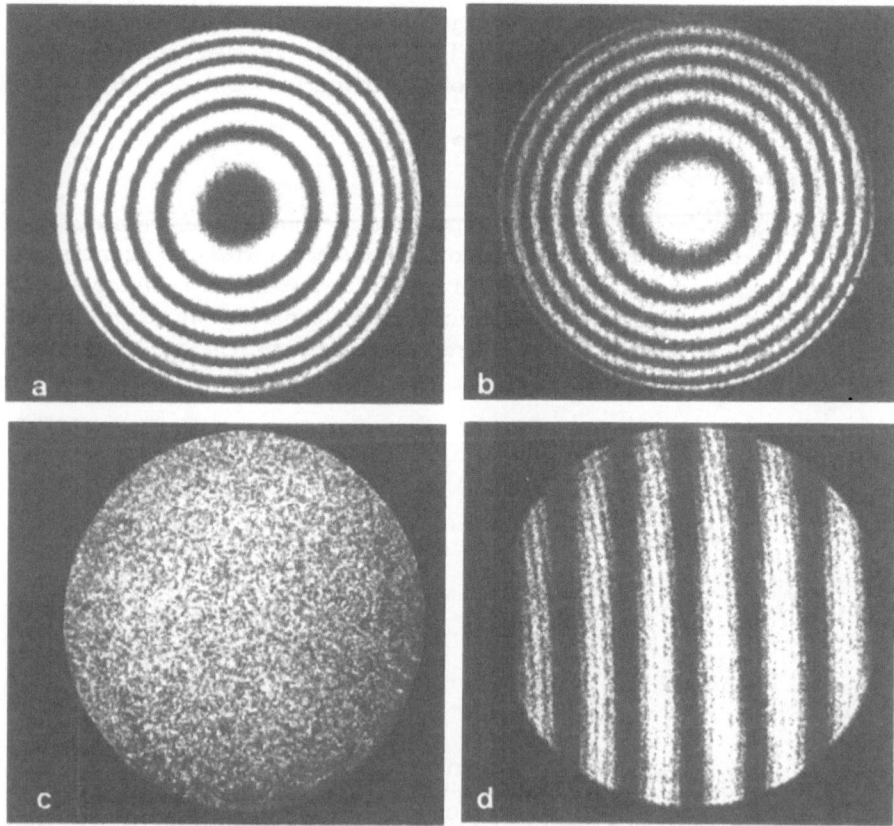

Fig. 5.26a–d. Interference patterns using the phase-conjugating interferometer shown in Fig. 5.25: **(a)** self-pumped phase conjugator in one arm, **(b)** distorter in front of the phase conjugator, (note that the interference pattern is still quite visible, although it has shifted due to the nonzero average thickness of the distorter), **(c)** distorter in front of the ordinary mirror, **(d)** no distorter and two ordinary mirrors, one tilted slightly off axis [5.97]

reference, it cannot compensate uniform phase changes, and so will preserve the distance information when used in an interferometer.

Interferometers with self-pumped phase conjugators have been demonstrated that correct for optical distortion in free space propagation [5.97] as shown in Fig. 5.26, and in transmission through an optical fiber [5.98]. *Ewbank et al.* demonstrated [5.71] a Michelson interferometer with *two* self-pumped phase conjugators, one in each arm. Due to the frequency shifts inherent in self-pumped phase conjugators, the relative phase of the two phase-conjugate beams varies with time, and the recombined beams alternately interfere constructively and destructively. However, Fig. 5.27 shows that after a few seconds of operation a new self-generated beam appears *between the two phase conjugators!* This new beam locks the relative phase of the two phase-conjugate beams so that complete destructive interference occurs at the unused port of the interferometer's beamsplitter. This experiment demonstrates that, given the chance, self-pumped phase conjugators tend to "self-couple" and act as one large phase

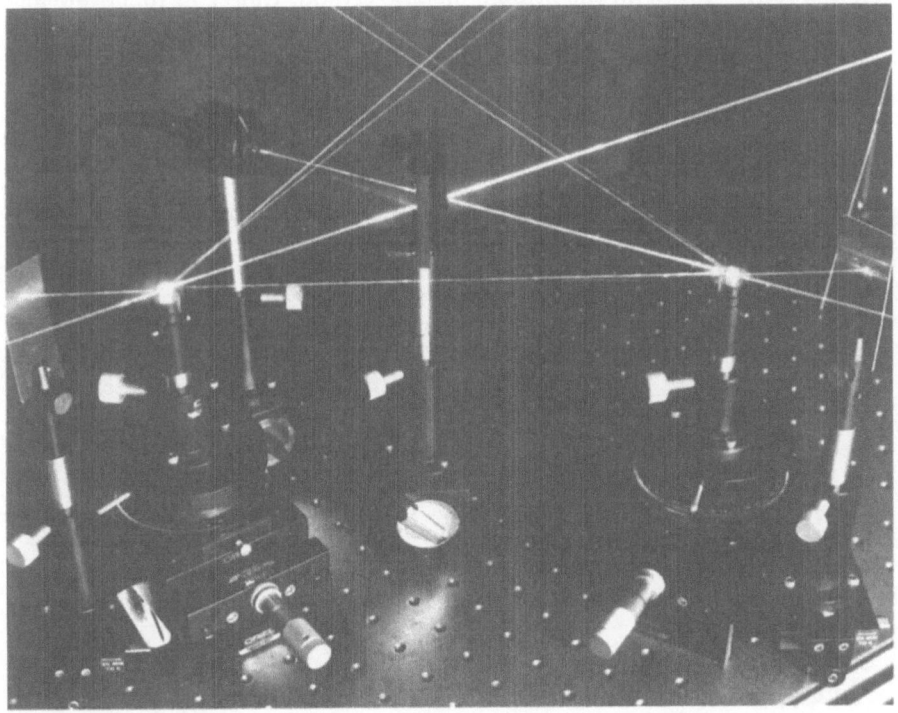

Fig. 5.27. Photograph of two phase conjugators coupled by a self-generated beam between them. The beam splitter in the center of the photograph directs light into the two BaTiO$_3$ crystals. The self-generated beam is the horizontal line in the center of the photograph. The "searchlight" beams seen crossing at the top center of the photograph are reflections from the entrance and exit faces of the two crystals [5.71]

conjugator. An application of this device for the real-time subtraction of two optical images is described in Sect. 5.8.5 below.

5.8.4 Photolithography

In conventional photolithography, a very well-corrected – and very expensive – lens system images a highly detailed mask onto a substrate. (The substrate is previously coated with a light-sensitive material. After light exposure it is chemically treated to reproduce the mask's features.) In order to reproduce the micron-sized features of the mask the optical system must be free of aberrations. *Levenson* suggested that phase conjugation can provide the required distortion-free imaging, with the only critical optical component being an accurately symmetric cube-type beamsplitter [5.99–101]. Using an externally pumped photorefractive crystal of lithium niobate as a phase conjugator, he demonstrated imaging of submicron features free of speckle noise, as shown in Fig. 5.28. Some drawbacks of this system are the inconvenience of using external pumping beams and the long exposure time needed to establish the photorefractive grating in the lithium niobate crystal. *Gower* [5.102] overcame these problems by using a CAT self-pumped conjugator in a barium titanate crystal to produce speckle-free images with micron-sized features.

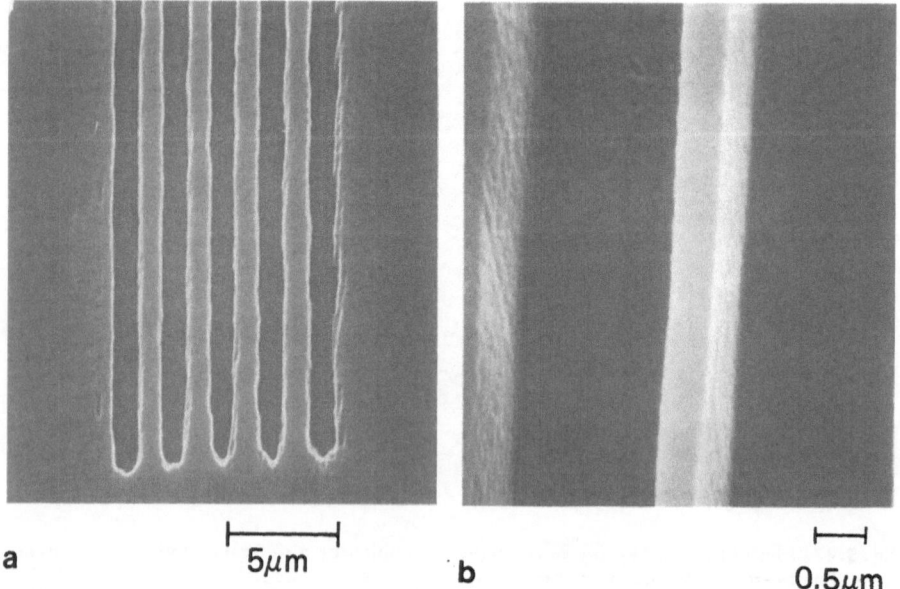

a 5μm   b 0.5μm

Fig. 5.28. High resolution photolithography using phase conjugation in photorefractive LiNbO$_3$. Adjacent stripes are about 2 microns apart. Note the absence of speckle noise [5.100]

5.8.5 Parallel Optical Processing: Image Correlation, Convolution, and Subtraction

A vast amount of information can be stored in an optical image: a single page from this book can store about 10^{12} visible-wavelength-sized pixels. Using parallel rather than sequential processing of such an optical image yields a nontrivial increase of 10^{12} in processing speed (from one picture a millenium to one picture a millisecond). *Lee* [5.103] has reviewed a number of techniques for optical image processing. Here we discuss the application of photorefractive materials to optical image processing.

White et al. [5.104] demonstrated parallel processing of an optical image using a four-wave-mixing geometry. The correlation or convolution of two image-bearing beams was optically produced within a few milliseconds using photorefractive BSO. One drawback of this geometry is the smearing of pixels due to the nonzero crossing angle of the beams [5.105]. *Brody* [5.106] has demonstrated the detection of correlations in an RF signal using an acousto-optic modulator and a self-pumped crystal of barium titanate.

Two optical images can be subtracted by arranging two coherent beams to be exactly out of phase across the entire wavefront. An optical image inserted into one wavefront will then be subtracted from an image placed in the other wavefront. Figure 5.29 shows how a phase-conjugating interferometer was used by *Chiou* et al. [5.107, 108] to generate two wavefronts 180° out of phase, and thereby subtract the amplitude images placed in the interferometer arms, as seen in Fig. 5.30. Image subtraction was also demonstrated by *Kwong* et al. using a barium titanate crystal [5.109] in the CAT self-pumped geometry [5.22].

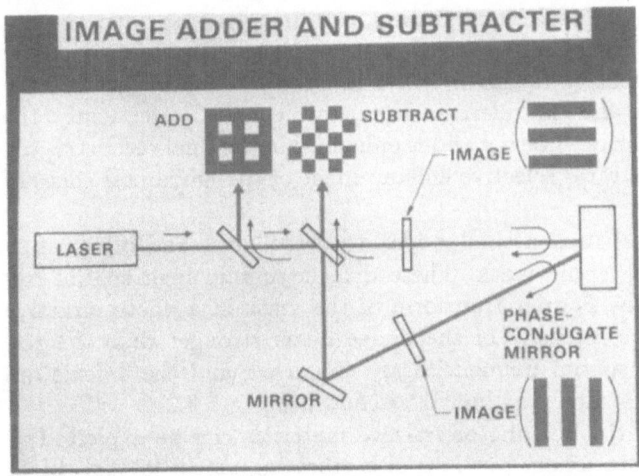

Fig. 5.29. Optical set-up for subtracting or adding two optical images using a phase conjugator [5.107]

191

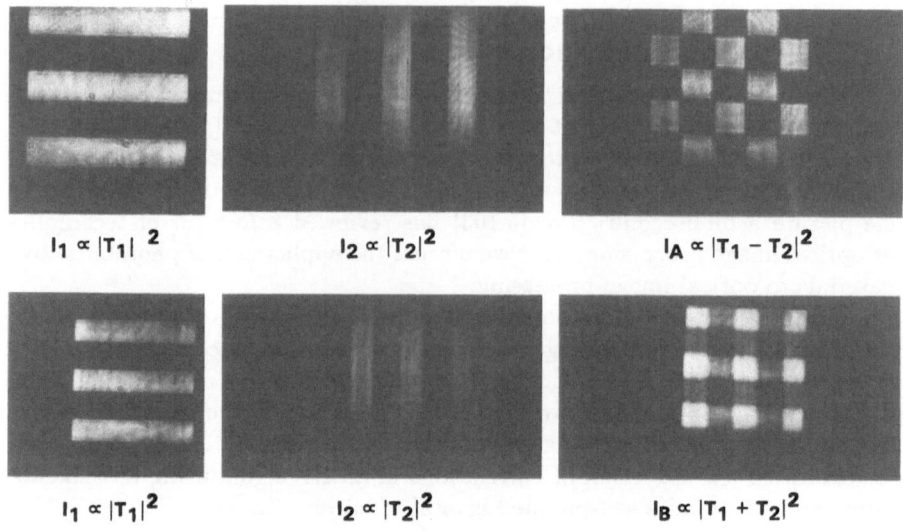

$I_1 \propto |T_1|^2$ $I_2 \propto |T_2|^2$ $I_A \propto |T_1 - T_2|^2$

$I_1 \propto |T_1|^2$ $I_2 \propto |T_2|^2$ $I_B \propto |T_1 + T_2|^2$

Fig. 5.30. Real-time subtraction (*top right*) and addition (*bottom right*) of two optical images (*left and center*) using the setup of Fig. 5.29 [5.107]

5.8.6 Edge and Defect Enhancement; Vibrational Modes

Photorefractive materials are ideally suited for performing real-time edge enhancement of an optical image [5.110, 111]. Edge enhancement occurs with the usual four-wave mixing geometry if the average intensity of the image-bearing beam is much larger than the intensity of the other beams. Because the photorefractive grating is largest when the intensities of the two writing beams are equal (see Sect. 5.3.1), both the weak and the strong parts of the image are ineffective in writing a high-efficiency grating, one being weaker than the reference beam and the other being stronger. At some place on an intensity edge, however, where the intensity ranges from its minimum to its maximum, the intensities of the image and the reference beams are equal. Consequently, the edge writes the most efficient grating and is enhanced in the final reconstructed image. Figure 5.31 illustrates selective enhancement of the horizontal edges of an image.

Ochoa et al. [5.112] used a similar idea to detect dust and other small defects in a photolithographic mask. These defects contain high spatial frequencies. By imaging the Fourier transform of the mask in a photorefractive crystal, and making the intensity of the image beam stronger than the reference beam, the high spatial frequencies are enhanced, and the defects are plainly visible in the reconstructed image, as shown in Fig. 5.32.

The slow response time of photorefractive materials can be exploited to produce images of the stationary portions of a vibrating object [5.113]. Light reflected from the vibrating surface of an acoustic loudspeaker is used to write a photorefractive grating in a crystal of BSO. The moving portions of the

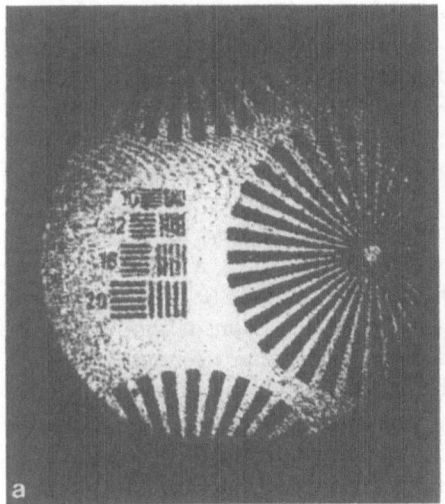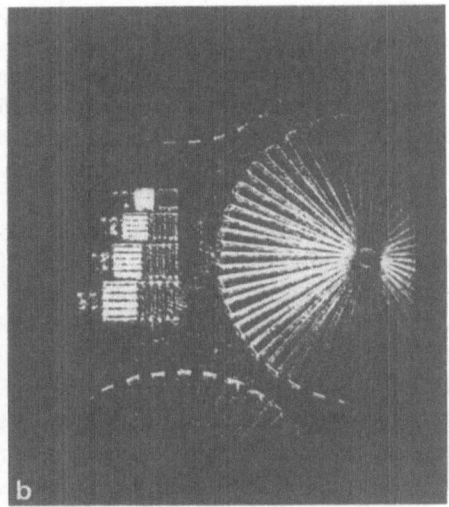

Fig. 5.31a,b. Selective edge enhancement of only the horizontal stripes of a resolution chart. (a) original image, (b) edge-enhanced image [5.111]

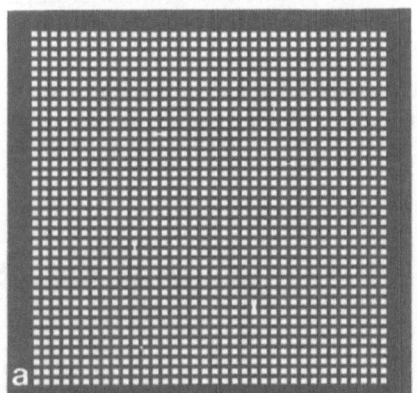

Fig. 5.32a,b. Detection of defects in a mask. (a) Original image, (b) defect-enhanced image [5.112]

loudspeaker cause a frequency shift in the reflected light, smearing out the corresponding portion of the hologram in the BSO crystal. In the resulting reconstructed image, only the stationary portions of the loudspeaker are visible, so that the standing-wave pattern of the loudspeaker cone is clearly seen.

5.8.7 Associative Memory

Photorefractive materials have recently been used to demonstrate an entirely new kind of optical computing. By combining holography and an optical resonator, one can retrieve a complete optical image from a partial version of the image. The numerical algorithms for creating and operating an associative memory [5.114] bear a striking resemblance to the procedure of storing and reconstructing a holographic image [5.115–117]. *Kohonen* [5.118] has described a number of geometries for an optical associative memory, and *Cohen* [5.119, 120] has proposed a holographic model for the behavior of complicated neural nets, such as those found in the brain.

The field of optical associative memory is in its infancy; there have been only two demonstrations at the time of writing. In the first, a set of images is stored in a holographic medium, each image with its own angularly encoded reference beam. To reconstruct an image, a partial version of that image illuminates the hologram, which reconstructs (most of) the reference beam for that image. The reconstructed reference beam enters a phase conjugator, which sends the reference beam back toward the hologram, where it reconstructs the original image. Because only a portion of the original image is supplied, in general some of the wrong reference beam is generated along with the correct reference beam, and therefore the phase conjugator must have a thresholding device to make the strongest reference beam dominant. *Soffer* et al. [5.121] have demonstrated recall of an image from a partial version of the image using a barium titanate phase conjugator and a thermoplastic holographic storage medium. Their preliminary demonstration had only one image and no thresholding device.

Anderson demonstrated an alternate approach using a ring resonator with a holographic mirror [5.122], as shown in Fig. 5.33. The information is stored by

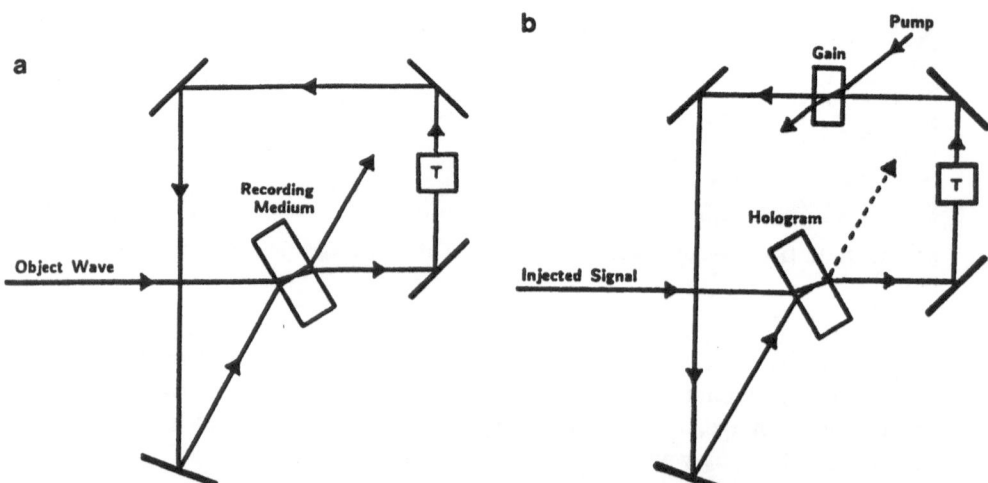

Fig. 5.33a,b. Optical associative memory with a generalized transformation (T) in a ring resonator cavity. (a) Recording the hologram. (b) Recalling the entire image by injecting a partial image. Optical gain is provided by two-wave mixing with an external pump beam in a photorefractive crystal [5.122]

injecting an image into the resonator, and exposing the hologram to the injected image (as the object beam) and its round-trip version (as the reference beam). This is repeated for each of the images to be stored. The hologram is then stabilized, so that it is not erased by light. The injected light is removed and the gain of the cavity is increased until it self-oscillates. The spatial modes of this cavity correspond to the images previously stored in the hologram, and the beam in the cavity hops from mode to mode. Injecting a partial version of one of the stored images increases that image's gain over the others, and only that image resonates, as seen in Fig. 5.34. Because the hologram contains the entire image, the entire image is displayed from only a partial input. The best results are obtained when the different modes compete for the cavity gain, which is provided by two-wave mixing in a separate barium titanate crystal.

The ideal implementation of these optical associative memories requires a holographic medium that (1) has gain, and (2) can be made temporarily insensitive to light, so that images can be semi-permanently stored. Photorefractive materials can provide the gain, but at present there is no easy way to temporarily disable their sensitivity to light.

5.8.8 Optical Novelty Filter

Recently *Anderson* et al. demonstrated a novelty filter: it displays only what is new in a scene [5.123]. This device uses a photorefractive crystal to perform the time derivative of an entire image in parallel. An inexpensive liquid-crystal television, modified for use as a spatial light phase modulator and connected to a video camera, impresses a live scene onto a laser beam. In one implementation, this beam is phase conjugated by the crystal and passed back through the television; if any portion of the scene changes during the response time of the photorefractive crystal, light corresponding to those pixels will not be returned to the source but deflected onto a screen, as shown in Fig. 5.35. Another implementation uses two-beam coupling to deplete light from stationary portions of the image. The crystal is oriented so that the transmitted object beam and the deflected pump beam *destructively* interfere, thereby eliminating stationary portions of the image (mountains, for example) but not moving portions (flying birds, for example). When a scene is viewed with these novelty filters, only those portions that are moving or changing appear, and the stationary background is removed.

5.9 Conclusion

This chapter has reviewed some of the applications of photorefractive materials for optical phase conjugation and optical resonators. In order for photorefractive materials to become truly practical for nonlinear optical applications, they will have to be durable and inexpensive. The demonstration of an infrared photorefractive effect in GaAs and in other semiconductors is encouraging, for

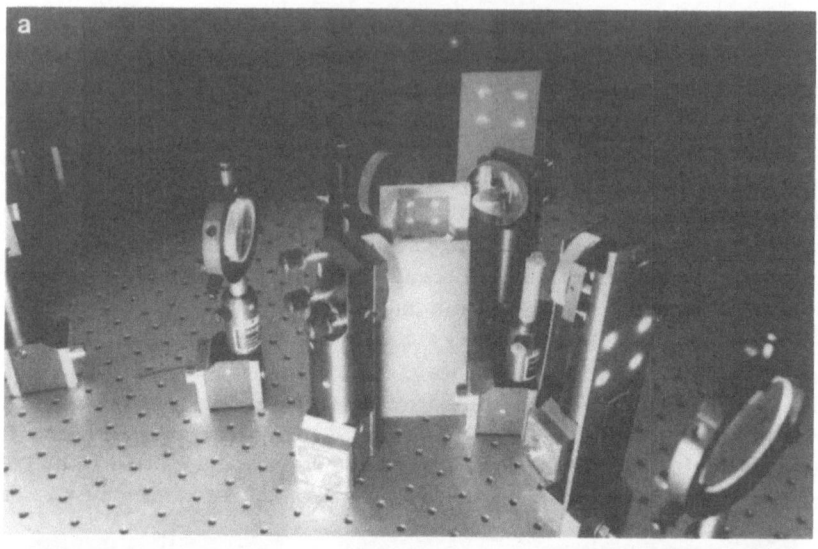

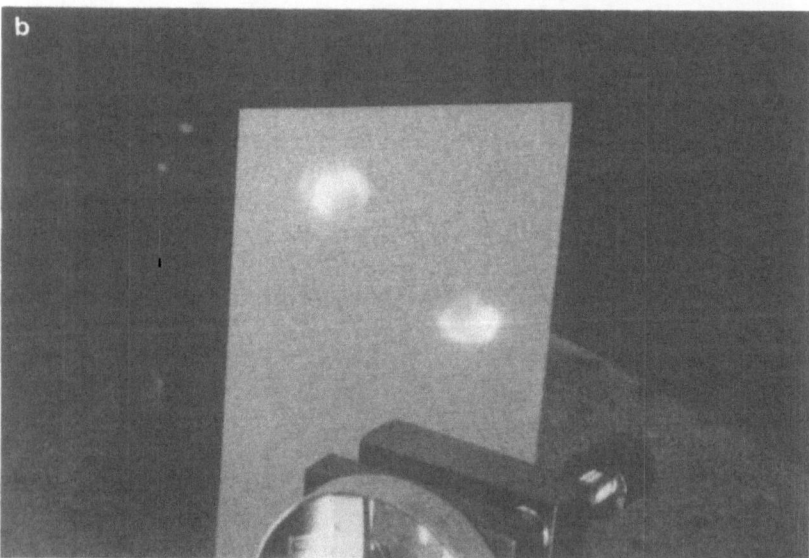

Fig. 5.34a–c. Two separate images of diagonal dots have been previously stored in the holographic medium. (a) The resonator is above threshold and both sets of dots oscillate. (b) The resonator reconstructs only one image (*two dots*) when seeded with a partial image (*lower right-hand dot*). (c) The resonator reconstructs the other image (*two dots*) when seeded with a partial image (*lower left-hand dot*) [5.122]

Fig. 5.34c. Caption see opposite page

these electronics-industry materials are inexpensive, well-studied, and manufactured with high quality. We hope that further research into these and related materials will eventually enable us to produce custom photorefractive materials containing an optimum density of photosensitive charge carriers embedded in a highly polarizable lattice.

The recent marriage of photorefractive materials and nonlinear optics has yielded a host of new effects and devices that need only low-power lasers. The relative ease of construction of these devices encourages creativity, because new ideas can be explored relatively quickly and with only a moderate financial investment. It is the unexpected effect that is often the most interesting, and photorefractive materials have provided a wealth of unexpected and interesting new effects.

Acknowledgements. This work was supported by contract no. F49620-85-C-0110 of the United States Air Force Office of Scientific Research and contract no. F49620-85-C-0071 of the Joint Services Electronics Program.

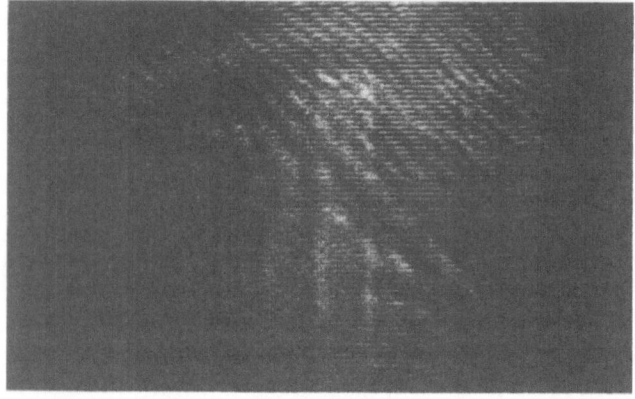

Fig. 5.35. Output of a novelty filter, which uses a barium titanate crystal to form the time derivative, in parallel, of an entire optical scene. *Top photograph*: The words "Novelty Filter" have been suddenly input into the device; the output shows this sudden change. *Bottom photograph*: A few seconds later, the input has not changed and so the image has faded from the screen [5.123]

References

5.1 Opt. Eng. **21**, March/April, 1982, issue on nonlinear optical phase conjugation
5.2 J. Opt. Soc. Am. **73**, May, 1983, issue on optical phase conjugation
5.3 R.A. Fisher (Ed.): *Optical Phase Conjugation* (Academic, New York 1983)
5.4 D. Gabor: "Microscopy by reconstructed wave-fronts," Proc. Roy. Soc. **A197**, 454 (1949)
5.5 D. Gabor: "Microscopy by reconstructed wave-fronts: II," Proc. Phys. Soc. **B64**, 449 (1951)
5.6 E.N. Leith, J.Upatnieks: "Wavefront reconstruction with diffused illumination and three-dimensional objects," J. Opt. Soc. Am. **54**, 1295 (1964)
5.7 H. Kogelnik: "Holographic image projection through inhomogeneous media," Bell Syst. Tech. J. **44**, 2451 (1965)
5.8 A. Yariv: "Three-dimensional pictorial transmission in optical fibers," Appl. Phys. Lett. **28**, 88 (1976)

5.9   A. Yariv: "On transmission and recovery of three-dimensional image information in optical waveguides," J. Opt. Soc. Am. **66**, 301 (1976)

5.10   R.W. Hellwarth: "Generation of time-reversed wave fronts by nonlinear refraction," J. Opt. Soc. Am. **67**, 1 (1977)

5.11   D.M. Bloom, G.C. Bjorklund: "Conjugate wave-front generation and image reconstruction by four-wave mixing," Appl. Phys. Lett. **31**, 592 (1977)

5.12   S.L. Jensen, R.W. Hellwarth: "Observation of the time-reversed replica of a monochromatic optical wave," Appl. Phys. Lett. **32**, 166 (1978)

5.13   D.M. Bloom, P.F. Liao, N.P. Economou: "Observation of amplified reflection by degenerate four-wave mixing in atomic sodium vapor," Opt. Lett. **2**, 58 (1978)

5.14   P.F. Liao, D.M. Bloom: "Cw optical wave-front conjugation by saturated absorption in atomic sodium vapor," Appl. Phys. Lett. **32**, 813 (1978)

5.15   R.L. Abrams, R.C.Lind: "Degenerate four-wave mixing in absorbing media," Opt. Lett. **2**, 94 (1978); **3**, 205 (1979)

5.16   P.F. Liao, D.M. Bloom: "Continuous-wave backward-wave generation by degenerate four-wave mixing in ruby," Opt. Lett. **3**, 4 (1978)

5.17   J. Feinberg, R.W. Hellwarth: "Cw phase conjugation of a nonuniformly polarized optical beam," 11th IQEC Digest of Technical Papers, Boston, 1980, Paper E9

5.18   J. Feinberg, D. Heiman, A.R. Tanguay, Jr., R.W. Hellwarth: "Photorefractive effects and light-induced charge migration in barium titanate," J. Appl. Phys. **51**, 1297 (1980); errata **52**, 537 (1981)

5.19   J. Feinberg, R.W. Hellwarth: "Phase-conjugating mirror with continuous-wave gain," Opt. Lett. **5**, 519 (1980); errata **6**, 257 (1981)

5.20   J.O. White, M. Cronin-Golomb, B. Fischer, A. Yariv: "Coherent oscillation by self-induced gratings in the photorefractive crystal $BaTiO_3$," Appl. Phys. Lett. **40**, 450 (1982)

5.21   M. Cronin-Golomb, B. Fischer, J.O. White, A. Yariv: "Passive (self-pumped) phase conjugate mirror: theoretical and experimental investigation," Appl. Phys. Lett. **41**, 689 (1982)

5.22   J. Feinberg: "Self-pumped, continuous-wave phase-conjugator using internal reflection," Opt. Lett. **7**, 486 (1982)

5.23   F.C. Jahoda, P.G. Weber, J. Feinberg: "Optical feedback, wavelength response, and interference effects of self-pumped phase conjugation in $BaTiO_3$," Opt. Lett. **9**, 362 (1984)

5.24   M. Cronin-Golomb, B. Fischer, J.O. White, A. Yariv: "Passive phase conjugate mirror based on self-induced oscillation in an optical ring cavity," Appl. Phys. Lett. **42**, 919 (1983)

5.25   T.Y. Chang, R.W. Hellwarth: "Optical phase conjugation by backscattering in barium titanate," Opt. Lett. **10**, 408 (1985)

5.26   B.Ya. Zel'dovich, V.I. Popovichev, V.V. Ragul'skii, F.S. Faizullov: "Connection between the wave fronts of the reflected and exciting light in stimulated Mandel'shtam-Brillouin scattering," Sov. Phys. JETP **15**, 109 (1972)

5.27   J.F. Lam: "Origin of phase conjugate waves in self-pumped photorefractive mirrors," Appl. Phys. Lett. **46**, 909 (1985)

5.28   C.-T. Chen, D.M. Kim, D. von der Linden: "Efficient pulsed photorefractive process in $LiNbO_3$: Fe for optical storage and deflection," IEEE J. Quantum Electron. QE-16, 126 (1980)

5.29   L.K. Lam, T.Y. Chang, J. Feinberg, R.W. Hellwarth: "Photorefractive-index gratings formed by nanosecond optical pulses in $BaTiO_3$," Opt. Lett. **6**, 475 (1981)

5.30   E. Kratzig, F. Welz, R. Orlowski, V. Doorman, M. Rosenkranz: "Holographic storage properties of $BaTiO_3$," Solid State Commun. **34**, 817 (1980)

5.31   S.P. Ducharme, J. Feinberg: "Speed of the photorefractive effect in a $BaTiO_3$ single crystal," J. Appl. Phys. **56**, 839 (1984)

5.32   N.V. Kukhtarev: "Kinetics of hologram recording and erasure in electrooptic crystals," Sov. Phys. Tech. Lett. **2**, 438 (1976)

5.33   F.P. Strohkendl, J.M.C. Jonathan, R.W. Hellwarth: "Hole-electron competition in photorefractive gratings," Opt. Lett. **11**, 312 (1986)

5.33a G.C. Valley: "Simultaneous electron/hole transport in photorefractive materials," J. Appl. Phys. **59**, 3363 (1986)

5.34 G.C. Valley: "Two-wave mixing with an applied field and a moving grating," J. Opt. Soc. Am. B1, 868 (1984)

5.35 H. Kogelnik: "Coupled wave theory for thick hologram gratings," Bell Syst. Tech. J. 48, 2909 (1969)

5.36 Y.H. Ja: "Energy transfer between two beams in writing a reflection volume hologram in a dynamic medium," Optical and Quantum Electronics 14, 547 (1982)

5.37 P. Yeh: "Contra-directional two-wave mixing in photorefractive media," Opt. Comm. 45, 323 (1983)

5.38 V.L. Vinetskii, N.V. Kukhtarev, S.G. Odulov, M.S. Soskin: "Dynamic self-diffraction of coherent light beams," Sov. Phys. Usp. 22, 742 (1979)

5.39 N.V. Kukhtarev, V.B. Markov, S.G. Odulov, M.S. Soskin, V.L. Vinetskii: "Holographic storage in electrooptic crystals. I. Steady state," Ferroelectrics 22, 949 (1979)

5.40 N.V. Kukhtarev, V.B. Markov, S.G. Odulov, M.S. Soskin, V.L. Vinetskii: "Holographic storage in electrooptic crystals. II. Beam coupling–light amplification," Ferroelectrics 22, 961 (1979)

5.41 D.W. Vahey: "A nonlinear coupled-wave theory of holographic storage in ferroelectric materials," J. Appl. Phys. 46, 3510 (1975)

5.42 R.A. Mullen, R.W. Hellwarth: "Optical measurement of the photorefractive parameters of $Bi_{12}SiO_{20}$," J. Appl. Phys. 58, 40 (1985)

5.43 P. Günter: "Holography, coherent light amplification and optical phase conjugation with photorefractive materials," Phys. Rep. 93, 199 (1982)

5.44 J.P. Huignard, A. Marrakchi: "Coherent signal beam amplification in two-wave mixing experiments with photorefractive $Bi_{12}SiO_{20}$ crystals," Opt. Comm. 38, 249 (1981)

5.45 H. Rajbenbach, J.P. Huignard, Ph. Refrégier: "Amplified phase-conjugate reflection by four-wave mixing with photorefractive $Bi_{12}SiO_{20}$ crystals," Opt. Lett. 9, 558 (1984)

5.46 H. Rajbenbach, J.P. Huignard: "Self-induced coherent oscillations with photorefractive $Bi_{12}SiO_{20}$ amplifier," Opt. Lett. 10, 137 (1985)

5.47 B. Fischer, M. Cronin-Golomb, J.O. White, A. Yariv: "Amplified reflection, transmission, and self-oscillation in real-time holography," Opt. Lett. 6, 519 (1981)

5.48 M. Cronin-Golomb, J.O. White, B. Fischer, A. Yariv: "Exact solution of a nonlinear model of four-wave mixing and phase conjugation," Opt. Lett. 7, 313 (1982)

5.49 J. Feinberg, G.D. Bacher: "Self-scanning of a continuous-wave dye laser having a phase-conjugating resonator cavity," Opt. Lett. 9, 420 (1985)

5.50 P. Yeh: "Theory of unidirectional photorefractive ring oscillators," J. Opt. Soc. Am. B2, 1924 (1985)

5.51 A. Yariv, S.K. Kwong: "Theory of laser oscillation in resonators with photorefractive gain," Opt. Lett. 10, 454 (1985)

5.52a M.D. Ewbank, P. Yeh: "Frequency shift and cavity length in photorefractive resonators," Opt. Lett. 10, 496 (1985)

5.52b D.Z. Anderson, R. Saxena: "Theory of multimode operation of a unidirectional ring oscillator having photorefractive gain: weak field limit," J. Opt. Soc. Am. 4, 164 (1987)

5.53 P. Yeh: "Photorefractive coupling in ring resonators," Appl. Opt. 23, 2974 (1984)

5.54 P. Yeh: "Electromagnetic propagation in a photorefractive layered medium," J. Opt. Soc. Am. 73, 1268 (1983)

5.54a K.R. MacDonald, J. Feinberg, M.Z. Zha, P. Günter: "Asymmetric Transmission through a photorefractive crystal of barium titanate", Opt. Comm. 50, 146 (1984)

5.55 P. Yeh: "Theory of phase-conjugate oscillators," J. Opt. Soc. Am. A2, 727 (1985)

5.56 E. Le Bihan, P. Verkerk, M. Pinard, G. Grynberg: "Observation of self-oscillation in ring cavities with sodium vapor phase conjugate mirrors," Opt. Comm. 56, 202 (1985)

5.57 M. Cronin-Golomb, K.Y. Lau, A. Yariv: "Infrared photorefractive passive phase conjugation with $BaTiO_3$: demonstrations with GaAlAs and 1.09-μm Ar^+ lasers," Appl. Phys. Lett. 47, 567 (1985)

5.58 B.T. Anderson, P.R. Forman, F.C. Jahoda: "Self-pumped phase conjugation in $BaTiO_3$," Opt. Lett. 10, 627 (1985)

5.59 K.R. MacDonald, J. Feinberg: "Theory of a self-pumped phase conjugator with two interaction regions," J. Opt. Soc. Am. 73, 548 (1983)

5.60 J. Feinberg: "Continuous-wave, self-pumped phase conjugator with wide field of view," Opt. Lett. 8, 480 (1983)

5.61 R.W. Hellwarth: "Theory of phase conjugation by stimulated scattering in a waveguide," J. Opt. Soc. Am. **68**, 1050 (1978)

5.62 S. Sternklar, S. Weiss, M. Segev, B. Fischer: "Beam coupling and locking of lasers using photorefractive four-wave mixing," Opt. Lett. **11**, 528 (1986)

5.63 A.M.C. Smout, R.W. Eason: "Analysis of mutually incoherent beam coupling in BaTiO$_3$," Opt. Lett. **12**, 498 (1987)

5.63a M.D. Ewbank: "Mechanism for photorefractive phase conjugation using incoherent beams," Opt. Lett. **13**, 47 (1988)

5.64 M. Ewbank, P. Yeh: "Fidelity of passive phase conjugators," SPIE's E-O Lase '86, Los Angeles, January, 1986, Paper 613-02

5.65 R.A. MacFarlane, D.G. Steel: "Laser oscillator using resonator with self-pumped phase-conjugate mirror," Opt. Lett. **8**, 208 (1983)

5.66 A.M.C. Smout, R.W. Eason, M.C. Gower: "Regular oscillations and self-pulsating in self-pumped BaTiO$_3$," Opt. Commun. **59**, 77 (1986)

5.67 P. Günter, E. Voit, M.Z. Zha, J. Albers: "Self-pulsation and optical chaos in self-pumped photorefractive BaTiO$_3$," Opt. Commun. **55**, 210 (1985)

5.68 M.C. Gower: "Photoinduced voltages and frequency shifts in a self-pumped phase conjugating BaTiO$_3$ crystal," Opt. Lett. **11**, 458 (1986)

5.69 P.Narum, D.J. Gauthier, R.W. Boyd: "Instabilities in a self-pumped barium titanate phase conjugate mirror," in *Optically Bistability III*, Proceeding of the Topical Meeting, Tucson, December, 1985, H.M. Gibbs et al., eds. (Springer, Berlin, Heidelberg 1986)

5.70 J.M. Ramsey, W.B. Whitten: "High-resolution self-scanning continuous wave dye laser," Anal. Chem. **56**, 2979 (1984)

5.71 M.D. Ewbank, P. Yeh, M. Khoshnevisan, J. Feinberg: "Time reversal by an interferometer with coupled phase-conjugate reflections," Opt. Lett. **10**, 282 (1985)

5.72 J.P. Huignard, A. Marrakchi: "Two-wave mixing and energy transfer in Bi$_{12}$SiO$_{20}$ crystals: application to image amplification and vibration analysis," Opt. Lett. **6**, 622 (1981)

5.73 M. Cronin-Golomb, A. Yariv: "Plane-wave theory of nondegenerate oscillation in the linear photorefractive passive phase-conjugate mirror," Opt. Lett. **11**, 242 (1986)

5.74 K.R. MacDonald, J. Feinberg: "Enhanced four-wave mixing by use of frequency-shifted optical waves in photorefractive BaTiO$_3$," Phys. Rev. Lett. **55**, 821 (1985)

5.75 S. Sternklar, S. Weiss, B. Fischer: "Tunable frequency shift of photorefractive oscillators," Opt. Lett. **11**, 165 (1986)

5.76 B. Fischer, S. Sternklar: "New optical gyroscope based on the ring passive phase conjugator," Appl. Phys. Lett. **47**, 1 (1985)

5.77 J.P. Jiang, J. Feinberg: "Dancing modes and frequency shifts in a phase conjugator," Opt. Lett. **12**, 266 (1987); also "Dancing Modes," 16 mm film or VHS video, copyright Jack Feinberg (Marmot Productions, Los Angeles, 1986)

5.78 V.V. Voronov, I.R. Dorosh, Yu.S. Kuz'minov, N.V. Tkachenko: "Photoinduced light scattering in cerium-doped barium strontium niobate crystals," Sov. J. Quantum Electron. **10**, 1346 (1981)

5.79 J. Feinberg: "Asymmetric self-defocusing of an optical beam from the photorefractive effect," J. Opt. Soc. Am. **72**, 46 (1982)

5.80 D.A. Temple, C. Warde: "Anisotropic scattering in photorefractive crystals," J. Opt. Soc. Am. **B3**, 337 (1986)

5.81 M.D. Ewbank, P. Yeh, J. Feinberg: "Photorefractive conical diffraction in BaTiO$_3$," Opt. Comm. **59**, 423 (1986)

5.82 S. Odulov, K. Belaev, I. Kisaleva: "Degenerate stimulated parametric scattering in LiTaO$_3$," Opt. Lett. **10**, 31 (1985)

5.83 E.M. Avakyan, S.A. Alaverdyan, K.G. Belabaev, V.Kh. Sarkisov, K.M. Tumanyan: "Charateristics of the induced optical inhomogeneity of LiNbO$_3$ crystals doped with iron ions," Sov. Phys. Solid State **20**, 1401 (1978)

5.84 E.M. Avakyan, K.G. Belabaev, S.G. Odulov: "Polarization-anisotropic light-induced scattering in LiNbO$_3$: Fe crystals," Sov. Phys. Solid State **25**, 1887 (1983)

5.85 V.I. Belinicher: "Space-oscillating photocurrents in crystals without symmetry center," Phys. Lett. **A66**, 213 (1978)

5.86 E.L. Ivchenko, G.E. Pikus: "New photogalvanic effect in gyrotropic crystals," JETP Lett. **27**, 604 (1978)

5.87 B.I. Sturman: "The photogalvanic effect – a new mechanism of nonlinear wave mixing in electrooptic crystals," Sov. J. Quantum Electron. **10**, 276 (1980)

5.88 V.M. Asnin, A.A. Bakun, A.M. Danishevskii, E.L. Ivchenko, G.E. Picus, A.A. Rogachev: "Observation of a photo-emf that depends on the sign of the circular polarization of the light," JETP Lett. **28**, 74 (1978)

5.89 S.G. Odulov: "Spatially oscillating photovoltaic currents in iron-doped lithium niobate crystals," JETP Lett. **35**, 10 (1982)

5.90 R. Grousson, S. Mallick, S. Odoulov: "Amplified backward scattering in $LiNbO_3 : Fe$," Opt. Commun. **51**, 342 (1984)

5.91 W.B. Whitten, J.M. Ramsey: "Self-scanning of a dye laser due to feedback from a $BaTiO_3$ phase-conjugate reflector," Opt. Lett. **9**, 44 (1984)

5.92 J.M. Ramsey, W.B. Whitten: "Phase-conjugate feedback into a continuous-wave ring dye laser," Opt. Lett. **10**, 362 (1985)

5.93 F.C. Jahoda, P.G. Weber, J. Feinberg: "Optical feedback, wavelength response and interference effects of self-pumped phase conjugation in $BaTiO_3$," Opt. Lett. **9**, 362 (1984)

5.94 K.R. MacDonald, J. Feinberg: 1984 Annual Meeting of the Optical Society of America, San Diego, October, 1984, Paper TuA7 [J. Opt. Soc. Am. **1**, 1213 (1984)]

5.95 J. Feinberg, G.D. Bacher: "Phase-locking lasers with phase conjugation," Appl. Phys. Lett. **48**, 570 (1986)

5.96 M. Cronin-Golomb, A. Yariv: "Self-induced frequency scanning and distributed Bragg reflection in semiconductor lasers with phase-conjugate feedback," Opt. Lett. **11**, 455 (1986)

5.97 J. Feinberg: "Interferometer with a self-pumped phase conjugator," Opt. Lett. **8**, 569 (1983)

5.98 B. Fischer, S. Sternklar: "Image transmission and interferometry with multimode fibers using self-pumped phase conjugation," Appl. Phys. Lett. **46**, 113 (1985)

5.99 M.D. Levenson: "High-resolution imaging by wave-front conjugation," Opt. Lett. **5**, 182 (1980)

5.100 M.D. Levenson, K.M. Johnson, V.C. Hanchett, K. Chiang: "Projection photolithography by wave-front conjugation," J. Opt. Soc. Am. **71**, 737 (1981)

5.101 M.D. Levenson, K. Chiang: "Image projection with nonlinear optics," IBM J. Res. Dev. **26**, 160 (1982)

5.102 M.C. Gower: in the "Annual Report to the Laser Facility Committee 1985, Laser Division, Rutherford Appleton Laboratory, Chilton, Didcot, Oxon, U.K."

5.103 S.H. Lee (ed.): *Optical Information Processing, Fundamentals*, Topics Appl. Phys., Vol. 48 (Springer, Berlin, Heidelberg 1981)

5.104 J.O. White, A. Yariv: "Real-time image processing via four-wave mixing in a photorefractive medium," Appl. Phys. Lett. **37**, 5 (1980)

5.105 C. Sexton, W.H. Steier: "Real-time correlations via degenerate optical four-wave mixing," CLEO '83, Baltimore, May, 1983

5.106 P.S. Brody: "Signal correlation with phase-conjugate holographic reconstruction using a $BaTiO_3$ crystal," SPIE's O-E Lase '86, Los Angeles, January, 1986, Paper 613-03

5.107 A.E.T. Chiou, P. Yeh, M. Khoshnevisan: "Coherent image subtraction using phase conjugate interferometry," SPIE's O-E Lase '86, Los Angeles, January, 1986, Paper 613-34

5.108 A.E.T. Chiou, P. Yeh: "Parallel image subtraction using a phase-conjugate Michelson interferometer," Opt. Lett. **11**, 306 (1986)

5.109 S.-K. Kwong, G. Rakuljic, V. Leyva, A. Yariv: "Image processing using self-pumped phase-conjugate mirrors," SPIE's O-E Lase '86, Los Angeles, January, 1986, Paper 613-07

5.110 J.P. Huignard, J.P. Herriau: "Real-time coherent object edge reconstruction with $Bi_{12}SiO_{20}$ crystals," Appl. Opt. **17**, 2671 (1978)

5.111 J. Feinberg: "Real-time edge enhancement using the photorefractive effect," Opt. Lett. **5**, 330 (1980)

5.112 E. Ochoa, J.W. Goodman, L. Hesselink: "Real-time enhancement of defects using BSO," Opt. Lett. **10**, 621 (1985)

5.113 J.P. Huignard, A. Marrakchi: "Two-wave mixing and energy transfer in $Bi_{12}SiO_{20}$: application to image amplification and vibration analysis," Opt. Lett. **6**, 622 (1981)
5.114 J.J. Hopfield: "Neural networks and physical systems with emergent collective computational abilities," Proc. Natl. Acad. Sci. USA **79**, 2554 (1982)
5.115 D. Gabor: "Associative holographic memories," IBM J. Res. Dev. **13**, 156 (1969)
5.116 N.H. Farhat, D. Psaltis, A. Prada, E. Paek: "Optical implementation of the Hopfield model," Appl. Opt. **24**, 1469 (1985)
5.117 D. Psaltis, N. Farhat: "Optical information processing based on an associative-memory model of neural nets with thresholding and feedback," Opt. Lett. **10**, 98 (1985)
5.118 T. Kohonen: *Self Organization and Associative Memory*, Springer Ser. Inf. Sci. Vol. 8 (Springer, Berlin, Heidelberg 1984)
5.119 M.S. Cohen: "Self-organization in neural networks and optical analogs," SPIE's O-E Lase '86, Los Angeles, January, 1986, Paper 625-30
5.120 M.S. Cohen: "Design of a new medium for volume holographic information processing," Appl. Opt. **25**, 2288 (1986)
5.121 B.H. Soffer, G.J. Dunning, Y. Owechko, E. Marom: "Associative holographic memory using phase-conjugate mirrors," Opt. Lett. **11**, 118 (1986)
5.122 D.Z. Anderson: "Coherent optical eigenstate memory," Opt. Lett. **11**, 56 (1986)
5.123 D.Z. Anderson, D.M. Lininger, J. Feinberg: "Optical tracking novelty filter," Opt. Lett. **12**, 123 (1987)

6. Optical Processing Using Wave Mixing in Photorefractive Crystals

Jean-Pierre Huignard and Peter Günter

With 56 Figures

Over the past several years, interest has arisen in nonlinear optical materials allowing operations in real time on the phase and on the amplitude of coherent wavefronts. The use of such nonlinear interactions has resulted in numerous new applications, such as dynamic holography, information storage, parallel signal processing, phase conjugation and self-induced optical resonators. It was recognized early that the photoinduced index change in electro-optic crystals (photorefractive effect) exhibits unique capabilities for these applications. Indeed, large index modulation can be created in different crystals with low power laser beams emitting at visible and near infra-red wavelengths. The effect was applied to real-time volume holography with parallel theoretical and experimental developments on the origin of the effect. The investigation of beam coupling effects during grating recording resulted in the amplification of a low intensity probe beam due to energy transfer from a pump beam [6.1–3]. The implication of these phenomena for applications in signal processing are therefore the subject of active research in different laboratories and are also stimulating further basic studies on the nonlinear optical properties of electro-optic crystals.

6.1 Photoinduced Space-Charge Field in Photorefractive Crystals

The mechanism of the photorefractive grating recording and erasure in electro-optic crystals is discussed in detail in TAP61. We will thus restrict this section to the presentation of the most important parameters involved in applications of photorefractive crystals, i.e., the crystal sensitivity [6.4, 5], the grating buildup time constants, the steady-state diffraction efficiency and the spatial frequency response [6.3]. The purpose of this section is thus to compare the different photorefractive crystals and to emphasize some of the optimal properties required for their application in optical signal processing, dynamic holography, and optical phase conjugation.

6.1.1 Photorefractive Sensitivity

The photorefractive sensitivity is defined, as for other holographic recording materials, as the refractive index change Δn per unit absorbed energy density:

$$S = \frac{\Delta n}{\alpha I_0 \tau} \tag{6.1}$$

where α is the crystal absorption coefficient at recording wavelength λ; τ is the crystal response time and I_0 the incident power density (also called incident intensity). This definition of S is a useful figure of merit since it allows one to compare, on an equal basis, materials having different absorption coefficients at a given wavelength [6.6]. The response time of a photorefractive crystal is the dielectric relaxation time multiplied by a function of different parameters such as applied field E_0, grating spacing Λ, drift and diffusion lengths (respectively r_E and r_D) of the photocarriers:

$$S = \frac{1}{2}n_0^3\frac{r}{\varepsilon\varepsilon_0}F(E_0,\Lambda,r_E,r_D) \quad . \tag{6.2}$$

Since n_0 and r/ε are nearly constant for all electro-optic crystals, we show in the following, that the photorefractive sensitivity is mainly determined by the recording conditions and by the relative values of the drift and diffusion lengths of the photocarriers compared to the grating spacing. The photorefractive sensitivity has been related, in a straightforward model, to the microscopic charge displacement in the crystal in [6.4]. For a sinusoidal variation of the photocarrier excitation of spatial frequency $K = 2\pi\Lambda^{-1}$ and of modulation depth m, S is given in the drift and diffusion regimes by the following expressions, only valid in the short-time limit:

For diffusion recording ($E_0 = 0$)

$$S = \frac{1}{2}n_0^3\frac{r}{\varepsilon\varepsilon_0}e\frac{\Phi}{h\nu}m\frac{Kr_D^2}{1+K^2r_D^2} \quad . \tag{6.3}$$

For drift recording ($E_0 \neq 0$)

$$S = \frac{1}{2}n_0^3\frac{r}{\varepsilon\varepsilon_0}e\frac{\Phi}{h\nu}m\frac{r_E}{(1+K^2r_E^2)^{1/2}} \quad . \tag{6.4}$$

S reaches a maximum value when the excited photocarries drift or diffuse over distances equal to or larger than the grating spacing ($Kr \geq 1$). The upper limit of S for an elementary grating ($m = 1$) and unit quantum efficiency ($\Phi = 1$) is

$$S_{\mathrm{max}} = \frac{1}{2}n_0^3\frac{r}{\varepsilon\varepsilon_0}\cdot\frac{e}{h\nu}\cdot\frac{1}{K} \quad . \tag{6.5}$$

This can be estimated as $0.1\,\mathrm{cm}^3\,\mathrm{J}^{-1}$ for $\lambda = 0.5\,\mu\mathrm{m}$. This optimim photorefractive sensitivity is reached in efficient photoconductive crystals such as KTN (two-photon absorption) and $Bi_{12}SiO_{20}$, $Bi_{12}GeO_{20}$, GaAs (linear absorption).

Another figure of merit which is commonly used for experiments with photorefraction is the energy per unit area W to write an elementary grating ($m = 1$) having 1% (or a few percent) efficiency in a crystal having $1\,\mathrm{mm}$ (or a few mm) thickness. This figure of merit enables a ready comparison of slow and fast materials illuminated with the same incident beam intensity I_0 [6.6]. The only problem with this definition is that some materials do not reach a 1% efficiency for a crystal thickness of few millimeters. For example in BSO-BGO,

$\eta = 10^{-3}$ when recording by diffusion. The relation between W and S is the following:

$$W = \sqrt{\eta}\frac{\lambda}{\alpha d}\frac{1}{\pi S}\exp\left(\frac{\alpha d}{2}\right) \quad . \tag{6.6}$$

From the value of $S_{\max}$ derived in [6.4, 5], the optimum grating recording energy would be of the order of $W = 50\,\mu\mathrm{J\,cm^{-2}}$ for a crystal thickness $d = 10\,\mathrm{mm}$ ($\alpha d \cong 1$). It must be noted that these values of recording energy in dynamic materials are nearly equivalent to high resolution silver halide plates.

The problem encountered with photorefractive crystals with respect to their applications in optical signal processing is obtaining a low writing energy while maintaining a high value of the photoinduced index modulation for efficient interaction between the recording beams. Since both of these requirements cannot be met, in general, with the same crystal, further research into improving certain properties of photorefractive crystals is presently being actively pursued.

6.1.2 Steady-State Diffraction Efficiency

The complete mathematical description of grating formation in photorefractive crystals has been derived by *Kukhtarev* et al. [6.3]. From this model, the photoinduced index modulation at saturation regime is given by the following expression valid for small phase changes [6.3, 8]

$$\Delta n = 2n_0^3 r \frac{\sqrt{\beta}}{(1+\beta)}\left[\frac{E_0^2 + E_D^2}{(1 + E_D/E_q)^2 + (E_0^2/E_q^2)}\right]^{1/2} \tag{6.7}$$

and the phase shift Φ_g between the incident fringe pattern and the photoinduced index modulation is given by:

$$\tan\Phi_g = \frac{E_D}{E_0}\left(1 + \frac{E_D}{E_q} + \frac{E_0^2}{E_D E_q}\right) \quad , \tag{6.8}$$

where β is the incident intensity ratio of the two interfering beams. E_D is the diffusion field and E_q is the maximum field which would correspond to a complete separation of the posivite and negative charges by one grating period. The expressions for these fields are the following:

$$E_D = \frac{2\pi kT}{e\Lambda} \quad ; \quad E_q = \frac{e}{2\pi\varepsilon_0\varepsilon}N_A\Lambda \quad ,$$

where N_A is the trap density in the crystal volume. From relations (6.7, 8) it is thus expected that the photoinduced index modulation may vary significantly with grating spacing and this point is of particular importance for imaging applications.

The diffraction efficiency η of a thick phase transmission grating (Fig. 6.1a) with a peak to peak index modulation $2\Delta n$ is derived from the *Kogelnik* formula [6.9]

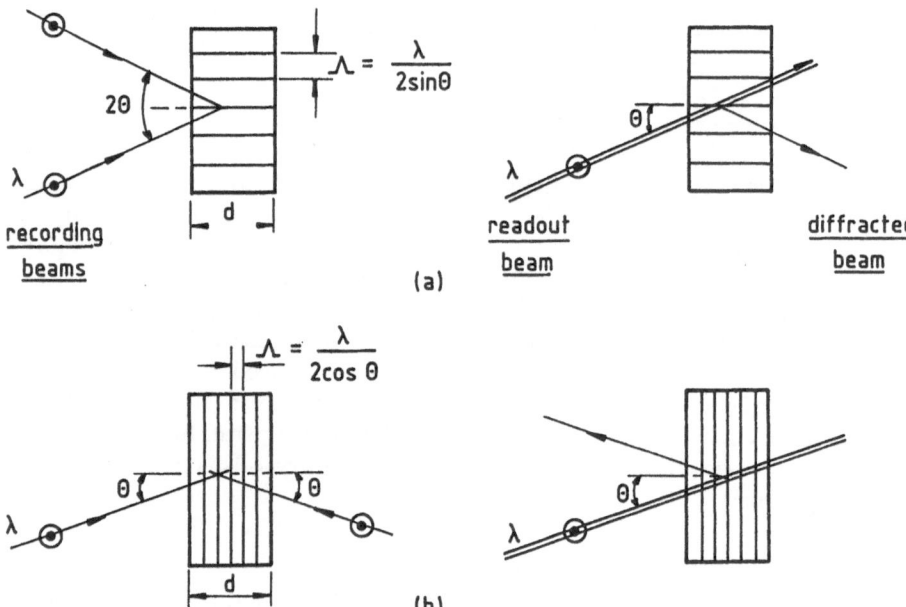

Fig. 6.1a,b. Volume grating recording and readout in photorefractive crystals. **(a)**: transmission type grating; **(b)**: reflection type grating

$$\eta = \exp\frac{(-\alpha d)}{\cos\theta}\,\sin^2\frac{\pi d\Delta n}{\lambda\cos\theta}\quad, \tag{6.9}$$

where θ is the Bragg angle inside the crystal.

The readout of the elementary grating (just one spatial frequency K) with a wavelength λ' different from the one used for recording λ is possible provided the Bragg conditions in the material is satisfied, i.e.

$$\sin\theta' = \frac{\lambda'}{4\pi}K\quad. \tag{6.10}$$

However, for the reconstruction of a wavefront which contains spatial information, readout of the phase volume hologram with the same wavelength is generally required for full image restoration. Although the transmission gratings are dominant in most experiments, the reflection-type gratings arising from the interference between two writing beams of opposite direction (Fig. 6.1b) may also be efficiently recorded due to the high spatial frequency response of the photorefractive effect. For such a configuration (holographic fringes parallel to the crystal faces) the grating efficiency is [6.9]

$$\eta = \exp\frac{(-\alpha d)}{\sin\theta}\,\tanh^2\frac{\pi d\Delta n}{\lambda\sin\theta}\quad. \tag{6.11}$$

After this reminder of the basic properties of volume phase gratings, we consider in the next section the evolution of the steady-state diffraction efficiency

η as a function of the spatial frequency of the photoinduced grating in the photorefractive crystal.

6.1.3 Spatial Frequency Response

A precise knowledge of the spatial frequency response of the photorefractive crystals is important for applications to high resolution imaging and optical signal processing [6.6, 7]. Indeed, the modulation transfer function represents the ability of the crystal to restore with high fidelity a wavefront containing two-dimensional spatial information. We present an analysis of the diffraction efficiency of an elementary grating which is based on the expression for the steady-state index modulation and which applies to crystals with a long transport length of the photocarriers. In the limit of low photoinduced index modulation, the grating diffraction efficiency is given by:

$$\eta = \frac{4\beta}{(1+\beta)^2}\left(\frac{\pi n_0^3 r}{\lambda \cos\theta}\right)^2 d^2 F^2 \frac{E_0^2 + E_D^2}{(1 + E_D/E_q)^2 + (E_0^2/E_q^2)} e^{-\alpha d} \qquad (6.12)$$

where $F = 1$ or 0.5 depending on the linear or quadratic recombination of the photocarriers, and θ is the Bragg angle inside the photorefractive crystal. This expression for the diffraction efficiency neglects the coupling between the two writing beams. This arises from self-diffraction effects in dynamic holographic materials when there is a phase shift between the intensity pattern and the photoinduced index modulation [6.1, 3]. A consequence of this beam coupling is the ability of the crystal to amplify a low intensity signal beam through energy transfer from a pump beam [6.10]. These effects will be considered in the last part of this chapter since they lead to the demonstration of new types of parametric image amplifiers and oscillators. The evolution of the diffraction efficiency in a photorefractive crystal such as BSO (or BGO) as a function of the grating spatial frequency is given in Fig. 6.2 for different values of the applied electric field E_0 [6.7, 8]. The grating is recorded with the green line of the argon laser ($\lambda = 514\,\mathrm{mm}$) – crystal thickness $d = 3\,\mathrm{mm}$ – and the diffraction efficiency is monitored in real time with a low power He-Ne laser, placed under Bragg incidence. The fields E_D and E_q will be written as follows:

$$E_D = A\Lambda^{-1} \quad ; \quad E_q = B\Lambda \quad . \qquad (6.13)$$

If in relation (6.12) we just retain the terms which depend on spatial frequency, we find the following expression for the diffraction efficiency:

$$\eta = C \times \frac{E_0^2 + A^2\Lambda^{-2}}{[1 + (A/B)\Lambda^{-2}]^2 + (E_0^2\Lambda^{-2}/B^2)} \frac{4n_0^2}{4n_0^2 - \lambda^2\Lambda^{-2}} \qquad (6.14)$$

where C is a constant characteristic of the photorefractive crystal used.

Depending on the respective values of the applied field E_0 and fringe spacing Λ, either drift or diffusion recording mechanisms will dominate the grating fromation and will therefore affect the grating efficiency differently.

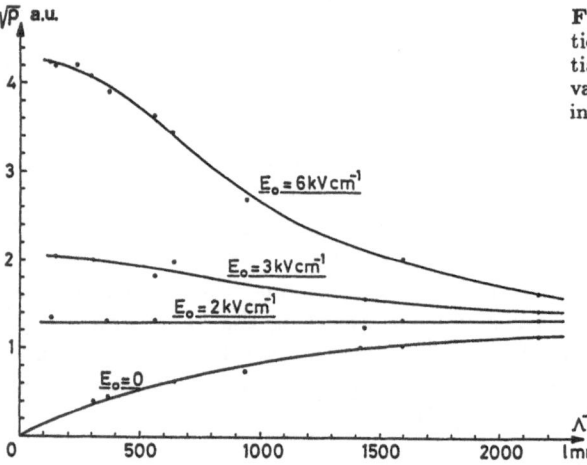

Fig. 6.2. Square root of diffraction efficiency versus grating spatial frequency Λ_g^{-1} for different values of the applied electric field in photorefractive BSO crystals

In Fig. 6.2, the theoretical curves have been fitted to the experimental points with the following data:

$$A = 0.16\,\text{V} \quad ; \quad B = 7 \times 10^7\,\text{V cm}^{-2} \quad ; \quad T = 300\,\text{K} \quad ; \quad \varepsilon = 56 \quad .$$

In agreement with the theory, the diffracted beam intensity increases for the diffusion recording ($E_0 = 0$) as the square of the grating spatial frequency. For large applied fields, the space-charge field is enhanced and η increases until saturation occurs when E_0 becomes comparable to E_q. Consequently, the drift recording mode ($E_0 \neq 0$) provides much higher diffraction efficiency but for gratings having low spatial frequencies. In that domain ($\Lambda^{-1} < 300\,\text{mm}^{-1}$) the efficiency increases as the square of the grating spacing. For $E_0 = 2\,\text{kV cm}^{-1}$, the diffusion and drift recording mechanisms add in such a way that the grating efficiency remains constant for all spatial frequencies. The results presented in Fig. 6.2 enable a determination of the maximum space-charge field E_q which corresponds to a complete separation of positive and negative charges by one grating period. The fitted values for E_q and E_D are the following:

$$E_q = 7\,\text{kV cm}^{-1} \quad ; \quad E_D = 1.6\,\text{kV cm}^{-1} \qquad \text{for} \quad \Lambda = 1\,\mu\text{m}$$
$$E_q = 70\,\text{kV cm}^{-1} \quad ; \quad E_D = 0.16\,\text{kV cm}^{-1} \qquad \text{for} \quad \Lambda = 10\,\mu\text{m} \quad .$$

From relations (6.13) we can derive the trap density $N_A \cong 1.2 \times 10^{16}\,\text{cm}^{-3}$ [6.7, 11, 12]. The results presented in Fig. 6.2 are a general characteristic of the photorefractive effect in electro-optic crystals. A similar dependence is obtained in crystals such as $KNbO_3$, $BaTiO_3$, SBN, $LiNbO_3$ and GaAs [6.8]. However, since these materials exhibit different electro-optic coefficients and trap densities, a noticeable difference may result in the value of the steady diffraction efficiencies for given recording conditions. Moreover, ferroelectric crystals such as $BaTiO_3$, SBN etc. are generally used with diffusion ($E_0 = 0$) since the application of a high electric field may have unwanted consequences such as crystal damage and depolarization. To summarize, the study of the spatial frequency response of a photorefractive crystal provides important information. Indeed, the experimental determination of the field E_q leads to a knowledge of the trap density, and it is shown that the crystal modulation transfer function can be controlled by the amplitude of the externally applied field. Recording by drift with $E_0 > 6\,\mathrm{kV\,cm^{-1}}$ gives a higher response in the low spatial frequency domain (low-pass filtering) while recording by diffusion at zero applied field ($E_0 = 0$) favors the high spatial frequencies (high-pass filtering).

6.1.4 Response Time of the Photorefractive Effect

The time constant for buildup of a grating is also an important characteristic of the photorefractive effect. The refractive index changes are due to electro-optic effects driven by space-charge fields and the time required to record a grating depends on the efficiency of the charge generation and transport process. The inertia in the nonlinear response of photorefractive media constitutes an important difference from other nonlinear media where the refractive index change is of electronic origin and thus occurs instantaneously. A complete analysis of the time evolution of the grating formation is presented by *Kukhtarev* in [6.13] for continuous wave illumination and by *Valley* for high irradiance nanosecond pulses [6.14]. In the following we give the expression for the grating time constant τ valid for cw illumination and for charge transport lengths which may be larger than or comparable to the grating. Under such conditions the crystal response has an overdamped oscillatory behavior with a response time given by [6.8]:

$$\tau = \tau_{\mathrm{di}} \frac{(1 + \tau_R/\tau_D)^2 + (\tau_R/\tau_E)^2}{[1 + (\tau_R\tau_{\mathrm{di}}/\tau_D\tau_I)](1 + \tau_R/\tau_D) + (\tau_R/\tau_E)^2(\tau_{\mathrm{di}}/\tau_I)} \quad , \qquad (6.15)$$

where τ_{di} is the dielectric relaxation of the crystal:

$$\tau_{\mathrm{di}} = \frac{\varepsilon\varepsilon_0}{n_0\mu e} \quad , \qquad\qquad\qquad (6.16)$$

n_0 is the free carrier concentration due to the incident illumination I_0:

$$n_0 = \tau_R \frac{\alpha\Phi}{h\nu} I_0$$

and μ is the mobility of the photocarriers. The charge recombination time τ_R may be written as:

$$\tau_R = \frac{1}{\gamma_R N_A} \quad ,$$

where γ_R is the recombination coefficient, τ_E and τ_D are the drift and diffusion time of the charges, given respectively by:

$$\tau_E = \frac{1}{K\mu E_0} \quad , \quad \tau_D = \frac{1}{\mu k T K^2} \quad .$$

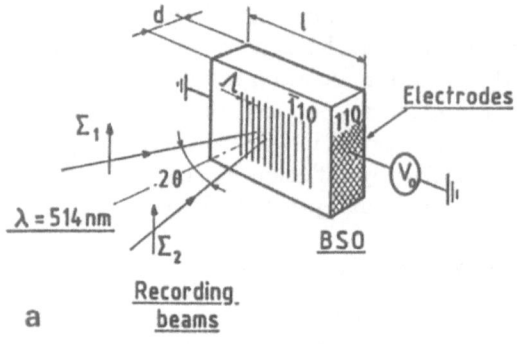

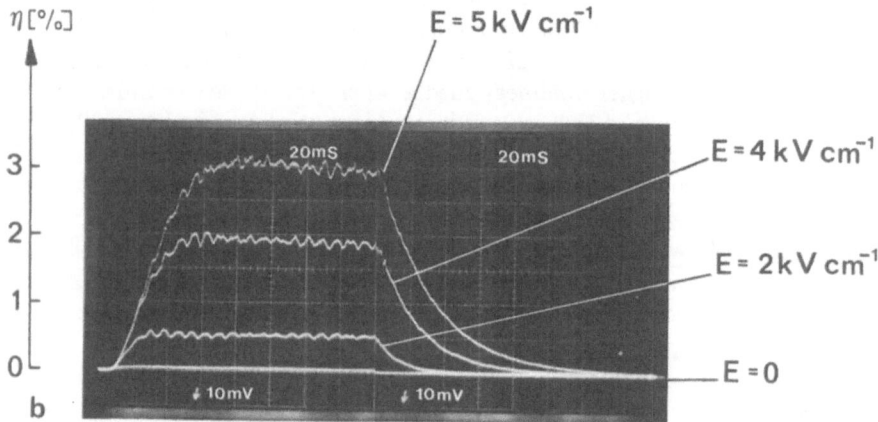

Fig. 6.3a,b. Grating recording and readout in photorefractive BSO. (a): crystallographic orientation; (b): recording-erasure cycles − Incident power = 13 mW cm^{-2} at $\lambda = 514$ nm. Applied voltages = 0−5 kV cm^{-1}. Grating period $\Lambda_g = 6\,\mu$m

τ_I is the inverse of the sum of photogeneration rate sI_0 and ion recombination rate $\gamma_R n_0$

$$\tau_I = \frac{1}{sI_0 + \gamma_R n_0} \quad .$$

A simple expression for the time dependence of the space-charge field during grating recording is the following:

$$\Delta E_{\text{sc}} = mE_{\text{sc}}[1 - e^{-t/\tau}] \quad . \tag{6.17}$$

During erasure by uniform illumination, the photoinduced space-charge field decreases according to the relation:

$$\Delta E_{\text{SC}} = mE_{\text{SC}}e^{-t/\tau} \tag{6.18}$$

where E_{SC} is the initial amplitude of the field and τ the same response time obtained for writing. The recording-erasure cycle is therefore symmetrical as shown in Fig. 6.3 for a photorefractive BSO crystal. The typical recording-erasure time for an elementary grating of 10–100 ms corresponds to an incident intensity of 10–100 mW cm^{-2} at the blue or green line of the Argon laser. BSO is known to be a fast and sensitive material with a relatively small electro-optic coefficient [6.15], while crystals such as BaTiO$_3$ have a large electro-optic coefficient but respond rather slowly, i.e., response times of few seconds for the conditions of Fig. 6.3 [6.6]. Therefore, another important figure of merit of a photorefractive crystal will be the energy required to reach the steady-state diffraction efficiency and this parameter often determines the crystal chosen for a particular application.

6.1.5 Isotropic Bragg Diffraction

a) **Ferroelectric Crystals.** In general, photorefractive crystals are used in the configuration of a transverse electro-optic modulator, i.e., the space-charge field is photoinduced along the direction of the electro-optic axis (Fig. 6.4a). For example, for ferroelectric LiNbO$_3$ and LiTaO$_3$ crystals, it is the $\vec{c}$ axis with the following refractive index modulation in that direction:

$$\Delta n = \tfrac{1}{2} n_e^3 r_{33} E_{SC} \quad .$$

Therefore, a readout beam whose polarization is parallel to the $\vec{c}$ axis is diffract-ed with the same polarization (isotropic Bragg diffraction).

When the highest electro-optic coefficient is not along the polar axis, the crystal is inclined with respect to the incident recording beams. Under such conditions, the grating K vector, and the wave vector of the readout beam of the grating have a component along the crystallographic axis. This geome-try exploits the large r_{42} electro-optic coefficient of single domain ferroelectric BaTiO$_3$, SBN or KNbO$_3$ crystals [6.16, 17].

b) **Cubic Crystals.** If we consider photorefractive crystals such as BSO, BGO, GaAs or InP, these crystals are cubic and have fourfold symmetry axes along the cube edges. An electric field along the $\langle 001 \rangle$ direction, either photoinduced or externally applied, induces in the x direction the following change of the index of refraction:

$$\Delta n_x = \tfrac{1}{2} n_0^3 r_{41} E_{SC} \tag{6.19}$$

while there is no change of refractive index along the z direction (see Fig. 6.4b). Thus, the maximum efficiency is obtained with a readout beam vertically polar-ized along the x direction, and the diffracted beam with the same polarization direction.

6.1.6 Anisotropic Bragg Diffraction

a) **Cubic Crystals.** When the electric field is applied along the $\langle 110 \rangle$ direction, it induces in this class of materials two new principal electro-optic axes x', y' with the refractive index modulations given by:

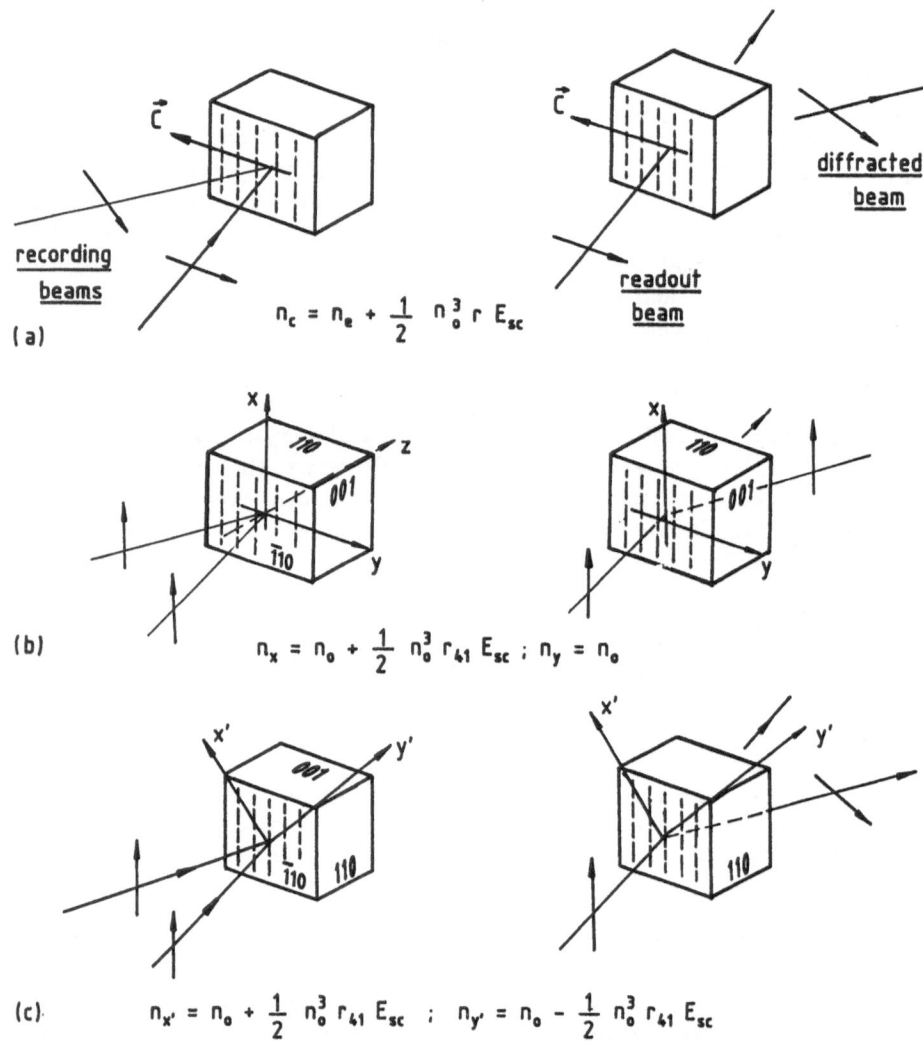

Fig. 6.4a–c. Crystallographic orientation and related index changes and beam polarizations for grating recording, readout in photorefractive crystals. Isotropic Bragg diffraction in ferroelectric (a) and in cubic (b) crystals. (c) anisotropic Bragg diffraction in cubic crystals

$$\Delta n_{x'} = +\tfrac{1}{2}n_0^3 r_{41} E_{SC} \quad , \quad \Delta n_{y'} = -\tfrac{1}{2}n_0^3 r_{41} E_{SC} \quad . \tag{6.20}$$

When reading a grating photoinduced in the photorefractive crystal with a vertical beam polarization such as shown in Fig. 6.4c, the diffracted beam is linearly polarized but its polarization is rotated by 90° with respect to the incident reading beam (anisotropic Bragg diffraction) [6.18, 19]. Moreover, with this configuration, the total index modulation seen by the reading beam is $\Delta n = n_0^3 r_{41} E_{SC}$, thus giving a diffraction efficiency 4 times higher than in the isotropic configuration. The explanation of this 90° polarization switching is

214

the following: The incident vertical beam polarization may be projected onto the two principal axes x' and y' and, therefore, each component gives rise to a diffracted beam polarized along the x' and y' directions, respectively. Since a diffracted beam is phase delayed by $\frac{\pi}{2}$ with respect to the reading beam, the phase difference between the two orthogonal components of diffraction is $(+\frac{\pi}{2}) - (-\frac{\pi}{2}) = \pi$. These components have the same amplitude and are phase shifted by π, therefore the polarization of the diffracted beam is switched by $90°$ with respect to the incident readout beam (the diffraction by the grating is equivalent to a half-wave plate). This polarization switching allows a significant increase in the signal to noise ratio of images in holographic experiments [6.18]. A correctly oriented polarizer suppresses the scattered light generated by the coherent illumination of the crystal and by the optical elements of the set-up. Experimental demonstration of high quality image reconstruction by polarization filtering will be reported in Sect. 6.3.2.

b) Optically Active Crystals. A difficult problem encountered with selenite type crystals (BSO, BGO, BTO) is related to their strong optical activity (for example $\varrho_0 = 45° \, \text{mm}^{-1}$ at $\lambda = 514 \, \text{nm}$ for BSO-BGO crystals). This circular birefringence is superimposed on the linear birefringence due to the Pockels effect. When a beam enters the crystal with the two polarization components E_1 and E_2 along the x' and y' axes, the beam emerging from the crystal has two components of polarization E_1', E_2' determined by the following transfer matrix:

$$\begin{pmatrix} E_1' \\ E_2' \end{pmatrix} = \begin{pmatrix} \cos \Delta d - \text{i} \sin \psi \sin \Delta d & -\cos \psi \sin \Delta d \\ +\cos \psi \sin \Delta d & \cos \Delta d + \text{i} \sin \psi \sin \Delta d \end{pmatrix} \begin{pmatrix} E_1 \\ E_2 \end{pmatrix}$$

(6.21)

where

$$\Delta = (\varrho_0 + \varphi_0^2/4)^{1/2} \quad ; \quad \psi = \tan^{-1} \frac{\varphi_0}{2\varrho_0}$$

in which ϱ_0 is the optical activity per unit length, φ_0 the birefringence per unit length, and d is the crystal thickness.

In the general case, the light emerging from an optically active and electro-optic crystal will be elliptically polarized. When a grating photoinduced in the crystal volume is read, these materials will exhibit very specific polarization properties which to date have been difficult to derive by analytical methods. As an example, the directly transmitted and diffracted beams have different elliptical polarizations and the grating efficiency is a complex function of the orientation of the reading beam polarization [6.20, 23]. These properties of selenite-type crystals are the subject of theoretical and experimental studies in various laboratories, since they have direct implications for the diffracted image intensity and signal-to-noise ratio in wave mixing experiments.

6.1.7 Space-Charge Field Nonlinearities

The expressions for the photoinduced space-charge field in electro-optic crystals given in the literature are essentially valid when the incident fringe pattern illuminating the crystal has a low contrast ($m \ll 1$). For these recording conditions, the solution of the *Kukhtarev* equations are linear and the amplitude of the space charge field E_{SC} is derived from (6.7). For example, for a drift dominant recording mechanism ($E_0 \gg E_D$; and $E_0 < E_q$) as often used in experiments with BSO, we have the following expression for E_{SC}:

$$E_{SC} = mE_0 \cos Kx \quad m \ll 1 \quad . \tag{6.22}$$

However, when the incident fringe modulation is nearly unity, the analytic expression for the photoinduced space-charge field is the following [6.3, 24]:

$$E_{SC} = \frac{E_0(1 - m_1^2)^{1/2}}{1 + m_1 \cos Kx} \quad , \tag{6.23}$$

where $m_1 = m[\tau g_0/(n_D + n_0)]$; n_D is the free carrier concentration in the dark and n_0 is the free carrier concentration under illumination.

A more general expression including the contribution of the diffusion field E_D and the photovoltaic field E_{ph} can be found in [6.24]. This expression applies, in particular, to BaTiO$_3$ and LiNbO$_3$ crystals, respectively, where one of these fields is dominant for grating recording. The validity domain of relation (6.23) is, however, restricted since it does not account for saturation effects in the space-charge field amplitude versus the applied field E_0 in photoconductive BSO-BGO crystals. For a given fringe spacing, saturation occurs due to complete charge separation by one grating period and trap filling. We may nonetheless use the closed-form expression for the steady-state photoinduced field in order to point out the nonlinearities of the crystal response by expanding (6.13) into its Fourier components:

$$E_{SC}(x) = 2E_0 \sum_{h=1}^{+\infty} \frac{1}{m_1^h} [(1 - m_1^2)^{1/2} - 1]^h \cos hKx \quad . \tag{6.24}$$

This expansion shows that harmonics of the fundamental grating wave vector K are created when the crystal is exposed with a fringe contrast equal to unity (interference of two beams of equal intensity). This ability of the crystal to generate grating harmonics will be used in experiments on nonlinear mixing of grating $\boldsymbol{K}$ vectors written with different wavelengths.

6.1.8 Collinear Bragg Diffraction

Two gratings wave vectors K_1 and K_2 are photoinduced in BSO with the two collinear writing beams at wavelengths λ_1 and λ_2 (Fig. 6.5). The space-charge field resulting from the successive (or simultaneous) illumination of the crystal with the two interference patterns having grating vectors K_1 and K_2 may be written as:

$$E_{SC}(x) = E_0 \frac{(1 - m_1^2)^{1/2}}{1 + \frac{1}{2}m_1(\cos K_1 x + \cos K_2 x)} \quad . \tag{6.25}$$

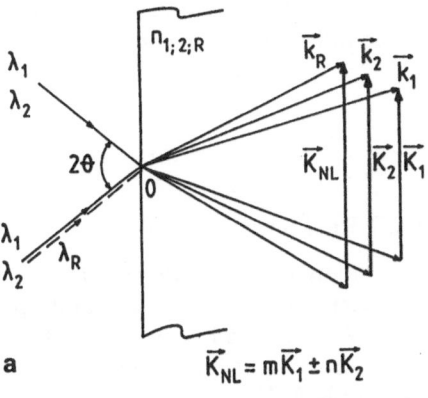

Fig. 6.5. (a) Wave-vector diagram illustrating the collinear Bragg diffraction. K_{NL} is induced by the nonlinear mixing of grating wave vectors K_1 and K_2. (b): collinear Bragg diffraction in BSO. Writing wavelengths $\lambda_1 = 633$ nm; $\lambda_2 = 514$ nm; readout wavelength $\lambda_R \approx 840$ nm

$$\vec{K}_{NL} = m\vec{K}_1 \pm n\vec{K}_2$$

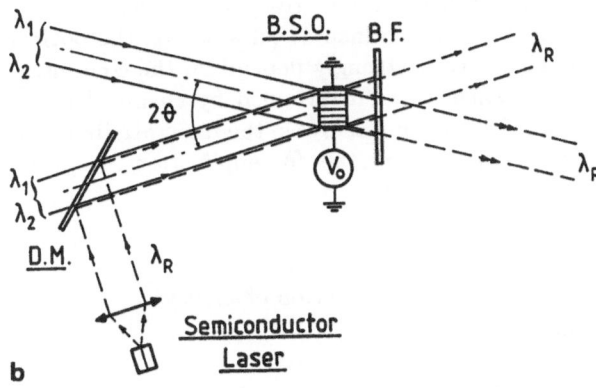

As analyzed in Sect. 6.1.7, this relation takes into account the space-charge field nonlinearities arising when recording with high contrast interference fringes. Expanded into Fourier components this relation takes the form:

$$E_{SC}(x) = \frac{E_0}{2} \sum_{n,m=0}^{+\infty} \alpha_{mn} \cos{[mK_1 + nK_2]x}$$
$$+ \alpha_{mn} \cos{[mK_1 + nK_2]x} \quad . \tag{6.26}$$

Therefore, exposing the crystal with two collinear grating vectors generates, through nonlinear mixing, the new wave vectors $K_{NL} = mK_1 \pm nK_2$ [6.25, 26] which satisfy the Bragg condition for a readout wavelength λ_R such that

$$\frac{1}{\lambda_R} = \frac{m}{\lambda_1} \pm \frac{n}{\lambda_2} \quad . \tag{6.27}$$

The experimental demonstration of the method is described in reference [6.26] with the two recording wavelengths $\lambda_1 = 633$ nm (He-Ne laser) and $\lambda_2 = 514$ nm (Argon laser). Choosing $m = 2$, $n = 1$, ensures a collinear Bragg diffraction at the semiconductor laser wavelength $\lambda_R = 822$ nm (Fig. 6.5b) [6.26].

Since this method is based on the nonlinear mixing of higher order nonlinearities, it consequently results in low values of the diffraction efficiency (typically $\eta \approx 10^{-3}$ for grating spacings $\Lambda \approx 3\,\mu m$).

6.2 Selection of Materials

6.2.1 Review of Photorefractive Crystal Performance

The photorefractive effect has been studied in many electro-optic crystals and a great deal of research is devoted to the understanding of the effect and optimizing the materials for various applications. As discussed in detail in other chapters, the source of photocarriers in photorefractive crystals was found to be a small concentration of impurities (for example Fe^{2+} in $LiNbO_3$) incorporated into the lattice during crystal growth. The electro-optic properties do not depend on these impurities, but the charge transport parameters, the optical absorption, and thus the crystal sensitivity strongly depend on the concentration of impurities and on their valence states determined by chemical treatment (oxidation-reduction). From the previous analysis, the main attributes to be considered for selecting photorefractive crystals for application in optical signal processing may be summarized as follows:

a)   Photorefractive sensitivity
b)   Photorefractive recording and erasure time constants
c)   Refractive index change (or maximum diffraction efficiency)
d)   Spatial frequency response
e)   Laser wavelength
f)   Crystal quality and availability
g)   Dark storage of the information

Starting from the results of the *Kukhtarev* equations, points (a), (b), (c), and (d) have already been analyzed in detail in the above sections. We just summarize in the following the main results:

− The ultimate photorefractive sensitivity ($W_0 \approx 100\,\mu J\,cm^{-2}$ for cw laser illumination) is reached for materials with large drift or diffusion lengths of the photocarriers i.e. materials that are efficient photoconductors at the recording wavelength (Kr_E, $Kr_D \geq 1$). The crystal is an efficient photoconductor, when the quantum efficiency Φ of the charge generation process is close to unity and when the crystal absorption coefficient at the recording wavelength is sufficiently high ($\alpha \cong 1\text{–}3\,cm^{-1}$ in highly reduced crystals).

− Under such conditions, the optimum grating time constant is obtained for an incident recording beam intensity I_0. These optimal recording conditions are encountered in photorefractive BSO, BGO, GaAs, and highly reduced $KNbO_3$ crystals.

− The high values of the photoinduced index change are obtained in materials with high electro-optic coefficients i.e. ferroelectric crystals such as $BaTiO_3$, $KNbO_3$, $LiNbO_3$ and SBN. In other materials having a low electro-optic coefficient, Δn can be increased with an externally applied electric field until saturation occurs for $E_0 \approx E_q$. In this saturation regime, The steady-state index change is proportional to the trap density N_A. However, making N_A too large by appropriate crystal doping may not be desirable because it increases the response time. From the above considerations, it clearly appears that up to now two classes of materials have to be considered: highly sensitive and fast materials such as BSO, BGO, GaAs with a relatively low value of the photoinduced index modulation, and on the other hand, crystals like BaTiO$_3$ and LiNbO$_3$, which are slow but highly efficient in terms of Δn. These two opposite situations are clearly seen on Figs. 6.6 and 6.7 which respectively show the steady-state index changes and the response times as a function of the grating period for BSO and BaTiO$_3$ for applied fields of $E_0 = 0$ and $E_0 = 10 \, \mathrm{kV \, cm^{-1}}$ [6.6]. The problem of crystal optimization for both fast response and high steady-state index change is an important objective presently under study in different laboratories. Success in achieving this goal will require a more precise analysis of the charge transport process as well as a precise identification and control of the donor and trapping centers in the crystal. Impurity doping is already known to be efficient in some materials (LiNbO$_3$-Fe^{2+}, KNbO$_3$-Fe^{2+}) and should be extended to other crystals (BSO, BaTiO$_3$ ...).

However, under special recording conditions, the photoinduced index modulation in photorefractive BSO, may be increased by a factor of 10 while retaining a short response time. This enhancement of Δn is obtained by a simple technique which involves a moving interference pattern at a particular grating spacing [6.27, 28]. A detailed analysis of this interaction with moving fringes will be presented, in Sect. 6.6 together with related implications for image amplification and self-induced optical resonators.

Concerning the spatial resolution, all photorefractive crystals have a high spatial frequency response and this can be controlled with the amplitude of the

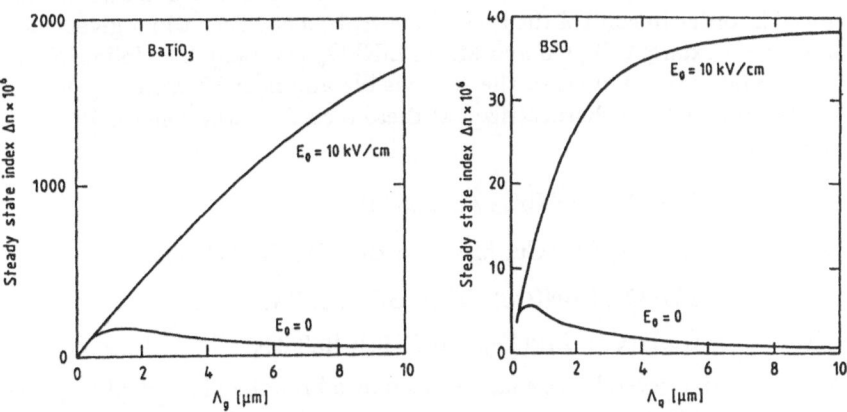

Fig. 6.6. Steady-state index change in photorefractive BaTiO$_3$ and BSO as a function of grating period Λ_g. Applied field $E_0 = 0$ and $10 \, \mathrm{kV \, cm^{-1}}$; mean irradiance $I_0 = 1 \, \mathrm{W \, cm^{-2}}$; $\lambda = 0.5 \, \mu\mathrm{m}$. From [6.6]

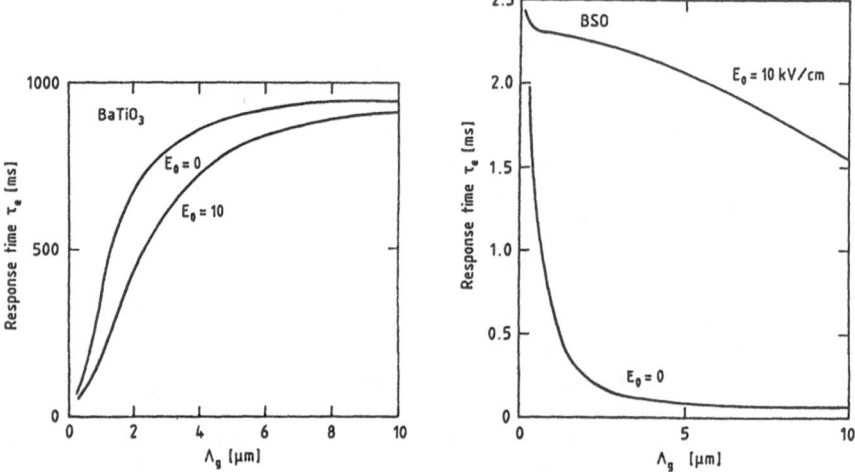

Fig. 6.7. Response time of photorefractive $BaTiO_3$ and BSO as a function of grating period Λ_g. Applied fields $E_0 = 0$ and $10\,kV\,cm^{-1}$; mean irradiance $I_0 = 1\,W\,cm^{-2}$; $\lambda = 0.5\,\mu m$. From [6.6]

externally applied electric field (see Fig. 6.2). Therefore, these crystals exhibit a high spatial frequency response and are perfectly adapted to high resolution imaging. In particular, reflection-type holograms or gratings (grating spacing $\Lambda_g \cong 0.1\,\mu m$) can be efficiently recorded. The only materials which have a limited spatial resolution are the electro-optic ceramics such as PLZT. These materials have a rather low sensitivity and can be prepared with large dimensions, but the granular nature of the ceramic limits the resolution to about $\Lambda_g \cong 1\,\mu m$ [6.29]. The electro-optic crystals usually considered for photorefractive recording are photosensitive at the common laser wavelengths from the UV to the near IR (semiconductor; Nd : YAG Lasers). Suitable dopants are also introduced in order to control the crystal absorption coefficient in a given spectral range. (For example Fe, Cu and Mn in $LiNbO_3$ crystals). The following list gives those materials adapted to the UV, visible and near IR wavelengths and which have already been characterized at these recording wavelengths [Ref. 6.8; pp. 233–243]:

UV	KH_2PO_4, $LiNbO_3$ (undoped)
Visible	$LiNbO_3$ (Fe, Cu, Mn ...); $BaTiO_3$, $KNbO_3$ (Fe) ;
	$LiTaO_3$, SBN(Ce); BSO; BGO; KTN.
Near IR	GaAs(Cr); InP(Fe); CdTe(In); $BaTiO_3$.

The photorefractive crystals are characterized by a broadband spectral response compared to the resonant interaction in other nonlinear media (gases, multi-quantum well structures in semiconductors ...). However, for each crystal there exists an optimum wavelength recording range for which the crystal sensitivity, and therefore the crystal speed are optimum. For example, in BSO and BGO the crystal sensitivity is close to the theoretical limit for $\lambda \cong 450$–$550\,\text{nm}$ because of the high value of the quantum efficiency in that domain ($\Phi > 0.5$). The recording energy, and thus the time response, are increased by two orders of magnitude in the red ($\Phi = 0.06$ at $\lambda = 633\,\text{nm}$).

6.2.2 Crystal Quality and Availability

Another problem related to the illumination with coherent light is the crystal quality and homogeneity. A defect that scatters light generates a wave that interferes with the original beam and the resulting interference pattern induces an index modulation which, in turn, leads to more scattering [6.30]. This generates a scattering pattern which limits the signal-to-noise ratio of the holograms. Another source of noise is the reflection of the recording beams from the front and back surfaces of the crystal. These effects may be minimized by recording a hologram with a low efficiency in order to prevent build-up of scattering. A more complete analysis of the different sources of noise in photorefractive crystals is presented in [Ref. 6.1; pp. 129–130]. The anisotropic Bragg diffraction properties presented in Sect. 6.1.6 are also extremely helpful for noise filtering and therefore for the enhancement of the signal-to-noise ratio in the diffracted image plane. Several of the photorefractive crystals listed in Sect. 6.2.1 are now available with an excellent optical quality and large dimensions ($LiNbO_3$, BSO, BGO ...) from a variety of suppliers. The problem of crystal quality and size is mainly encountered in ferroelectrics such as $BaTiO_3$, SBN and KTN, which show inhomogeneities and striations. Recently, however, progress in crystal-growing methods has resulted in crystals of limited size but essentially striation free and which exhibit only a minimal amount of scattering.

6.2.3 Dark Storage Time

One of the peculiarities of a photorefractive crystal is its ability to store information in the dark. In the absence of illumination, and at room temperature, the crystal dielectric relaxation time is simply given by $\tau_{di} = \varepsilon_0 \varepsilon / \sigma_0$ where σ_0 is the dark conductivity. Most of the ferroelectric crystals have a low conductivity in the dark $\sigma_0 < 10^{-18}\,(\Omega\,\text{cm})^{-1}$ and the related memory times range from 10 h for KTN, to the weeks for SBN, $BaTiO_3$ and several months for $LiNbO_3$. In photorefractive BSO-BGO, the measured dark crystal conductivity is $\sigma_0 \approx 10^{-14}\,(\Omega\,\text{cm})^{-1}$ and the observed memory time is typically 10–20 h. The dark storage time is considerably reduced in low bandgap semi-insulating semiconductor materials sensitive in the near IR and is typically equal to 10^{-4} s in InP-Fe [$\sigma_0 = 10^{-8}\,(\Omega\,\text{cm})^{-1}$]. These materials have nearly no memory time, which excludes their application to information storage, but are perfectly adapted to short-time photorefractive recording and erasure in dynamic holographic experiments. For each of the crystals mentioned above, the dark conductivity mechanisms are the subject of investigations; in particular, the dark conductivity can be ionic, whereas it is electronic under illumination (e.g. in BSO, BGO, LiNbO$_3$).

This chapter has given a general overview of the most important features of photorefractive crystals with a view to their applications in dynamic holography and optical signal processing. Several of these points, in particular the basic mechanism of the photorefractive effect and the role of impurities are discussed in detail in other chapters in this book. We have concentrated on parameters such as photorefractive sensitivity and steady-state index change which govern the crystal speed and the grating diffraction efficiency. The other important parameter which will be now considered is the phase shift between incident fringe pattern and the photoinduced index modulation. The existence of this phase shift is the basis of the energy transfer in wave-mixing experiments and of related applications to self-induced resonators.

6.3 Holography with Photorefractive Crystals

6.3.1 General Introduction

Photorefractive crystals are well suited for recording in real time the interfernce field due to an object (or signal) wavefront containing spatial information and a plane (or spherical) reference beam. The intensity distribution within the volume of the recording medium is given by

$$I(r) = |S + R|^2$$
$$= S_0^2 + R_0^2 + SR^* + S^*R \quad \text{with} \quad S = \text{Re}\left\{S_0(r)e^{i(\omega t - kz + \varphi_S)}\right\} (6.28)$$

where $S_0(r)$ and $\varphi_S(r)$ are the respective amplitude and phase of the signal wavefront propagating in the $+z$ direction. R is the off-axis plane wave reference beam whose amplitude is R_0 and ω is the optical frequency of the recording beams. The recording medium changes its optical properties according to the illumination pattern in relation 6.28, (phase or amplitude volume hologram). As shown in [6.31, 33], the reconstruction of a signal beam can be performed in two ways:

i) By illuminating the hologram with the same reference beam, the term SR^* reconstructs the original signal wave S.

ii) By illuminating the hologram with a readout beam antiparallel to R, the term S^*R reconstructs the phase conjugate replica of the incident signal wave S.

In contrast to classical holographic recording materials (silver halide plates, photoresist etc.) which require recording, developing and readout of the hologram, all these steps occur simultaneously with a dynamic photorefractive crystal and any change of the object is continuously recorded. The two optical configurations which are of particular interest in holographic experiments with photorefractive crystals are given in Fig. 6.8. The configuration in Fig. 6.8a is

222

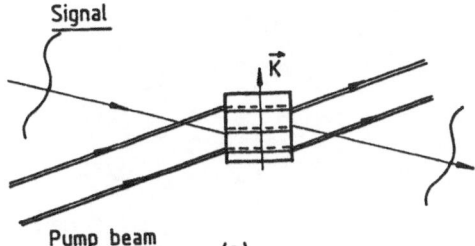

Signal

$\vec{K}$

Pump beam   (a)

Fig. 6.8a,b. Dynamic holography in photorefractive crystals: **(a)** degenerate two wave mixing. **(b)** phase conjugation via degenerate four wave mixing. Continuous lines: interfernce pattern. Dashed lines: index modulation. K grating vector

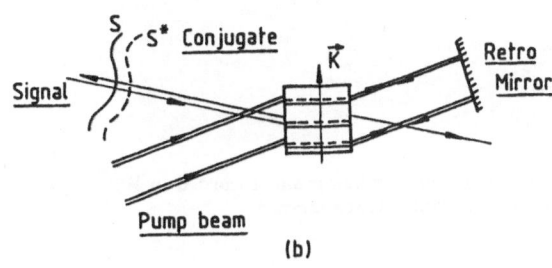

S   S* Conjugate $\vec{K}$ Retro Mirror

Signal

Pump beam

(b)

called two-wave mixing (2WM) while four-wave mixing (4WM) in Fig. 6.8b allows the generation in real time of the conjugate of the incident signal beam [6.34] (the readout beam of the hologram originates from the retroreflection of the reference beam). Both of these configurations reconstruct the waves under phase matching conditions (Bragg diffraction) with the following K vector relations:

$$2WM : k_S = k_R - K$$
$$4WM : k_C = -k_R + K \tag{6.29}$$

where k_S and k_C are the wave vectors of the incident (or diffracted) signal and conjugate waves, respectively.

6.3.2 Review of the Properties of Two- and Four-Wave Mixing Configurations

a) 2WM and Holography. This configuration is derived from the classical holographic recording but the new feature related to the use of photorefractive crystals is that the readout of the hologram is destructive since information erasure occurs through space-charge field relaxation by uniform illumination with the reference beam [6.15]. Examples of transient diffracted images from defocused Fourier holograms recorded in a photorefractive BSO crystal are shown in Fig. 6.9. Figure 6.10 illustrates the anisotropic Bragg diffraction properties discussed in Sect. 6.1.6. A polarizer correctly oriented in the plane of the diffracted image suppresses the scattered noise due to the reference beam and this polarization filtering thus significantly increases the signal-to-noise ratio in the image [6.18]. (For the linear recording of the spatial information, the

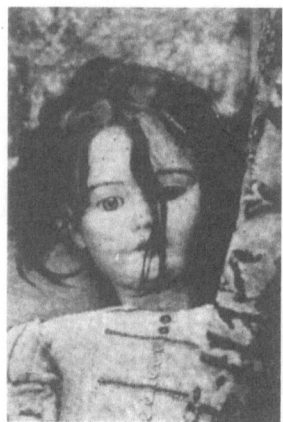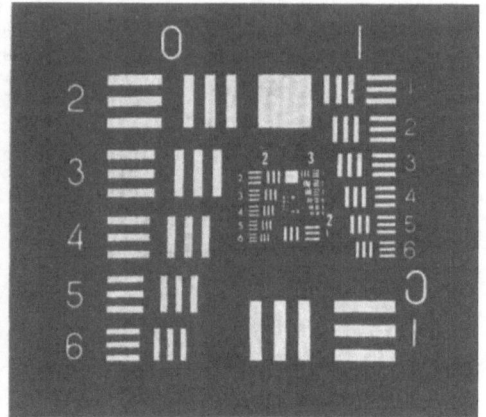

Fig. 6.9. Reconstructed images from out focus Fourier holograms recorded in BSO crystals. The object wavefront originates from a photographic transparency

reference beam is 10 times more intense than the signal beam.) If such a dynamic recording and erasure process is suitable for applications to real-time image processing, the applications of the photorefractive crystals to high density information storage would require that the hologram does not decay during the continuous readout of the crystal. Different techniques have been proposed to solve this problem of the nondestructive readout phase volume holograms recorded in photorefractive crystals [6.12, 35]:

a)   wavelength change for readout,
b)   crystals with asymmetric recording-erasure cycles,
c)   thermal and/or electrical fixing,
d)   multiphoton absorption.

One of the simplest method, which apparently leads to nondestructive readout is to choose a wavelength which differs significantly from that of the writing beams and which is outside of the spectral sensitivity range of the crystal (red or near IR region). While this technique is perfectly convenient for thin gratings, it does not apply to phase volume holograms which contain high density spatial information. However, some special configurations based on crystal birefringence have been found in LiNbO$_3$ by *Petrov* et al. which partially relax the constraint of the Bragg selectivity of volume holograms. The effectiveness of this method as well as the crystal orientation and beam incidence conditions are detailed in [6.36]. This method is also accompanied by a 90° rotation of the plane of polarization of the readout beam at $\lambda' = 633$ nm while recording is done $\lambda = 440$ nm. It is beyond the scope of this section to review the details of other methods allowing nondestructive readout of the stored information. These techniques were developed in the 1970s for applications of photorefractive crystals to read-only holographic memories by *Staebler* and *Phillips* [6.37] (low sensitivity to erasure in 0.1 % Fe^{3+}-LiNbO$_3$), *Staebler* and

Fig. 6.10a,b. Anisotropic Bragg diffraction in BSO. Noise suppression in the diffracted image plane by polarization filtering. (**a**) experimental evidence of the scattered noise in the diffracted image plane. (**b**) noise suppression by correct orientation of a polarizer sheet. Signal-to-noise ratio improvement: +25 dB

Amodei [6.38] (thermal fixing in LiNbO$_3$), *Micheron* et al. [6.29, 39] (electrical fixing in BaTiO$_3$ and SBN via field-assisted reversal of ferroelectric domains) and *von der Linde* et al. [6.40] (photorefractive recording at $\lambda = 1.06\,\mu$m by two-photon absorption in LiNbO$_3$ and KTN). A detailed description of the mechanism involved in these different techniques may be found in the review papers [6.1, 2, 35]. Although recent applications of photorefractive crystals are based on the mixing of optical beams of the same frequency, the problem of the nondestructive readout of the hologram is still an important point which at present cannot be considered as completely solved for practical applications.

b) 4WM and Phase Conjugation. In the configuration shown in Fig. 6.8b the photorefractive crystal can be recognized as a phase conjugate mirror whose reflectivity is $\varrho = I_C/I_{S_0}$. I_C and I_{S_0} are the respective conjugate and incident beam intensities. The conjugate wave S^* corresponds to a wave propagation in the $-z$ direction, with a reversed phase relative to the incident wave, and whose field is expressed by:

$$S^* = \mathrm{Re}\left\{S_0(\boldsymbol{r})\mathrm{exp}[\mathrm{i}(\omega t + kz - \varphi_S)]\right\} \quad . \tag{6.30}$$

The application of phase conjugation in adaptive optics, where aberrated wavefronts should be restored to their initial state, is illustrated in Fig. 6.11 [6.41, 43]. An aberrator placed in the signal beam path distorts the incident signal wave-

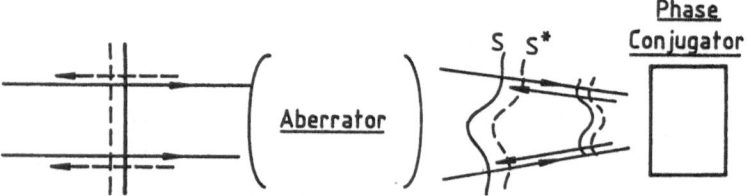

Fig. 6.11. Phase distortion compensation by wavefront reflection on a conjugate mirror

front. However, when the conjugate wave passes back through the aberrator it will emerge completely free of distortions. Several experiments illustrating these unique properties of the conjugate waves generated via 4WM in photorefractive crystals will be described in the following sections.

6.3.3 Imaging Through a Phase Disturbing Medium

The optical set-up for correction of the phase distortion in an imaging experiment based on 4WM in a photorefractive BSO (or BGO) crystal is shown in Fig. 6.12. This is the transposition, with a dynamic material, of the *Kogelnik* [6.44] imaging technique through an aberrating medium. Experimental results are shown in Fig. 6.13. Figure 6.13a is the direct imaging through an aberrating glass plate, while Fig. 6.13b is imaging via phase conjugation. Phase distortion can thus be removed by allowing the time reversed wavefront to travel back through the aberrating medium [6.45]. Sensitivity of the image reconstruction process to a misalignment of the retromirror by some milliradian can be appreciated from Fig. 6.13c. The conjugate beam reflectivity obtained in early experiments with BSO-BGO was $\varrho \cong 2 \times 10^{-3}$, but with recent improvements of the crystal response, reflectivities exceeding unity have been demonstrated ($\varrho \cong 1$–30). Besides static aberrations, dynamic phase distortions can also be compensated since the recording time of the hologram is of the order of 10 ms at the power levels used in the reference beam: $I_{R_0} = 50\,\mathrm{mW\,cm^{-2}}$ at $\lambda = 514\,\mathrm{nm}$. This response time can be further decreased by increasing the reference beam intensity.

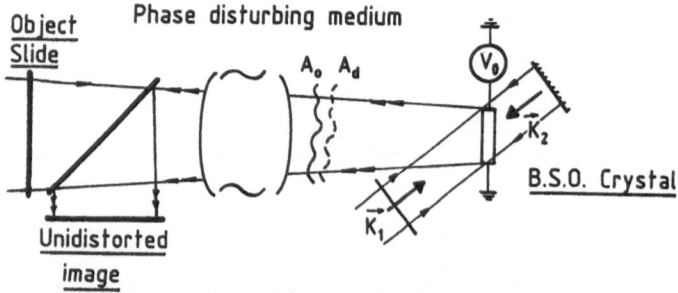

Fig. 6.12. Dynamic holography and 4WM in BSO. Application of phase conjugation to imaging through a phase disturbing media. From [6.51]

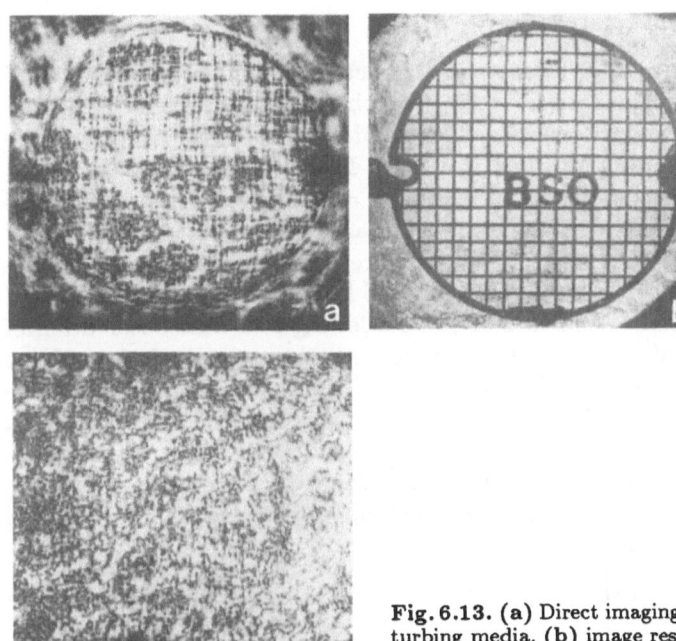

Fig. 6.13. (a) Direct imaging through the phase disturbing media. (b) image restoration via phase conjugation. (c) image distortions due to a misalignment of the retromirror

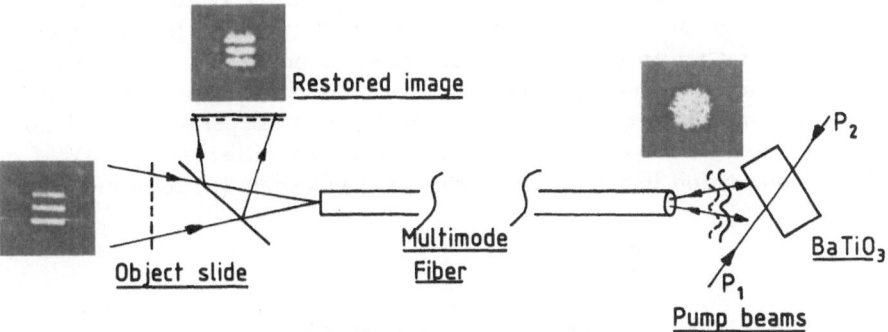

Fig. 6.14. Image restoration through phase conjugation in a multimode optical fiber

Similar techniques of wavefront correction by 4WM in photorefractive crystals such as $LiNbO_3$, $LiTaO_3$, $KNbO_3$, $BaTiO_3$ and SBN have also been reported by different authors [6.47–50]. With each of these crystals, amplified phase conjugation is now obtained ($\varrho \cong 1$–40) at low incident power levels. The transmission of images through multimode fibers by means of phase conjugation has been demonstrated in [6.51, 52] using a $BaTiO_3$ crystal in the diffusion recording mode. The wavefront reversal properties of the conjugate beam permit compensation, after double pass through the same fiber, of the modal dispersion of the fiber. Experimental results are shown in Fig. 6.14.

6.3.4 Real-Time Interferometry

Laser induced gratings in dynamic photorefractive crystals are ideally suited for real-time holographic interferometry. The obvious advantages of these materials are that in comparison with silver halide plates (or thermoplastic) no processing of the recording medium is needed. In photorefractive BSO, BGO, $KNbO_3(Fe^{2+})$, recording-erasure sensitivities comparable to those obtained with high resolution photographic plates are achieved ($S^{-1} \cong 100 \, \mu J \, cm^{-2}$ at $\lambda = 514 \, nm$). These materials are therefore very interesting for "double-exposure" or "time-average" interferometry of large size objects. The optical configuration, based on 4WM in BSO, and allowing real time observation of the vibration mode of a loudspeaker membrane is shown in Fig 6.15 [6.53]. A plane wave reference beam and a low intensity signal beam scattered by the vibrating object structure interfere in the crystal volume. Through 4WM in BSO, a conjugate real image of the loudspeaker is obtained after spatial separation of the incident and conjugate beams. When the structure is vibrating, one obtains a reconstructed image with superimposed dark fringes given by the zero order Bessel function:

$$I_D = I_S J_0^2 \left[\frac{4\pi}{\lambda} \delta(x, y) \right] \tag{6.31}$$

where $\delta(x, y)$ is the local amplitude of the periodic deformation of the membrane. This is obtained when the recording time τ of the holographic space-charge field is long compared to the vibration period T of the object $\tau \gg T$ for time-average intensity recording in the crystal volume [6.31] (typically $\tau \cong 10\text{--}100 \, ms$ for an incident reference beam intensity $I_{R_0} = 10 \, mW \, cm^{-2}$ at $\lambda = 514 \, nm$). The mode pattern visualization of the loudspeaker membrane excited at different frequencies is shown in Fig. 6.16. Experiments in double-exposure interferometry can be performed similarly either using a 4WM or 2WM configuration [6.6–54]. The cycle is the following: (a) recording of the

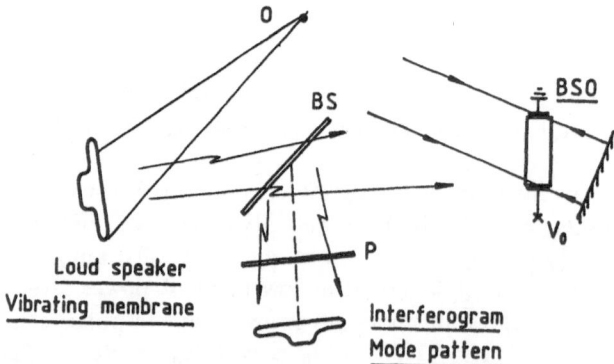

Fig. 6.15. Application of phase conjugation and 4WM in BSO to dynamic interferometry of vibrating object structures. BS: beam splitter. P: polarizer for noise filtering. $V_0 = 8 \, kV$. Reference beam intensity $I_0 = 20 \, mW \, cm^{-2}$ at $\lambda = 514 \, nm$

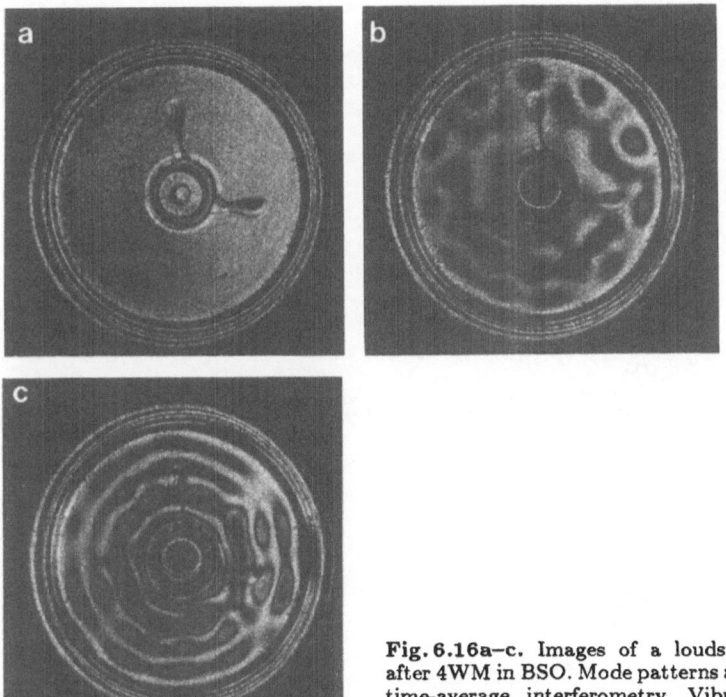

Fig. 6.16a–c. Images of a loudspeaker membrane after 4WM in BSO. Mode patterns are displayed using time-average interferometry. Vibration frequencies are respectively: (a) $F = 0$ (b) $F = 6\,\mathrm{kHz}$ (c) $F = 30\,\mathrm{kHz}$

original wavefront diffracted by the object (exposure time τ); (b) recording of the wavefront diffracted by the object deformed by a mechanical and/or thermal constraint (exposure time $\tau/2$); (c) readout with the reference beam – transient interferogram projected onto a vidicon memory tube. Experimental results following such a cycle are given in Fig. 6.17. With similar set-ups, other types of interferometric techniques have been successfully demonstrated such as: speckle interferometry and object contouring by two wavelength recording [6.55]. All of these experiments can also be performed with a crystal and beam orientation allowing the recording of reflection type holograms ("diffusion" recording with no applied field, but lower efficiency compared to the "drift" recording) [6.56].

6.3.5 Speckle-Free Imaging

Images reconstructed with coherent light are in general noisy because of the complex interference pattern (speckle) of the scattered light resulting from dust particles, surface defects or inhomogeneities in the recording medium or optical elements. This leads to a reduction of the spatial resolution and severely degrades the signal-to-noise ratio of the coherent imaging process. A method for speckle noise reduction, which applies to phase conjugate wavefronts generated in 4WM experiments with dynamic photorefractive crystals such as BSO, is

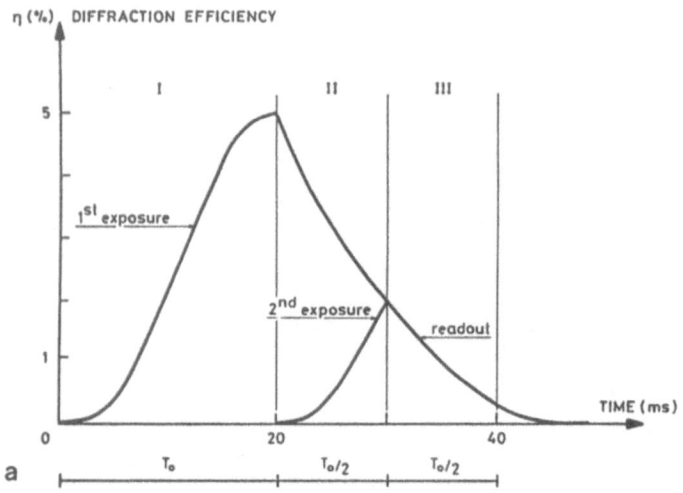

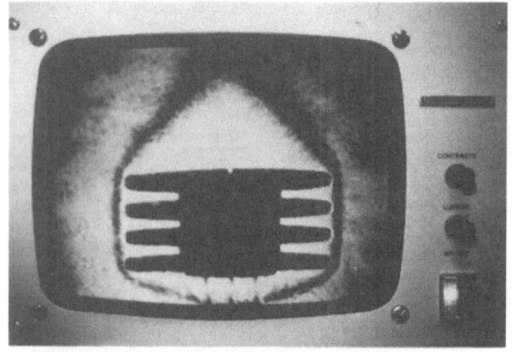

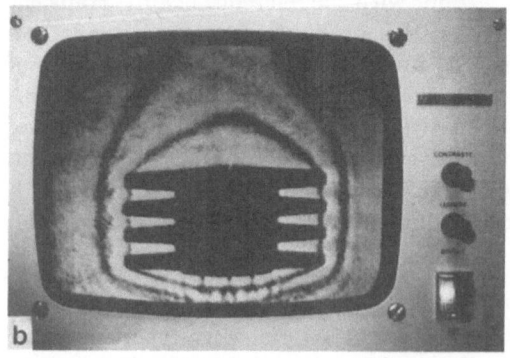

Fig. 6.17a,b. Double-exposure interferometry. Recording-readout cycle (**a**) and interferogram reconstruction (**b**) after 2WM in photorefractive BSO

shown in Fig. 6.18. It is based on the time integration in the detection plane of N coherent images having uncorrelated speckle noise pattern. These patterns are generated with a diffusing screen which is moved step by step between successive exposures of the camera [6.57]. Under such conditions the signal-to-

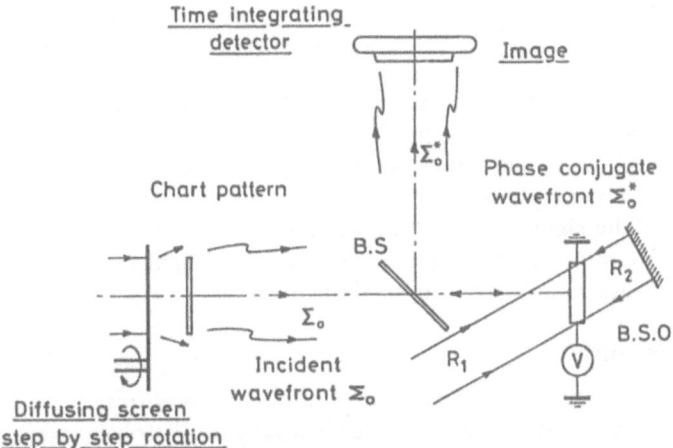

Fig. 6.18. 4WM experiment in BSO crystals showing speckle noise reduction by integration on an intensity basis of N independent coherent images

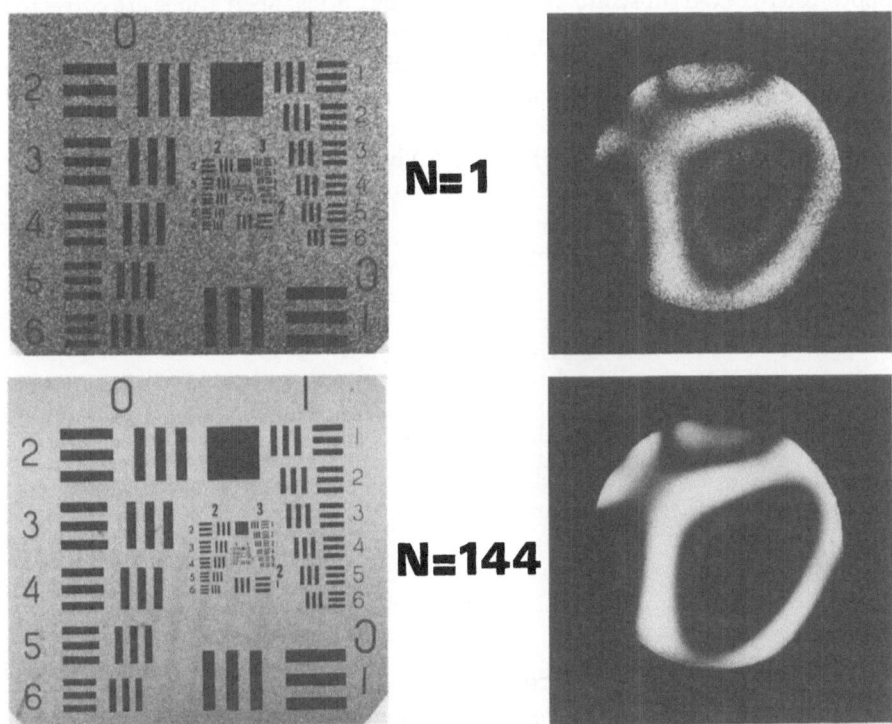

Fig. 6.19. Experimental demonstration of signal-to-noise ratio improvement in images with increasing the number N of incoherent superpositions in the detection plane. $N = 1$ (top) $N = 144$ (bottom). Binary chart pattern and interferogram of a vibrating membrane

noise ratio in the image plane is improved by $\sqrt{N}$ while the integrating time in the detector plane is $T = N\tau$ [6.58]. Experimental evidence of signal-to-noise improvement with an increase of the number of incoherent superpositions is shown in Fig. 6.19. The chart pattern may also be replaced by a vibrating object structure, thus allowing one to obtain a speckle-free time-average interferogram of the membrane. From these experiments, it can be concluded that images reconstructed in 4WM with photorefractive crystals may have an excellent signal-to-noise ratio. The signal-to-noise ratio of incoherent illumination may be approached while retaining all the capabilities of coherent light for recording and processing amplitude and phase of optical fields.

6.3.6 Photolithography

The principle of lensless imaging and focusing has already been presented in previous sections and it is now applied to mask-copying in photolithography. The experiment has been performed with a photorefractive $LiNbO_3$ phase conjugator and with the set-up shown in Fig. 6.20 [6.59]. The signal wave originated from an illuminated resolution test pattern drawn on a chromium mask. The retroreflected conjugate wave then forms an image on the surface of the photoresist-coated substrate. Since most of the optical aberrations are corrected in the projection process using phase conjugation, full spatial resolution over a large image field is possible. The prime difficulty is the low intensity in the conjugate image plane due to a low conjgate beam reflectivity with the material used in this early demonstration. The use of high reflectivity photorefractive conjugators such as $BaTiO_3$, SBN and $KNbO_3$ could overcome this problem.

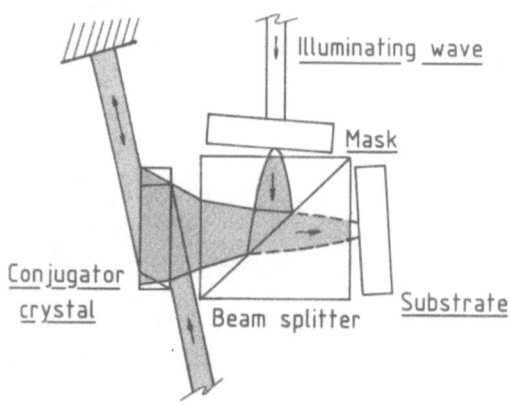

Illuminating wave

Mask

Conjugator crystal

Beam splitter

Substrate

Fig. 6.20. Application of 4WM in a $LiNbO_3$ crystal to mask-copying for high resolution photolithography. From [6.50]

6.3.7 Multiple Image Storage

Information recording in the volume of a dynamic photorefractive crystal potentially offers a high storage capacity (diffraction limit $\lambda^3 = 10^{12}$ bits cm^{-3}), contact-free parallel readout of an array of information and the possibility of erasing part of the whole memory for writing new data. These arguments promoted extensive studies on random-access high capacity holographic memories in the 1970s. It is not the objective of this section to review the proposed architectures for holographic read-only and read-write memories using electro-optic crystals as storage media, since no practical system has been developed (for a review see [6.35]). Therefore, we merely present some properties which are unique to the holographic recording in photorefractive crystals. The attractive feature is the possibility of superimposing many holograms in the same crystal volume. This is accomplished, for example, by varying the angle of incidence of the reference beam, each individual hologram being associated with a well-defined angle (Fig. 6.21). Selective reconstruction of the superimposed holograms is achieved because efficient diffraction occurs only when the hologram is addressed at the right Bragg angle. This multiplexing technique requires a photorefractive crystal with an asymmetric recording-erasure cycle (for example $LiNbO_3 - 0.1\%\ Fe^{3+}$) in order to prevent erasure of previously recorded data pages during the recording of a new one in the same crystal volume. At present, up to 500 holograms have been superimposed and fixed in a cube of $LiNbO_3$ crystal giving a total capacity of 0.5 Gbits cm^{-3} [6.60]. A method for selective erasure of any information block, based on a coherent image subtraction technique, has also been demonstrated by *Huignard* et al. [6.61].

The possibility of electrically controlled volume hologram written in $LiNbO_3$ crystals has been shown by *Petrov* et al. [6.62] in $LiNbO_3$. Due to the large electro-optic effect in this crystal, it is possible to control the Bragg conditions for image reconstruction by the bias voltage applied to the crystal. The independent reconstruction of two or three holograms written in the same crystal under different voltages is demonstrated in [6.62]. The maximum efficiency is reached at the same voltage as was used for the recording.

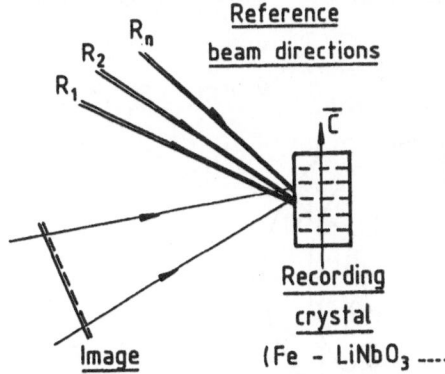

Fig. 6.21. Multiple image storage in the volume of a photorefractive crystal by angular coding of the reference beam direction

6.3.8 Beam Deflection and Interconnection

A photorefractive grating may also be considered as an active electro-optical element allowing angular deflection of a third beam of different wavelength. The dynamic grating is recorded in the visible and may be read out at the wavelength of a semiconductor laser. Potentially large deflection angles are possible with millisecond random access time. These techniques could therefore provide attractive solutions for a reconfigurable optical interconnection between fiber matrices [6.62, 66]. The deflection of a near IR beam is achieved over a large scan angle ($\Delta\theta \cong 10°$) by changing the spacing of a dynamic grating photoinduced in a photorefractive BSO crystal. The Bragg diffraction conditions for a fixed incidence of the IR beam are obtained either by taking account of the space-charge field nonlinearities (collinear Bragg diffraction [6.26]) or by recording with a dye laser source. In this latter method demonstrated by *Sincerbox* and *Roosen* [6.64] and *Pauliat* et al. [6.67], the grating is induced in the photorefractive crystal by the interference of two coherent plane waves issued from a dye laser, and one of the recording beams is passed through a fixed dispersive element. Changing the recording wavelength of the laser will produce both a modification of the period and of the tilt angle of the photoinduced grating in the crystal volume thus ensuring efficient diffraction of a near IR beam over a large scan angle.

In another experiment *Voigt* et al. [6.67] utilizes the anisotropic Bragg diffraction in $KNbO_3$ for achieving the angular deflection of a third beam of different wavelength. In anisotropic crystals the diffracted light wave can have a polarization direction normal to the incident readout beam polarization. This occurs in materials having nonzero off-diagonal elements of the electro-optic tensor [6.68]. For optimum diffraction efficiency, a generalized Bragg condition has to be fulfilled. Since incident and diffracted wavevectors in anisotropic materials can have different lengths, the incidence and diffraction angles may be different. Figure 6.22 shows examples of wavevector surfaces in the x-y plane of a biaxial crystal and possible diffraction geometries for photoinduced wavevectors parallel to the y-axis. Figure 6.23 shows the wavevector dependence of the incidence (θ_i) and diffraction angle (θ_d) for biaxial $KNbO_3$ crystals.

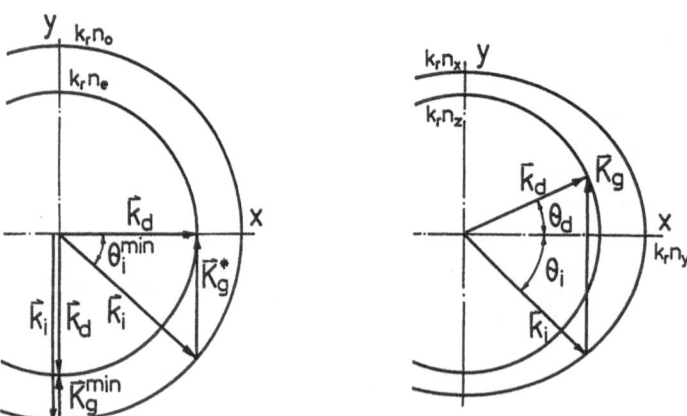

Fig. 6.22. Wave-vector diagrams for anisotropic Bragg diffraction in $KNbO_3$ with different grating wavevectors K_g^{min}, K_g^* and K_g

From Fig. 6.22 we see that there exists a lower and upper limit for the grating wavenumber:

$$K_g^{min} = \frac{2\pi}{\Lambda^{min}} = |n_0 - n_e|k$$

$$K_g^{max} = \frac{2\pi}{\Lambda^{max}} = |n_0 + n_e|k$$

below which no diffraction is possible. k_i and k_d are in the same direction whereas k_g is antiparallel to k_i. $k = 2\pi/\lambda$ is the vacuum wavevector of the optical wave. The diffracted wave travels in the same direction as the incident wave but the polarizations of the two waves are different.

Another special configuration of great practical interest is that of minimal input angle θ_i^{min}, where the diffraction angle is exactly zero degrees. In this case the grating wavenumber K_g^* is given by

$$K_g^* = \frac{2\pi}{\Lambda^*} = \sqrt{|n_0^2 - n_e^2|}\,k \quad .$$

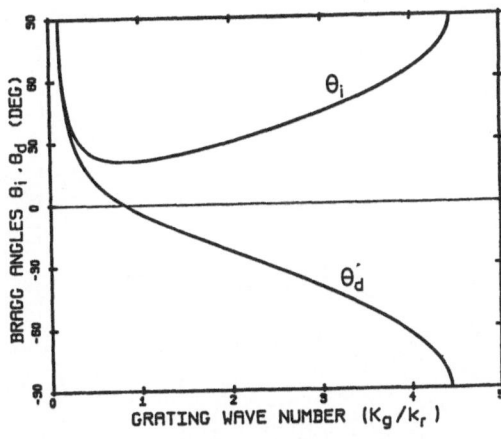

Fig. 6.23. Wave-vector dependence of incidence and diffraction angles θ_i and θ_d for anisotropic Bragg diffraction in KNbO$_3$ for grating wave vectors parallel to the b-axis

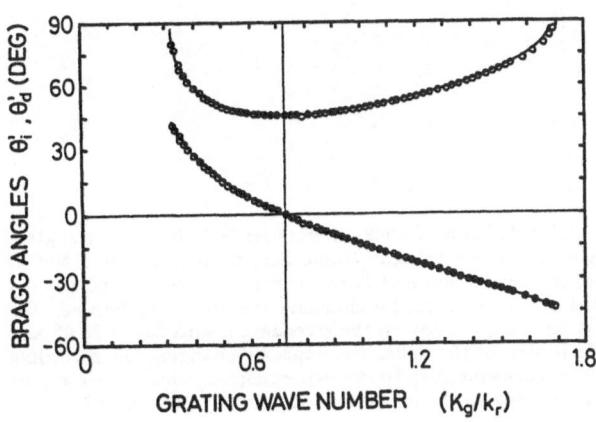

Fig. 6.24. Grating wave number dependence of incident and diffraction angles θ_i' and θ_d' for anisotropic Bragg diffraction in KNbO$_3$ with grating wave vector along the b-axis

This configuration is also depicted in Fig. 6.22. This arrangement is of basic importance for realizing optically controllable light deflection by means of anisotropic photorefractive deflection. Figure 6.24 shows experimental results for the Bragg angles θ_i and θ_d for the configurations using the electro-optic tensor element r_{42}. The grating wavenumber in these experiments was altered by changing the angle between writing argon laser beams. Readout was performed by a weak He-Ne laser beam. The broad minimum in the angle of incidence around K_g^* (see Fig. 6.23) is very useful in light deflectors because the light in-

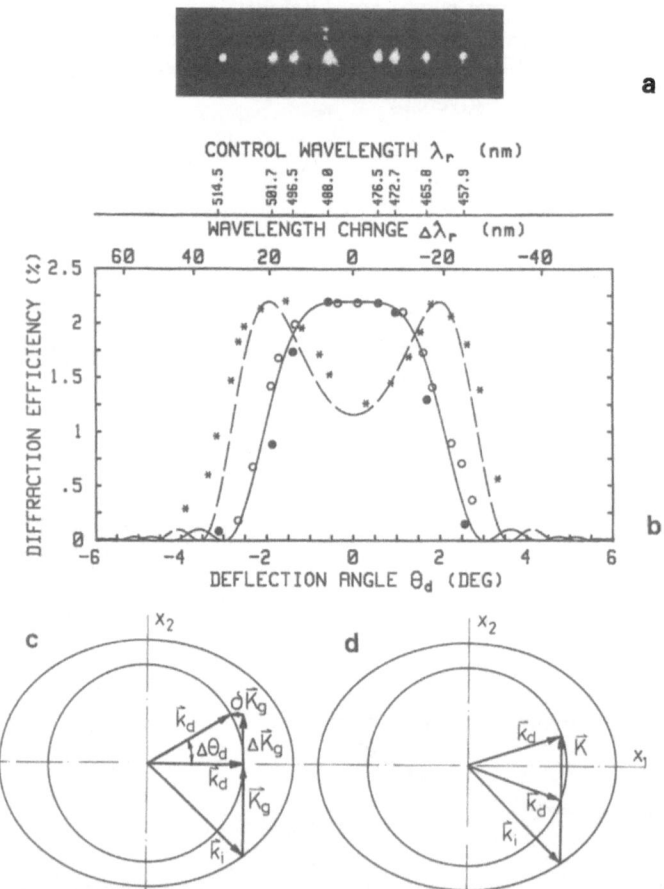

Fig. 6.25. (a) Photograph of the diffracted spots for six different Ar+-ion laser wavelengths. (b) Diffraction efficiency and deflection angle for anisotropic Bragg-diffraction in KNbO₃. (o, *) Recording at fixed wavelength $\lambda = 488$ nm and fixed incidence angle $\theta_i = 56.23°$ (o) and $\theta_i = 56.30°$ (*); tuning of the deflection angle by changing the angle $2\theta_0$ between the recording beams. (•) Recording at fixed angle between the recording beams $2\theta_0 = 36.9°$ and fixed incidence angle $\theta_i = 56.23°$: tuning of the deflection angle by changing the recording wavelength. (c) Wavevector diagram corresponding to the experimental points (o, •) in part b). (c) Wavevector diagram corresponding to the experimental points (*) in part b)

236

cident at a fixed angle can be effectively deflected by a broad range of grating wavenumbers K_g (or recording wavelengths λ_0), resulting in a large range of deflection angles. Figure 6.25 displays experimental data for the diffraction angle and the diffraction efficiency obtained with a 2.55 mm thick crystal of $KNbO_3$. Variation of the deflection was induced by changing the angle between recording beams at fixed recording wavelength $\lambda = 488$ nm. In another experiment performed in $KNbO_3$, a He-Ne laser beam incident under the Bragg angle was deflected in several directions by varying the wavelength of the writing beams between 457.9 nm and 514.5 nm (Ar^+-laser lines). Figure 6.25a is a photograph of the diffracted spots for the different wavelengths. The variation of the deflection angle $\Delta\theta_d$ was 5.7° for $\Delta\lambda_0 = 57.7$ nm. No adjustment of incidence angles was required in these experiments. The number of resolvable spots for a beam divergence angle of $\delta\theta = 1.0$ mrad is $N = 100$. This number could be further increased by using optical beams of smaller divergence angle, yielding results comparable with those obtained from acousto-optic deflectors. To remain within 50 % of the maximum diffraction efficiency for all wavelengths used, a slight increase of the Bragg angle is needed. This configuration was employed for the measurement of the dashed curve in Fig. 6.25b. The dip at $\theta_d = 0$ occurs because of the slight off-Bragg adjustment.

6.4 Image and Signal Processing

6.4.1 Image Convolution and Correlation

Dynamic cross-correlation or spatial convolution with a classical Fourier transform lens configuration can be achieved by two- or four-wave mixing of optical fields in photorefractive crystals. The four-wave mixing geometry used in these experiments of *White* and *Yariv* [6.68] is shwon in Fig. 6.26. All the beams have the same wavelength, and the amplitudes $u_1(x,z)$, $u_2(x,z)$ and $u_4(x,z)$ in the outer focal planes are Fourier transformed by propagating to the common focal plane. The transform fields mixed in the photorefractive crystal are the following:

$$U_1 = \text{FT}[u_1(x,z)] \quad ; \quad U_2 = \text{FT}[u_2(x,z)] \quad ; \quad U_4 = \text{FT}[u_4(x,z)] . \quad (6.32)$$

The backward wave generated through 4WM in the crystal, $u_3(x,z)$, evaluated at a distance f from lens L, is of the form:

$$u_4(x_0, z_0) = \alpha_0 u_1(-x, -z) \cdot u_2(-x, -z) \otimes u_4(-x, -z) \qquad (6.33)$$

where α_0 depends on the amplitude of the photoinduced index modulation: · and $\otimes$ denote respectively the product and the convolution product of the optical fields. Figure 6.27 illustrates another configuration based on 2WM in BSO [6.69]. The intereference pattern recorded in the Fourier plane of lens L is read out by an auxiliary low power laser (He-Ne; $\lambda_R = 633$ nm). The thickness of the crystal implies that this readout beam of different wavelength has to be positioned at the correct Bragg angle for obtaining the maximum intensity in

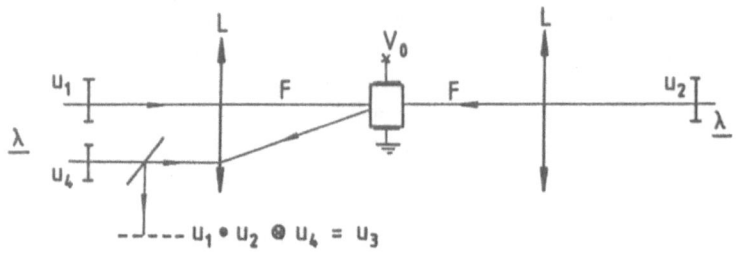

$$U_1 \bullet U_2 \otimes U_4 = U_3$$

U_1	U_2	U_4	U_3
• • •	DELTA FUNCTION	• • • •	
• • •	DELTA FUNCTION	E	
C	DELTA FUNCTION	CAL TECH	
C	• • • •	DELTA FUNCTION	

Fig. 6.26. Application of 4WM in photorefractive crystals to image convolution or correlation. Experimental results for different input data masks U_1-U_4. From [6.68]

the diffracted cross-correlation peak. Experimental results for different input signals, and using BSO as a dynamic matched filter are shown in Fig. 6.27 (typical response time $\tau = 50\,\text{ms}$ for $1\,\text{mW}$ incident intensity on the photorefractive crystal). For both configurations, the Bragg angular selectivity of the phase volume hologram may limit the number of pixels that can be processed, and futher studies should quantitatively estimate the capabilities of these architectures for high capacity parallel optical processing of analog or digital images. In these experiments, an auxiliary beam incident on the photorefractive crystal may also be used to modify the output of the processor in real time; thus provid-

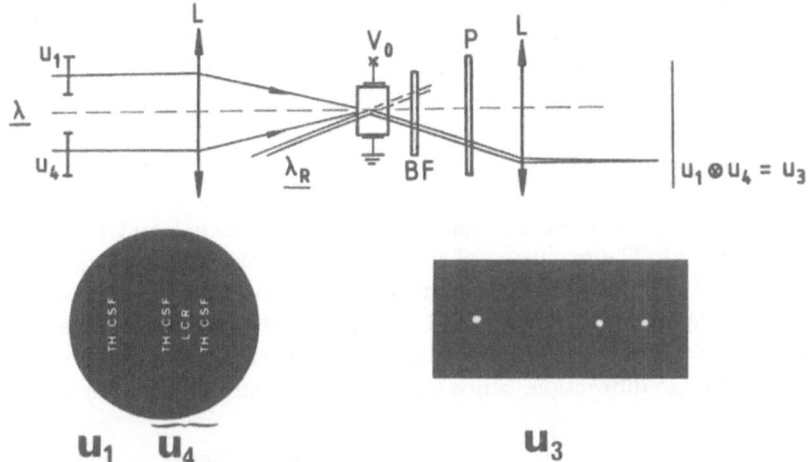

Fig. 6.27. Application of 2WM in photorefractive crystals to image convolution or correlation. (BF) blocking filter for λ_R; (P) Polarizer for noise filtering. From [6.69]

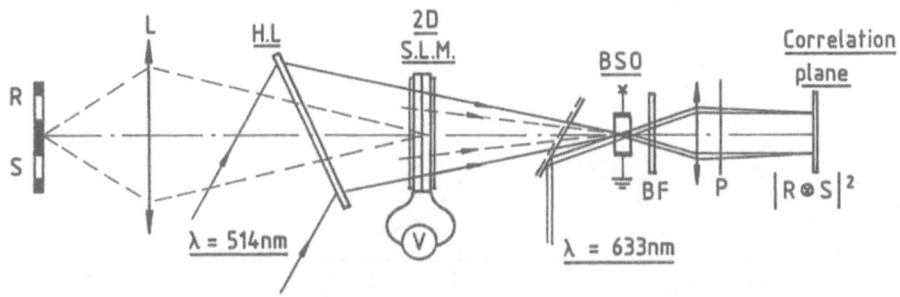

Fig. 6.28. Schematic of a complete dynamic optical processor. Spatial light modulator (2D-SLM) for data input, and photorefractive crystal (BSO) in the Fourier plane. (BF) blocking filter, (P) polarizer, (H.L) holographic lens

ing a means of weighting the correlation product in favor of specified spatial frequencies [6.70]. Moreover, a two-dimensional spatial light modulator is used for real-time data input (Fig. 6.28) in the optical processor setup demonstrated by *B. Loiseaux* et al. in [6.71]. Considering the technological progress on input-output interfaces and in nonlinear media, these parallel processors should produce attractive new developments for application to pattern recognition, and analog optical computing.

6.4.2 Image Edge Enhancement and Inversion

The edge enhancement of an image transparency is based on the fact that in holographic recording, the reconstructed image is only a faithful replica of the original one if the intensity of the signal beam is less than the intensity of the reference beam. If this condition is violated, edge enhancement can be produced in the diffracted image. These effects are also known in recording with classical photographic films, and they can now be applied in real time with photorefractive crystals using either a 2WM or 4WM interaction. The modulation ratio m of the interference pattern can be written as:

$$m = \frac{2\sqrt{I_R I_S}}{I_S + I_R} \tag{6.34}$$

where I_S and I_R are the intensity distributions due to signal and reference beams on the crystal. The evolution of m versus the beam ratio $\beta = I_R/I_S$ is shown in Fig. 6.29. The region $\beta \gg 1$ corresponds to linear image reconstruction ("ordinary" holography) while $\beta \ll 1$ provides edge enhancement. In this region we have $m = 2(I_R)^{1/2} \times (I_S)^{-1/2}$ and therefore the bright region of the image gives negligible diffraction. Since for $I_S \approx 0$, no diffraction occurs, only

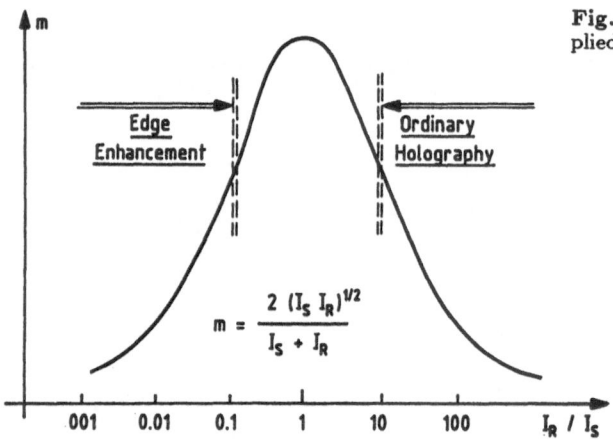

Fig. 6.29. Four-wave mixing applied to image edge enhancement

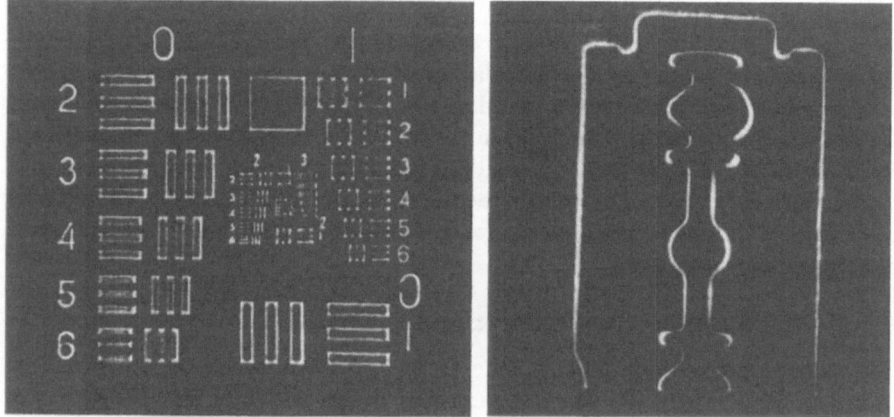

Fig. 6.30. Examples of reconstructed images with edge enhancement

the transition regions between light and dark regions of the input object beam will be efficiently reconstructed thus giving rise to an edge enhancement in the image. Such experiments in dynamic photorefractive crystals were first successfully performed with $Bi_{12}SiO_{20}$ crystals using a 4WM configuration and then in $BaTiO_3$ [6.72, 73]. Figure 6.30 shows experimental results for edge enhancement. A general expression for the beam ratio dependence of the diffraction efficiency has been derived in [6.74]. It shows in particular the range of beam ratios $\beta < 1$ over which the diffracted image can be exactly proportional to the inverse of the object beam intensity. Using this inversion of dynamic range, real-time optical intensity inversion of gray scale images has been obtained, as well as enhancement of defects in a periodic mask [6.75] and real-time image division and deblurring [6.76, 77].

6.4.3 Image Subtraction and Parallel Optical Logic

A further application of dynamic holography in photorefractive crystals is in image subtraction and logical operations between two-dimensional data planes. A series of different Boolean operations can be demonstrated. The principle of image subtraction (or partial image erasing) consists in superimposing in the crystal volume a complementary refractive index modulation. Let us consider Δn_A and Δn_B, the photoinduced index modulation in the hologram corresponding to the image fields $A(x,y)$ and $B(x,y)$, successively recorded in the same crystal volume

$$\Delta n_A = \delta n_A \cos [Kx + \Phi_A(x,y,z)] \quad ,$$
$$\Delta n_B = \delta n_B \cos [Kx + \Phi_B(x,y,z)] \quad . \tag{6.35}$$

Subtraction is achieved when $\Delta n_A + \Delta n_B = 0$. The feasibility of this process was first demonstrated for Fourier holograms recorded in iron-doped $LiNbO_3$ crystals [6.78]. Selective erasure of a page of information (complete, partial or single bit) is achieved by a second recording with the initial object transparency partially masked and with a phase shift of π on the reference beam. Therefore, the superposition of these two holograms in the crystal volume reconstructs the image difference $A - B$. This binary image subtraction corresponds to the logic Boolean operation "Exclusive OR" (Fig. 6.31a). In the particular situation where B is a uniform object transparency, ($B = 1$) the reconstructed image is the initial object A but with a reversed contrast (Fig. 6.31b) [6.78, 79]. These operations of optical logic can also be performed in data planes superimposed in the same crystal volume by angular coding of the reference beam. Based on the same principles, real-time subtraction and differentiation of digital images using degenerate four-wave mixing in highly sensitive BGO crystals (reflection type holograms) is demonstrated in [6.80, 81].

To summarize, the experimental results prove the capabilities of optics for performing in parallel numerical operations on digital images. Since there is a need for increased computing power in the future, optical techniques may be a very efficient way of handling parallel data channels for very high data rate computation.

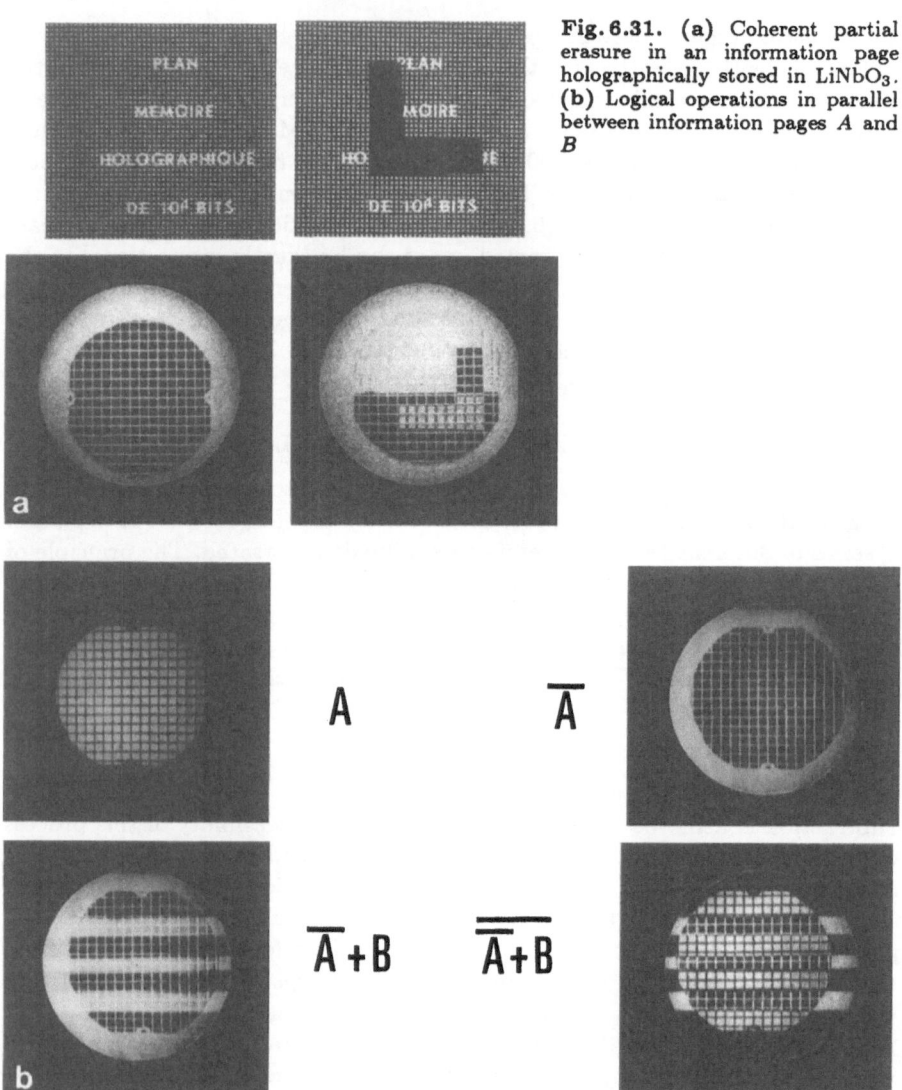

Fig. 6.31. (a) Coherent partial erasure in an information page holographically stored in LiNbO₃. (b) Logical operations in parallel between information pages A and B

6.4.4 Acousto-Photorefractive Effect

The acousto-photorefractive effect results from the interaction of a high intensity short laser pulse with an acoustic wave (volume or surface wave propagation) propagating in a photorefractive crystal such a LiNbO₃. It results in semipermanent refractive index changes that replicates the acoustic signal at a given instant of time. The mechanisms involved for the storage of the acoustic signal include: (i) modulation of the photorefractive effect by the acoustically induced pressure or density variations; (ii) modulation of the photorefractive effect by the electric fields associated with an acoustic signal propagating in piezoelectric media; (iii) light diffracted by the acoustic signal can produce photorefractive storage. These three factors contribute to either bulk or surface acoustic signal wave storage. As shown in Fig. 6.32, the demonstration of an acousto-optic memory correlator based on this effect by *Berg* et al. [6.82, 83], implies the following steps: (i) writing, in which a photorefractive image of the acoustic signal is stored in a LiNbO₃ surface acoustic wave device; (ii) reading, in which a signal is acousto-optically correlated with the stored reference signal, and (iii) erasure by UV exposure or by annealing the crystal at 250° C for one hour. This acousto-optic processor has been successfully used for correlation of linear FM chirp signals with the stored image of the same signal. The laser pulse writing source was a frequency doubled Nd : YAG laser and, by illuminating the acoustic delay line with both green and IR radiation, ($0.53\,\mu$m and $1.06\,\mu$m respectively), memory times as long as two months have been demonstrated [6.83]. Based on these techniques, filters that use the reflections of the bulk acoustic wave from holographically created gratings in LiNbO₃ have been demonstrated [6.84] with possible extension to resonators and very wide bandwidth dispersive delay lines. Attractive additional areas for acousto-photorefractive memory devices, other than signal processing, may also include high speed and high density data recording, and 1D or 2D spatial light modulation.

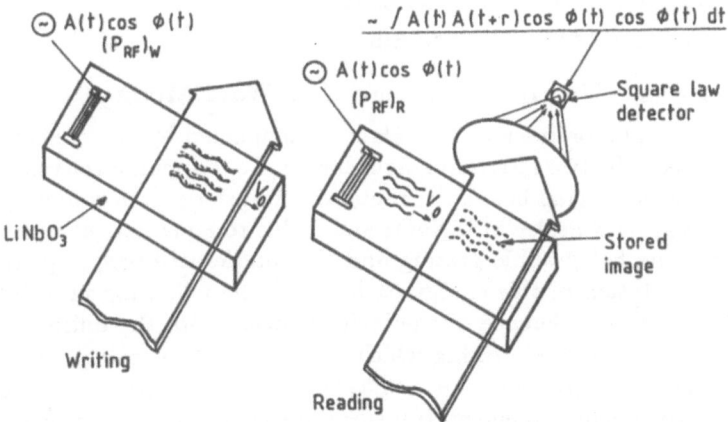

Fig. 6.32. Basic principle of an acousto-optic correlator based on the acousto-photorefractive effect in LiNbO₃. From [6.83]

6.5 Summary of Crystal Properties

Photorefractive crystals appear to be perfectly adapted recording materials for application to dynamic holography and coherent image processing. Several of these crystals have already shown a sensitivity comparable to high resolution photographic plates, thus allowing one to record and process information in parallel at millisecond speed with low power cw lasers. Moreover, shorter response time is obtained by increasing the incident light intensity and for example nanosecond response is achieved with pulse lasers emitting in the blue-green spectral range (frequency doubled Nd : YAG delivering $10\,\mathrm{kW\,cm^{-2}}$). All these features, including availability of crystals with the required optical quality, provide attractive features for the proposed applications in real-time interferometry, adaptive optics, image storage, optical interconnections, etc. For all the experiments, the photorefractive crystal has been considered merely as a real-time holographic media. However, we will show in the following sections that a new class of applications emerge when the effects of self-interference between the recording beams inside the crystal volume are also considered. This leads to energy transfer from the reference to the signal beam, and in such conditions, the photorefractive crystal may be regarded as a nonlinear material allowing parametric amplification of a coherent signal wave containing spatial information.

A discussion of the interactions based on 2WM and 4WM in different types of photorefractive crystals will be developed in the following as well as applications to image amplification, self-induced cavities and phase conjugation with gain.

6.6 Energy Transfer in Wave Mixing with Photorefractive Crystals

6.6.1 Degenerate and Nearly Degenerate Two-Wave Mixing

Grating fromation in photorefractive crystals is accompanied by an intensity redistribution between the two interfering light beams, i.e. the pump (or reference) and a low intensity signal beam. This intensity transfer (beam coupling) was initially observed by *Staebler* in LiNbO$_3$ and is due to a permanent phase mismatch between the holographic grating and the interference fringes [6.1]. The physical interpretation of this energy exchange is the following: in a dynamic material, the self-interference of the incident beam with the diffracted beam creates a new holographic grating which can add to (or subtract from) the initial one. Since the diffracted wave is phase delayed by $\frac{\pi}{2}$ with respect to the reading beam, the maximum energy transfer is obtained when the incident fringe pattern and the photoinduced index modulation are shifted by $\frac{\pi}{2}$ thus corresponding to a spatial shift by $\Lambda/4$ [6.3, 85]. In photorefractive crystals, such a $\frac{\pi}{2}$ phase shift exists when the recording is by "diffusion" of photocarriers (no external applied electric field) as shown in Fig. 6.33. As a consequence, a permanent and efficient amplification of a low intensity signal beam has been observed in crystals like BaTiO$_3$, LiNbO$_3$, or KNbO$_3$ [6.8]. If we now apply the coupled wave equations to the $\frac{\pi}{2}$ phase shifted component of the photoinduced index modulation, the coherent interaction between the two waves of respective amplitude R and S is the solution of:

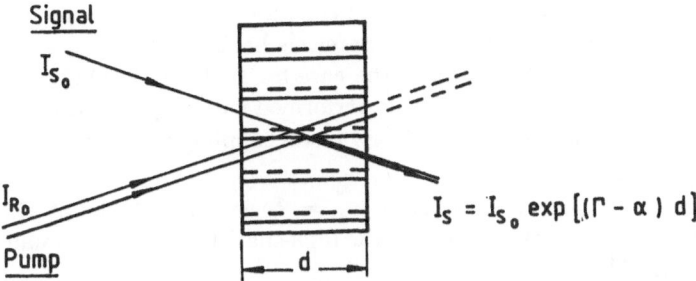

Signal

I_{S_0}

I_{R_0}

Pump

$\longleftarrow d \longrightarrow$

$I_S = I_{S_0} \exp\left[(\Gamma - \alpha)\, d\right]$

Fig. 6.33. Degenerate two-wave mixing in photorefractive crystals. *(Continuous line)* intensity pattern. *(Dashed line)* index modulation

$$\frac{dS}{dz} = \frac{1}{2}\Gamma\frac{R^2 S}{R^2 + S^2} - \frac{1}{2}\alpha R \quad , \quad \frac{dR}{dz} = \frac{1}{2}\Gamma\frac{R S^2}{R^2 + S^2} - \frac{1}{2}\alpha S \quad . \tag{6.36}$$

The transmitted signal beam intensity resulting from the dynamic two-beam coupling takes the form

$$I_S = I_{S_0}\frac{\beta + 1}{\beta + \exp\left(\Gamma l\right)} \times \exp\left[(\Gamma - \alpha)d\right] \tag{6.37}$$

where β is the incident intensity ratio of the two interfering waves and Γ is the exponential gain coefficient of the interaction. Γ is related to the maximum amplitude of the photoinduced index modulation Δn_S through the relation

$$\Gamma = \frac{4\pi\Delta n_S}{\lambda\cos\theta}\sin\Phi_g \quad . \tag{6.38}$$

Φ_g is the spatial phase shift of the grating and, in agreement with the previous arguments, Γ is maximum when $\Phi_g = \frac{\pi}{2}$. In the case of a negligible pump beam depletion, the transmitted signal beam intensity is simply given by

$$I_S = I_{S_0}\exp\left[(\Gamma - \alpha)d\right] \tag{6.39}$$

and therefore, when the condition $\Gamma > \alpha$ is fulfilled, the incident signal exhibits gain and the photorefractive crystal may be regarded as a parametric amplifier. A practical parameter for characterizing the energy transfer due to the two-beam coupling is the effective gain γ_0 defined [6.86] as

$$\gamma_0 = \frac{I_S\,(\text{with pump})}{I_S\,(\text{without pump})} \tag{6.40}$$

and for the undepleted pump beam approximation we have

$$\gamma_0 = \exp\left(\Gamma d\right) \quad . \tag{6.41}$$

Therefore, from the measurement of γ_0, the value of the exponential gain coefficient Γ of the interaction can be easily deduced. Γ is derived from the Kukhtarev relations given in Sect. 6.1.2 which express the steady-state index modulation Δn_S and the phase shift Φ_g as a function of the grating recording conditions (applied field, grating spacing) and of the crystal parameters (electro-optic coefficient, trap density, dielectric constant ...). Thus large values of Γ ($\Gamma \cong 12\,\mathrm{cm}^{-1}$) may be obtained in materials having large Δn_S when recording by diffusion ($\Phi_g = \frac{\pi}{2}$) and this is the case for BaTiO$_3$, LiNbO$_3$ and KNbO$_3$. However, the same 2WM experiment performed with highly photoconductive BSO (or BGO) crystals leads to very low beam coupling ($\Gamma \cong 1.5\,\mathrm{cm}^{-1}$) for the following reasons:

— For diffusion, the required phase shift $\Phi_g = \frac{\pi}{2}$ is established but the steady-state index modulation is low in the high spatial frequency domain ($\Lambda^{-1} > 1000\,\mathrm{mm}^{-1}$).
— For drift, the index modulation is much higher at $\Lambda^{-1} < 300\,\mathrm{mm}^{-1}$ but the corresponding phase shift Φ_g is nelgigible [6.7].

However, an efficient beam coupling can be obtained if the fringe pattern (or the crystal) is moved at a constant velocity with an electric field applied to the crystal. The speed is adjusted such that the index modulation is recorded at all times, but with a spatial phase shift with respect to the interference fringes. Clearly, the optimum fringe velocity will depend on the recording-erasure time constant τ and when an interference pattern moving at velocity v is introduced into the coupled wave equations [6.87], the resulting gain coefficient Γ is given by

$$\Gamma = \frac{4\pi \Delta n_S}{\lambda \cos \theta} \frac{K v \tau}{1 + K^2 v^2 \tau^2} \ . \tag{6.42}$$

In a 2WM configuration such as that shown in Fig. 6.34 the fringe displacement increases the amplitude of the $\frac{\pi}{2}$ phase shifted component of the index modulation and consequently efficient energy transfer can now be obtained in photorefractive crystals like BSO. From (6.42) it is thus expected that the optimum fringe velocity is $v_0 = \Lambda(2\pi\tau)^{-1}$ which corresponds to a frequency detuning by $\delta\omega = \tau^{-1}$ of the reference beam. The time constant τ (and Δn_S) depends on the grating recording conditions as well as on the physical parameters of the crystal. This interaction with moving fringes is consequently named "nearly degenerate two wave-mixing" and a simple method for frequency de-

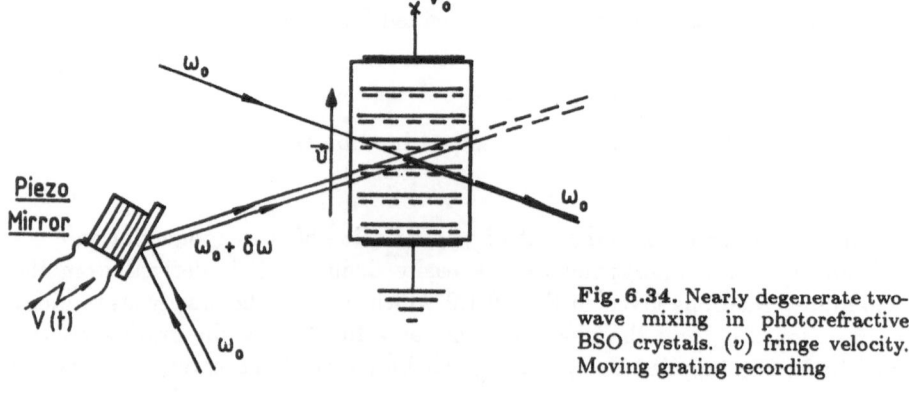

Fig. 6.34. Nearly degenerate two-wave mixing in photorefractive BSO crystals. (v) fringe velocity. Moving grating recording

tuning the reference beam by $\delta\omega$ is to use a piezomirror driven by a saw-tooth voltage.

6.6.2 Degenerate and Nearly Degenerate Four-Wave Mixing

In the 4WM configuration shown in Fig. 6.35, the two counterpropagating pump beams interact in the crystal volume. Using the holographic approach [6.43], the generation of the conjugate beam arises from the diffraction of one of the pump beams by the index grating recorded by the other two interfering beams. If the transmission-type grating is dominant, R_1 and S_1 are the recording beams, while R_2 is the readout beam of the photoinduced grating. As in 2WM, the self-diffraction process plays a dominent role for obtaining an efficient interaction between the recording beams. When a $\frac{\pi}{2}$ phase shifted component of the index modulation is present, the wavefront reflectivity is simply given by:

$$\varrho = \frac{I_2}{I_1}\left[\exp\left(\frac{\Gamma d}{2}\right) - 1\right]^2 . \tag{6.43}$$

Depending on the crystal orientation $(\Gamma > 0)$, the conjugate mirror reflectivity can be much larger than unity, thus allowing phase conjugation with gain. In a first approximation, ϱ depends linearly on the pump beam ratio $r = I_2/I_1$, but in a more general formalism derived by *Fischer* et al. [6.88], an optimum pump-beam ratio also exists as well as an optimum value of the phase shift Φ_g which may be different from $\frac{\pi}{2}$ in this 4WM interaction. With photorefractive BSO, this phase shift, as in 2WM, is induced by a frequency detuning of the pump beam R_1. In the nearly degenerate 4WM interaction shown in Fig. 6.31, the generated conjugate beam is thus frequency shifted by $\delta\omega$. However, it will be shown in the following sections that the applicability domain of (6.42) is limited in photorefractive BSO and in the general formalism it will result in a resonance of Γ with both the fringe velocity and the grating spacing. The optimization of these two parameters will also result in large values of Γ $(\Gamma \cong 5\text{–}12\,\text{cm}^{-1})$.

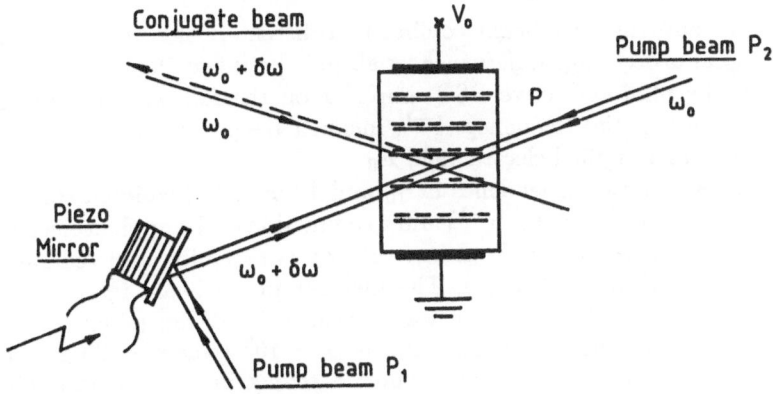

Fig. 6.35. Nearly degenerate four-wave mixing

6.6.3 Transient Energy Transfer

This is achieved in photorefractive materials with local response (no phase shift between intensity and index grating) when the recording time is less than or comparable to the grating erasure time. In the initial stage of holographic recording, the light fringes are normal to the crystal surface (Fig. 6.36a). Then, due to self-diffraction, the phases of the two beams vary and these phase changes depend on the relative intensities of the two recording beams (Fig. 6.36b). As a result, light interference fringes become tilted and curved and, due to crystal inertia, a spatial mismatch occurs between the interference fringes and the index grating. This gives rise to an energy redistribution between the two output beams [6.89]. In the steady state, the index modulation coincides with the light intensity pattern (Fig. 6.36c) and thus the energy transfer disappears. In agreement with the theory [6.90, 91], the probe beam intensity oscillates with time and the whole pump beam intensity can be temporarily transferred to the signal beam. These transient phenomena are clearly observed in iron-doped $LiNbO_3$ (recording by photovoltaic effect) and in BSO-BGO when recording with high applied electric field at large grating spacings ($\Lambda \cong 10$–$20\,\mu m$).

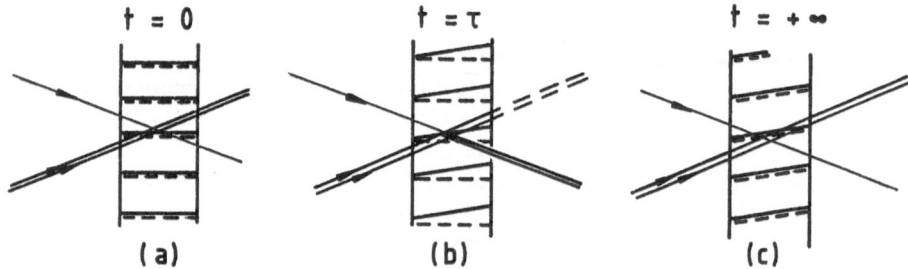

Fig. 6.36a–c. Transient beam coupling during recording in photorefractive $LiNbO_3$-Fe^{2+}. (a) initial state. (b) transient signal beam amplification. (c) steady-state

6.6.4 Spatial Frequency Dependence of the Gain

With a view to applying two-beam couling to coherent image amplification and optical signal processing, *Rajbenbach* et al. [6.92] studied the dependence of the gain of the photorefractive BSO amplifier on the following recording parameters: grating spatial frequency Λ^{-1}, incident beam ratio β of the two interfering waves, and applied electric field E_0.

Figure 6.37 shows the dependence of γ_0 and Γ on Λ^{-1} (explored spatial frequency range $10 < \Lambda^{-1} < 400\,mm^{-1}$) and as a function of the applied electric field E_0. For each measurement the fringe velocity is carefully adjusted such that the maximum of γ_0 is obtained. The incident pump beam intensity is $I_{R_0} = 140\,mW\,cm^{-2}$ at the recording wavelength $\lambda = 568\,nm$ (single-mode krypton laser), and the incident beam ratio is $\beta = 10^3$ (corresponding to a time constant $\tau \cong 120\,ms$). These curves show a strong increase in the gain for $\Lambda^{-1} < 100\,mm^{-1}$ with a sharp maximum at $\Lambda^{-1} \cong 45\,mm^{-1}$. Figure 6.38

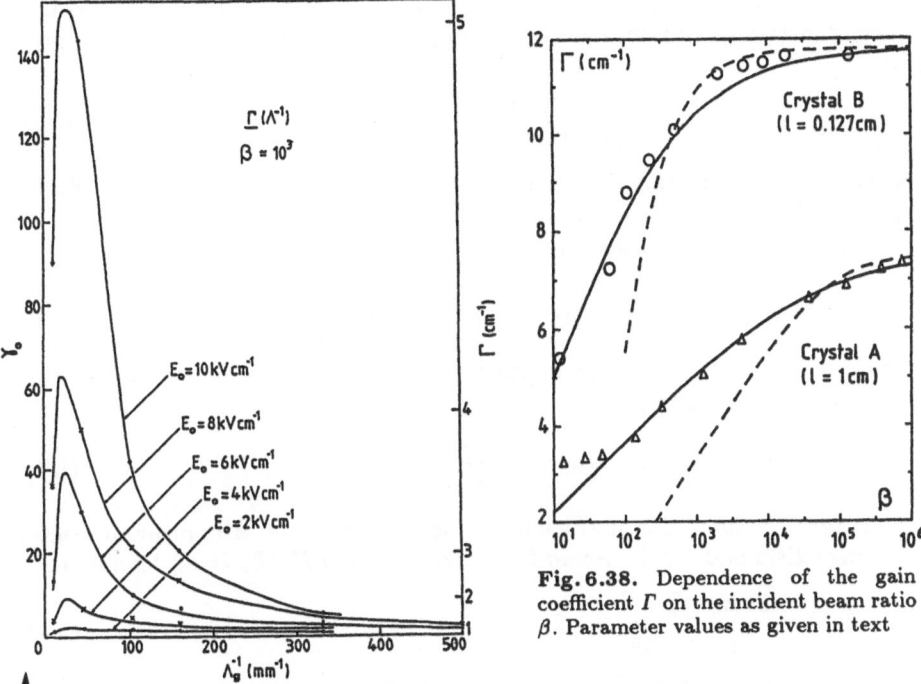

Fig. 6.38. Dependence of the gain coefficient Γ on the incident beam ratio β. Parameter values as given in text

Fig. 6.37. Dependence of the gain coefficient Γ on the grating spatial frequency and applied field in nearly degenerate two-wave mixing with BSO

represents the variation of Γ as a function of the incident beam ratio β for the optimum spatial frequency ($\Lambda_{opt}^{-1} = 45\,\mathrm{mm}^{-1}$; $E_0 = 10\,\mathrm{kV\,cm}^{-1}$) and for two different crystal lengths ($d = 10\,\mathrm{mm}$ and $d = 1.27\,\mathrm{mm}$).

The following points summarize the main conclusions that can be drawn from these curves: (i) High gain is available in photorefractive BSO when recording with a high electric field and moving the fringes at the optimum velocity. (ii) The gain of the amplifier is strongly dependent on the grating spatial frequency. (iii) The gain reaches saturation at high beam ratio. Consequently, a wide range of experimental conditions allows one to obtain a value of Γ in excess of the crystal absorption losses ($\alpha \cong 0.6\,\mathrm{cm}^{-1}$ at $\lambda = 568\,\mathrm{nm}$, and $\Gamma \cong 8\text{--}12\,\mathrm{cm}^{-1}$ for optimized recording conditions), thus suggesting applications to coherent image amplification and self-induced ring cavities.

Kukhtarev's equations can still explain the dependence of Γ on the grating spatial frequency Λ^{-1} and the incident beam ratio β. The starting point is the set of differential equations that describe the charge transport and trapping and for which the incident fringe illumination is

$$I(x,t) = I_0[1 + m \cos K(x - vt)] \quad . \tag{6.44}$$

The details of the theory may be found in related publications on this subject [6.27, 28, 94] and therefore we just emphasize in this section the important features related to the fringe displacement in crystals like BSO. Derived from these references, the velocity that maximizes the imaginary part of $E_{\rm SC}$ is given by

$$V_{\rm opt} = s(I_{\rm R_0} + I_{\rm S_0})\frac{N_{\rm D}}{N_{\rm A}}\frac{E_q}{E_0}\frac{\Lambda}{2\pi} \quad , \tag{6.45}$$

where s is the ionization cross-section, $N_{\rm D}$ the density of donor atoms, $N_{\rm A}$ the density of acceptor atoms, and $E_q = eN_{\rm A}\Lambda/2\pi\varepsilon_0\varepsilon$.

Under these conditions, we have:

$$n_1 = -{\rm i}\frac{n_{\rm av}}{D}m\frac{E_q}{E_0} \quad , \quad N_{\rm D^+} = \frac{m}{D}N_{\rm A} \quad , \quad E_{\rm S1} = {\rm i}\frac{m}{D}E_q \quad , \tag{6.46}$$

where n_1 is the free electron density, $N_{\rm D}^+$ the ionized donor density, $n_{\rm av}$ the average electron density.

$$D = 1 + \frac{E_q E_M}{E_0^2} \quad ; \quad E_M = \gamma_{\rm R}\frac{N_{\rm A}\Lambda}{2\pi\mu} \tag{6.47}$$

$\gamma_{\rm R}$ is the recombination coefficient of the photocarriers. An optimum grating spacing exists and can be found from the condition $E_M E_q = E_0^2$ leading to:

$$\Lambda_{\rm opt} = \frac{2\pi E_0}{N_{\rm A}}\sqrt{\frac{\mu\varepsilon_0\varepsilon}{\gamma_{\rm R}e}} \quad , \tag{6.48}$$

and to a frequency detuning of the pump beam by

$$\delta\omega = 2\pi v_{\rm opt}\cdot(\Lambda_{\rm opt}^{-1}) = sI_{\rm R_0}\frac{N_{\rm D}}{N_{\rm A}}\cdot\frac{E_q}{E_0} \quad . \tag{6.49}$$

It may be concluded that recording in BSO with a moving grating has two consequences: First, under optimum conditions ($v_{\rm opt}$ and $\Lambda_{\rm opt}$) the space-charge field is $\frac{\pi}{2}$ out of phase with the interference pattern, i.e., all of the electric field is useful for promoting the energy transfer from the reference beam to the low intensity probe beam. Secondly, the modulus of the space-charge field is increased from a value of mE_0 in the absence of fringe movement to $m(E_q/2)$ at the velocity $v_{\rm opt}$ (typically a tenfold increase for $E_0 \cong 10\,{\rm kV\,cm^{-1}}$).

6.6.5 Beam Ratio Dependence of the Gain

The dependence of the gain Γ on the incident beam ratio β shown in Fig. 6.38 is interpreted by introducing the second-order terms in the expansion of the space charge field (second-order perturbation):

$$E_{\rm S} = \tfrac{1}{2}E_{\rm S1}\exp[{\rm i}K(x - vt)] + \tfrac{1}{2}E_{\rm S2}\exp[2{\rm i}K(x - vt)] + {\rm c.c.} \tag{6.50}$$

It follows that the fundamental component of $E_{\rm S}$ will increase at a less than linear rate as m increases. For example, for the second-order calculations the space-charge field at optimum velocity and grating period may be written as

$$E_{\rm S1} = {\rm i}\frac{m}{2}E_q(1 - \chi m^2) \tag{6.51}$$

where χ is a correcting factor. From this analysis it can be seen that E_{S1} is no longer proportional to m, and using a phenomenological approach, we can write

$$E_{S1} = \frac{i}{2}E_q f(m) \quad . \tag{6.52}$$

The signal beam intensity is thus the solution of

$$\frac{dI_S}{dz} = \Gamma \frac{f(m)}{m} I_S \quad . \tag{6.53}$$

The problem then is to find a function which for $m \ll 1$ is equal to m and for higher values of m will display slower growth. We obtained the best results with $f(m) = (1/a)[1 - \exp(-am)]$ with $a = 2.8$, and the agreement as shown by the continuous lines in Fig. 6.38 is quite good. We also note on this figure that a higher value of Γ is measured for a thinner crystal. The reason for this may be that a thinner crystal causes less polarization rotation and also provides a better uniformity of the optical intensity through the crystal volume.

In conclusion, moving the grating at the optimum grating velocity increases the $\frac{\pi}{2}$ phase-shifted amplitude of the space-charge field. In accordance with the experimental results, an optimum value of the grating spacing also exists from the condition $E_M E_q = E_0^2$. The dependence of the gain coefficient Γ on the beam ratio β is explained by adding a second term in the Fourier expansion of the space-charge field, and for high values of β (typically $\beta < 10^4$), gain coefficients Γ on the order of 8–12 cm^{-1} are obtained. Such values of the gain coefficient are equivalent to those of other photorefractive crystals such as BaTiO$_3$ and LiNbO$_3$ and are now compatible with the realization of BSO phase conjugate mirrors with gain and self-starting optical resonators with an optical feedback to the photorefractive BSO amplifier. These applications based on nearly degenerate two-wave mixing and four-wave mixing are presented in the next sections.

6.6.6 Further Comments on the Beam Coupling

Since energy transfer is related to a self-interference process, the maximum beam coupling is obtained when the incident and diffracted waves have the same polarization. This occurs in BaTiO$_3$, LiNbO$_3$, KNbO$_3$ when the incident beams have a polarization parallel to the polar c-axis or in BSO with the configuration shown in Fig. 6.4b (applied field along $\langle 001 \rangle$ direction). With this latter crystal, the configuration of anisotropic diffraction, shown in Fig. 6.4c is not adapted to the beam coupling experiments, since the incident and diffracted waves have orthogonal polarization states and thus cannot interfere. These arguments justify in fact that the two configurations may be respectively named γ_{max} (maximum coupling) and η_{max} (maximum diffraction efficiency) [6.86]. However the coexistence in BSO of optical activity and linear birefringence, renders in general the direct and diffracted beams elliptically polarized. Therefore, the coupled wave analysis presented here for linear beam polarization, should be extended to such complex media having both linear and circular birefringence. These polarization properties are being investigated by various authors on a theoretical and/or experimental basis [6.22, 94].

6.6.7 Frequency Shifters for Photorefractive Crystals

As shown in previous sections, the required frequency shift of the pump beam is dependent on the grating recording parameters as well as on the physical parameters of the crystal. In photorefractive crystals where the usual intensity levels are typically $10-500\,\mathrm{mW\,cm^{-2}}$, the deduced value of the frequency detuning is in the range of $1\,\mathrm{Hz}$ to few $100\,\mathrm{Hz}$. Such low frequency shifts can be obtained and precisely controlled with the following two techniques:

— A piezo mirror driven by a saw-tooth voltage; precise adjustment of the voltage period and amplitude ensures that all the incident pump beam intensity is transferred into the sideband at frequency $\omega + \delta\omega$.
— A rotating half-wave plate placed between two fixed quarter-wave plates; the optical field transmitted by these wave plates is frequency shifted by $\delta\omega$ when the half-wave plate rotates at angular velocity $\delta\omega/2$.

Both of these techniques, for which more details can be found in [6.95], are well suited for frequency shifting the pump (or signal) beam of a photorefractive amplifier.

Table 6.1. Properties of ferroelectric photorefractive crystals [6.8]

Ferroelectrics	I_0	τ	Γ	ϱ
LiNbO$_3$; BaTiO$_3$	$100\ \mathrm{mW\,cm^{-2}}$	seconds	$10-20\ \mathrm{cm^{-1}}$	$1-50$
SBN; KNbO$^*{}_3$; KTN	$\lambda=514\ \mathrm{nm}$			

Comments:
– Single domain crystals
– No applied field in general
– Optimum grating spacing $\Lambda \cong 1\,\mu\mathrm{m}$
– KNbO$^*{}_3$ = Faster speed demonstrated with highly reduced crystals

Table 6.2. Properties of nonferroelectric photorefractive crystals

Non-ferroelectrics	I_0	τ	Γ	ϱ
Bi$_{12}$(Si,Ge,Ti)O$_{20}$	$10-100\ \mathrm{mW\,cm^{-2}}$	$10\ \mathrm{ms}$	$8-12\ \mathrm{cm^{-1}}$	$1-30$
	$\lambda=514\ \mathrm{nm}$			
GaAs-Cr; InP-Fe	$10-100\ \mathrm{mW\,cm^{-2}}$	$10\,\mathrm{ms}$	$1-6\,\mathrm{cm^{-1}}$	$0.1-1$
[6.96, 97]	$\lambda=1.06\,\mu\mathrm{m}$			

Comments:
– Applied field and moving grating or A.C electric field
– Optimum grating spacing (for BSO) $\Lambda=20\,\mu\mathrm{m}$

6.6.8 Summary of Crystal Performance

Tables 6.1 and 6.2 summarize some of the properties of different photorefractive crystals which are of interest for optical signal processing applications. These include the typical grating writing time constant τ, the two-wave coupling gain coefficient Γ and the conjugate beam reflectivity ϱ (recording beam intensity I_0).

6.7 Applications of the Energy Transfer

6.7.1 Image Amplification

The large values of the gain coefficient Γ in photorefractive crystals permit the amplification of a low intensity signal beam containing spatial information [6.10]. The optical set-up for image amplification of a signal wavefront modulated by a photographic transparency is shown in Fig. 6.39a. With this configuration, the energy transfer from the pump beam allows to amplify the image in the detection plane. When using a photorefractive amplifier such as BSO, the fringe velocity and the reference-signal beam angle are adjusted in order to receive maximum gain and, according to the curves shown in Figs. 6.37, 38, this takes place for a carrier spatial frequency of the hologram equal to $\Lambda^{-1} \approx 45\,\mathrm{mm}^{-1}$. However, since the spatial frequency response of the photorefractive BSO amplifier is of the bandpass type, the difference in

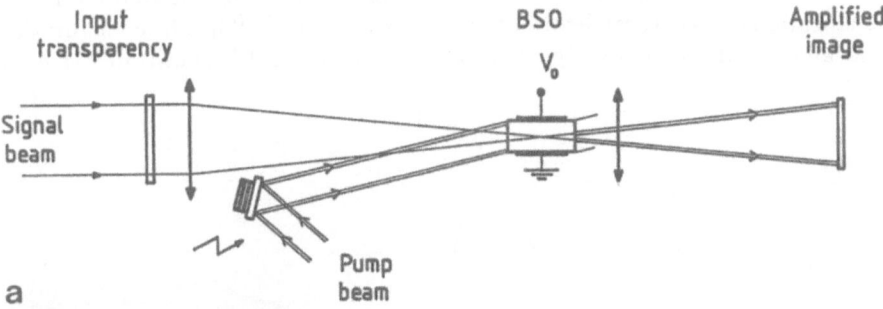

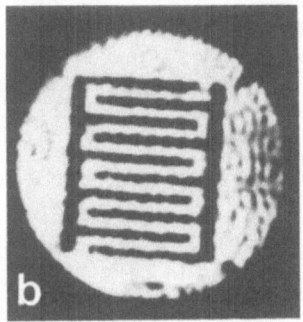

Fig. 6.39. (a) Image amplification through energy transfer from the pump beam. (b) 20 $\times$ amplified image. $E_0 = 6\,\mathrm{kV\,cm}^{-2}$

gain for the various spatial frequencies may be noticeable and can limit the size of the image to be amplified. Figure 6.39b shows a 20 × amplified image for an applied field of $E_0 = 6\,\mathrm{kV\,cm^{-1}}$. Higher values of the gain are possible when the electric field is increased, but this would correspond to a loss in image uniformity and quality [6.92]. Similar demonstrations of efficient image amplification have been performed with other photorefractive crystals such as $\mathrm{LiNbO_3}$, $\mathrm{BaTiO_3}$ and $\mathrm{KNbO_3}$, in which the phase-shifted volume hologram is recorded by diffusion (no applied field) with a carrier spatial frequency of the order of $\Lambda^{-1} \approx 1000\,\mathrm{mm^{-1}}$. When recording by the photovoltaic effect in iron-doped $\mathrm{LiNbO_3}$, only a transient image amplification is observed, since this mechanism provides a local response [6.10].

To summarize, the main limitations of these coherent image amplifiers stem from crystal inhomogeneities and from light-induced scattering, which limits both the resolution and the minimum intensity of the image to be amplified. This light-induced scattering is due to local fluctuations of the crystal dielectric constant, thus causing scattered waves which are efficiently amplified by 2WM with the pump beam. Clearly, in these experiments, a trade-off between the gain of the photorefractive amplifier and the signal-to-noise ratio of the amplified image has to be expected.

6.7.2 Interferometry of Large Objects

The above-mentioned technique of image amplification is not only limited to plane images but can also be applied to the amplification of diffuse three-dimensional objects, with related applications to dynamic interferometry for visualization of structural deformations and vibrations [6.98]. The configuration used and based on the nearly degenerate 2WM in BSO is shown in Fig. 6.40.

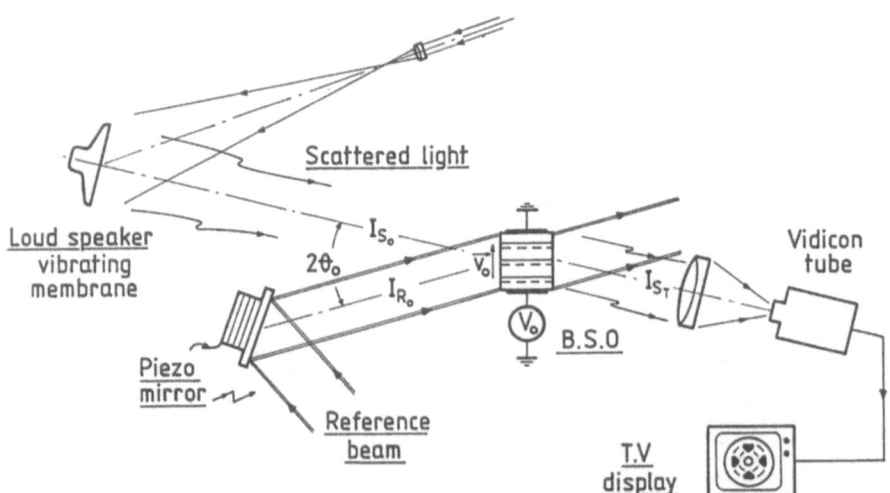

Fig. 6.40. Two-wave mixing in BSO applied to dynamic interferometry of a vibrating loud-speaker membrane

A frequency shifted pump beam and a low intensity signal beam interfere in the crystal volume − beam ratio of the two interfering waves $\beta \cong 10^3$. This signal beam carries the spatial and temporal information of a diffuse vibrating plate or membrane used as object. Therefore, due to the energy transfer of the pump beam, the transmitted image of the loudspeaker is amplified by a factor of 10 (see Fig. 6.41a). If the membrane vibrates at a frequency Ω, it induces in the crystal, fast moving fringes ($\Omega \gg 1/\tau$ condition of time-average intensity recording) and the amplified image displays bright and dark regions which correspond to the mode pattern (Fig. 6.41b). The related image intensity emerging from the crystal is given by:

$$I_S = I_{S_0} \exp\left[(\Gamma - \alpha)d\right] \quad \text{with} \tag{6.54}$$

$$\Gamma = \frac{4\pi}{\lambda} C E_0 \frac{K v \tau}{1 + K^2 v^2 \tau^2} J_0\left(\frac{4\pi\delta}{\lambda}\right) \quad .$$

Here C is a constant dependent on the recording conditions, J_0 is the zero-order bessel function, and δ the local amplitude of vibration of the object. The transmitted image exhibits a spatial modulation of the gain, and for optimum frequency detuning ($K v \tau \cong 1$ or $\delta\omega = 1/\tau$) of the pump beam we have:

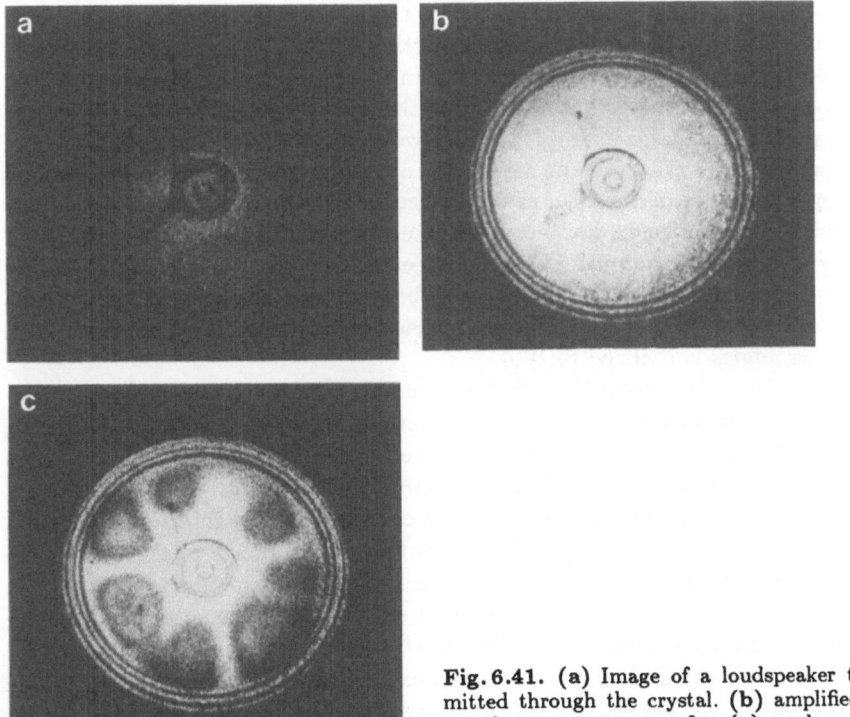

Fig. 6.41. (a) Image of a loudspeaker transmitted through the crystal. (b) amplified image due to energy transfer. (c) mode pattern of the vibrating membrane. Vibration frequency $F = 1.6\,\text{kHz}$

$$I_S = I_{S_0} \exp\left[\frac{2\pi}{\lambda} C E_0 J_0\left(\frac{4\pi\delta}{\lambda}\right) - \alpha\right] d \quad.$$

The related contrast of the mode pattern is simply given by:

$$M = \frac{\exp\left(\Gamma d\right) - 1}{\exp\left(\Gamma d\right) + 1} \quad. \tag{6.55}$$

In the pictures of interferograms shown in Fig. 6.41, the observed contrast of the mode pattern corresponds to a Γ value equal to $2.4\,\mathrm{cm}^{-1}$ (absorption losses at $\lambda = 514\,\mathrm{nm}$, $\alpha = 1\,\mathrm{cm}^{-1}$). Much higher Γ values could in principle be obtained, but due to the large object size used in these experiments, the hologram spatial carrier frequency is $\Lambda^{-1} \approx 300\,\mathrm{mm}^{-1}$ and therefore the coupling effect is not at its maximum. It can also be very advantageous to take account of the fact that the directly transmitted and self-diffracted waves may have different polarization states for a given crystallographic orientation. Thus polarization filtering of the scattered beams can be easily achieved by proper orientation of a polarizer, allowing one to attain higher signal-to-noise ratios for larger scale object structure under test [6.99].

6.7.3 Laser Beam Steering

This application relies upon the use of a two-dimensional spatial light modulator in combination with a photorefractive crystal (Fig. 6.42a). The basic principle is the following: The pump beam interferes in the crystal with the probe beam which is direction selected by the spatial light modulator (array of electro-optic shutters for example). After two-beam coupling a complete energy exchange from the pump to the selected probe beam direction can be obtained by using a photorefractive crystal with large gain coefficient. Therefore we can say that the pump beam has been deflected in the direction of the probe. If another direction of the probe beam is selected, the previous grating is erased, and rewriting a new one deflects the pump beam in another direction. By means to this principle, a new type of random-access digital laser beam deflector with large scan angles is realized [6.100].

A practical demonstration of this principle over a limited number of positions can be achieved with the experimental set-up shown in Fig. 6.42b. The low intensity signal beam is expanded and reflected by an array of piezo-mirrors (4×3). In the focal plane of lens L where a photorefractive $BaTiO_3$ crystal is placed, the pump beam and the array of signal beams interfere. Selection of one probe beam direction is achieved as follows: all of the piezo-mirrors are excited with a ramp generator except for one, which corresponds to the selected direction of deflection. Due to the Doppler shift δ induced by the moving mirrors, the interference fringes move. If $\delta \gg \tau^{-1}$ ($\tau \approx 1\,\mathrm{s}$ is the time constant for energy exchange in $BaTiO_3$), the corresponding index modulation cannot be recorded due to the crystal inertia. Therefore, the probe beam whose direction is selected by the nonexcited piezo-mirror is amplified. Figure 6.42c shows the experimental results obtained by driving the piezo-mirrors, and where the de-

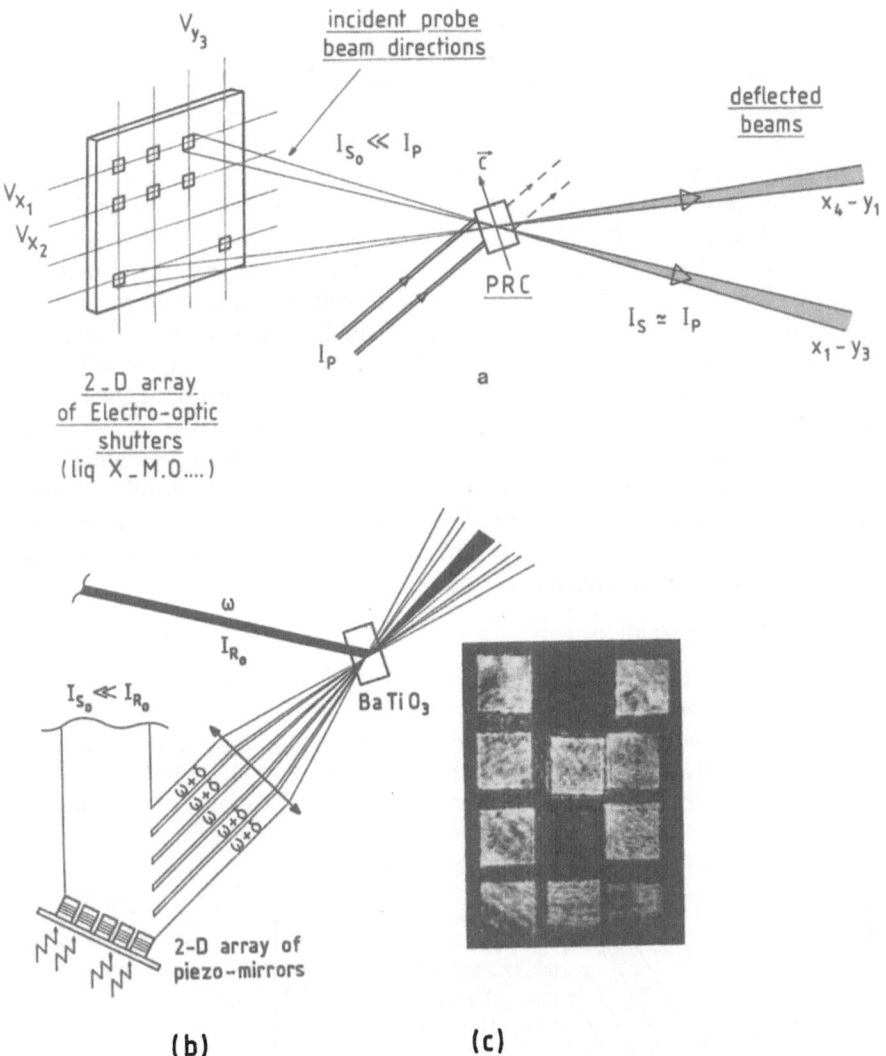

Fig. 6.42. (a) Application of the energy transfer in BaTiO$_3$ to 2D laser beam steering. (b) laser beam deflection obtained by driving an array of 4 × 3 piezo-mirrors. (c) generated pattern. The beam is randomly deflected by driving all the piezo-mirrors except the one corresponding to the deflection position

flected beam is about 10 % of the pump intensity. In this experiment, the use of a mirror array with a temporal phase modulation of the incident wavefront makes possible a perfect discrimination between the nonselected and the selected probe beam directions. This deflection principle can certainly be applied in laser cavities.

6.7.4 Amplified Phase Conjugation in Photorefractive BSO Crystals

The optical configuration used for phase conjugation by nearly degenerate 4WM is presented in Fig. 6.35. In this interaction, the conditions of high reflectivity closely depend on the same parameters as the exponential gain coefficient Γ previously considered in the 2WM interaction, i.e., the fringes in the crystal move at a constant velocity and the fringe spacing is adjusted at the optimum value $\Lambda_{\text{opt}} \cong 23\,\mu\text{m}$ [6.101, 102]. An extra parameter, in this 4WM interaction is the pump beam ratio $r = I_2/I_1$ and, as shown in Fig. 6.43, there is a noticeable dependence of the conjugate beam reflectivity on the parameter r. An amplified reflectivity $\varrho \cong 2.7$, is reached, for an incident beam ratio $\beta > 3 \times 10^4$ and a pump beam ratio $r = 0.2$. Such a maximum of the reflectivity obtained for asymmetric pump beam intensities is in accordance with the coupled mode theory of *Fischer* et al. [6.88]. Higher reflectivities are demonstrated with the configuration of Fig. 6.44 which uses two BSO crystals whose roles are respectively to generate (BSO1 $-$ 4WM) and then to amplify (BSO2 $-$ 2WM) the conjugate signal beam. For each crystal, the fringe velocity is carefully adjusted so that optimum 4WM and 2WM interactions are reached. The conjugate beam reflectivity is thus given by:

$$\varrho' = \varrho \exp\left[(\Gamma - \alpha_t)d\right] \tag{6.56}$$

ϱ is the conjugate beam reflectivity of BSO1 and Γ the gain coefficient of the BSO2 photorefractive amplifier. α_t is the loss coefficient that includes absorption and interface reflections. Based on this set-up, reflectivities as high as $\varrho' = 35$ have been measured, corresponding to $\varrho = 1.4$ for the generator and $\Gamma = 4\,\text{cm}^{-1}$ for the amplifier crystal ($\alpha_t = 0.6\,\text{cm}^{-1}$; $\lambda = 568\,\text{nm}$;

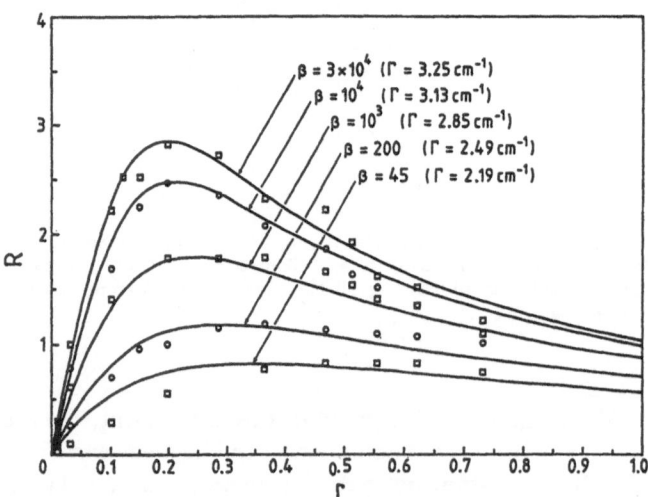

Fig. 6.43. Dependence of the conjugate beam reflectivity ϱ on the pump beam ratio r in nearly degenerate four-wave mixing with BSO

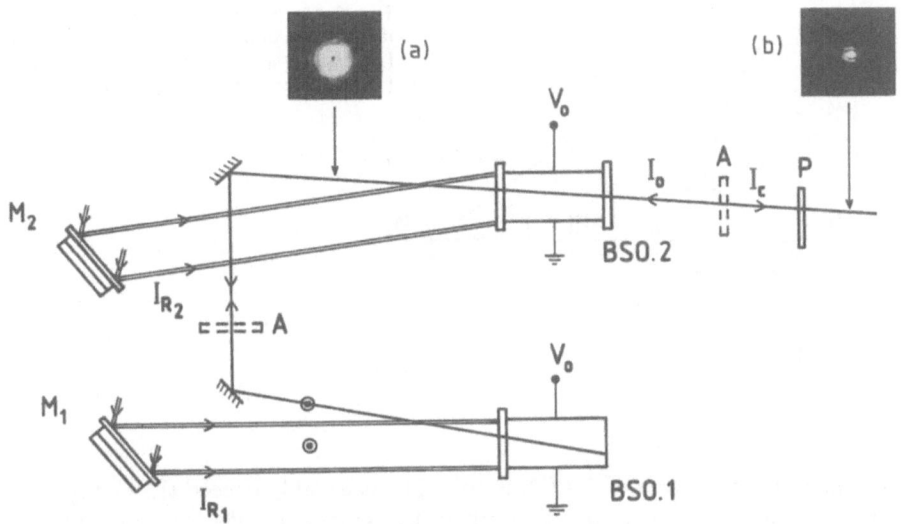

Fig. 6.44. High reflectivity phase conjugation ($\varrho \approx 35$) using two distinct photorefractive crystals. (BSO1) Phase conjugator (4WM). (BSO2) Amplifier (2WM). (A) Aberrator. (a) distorted probe beam; (b) corrected phase conjugate beam

$E_0 = 10\,\mathrm{kV\,cm^{-1}}$). Demonstration of phase distortion compensation is also shown in Fig. 6.44.

6.7.5 Self-Induced Optical Cavities

Due to the large gain coefficients of photorefractive amplifiers, different types of self-starting oscillators can be obtained by adding an optical feedback to the photorefractive amplifier [6.103]. These coherent oscillations have been reported with BaTiO$_3$, SBN and LiNbO$_3$ due to the high gain resulting from the high electro-optic coefficient, and they are also obtained with BSO-BGO because of the gain enhancement due to self-induced moving gratings when an electric field is applied to the crystal [6.104, 105]. Some of the characteristic properties of the ring and phase-conjugate oscillators are reviewed in the following sections.

a) Ring Oscillators. The optical set-up for obtaining a ring oscillator from a photorefractive amplifier is shown in Fig. 6.45. The photorefractive crystal is introduced into the beam path defined by the three mirrors $M_1 M_2 M_3$, and the angle between the pump beam and the $M_1 M_2$ direction is chosen so as to correspond to the optimum fringe spacing for the energy transfer of the pump beam. The condition for oscialltion is given by

$$(1 - R_{\mathrm{BS}})R^3 \exp\left[(\Gamma - \alpha_t)d\right] \geq 1 \tag{6.57}$$

where R and R_{BS} are the reflectivities of the cavity mirrors and beam splitter, Γ is the gain coefficient of the 2WM interaction and α_t represents the total

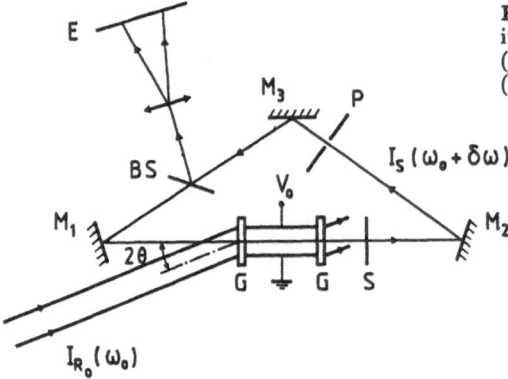

Fig. 6.45. Self-induced optical ring cavity with a photorefractive BSO crystal. (BS) beam splitter for mode observation; (P) pinhole for mode selection

losses. Since the values of Γ ($\Gamma > 8\,\mathrm{cm}^{-1}$) considerably exceed the cavity losses ($\alpha_t \cong 1\,\mathrm{cm}^{-1}$) oscillation builds up in the cavity [6.106]. The oscillation in the cavity is self-starting; the optical noise due to the pump beam is sufficient to generate a weak probe beam that is then amplified after each round trip in the cavity. The required detuning $\delta\omega$ between the pump and the cavity beam in the ring oscillator is also self-induced. In other words, the crystal chooses from the optical noise spectrum the frequency component shifted by $\delta\omega$ that will be optimally amplified in the cavity. In photorefractive BSO, for an applied field $E_0 = 10\,\mathrm{kV\,cm}^{-1}$, the beam in the cavity is typically frequency shifted by 30 Hz for $I_{\mathrm{R}_0} = 150\,\mathrm{mW\,cm}^{-2}$ at $\lambda = 568\,\mathrm{nm}$.

Figure 6.46 shows different transverse modes of the oscillator observed by inserting into the cavity a circular pinhole used as a mode selector; fundamental and higher order transverse modes of oscillations are observed, depending on the pinhole adjustment. A specific property of these photorefractive ring oscillators is that the gain is unidirectional and only one wave is amplified in the cavity. In particular, the residual coherent retrodiffused beams due to the mirrors M_1, M_2 and M_3 are not amplified: after interference with the pump beam they give reflection-type photoinduced gratings that are not efficiently recorded in the BSO with this configuration. The theory for oscillation in ring resonators with photorefractive gain has been developed by *Yariv* and *Kwong* [6.107], *Ewbank* and *Yeh* [6.108], and, in particular, it is shown that the amount of frequency shift depends on the length of the cavity. Consequently, the photorefractive resonators may be used in a new type of interferometry which directly converts optical path length changes into frequency shifts. The peculiarities of these ring cavities can also be applied for the conception of new gyroscopes based on the Sagnac effect. Other applications of the ring cavities can be found in reference [6.109] where the multimode functioning of a BSO ring oscillator is used for the determination of spheroidal wave functions for any value of the Fresnel number of the cavity. These specific properties could certainly be generalized to other operations in analog optical computing.

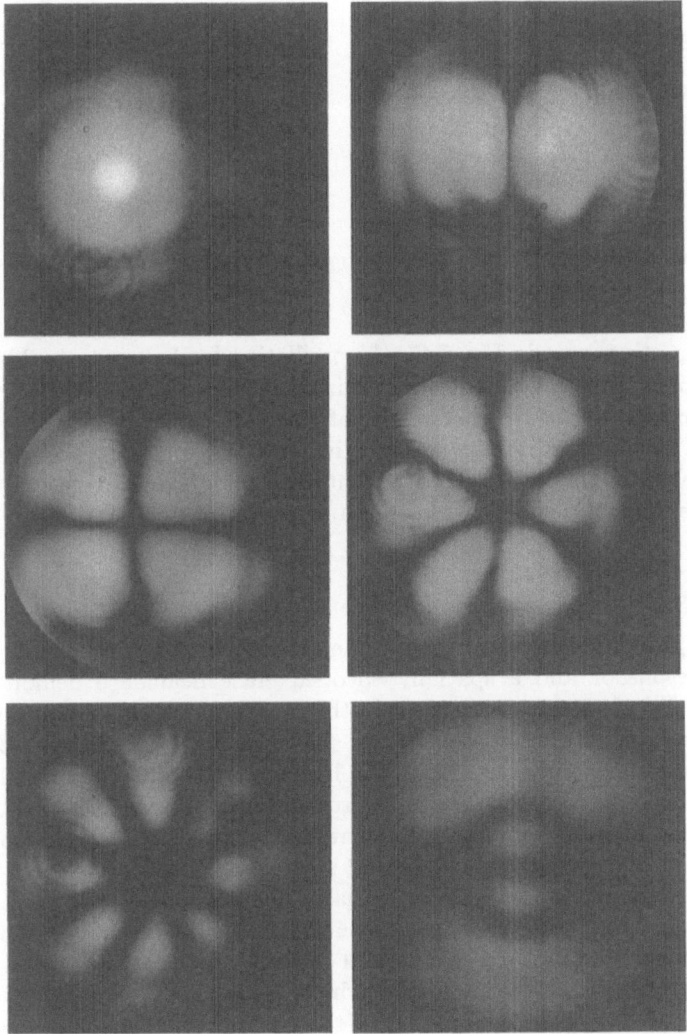

Fig. 6.46. Transverse mode structures of the beam in the self-induced ring cavity

b) Oscillator with Phase Conjugate Mirror. As reviewed in Sect. 6.7.4, the conjugate beam reflectivity in a 4WM interaction exceeds unity after optimization of the grating recording parameters. It is thus possible to induce an oscillation between a classical mirror and a photorefractive phase conjugate mirror. Since the first demonstration by *Feinberg* and *Hellwarth* [6.16] with a $BaTiO_3$ crystal, similar phase conjugate resonators have been obtained with $LiNbO_3$, and more recently with $Bi_{12}(Si,Ge)O_{20}$ crstals [6.110, 111]. The conditions which optimize the reflectivity in these latter crystals have been detailed previously: for $E_0 = 10\,kV\,cm^{-1}$, the optimum pump beam ratio is $r = 0.2$ and the grating

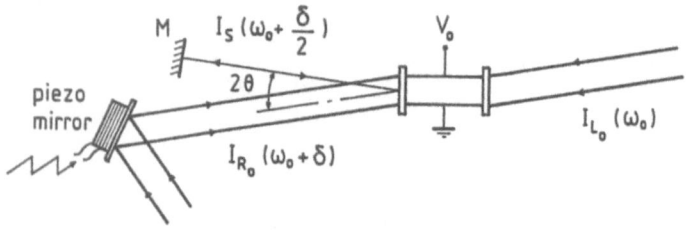

Fig. 6.47. Self-induced oscillation between a plane mirror M and photorefractive BSO conjugate mirror

spacing $\Lambda \cong 23\,\mu m$. As shown in Fig. 6.47, the oscillation in the cavity builds up from the noise only when a frequency shift is introduced with respect to the pump beam. In such conditions, the beam oscillating in the cavity is frequency shifted by $\delta/2$ and this frequency shift ensures, in the BSO, a grating moving at the optimum velocity. The oscillation is maintained even if an aberrator is placed between the mirror M and the phase conjugator crystal.

6.7.6 Image Threshold Detector
Using a Phase Conjugate Resonator

In this experiment [6.112] a nonlinear threshold detecting device using a BaTiO$_3$ phase conjugate resonator and a spatially encoded erase beam are demonstrated. The threshold of the resonator is set so that pixels above threshold are amplified while those below give no output. As shown in Fig. 6.48, the set-up combines the photorefractive oscillator with the incoherent to coherent spatial light modulator concept of [6.113]. Gain in the resonator is provided by 4WM in BaTiO$_3$, while the input is achieved by a spatially modulated image beam that is used to selectively erase the gratings that support the modes of the resonator. The experiment is done with an argon laser operating at $\lambda = 514\,nm$, while the erase beam is a He-Ne laser at $\lambda = 442\,nm$. The resolution in the processed image transmitted through the crystal depends on the crystal thickness and on the number of transverse modes which can be supported by the resonator. The modulation transfer function of the interaction is modeled by

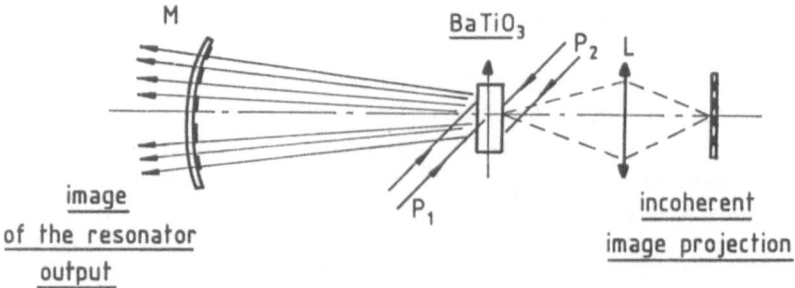

Fig. 6.48. Imaging threshold detector using a phase conjugate resonator in BaTiO$_3$. From [6.113]

modifying the gain Γ of the 4WM interaction in BaTiO$_3$ to

$$\Gamma' = \frac{\Gamma}{1 + I_e/I_0}$$

where I_e and I_0 are the erase and the pump beam intensities, respectively.

6.7.7 Optical Logic Using Two-Beam Coupling

Optical devices performing logic operations are highly desirable for applications in parallel computing. For this purpose, the nonlinear effects under study are mainly based on the properties of multi-quantum well structures in semiconductors, but more recently optical digital logic was also demonstrated by *Fainman* et al. [6.114] using two beam coupling in photorefractive crystals. In these materials, the nonlinear phenomena relies on the properties of the gain coefficient Γ in a classical two-beam interaction with gain. Three different effects are employed to perform optical logic operations: gain saturation, pump beam depletion, and optically controlled two-beam coupling. These interactions are detailed in [6.114] and examples of OR and NOR gates are given in Fig. 6.49. For example, in Fig. 6.49a, the weak signal beam is amplified by the same amount when one or both pump beams A_2', A_2'' are present with a high intensity level (logic 1). This property, related to the saturation of the

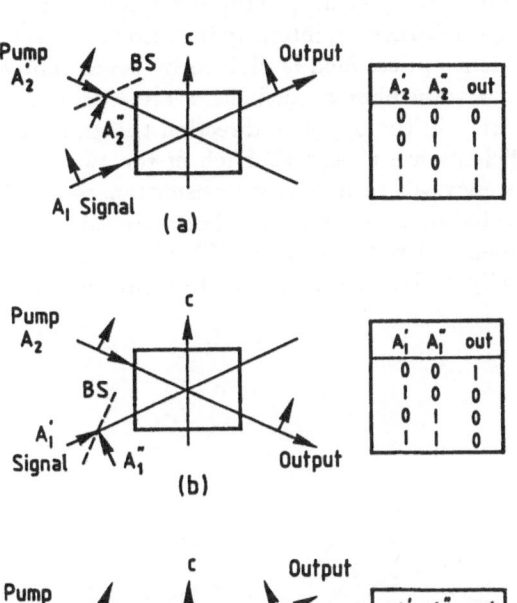

Fig. 6.49a–c. Optical logic by two beam coupling in photorefractive crystals. (a) OR gate – signal beam saturation. (b) NOR gate – pump depletion mode. (c) AND gate – signal beam saturation. From [6.114]

photorefractive gain versus the incident beam ratio, leads to the logic operation OR. In a similar way, in Figs. 6.49b, c, NOR and AND gates are achieved by using the pump beam depletion and the signal beam saturation modes, respectively. In other experiments, an additional incoherent beam ensures the control of the gain coefficient by erasing the interference grating formed by the signal and pump beams. All these optical logic operations rely on an efficient two-beam coupling with nonlinear photorefractive crystals such as $BaTiO_3$, $KNbO_3$ and BSO. Since the digital image is directly projected onto the crystal, it is necessary that a high gain is obtained while the crystal interaction length is reduced. This allows one to maintain optimum interactions between corresponding pixels and permits realization of optical logic on digital images of large space bandwidth product. Other parameters such as crystal orientation and beam incidence angles should also be considered in a careful optimization of the gain of the photorefractive amplifier with respect to the image characteristics [6.115].

6.7.8 Image Subtraction Using a Self-Pumped Phase Conjugate Mirror Interferometer

Figure 6.50 illustrates an interferometric set-up employing a self-pumped phase conjugate $BaTiO_3$ mirror as used by *Kwong* et al. [6.116] and *Chiou* and *Yeh* [6.117] for performing parallel image subtraction, intensity inversion and exclusive OR logic operation. The incident optical field is divided by beam splitter BS whose amplitude reflection and transmission coefficients are r and t, respectively. For the waves propagating in the opposite direction the amplitude reflection and transmission coefficients are r' and t'. Each of the two waves is then passed through a transparency whose intensity transmittances are T_1 and T_2. These two waves are reflected by a self-pumped photorefractive phase conjugate mirror with a nearly identical reflectivity R. The phase conjugate beams recombine interferometrically at beam splitter BS to form an output image intensity given by

$$I_{\text{out}} = I_0 R |t'r^*T_2 + t^*r'T_1|^2 \ .$$

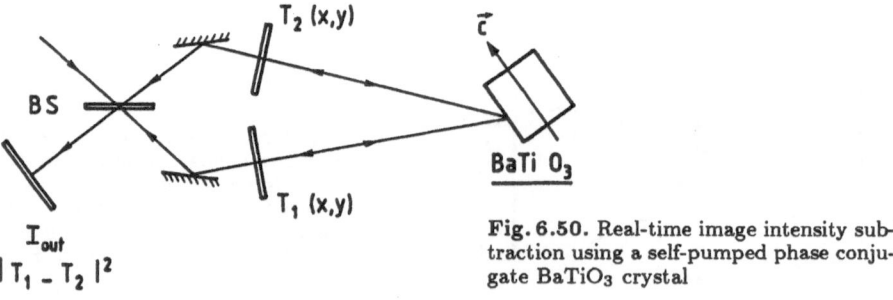

Fig. 6.50. Real-time image intensity subtraction using a self-pumped phase conjugate $BaTiO_3$ crystal

From the Stokes principle of reversibility of light it holds that

$$r't^* + t'r^* = 0$$

and therefore

$$I_{out} = I_0 R |r'r^*|^2 |T_2 - T_2|^2 \quad .$$

Consequently, the interferometer provides an image intensity subtraction proportional to the square of the intensity transmittance functions of the two input slides. This operation represents the Boolean exclusive OR achieved in parallel between the two images T_1 and T_2. The image intensity subtraction occurs in one step while usual methods employ two sequential holographic exposures phase shifted by π. Moreover, the interferometer is only sensitive to intensity difference and is independent of the phase information of the transparencies or optical path lengths of the two arms.

The properties of other types of interferometers including self-pumped photorefractive phase conjugate mirrors are currently being investigated [6.118–120]. The unique characteristics of these interferometers, which are reviewed in the Chap. 9 of this volume could be applied to the testing of optical elements or to producing new fibre sensor interferometers.

6.7.9 Associative Memories

Associative memory systems that use holographic data bases and phase conjugate mirrors to provide regenerative optical feedback, thresholding and gain have been recently reported by *Soffer* et al. [6.121] and *Anderson* [6.122]. The principles of information retrieval by association using parallel optical techniques, and in particular those based on holographic principles were recognized early by various authors [6.123, 124]. However, these first approaches were limited in their ability to faithfully reconstruct the output object from a partial input because of the large cross-talk which results when multiple objects are holographically stored in the memory. Nonlinear elements such as photorefractive crystals now permit these problems to be overcome, since they introduce optical feedback and gain, thus improving the selectivity and the stability of the memory. The principle of a holographic associative memory is shown in Fig. 6.51. Only a single hologram is used in this configuration and it is simultaneously addressed by the object as well as by the conjugate reference beam, the latter beam acting as the key that unlocks the associated information. A photorefractive $BaTiO_3$ phase conjugator is used both for reference beam retroreflection as well as for gain and thresholding. This provides the necessary nonlinearity emphasizing only the strongly correlated signals. The demonstration of total image reconstruction of an object image when only a partial image addressed the system, is also shown in Fig. 6.51. The illumination of the hologram by part of the object generates a diffracted beam propagating in the original direction of the reference beam. This beam is then phase conjugated and amplified by four-wave mixing in a photorefractive $BaTiO_3$ crystal. When

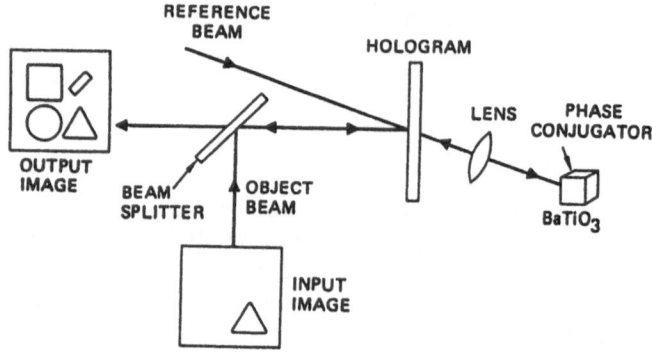

Fig. 6.51. Associative holographic memory. Complete object image reconstruction from a partial input image. From [6.121]

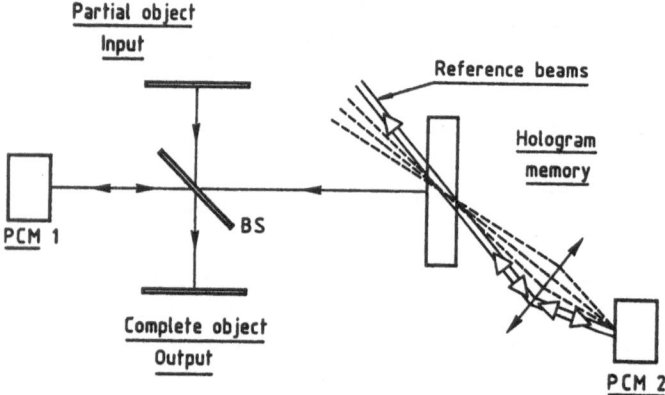

Fig. 6.52. Optical associative memory using a phase conjugate resonator and information storage in a volume holographic memory

this readout beam impinges on the hologram, it is diffracted and recreates the initial object beam. This recreated object beam contains all the information originally recorded in the hologram memory. This principle can be extended to different objects superimposed in the hologram memory by angular coding of the reference beam directions [6.121, 125].

A schematic implementation of such a nonlinear associative memory based on a hologram placed in a cavity formed by two phase conjugate mirrors is illustrated in Fig. 6.52 [6.126]. The phase conjugate mirrors provide beam retroreflection with gain and thresholding and give a self-alignment of the object and reference wavefronts with respect to the hologram memory. The optical feedback and the thresholding effects due to the nonlinear mirrors favours the strongly correlated signals and forces the system to converge to a stable state. The steady-state output signal thus consists of the image stored in the holographic memory and that which presents the highest degree of correlation with the input image. Real-time modification of the memory is also possible if holograms are stored in the volume of a photorefractive crystal. This would be a requirement for adaptive or learning behaviour of the system. This ability to reconstruct an image from a partial input information plane has important implications for pattern recognition, robotic vision and image processing operations. The nonlinear holographic associative memory also constitutes a first step towards the optical implementation of the neural network model recently proposed by *Hopfield* [6.127]. This model is based on the feasability of distributed and interconnected memory elements with a nonlinear feedback, and is thus analogous in many aspects to optical holography.

6.7.10 Laser Beam Cleanup

Atmospheric turbulence can give rise to highly distorted optical beams thus inducing a large beam divergence after propagation over long distances. It is therefore important to have the possibility of improving the spatial properties of optical beams in real time (beam cleanup). To this end, the basic idea is to use a two-wave-mixing interaction where the power of the aberrated pump beam can be transferred to a low intensity plane wave signal beam. This function can be achieved with photorefractive crystals since this interaction transfers the amplitude of the wave without transferring the phase. The self-cancellation of the phase information can be interpreted as the holographic reconstruction of the reference beam when the dynamic hologram is read out by a complex object wavefront. Laser beam cleanup has been demonstrated using two-wave-mixing in SBN and BaTiO$_3$ crystals with the set-up shown in Fig. 6.53 [6.128]. An aberrated pump beam interferes with a diffraction-limited signal wavefront of low intensity. After two-beam coupling the pump beam intensity can be transferred into the signal. It has been experimentally confirmed that this high intensity signal beam transmitted by the nonlinear crystal is still limited by diffraction and is therefore unaffected by the pump beam distortions. In another experiment shown in Fig. 6.54 [6.129], the beam cleanup results from the generation of a self-induced beam by a single mode photorefractive ring oscillator pumped by the aberrated beam. The response time for beam cleanup is, as in other experiments with photorefractive crystals, controlled by the pump

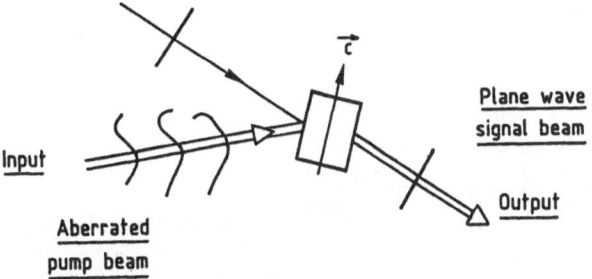

Fig. 6.53.
Laser beam cleanup using two-wave mixing in a photorefractive crystal

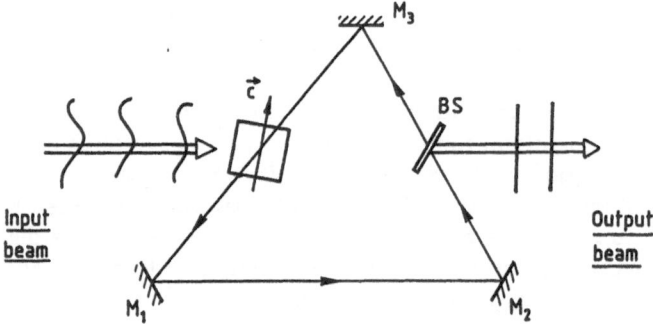

Fig. 6.54. Beam cleanup by wavefront generation using a unidirectional photorefractive ring oscillator

beam intensity. This crystal inertia will limit the speed at which time varying phase distortions of the beam can be converted into a clean beam.

6.7.11 Phase Locking of Lasers

Four-wave mixing techniques in photorefractive crystals can also be applied to phase locking of lasers to form a single-mode output beam arising from the coupling of several coherent sources. This coherence is achieved despite large differences in the individual amplifier characteristics and optical path lengths. Moreover, any changes in the cavity lengths and spatial mode structures of the optical beams are automatically compensated. The basic optical configuration used by different authors for the demonstration of the locking of two independent laser sources using phase conjugation is shown in Fig. 6.55 [6.130–132]. It consists of a stable master laser that supplies two counterpropagating pump beams in the photorefractive $BaTiO_3$ crystal. The slave laser is such that its output beam intersects the pumping beams in the nonlinear medium. With the output mirror of the slave laser being removed, the scattered light from the crystal initiates an oscillation between the mirror M_1 and the nonlinear material. This arises from four-wave mixing in the $BaTiO_3$ crystal, which behaves

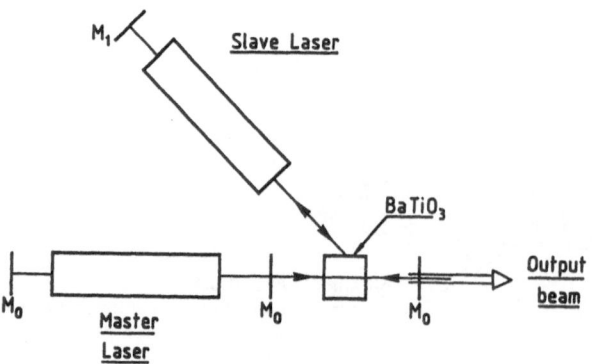

Fig. 6.55. Phase locking of lasers by four-wave mixing in a photorefractive crystal

as a phase conjugator with gain. The phase conjugating mirror thus directs a portion of the pumping beams into the slave laser cavity while automatically choosing the spatial and the spectral modes which optimize the gain of the system. The location of the rear mirror M_3 on the slave laser is arbitrary and the frequency of oscillation is unaffected by its position. Moreover, phase distortions can be introduced in the path of the slave laser due to phase conjugate reflection by four wave mixing in the photorefractive crystal. The steady-state gain condition requires that there is no change in the amplitude of the optical wave in the slave laser after one round trip in the cavity. This condition may be written as follows:

$$\exp\left(2\Gamma L\right) = -M_1 R$$

where M_1 is the intensity reflectivity of the mirror of the slave laser cavity, R the reflectivity of the phase conjugate mirror (photorefractive crystal) and ΓL the single-pass net intensity gain of the slave laser medium of length L. Figure 6.56 shows two different optical configurations [6.131, 132] in which a self-pumped photorefractive phase conjugator is used to lock the oscillation of a slave laser or several laser cavities. In these configurations, the pumping beams are perfectly phase conjugate waves thus allowing a high tolerance with respect to the optical alignments and the intra-cavity phase distortions of both master and slave lasers. The extension of these principles to the phase locking of many lasers can be achieved by using either several photorefractive crystals or by mixing all the beams of the slave laser into the same nonlinear crystal pumped

(a)

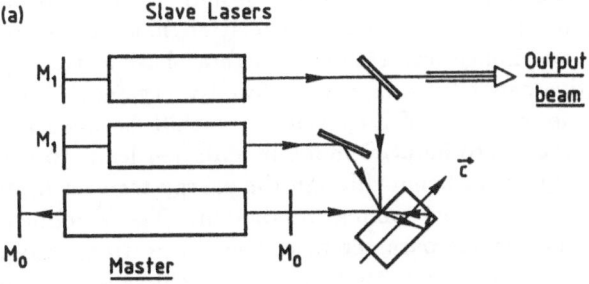

(b)

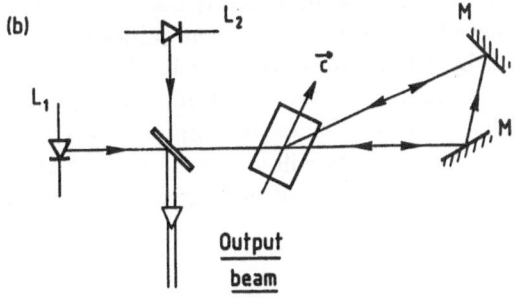

Fig. 6.56a,b. Phase locking of several lasers using a self-pumped phase conjugate photorefractive mirror (a); Ring self-pumped phase conjugate mirror to couple two semiconductor lasers (b). From [6.132, 133]

by the master laser. This last configuration requires a nonlinear medium with large gain since each individual beam should have a phase conjugate reflectivity close to unity. In this set-up, the conjugate mirror reflectivity may vary as a function of the incident angles of the various slave laser beams. In conclusion, all of these devices have the possibility of locking together several independent laser sources [6.133] and of combining the power of the individual laser beams in a linear laser cavity. These concepts are quite general and can be applied to a variety of laser media and, in particular, to arrays of semiconductor lasers.

6.8 Conclusion

In this chapter we have given a detailed description of photorefractive effects and their applications to optical signal processing. Presently, the basic physical mechanisms leading to the photo-induced index change in different crystals are well identified, but despite the great interest for applications, little is known about the charge transport processes or the species responsible for the photore-fraction in these crystals. Therefore, continued research into the microscopic origins of the photorefractive effect is essential for further optimization of the materials through impurity doping and thermal treatments. It is to be hoped that crystals with much higher sensitivity and displaying larger photo-induced nonlinearities in the visible and near IR wavelength range will be developed in the future. Several of the existing materials already work at millisecond speed with low power visible lasers, and thus offer promising applications in optical data processing, dynamic holography, phase conjugation and contact-free interconnection. A lot of experiments based on image correlation, holographic interferometry, beam steering and image restoration through a phase disturbing media illustrate the attractive capabilities of photorefractive crystals. Moreover, we have shown that the existence of a spatial phase shift between the incident fringe pattern and the photo-induced index modulation leads to the amplification of a low intensity signal beam through the energy transfer from a pump beam in two-wave or four-wave mixing interactions. The amplification factor depends on several recording parameters such as the grating period, the beam ratio of the interfering waves, the applied electric field as well as on the intrinsic material parameters. For different types of photorefractive crystals, the optimum conditions for obtaining high gain in nearly degenerate wave mixing experiments have been reviewed and new types of self-induced optical resonators have been demonstrated by adding an optical feedback to a photorefractive amplifier. To summarize, photorefractive crystals certainly have most of the characteristics needed for an initial demonstration of the power of the parallel optical processing. These capabilities will stimulate further studies on new mechanisms and materials for a variety of new applications using all-optical interactions.

Acknowledgements. The authors gratefuly acknowledge their colleagues for their very fruitful collaboration. They are also greatly indebted to J.P. Herriau (TH-CSF – LCR) for this expert contribution to the work presented in this chapter and to F. Micheron (TH-CSF – LCR) for stimulating contributions in the early studies of photorefractive crystals.

References

6.1   D.L. Staebler: In *Holographic Recording Materials*, Topics Appl. Phys., Vol. 20, ed. by H.M. Smith (Springer, Berlin, Heidelberg 1977) pp. 102–132

6.2   A.M. Glass: Opt. Eng. **17**, 470 (1978)

6.3   N.V. Kukhtarev, V.B. Markov, S.G. Odulov, M.S.Soskin, V.L. Vinetskii: Ferroelectrics **22**, 949 and 961 (1979)

6.4   L. Young, W.K.Y. Wong, M.L. Thewalt, W.D. Cornish: Appl. Phys. Lett. **24**, 264 (1974)

6.5   F. Micheron: Ferroelectrics **18**, 153 (1978)

6.6   G.C. Valley, M.B. Klein: Opt. Eng. **22**, 704 (1983)

6.7   J.P. Huignard, J.P. Herriau, G. Rivet, P. Günter: Opt. Lett. **5**, 102 (1980)

6.8   P. Günter: Phys. Reports **93** (1982)

6.9   H. Kogelnik: Bell Syst. Tech. J. **48**, 2909 (1969)

6.10 V. Markov, S. Odulov, M. Soskin: Opt. Laser Tech. **11**, 95 (1979)

6.11 J. Feinberg, D. Heiman, A.R. Tanguay, R.W. Hellwarth: J. Appl. PHys. **51**, 1297 (1980)

6.12 R.A. Mullen, R.W. Hellwarth: J. Appl. Phys. **58**, 40 (1985)

6.13 N.V. Kukhtarev: Sov. Tech. Phys. Lett. **2**, 438 (1976)

6.14 G.C. Valley: IEEE, J. Quantum Electron. QE **19**, 1637 (1983)

6.15 J.P. Huignard, F. Micheron: Appl. Phys. Lett. **29**, 591 (1976)

6.16 J. Feinberg, R.W. Hellwarth: Opt. Lett. **5**, 519 (1982)

6.17 J.O. White, M. Cronin-Golomb, B. Fischer, A. Yariv: Appl. Phys. Lett. **41**, 689 (1982)

6.18 J.P. Herriau, J.P. Huignard, P. Aubourg: Appl. Opt. **17**, 1851 (1978)

6.19 M.P. Petrov, T.G. Pencheva, S.I. Stepanov: J. Opt. **12**, 287 (1981)

6.20 M. Miteva, L. Nikolova: Opt. Commun. **42**, 307 (1982)

6.21 N.V. Kukhtarev, G.E. Dovgalenko, V.N. Starkov: Appl. Phys. A**33**, 227 (1984)

6.22 A. Marrakchi, R.V. Johnson, A.R. Tanguay: J. Opt. Soc. Am. B**3**, 321 (1986)

6.23 A.G. Apostolidis, S. Mallick, D. Rouede, J.P. Herriau, J.P. Huignard: Opt. Commun. **56**, 73 (1985)

6.24 M.G. Moharam, T.K. Gaylord, R. Magnusson, L. Young: J. Appl. Phys. **50**, 5642 (1979)

6.25 M.P. Petrov, S.V. Mirodonov, S.I. Stepanov, V.V. Kulikov: Opt. Commun. **31**, 301 (1979)

6.26 J.P. Huignard, B. Ledu: Opt. Lett. **7**, 310 (1982)

6.27 S.I. Stepanov, V. Kulikov, M.P. Petrov: Opt. Commun. **44**, 19 (1982)

6.28 Ph. Refregier, L. Solymar, H. Rajbenbach, J.P. Huignard: J. Appl. Phys. **58**, 45 (1985)

6.29 F. Micheron, C. Mayeux, J.C. Trotier: Appl. Opt. **13**, 784 (1974)

6.30 R. Magnusson, T.K. Gaylord: Appl. Opt. **13**, 1546 (1974)

6.31 R.J. Collier, C.B. Burkhardt, L.H. Lin: *Optical holography* (Academic, New York 1971)

6.32 L. Solymar, D.J. Cooke: *Volume Holography and Volume Gratings* (Academic, New york 1981)

6.33 A. Yariv: Opt. Commun. **25**, 23 (1978)

6.34 A. Yariv: IEEE J. Quantum Electron. **14**, 650 (1978)

6.35 J.P. Huignard, F. Micheron, E. Spitz: In *Optical Properties of Solids. New developments* ed. by B.O. Seraphin (North-Holland, Amsterdam 1976) Chap. 16

6.36 M.P. Petrov, S.I. Stepanov, A.A. Kamshilin: Opt. and Laser Tech. **149** (1979)

6.37 D.L. Staebler, W. Phillips: Appl. Opt. **13**, 789 (1974)

6.38 D.L. Staebler, J.J. Amodei: Ferroelectrics **3**, 107 (1972)

6.39 F. Micheron, G. Bismuth: Appl. Phys. Lett. **20**, 79 (1972) – **23**, 71 (1971)

6.40 D. von der Linde, A.M. Glass, K.F. Rodgers: Appl. Phys. Lett. **25**, 155 (1974)

6.41   B. Ya Zeldovich, N.F. Pillipetskii, V.V. Ragulskii, V.V. Shkunov: Sov. J. Quantum Electron. **8**, 1021 (1978)
6.42   R.W. Hellwarth: J. Opt. Soc. **67**, 1 (1977)
6.43   D.M. Pepper: Opt. Eng. **21**, 156 (1982)
6.44   H. Kogelnik: Bell. Syst. Tech. J. **44**, 2451 (1965)
6.45   E. Spitz, A. Werts: Compt. Rend. Acad. Sc. Paris **264**, 1015 (1967)
6.46   J.P. Huignard, J.P. Herriau, Ph. Aubourg, E. Spitz: Opt. Lett. **4**, 21 (1979)
6.47   S. Odulov, M. Soskin, V. Vasuetsov: Opt. Commun. **32**, 183 (1980)
6.48   P. Günter: Opt. Lett. **7**, 10 (1982)
6.49   J. Feinberg, R.W. Hellwarth: Opt. Lett. **5**, 519 (1980)
6.50   B. Fischer, M. Cronin-Goloomb, J.O. White, A. Yariv, R. Neurgaonkar: Appl. Phys. Lett. **40**, 863 (1982)
6.51   G.J. Dunning, R.C. Lind: Opt. Lett. **7**, 558 (1982)
6.52   B. Fischer, S. Sternklar: Appl. Phys. Lett. **46**, 113 (1985)
6.53   A. Marrakchi, J.P. Huignard, J.P. Herriau: Opt. Commun. **34**, 15 (1980)
6.54   J.P. Huignard, J.P. Herriau: Appl. Opt. **16**, 180 (1977)
6.55   H. Tiziani, J. Klenk: Appl. Opt. **20**, 1467 (1981)
6.56   Y.H. Ja: Appl. Opt. **21**, 3230 (1982)
6.57   J.P. Huignard, J.P. Herriau, L. Pichon, A. Marrakchi: Opt. Lett. **5**, 436 (1980)
6.58   J.W. Goodman: In *Laser Speckle*, Topics Appl. Phys., Vol. 9, ed. by J.C. Dainty (Springer, Berlin, Heidelberg 1975) pp. 10–75
6.59   M.D. Levenson, K.M. Johnson, V.C. Hanchett, and K. Chiang: J. Opt. Soc. Am. **71**, 737 (1981)
6.60   D.L. Staebler, W. Burke, W. Phillips, J.J. Amodei: Appl. Phys. Lett. **26**, 182 (1975)
6.61   J.P. Huignard, J.P. Herriau, F. Micheron: Appl. Phys. Lett. **26**, 256 (1975)
6.62   M.P. Petrov, S.I. Stepanov, A.A. Kamshilin: Opt. Commun. **21**, 297 (1977)
6.63   P.D. Henshaw: Appl. Opt. **21**, 2323 (1982)
6.64   G.T. Sincerbox, G. Roosen: Appl. Opt. **22**, 690 (1983)
6.65   J.Y. Moisan, P. Gravey, R. Lever, L. Bonnel: Opt. Eng. **25**, 151 (1986)
6.66   J.W. Goodman, F.J. Leonberger, S.Y. Kung, R.A. Athale: Proc. IEEE **72**, 850 (1984)
6.67   G. Pauliat, J.P. Herriau, A. Delboube, G. Roosen, J.P. Huignard: J. Opt. Soc. Am. **B3**, 306 (1986);
        E. Voit, C. Zaldo, P. Günter: Opt. Lett. **11**, 309 (1986)
6.68   J. White, A. Yariv: Appl. Phys. Lett. **37**, 5 (1980)
6.69   L. Pichon, J.P. Huignard: Opt. Commun. **36**, 277 (1981)
6.70   M.W. McCall, C.R. Petts: Opt. Commun. **53**, 7 (1985)
6.71   B. Loiseaux, G. Illiaquer, J.P. Huignard: Opt. Eng. **24**, 144 (1985)
6.72   J.P. Huignard, J.P. Herriau: Appl. Opt. **17**, 2671 (1978)
6.73   J. Feinberg: Opt. Lett. **5**, 330 (1980)
6.74   E. Ochoa, L. Hesselink, J.W. Goodman: Appl. Opt. **24**, 1826 (1985)
6.75   E. Ochoa, J.W. Goodman, L. Hesselink: Opt. Lett. **10**, 430 (1985)
6.76   Y.H. Ja: Opt. Commun. **44**, 24 (1982)
6.77   Y.H. Ja: Opt. Quant. Electron. **15**, 457 (1983)
6.78   J.P. Huignard, J.P. Herriau, F. Micheron: Ferroelectrics **11**, 393 (1976)
6.79   C.C. Guest, M.M. Mirsalehi, T.K. Gaylord: Appl. Opt. **23**, 3444 (1984)
6.80   Y.H. Ja: Opt. Commun. **42**, 377 (1982)
6.81   Y.H. Ja: Appl. Phys. **B36**, 21 (1985)
6.82   N.J. Berg, B.J. Udelson, J.N. Lee: Appl. Phys. Lett. **31**, 555 (1977)
6.83   J.J. Berg, J.N. Lee, M.W. Casseday, B.J. Udelson: Opt. Eng. **19**, 359 (1980)
6.84   D.E. Oates, P.G. Gottschalk, P.B. Wright: Appl. Phys. Lett. **46**, 1125 (1985)
6.85   V. Kondilenko, V. Markov, S. Odulov, M. Soskin: Optica Acta **26**, 238 (1979)
6.86   A. Marrakchi, J.P. Huignard, P. Günter: Appl. Phys. **24**, 131 (1981)
6.87   J.P. Huignard, A. Marrakchi: Opt. Commun. **38**, 249 (1981)
6.88   B. Fischer, M. Cronin-Golomb, J.O. White, A. Yariv: Opt. Lett. **6**, 519 (1981)
6.89   N. Kukhtarev, V. Markov, S. Odulov: Opt. Commun. **23**, 338 (1977)
6.90   N. Kukhtarev, V. Markov, S. Odulov: Sov. Phys. Tech. **25**, 1109 (1980)
6.91   J.M. Heaton, L. Solymar: Optica Acta **32**, 397 (1985)
6.92   H. Rajbenbach, J.P. Huignard, B. Loiseaux: Opt. Commun. **48**, 247 (1983)
6.93   G.C. Valley: J. Opt. Soc. Am. **B1**, 868 (1984)

6.94 A.G. Apostolidis, S. Mallick, D. Rouede, J.P. Herriau, J.P. Huignard: Opt. Commun. **56**, 73 (1985)
6.95 J.P. Huignard, J.P. Herriau: Appl. Opt. **24**, 4285 (1985)
6.96 A.M. Glass, A.M. Johnson, D.H. Olson, W. Simpson, A.A. Ballman: Appl. Phys. Lett. **44**, 948 (1984)
6.97 M.B. Klein: Opt. Lett. **9**, 350 (1984);
      B. Imbert, H. Rajbenbach, S. Mallick, J.P. Herriau, J.P. Huignard: Opt. Lett. **13**, 327 (1988)
6.98 J.P. Huignard, A. Marrakchi: Opt. Lett. **6**, 222 (1981)
6.99 J.P. Herriau, J.P. Huignard, A.G. Apostolidis, S. Mallick: Opt. Commun. **56**, 141 (1985)
6.100 D. Rak, I. Ledoux, J.P. Huignard: Opt. Commun. **49**, 302 (1984)
6.101 H. Rajbenbach, J.P. Huignard, Ph. Refregier: Opt. Lett. **9**, 558 (1984)
6.102 S.I. Stepanov, M.P. Petrov: Optica Acta **31**, 1335 (1984)
6.103 M. Cronin-Golomb, B. Fischer, J.O. White, A. Yariv: IEEE QE **20**, 12 (1984)
6.104 J.P. Huignard, H. Rajbenbach, Ph. Refregier, L. Solymar: Opt. Eng. **24**, 586 (1985)
6.105 J. Feinberg, G.D. Bacher: Opt. Lett. **9**, 420 (1984)
6.106 H. Rajbenbach, J.P. Huignard: Opt. Lett. **10**, 137 (1985)
6.107 A. Yariv, S.K. Kwong: Opt. Lett. **10**, 454 (1985)
6.108 M.D. Ewbank, P. Yeh: Opt. Lett. **10**, 496 (1985)
6.109 P. Pellat-Finet, J.L. de Bougrenet de la Tocnaye: Opt. Commun. **55**, 305 (1985)
6.110 S. Odulov, M. Soskin: JETP Lett. **37**, 289 (1983)
6.111 H. Rajbenbach, J.P. Huignard: Opt. Lett. **10**, 137 (1985)
6.112 M.B. Klein, G.J. Dunning, G.C. Valley, R.C. Lind, T.R.O. Meara: Opt. Lett. **11**, 575 (1986)
6.113 Y. Shi, D. Psaltis, A. Marrakchi, A.R. Tanguay: Appl. Opt. **22**, 3665 (1983)
6.114 Y. Fainman, C. Guest, Sh. Lee: Appl. Opt. **25**, 1598 (1986)
6.115 Y. Fainman, E. Klancnik, Sh. Lee: Opt. Eng. **25**, 228 (1986)
6.116 S.K. Kwong, G.A. Rakuljic, A. Yariv: Appl. Phys. Lett. **48**, 201 (1986)
6.117 A.E. Chiou, P. Yeh: Opt. Lett. **11**, 306 (1986)
6.118 J. Feinberg: Opt. Lett. **8**, 569 (1983)
6.119 B. Fischer, S. Sternklar: Appl. Phys. Lett. **46**, 113 (1985)
6.120 I.M. Michael, P. Yeh: Opt. Lett. **11**, 686 (1986)
6.121 B.H. Soffer, G.J. Dunning, Y. Owechko, E. Marom: Opt. Lett. **11**, 118 (1986)
6.122 D.Z. Anderson: Opt. Lett. **11**, 56 (1986)
6.123 R.J. Collier, K.S. Pennington: Appl. Phys. Lett. **8**, 44 (1966)
6.124 D. Gabor: IBM J. Res. Develop. **13**, 156 (1969)
6.125 A. Yariv, S. Kwong, K. Kyuma: Appl. Phys. Lett. **48**, 114 (1986)
6.126 B. Fischer, S. Sternklar, S. Weiss: Appl. Phys. Lett. **48**, 1567 (1986)
6.127 J.J. Hopfield: Proc. Natl. Acad. Sci. USA **79**, 554 (1982)
6.128 A. Chiou, P. Yeh: Opt. Lett. **10**, 621 (1985)
6.129 S. Kwong, A. Yariv: Appl. Phys. Lett. **48**, 564 (1986)
6.130 S. Sternklar, S. Weiss, M. Segev, B. Fischer: Opt. Lett. **11**, 528 (1986)
6.131 M. Croning-Golomb, A. Yariv, I. Ury: Appl. Phys. Lett. **48**, 1240 (1986)
6.132 J. Feinberg, D. Bacher: Appl. Phys. Lett. **48**, 570 (1986)
6.133 D. Rockwell, C. Guliano: Opt. Lett. **11**, 147 (1986)

7. The Photorefractive Incoherent-To-Coherent Optical Converter

Jeffrey W. Yu, Demetri Psaltis, Abdellatif Marrakchi, Armand R. Tanguay, Jr., and Richard V. Johnson

With 32 Figures

7.1 Overview

High performance spatial light modulators (SLMs) are essential in many optical information processing and computing applications for converting incoherent images to coherent replicas suitable for subsequent processing [7.1, 2]. A typical spatial light modulator consists of a photosensitive element to capture the incoherent light image and an optical modulator element to impress the incoherent input image content onto a coherent readout beam. A particularly important class of spatial light modulators employs photorefractive crystals that combine both photosensitive and electrooptic modulation functions within the same medium. Examples of electrooptic spatial light modulators that utilize photorefractive crystals include the Pockels Readout Optical Modulator (PROM) [7.3] and the PRIZ (a Soviet acronym for a crystallographically modified PROM) [7.4].

During operation of the Pockels Readout Optical Modulator, the input image-bearing beam creates photoinduced carriers which are longitudinally separated by an applied bias electric field. This charge separation produces a space-variant division of the applied field across the active electrooptic crystal and one or more dielectric blocking layers, as shown in the upper left quadrant of Fig. 7.1. The local birefringence of the medium depends on the longitudinal component of the local electric field, and hence can be sensed by a polarized readout beam observed through an exit analyzer. In the PRIZ, the same charge generation and separation process is utilized, with the difference that the crystallographic orientation is chosen to emphasize readout sensitivity to the transverse components of the induced electric field distribution, as shown in the upper right quadrant of Fig. 7.1.

The PROM and the PRIZ are typically limited in spatial frequency response to of order 10 cycles/mm at optimum optical exposure [7.5], and hence are limited to relatively modest bandwidth optical processing and computing applications. The physical configurations of the PROM and PRIZ devices do not lend themselves readily to exploitation of the remarkably high spatial bandwidths available in holographic configurations, in which the input image is encoded on a spatial carrier (as shown in the lower left quadrant of Fig. 7.1 and discussed extensively elsewhere in this book). Even when utilizing the same electrooptic crystal as in the PROM and PRIZ (typically bismuth silicon oxide), holographic recording configurations employing transverse applied elec-

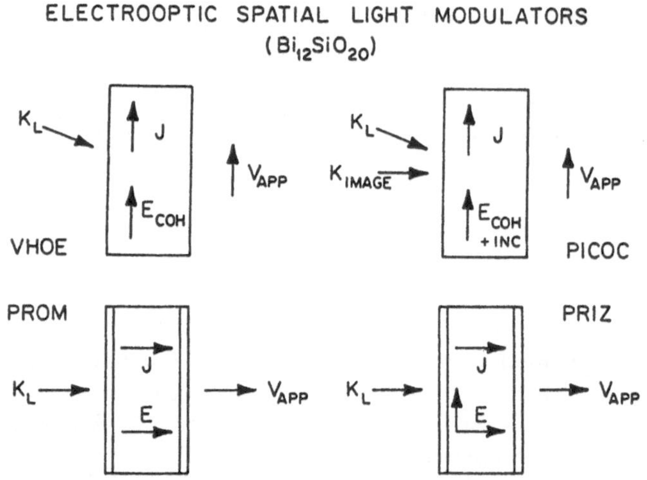

ELECTROOPTIC SPATIAL LIGHT MODULATORS
($Bi_{12}SiO_{20}$)

Fig. 7.1. Schematic diagram of the principal components of the applied voltage, charge transport, sensed electric field components, and input light wave vectors for four types of spatial light modulators that utilize single crystal bismuth silicon oxide ($Bi_{12}SiO_{20}$). The acronym VHOE stands for volume holographic optical element; likewise PROM stands for the Pockels Readout Optical Modulator, PRIZ is a Soviet acronym for a crystallographically modified PROM, and PICOC stands for the photorefractive incoherent-to-coherent optical converter, the subject of this chapter

tric fields and no blocking layers have been shown to exhibit spatial frequency bandwidths in excess of 2000 cycles/mm. However, such purely holographic recording requires input of image-dependent information on one of two coherent input beams, and as such cannot be directly utilized for performing the incoherent-to-coherent conversion function.

A fourth distinct type of photorefractive spatial light modulator has been independently proposed by *Kamshilin* and *Petrov* [7.6] and by the present authors [7.7, 8] which combines the incoherent-to-coherent conversion function with an essentially holographic recording process, and thereby exhibits several advantages of each. As in the holographic recording case, a transverse applied electric field is used in conjunction with two uniform coherent writing beams to produce a volume grating that is then selectively modified by a third beam encoded with the information content to be stored or converted. This configuration is shown schematically in the lower right quadrant of Fig. 7.1. The Photorefractive Incoherent-to-Coherent Optical Converter (PICOC) [7.7, 8] device is capable of recyclable real time operation, is characterized by an enhanced spatial frequency response, and is even simpler to construct than the PROM or PRIZ. In addition, the PICOC device configuration allows its use in many quasi-holographic techniques (such as optical phase conjugation), which in turn lead to potentially novel optical information processing and computing architectures [7.9–11].

The photorefractive incoherent-to-coherent optical conversion process is described in Sect. 7.2, as are alternative sequencing schemes and optical implementations. In Sect. 7.3, the limiting assumptions needed to derive a tractable analytical model of PICOC performance are specified, in preparation for detailed discussion of the recording stage in Sect. 7.4, and of the readout stage in Sect. 7.5. Conclusions and future research directions are offered in Sect. 7.6.

7.2 Physical Principles and Modes of Operation

The photorefractive incoherent-to-coherent optical conversion (PICOC) process is perhaps best understood as an extension of the more familiar holographic recording process in a photorefractive medium. The physical principles governing such recording are briefly reviewed in this section, and in much greater detail in Sect. 7.4. The extension of this recording process to include PICOC allows for at least three different temporal modes for sequencing the coherent grating with respect to the incoherent image. These modes are identified and compared in this section. In addition, two alternative optical architectures are defined, and converted images generated by one representative configuration are presented.

The high sensitivity of photoconductive and electrooptic crystals such as bismuth silicon oxide ($Bi_{12}SiO_{20}$, or BSO) in the visible portion of the spectrum has allowed the simultaneous recording and reading of volume holograms to be achieved with time constants amenable to real-time operation [7.12]. The holographic recording process in photorefractive materials involves photoexcitation, charge transport, and trapping mechanisms [7.13]. When two coherent beams are allowed to interfere within the volume of such a crystal, free carriers are nonuniformly generated by absorption and are redistributed by diffusion and/or drift under the influence of an externally applied electric field. Subsequent trapping of these charges at relatively immobile trapping sites generates a stored space-charge field, which in turn modulates the refractive index through the linear electrooptic (Pockels) effect and thus records a volume phase hologram. If both coherent writing beams are plane waves, the induced hologram consists of a uniform grating.

In the photorefractive incoherent-to-coherent optical conversion (PICOC) process, an incoherent image is focused in the volume of the photorefractive material in addition to the coherent grating beams, creating an additional spatial modulation of the charge distribution stored in the crystal. This spatial modulation can be transferred onto a coherent readout beam by reconstructing the holographic grating. The spatial modulation of the coherent reconstructed beam will then be a negative replica of the input incoherent image, as shown in Fig. 7.2. It should be noted here that a related image encoding process can be implemented nonholographically by premultiplication of the image with a grating [7.14].

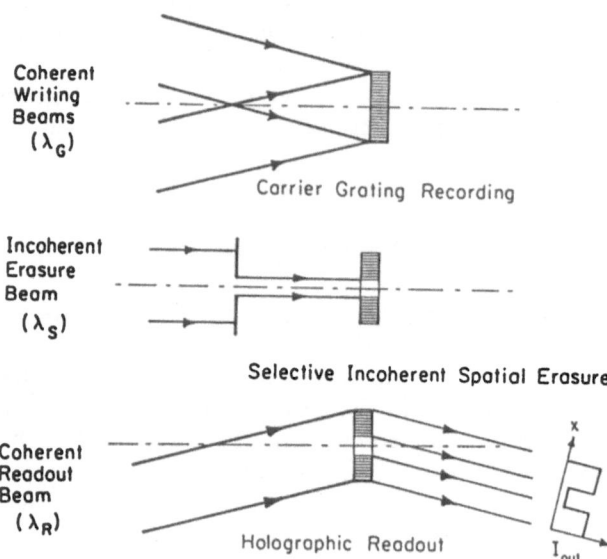

Selective Grating Erasure

Coherent Writing Beams (λ_G)

Carrier Grating Recording

Incoherent Erasure Beam (λ_S)

Selective Incoherent Spatial Erasure

Coherent Readout Beam (λ_R)

Holographic Readout

I_{out}

Fig. 7.2. Principle of operation of the photorefractive incoherent-to-coherent optical converter (hereinafter referred to as PICOC)

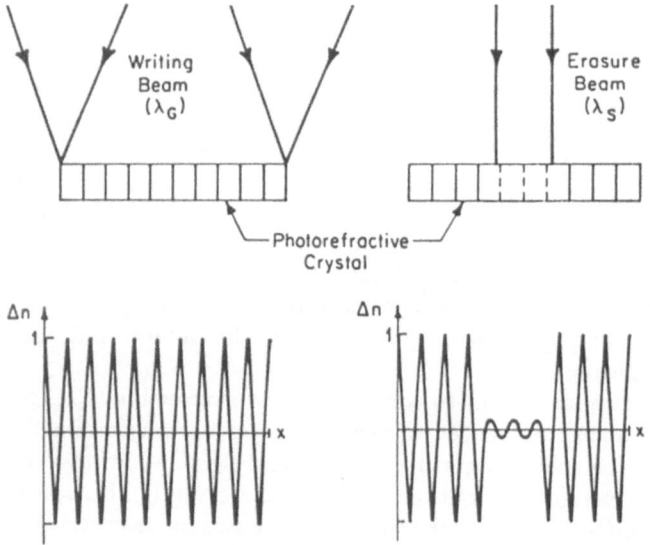

Writing Beam (λ_G)

Erasure Beam (λ_S)

—Photorefractive—Crystal

Δn

Δn

Fig. 7.3. PICOC in the grating erasure mode (GEM), in which the carrier grating is recorded before the incoherent image-bearing signal

In PICOC, the holographic grating can be recorded before, during, or after the crystal is exposed to the incoherent image. Therefore, several distinct operating modes are possible. These include the grating erasure mode (GEM; Fig. 7.3); the grating inhibition mode (GIM; Fig. 7.4), and the simultaneous erasure/writing mode (SEWM; Fig. 7.5).

In the *grating erasure mode* (GEM), shown schematically in Fig. 7.3, a uniform grating is first recorded by interfering two coherent writing beams in the photorefractive crystal. The writing beams are turned off, and this grating is then selectively erased by incoherent illumination of the crystal with an image-bearing beam. The incoherent image may be incident either on the same face of the crystal as the writing beams, or on the opposite face. When the absorption coefficients at the writing and image-bearing beam wavelengths give rise to significant depth nonuniformity within the crystal, these two cases will have distinct wavelength-matching conditions for response optimization [7.8].

In the *grating inhibition mode* (GIM), shown schematically in Fig. 7.4, the crystal is pre-illuminated with the incoherent image-bearing beam prior to grating formation. This serves to selectively decay (enhance) the applied transverse electric field in exposed (unexposed) regions of the crystal. After this pre-exposure, the coherent writing beams are then allowed to interfere within the crystal, causing grating formation with spatially varying efficiency due to significant differences in the local effective applied field.

In the *simultaneous erasure/writing mode* (SEWM), shown schematically in Fig. 7.5, the incoherent image modulation, the coherent grating formation

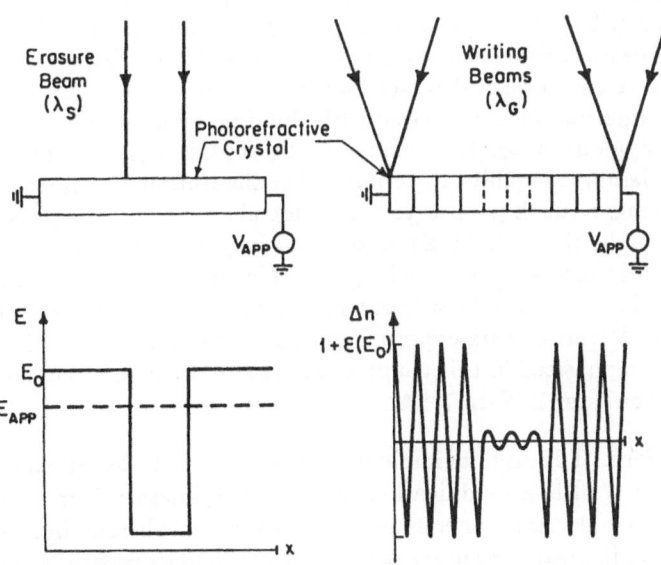

Fig. 7.4. PICOC in the grating inhibition mode (GIM), in which the carrier grating is recorded after the incoherent image-bearing signal

279

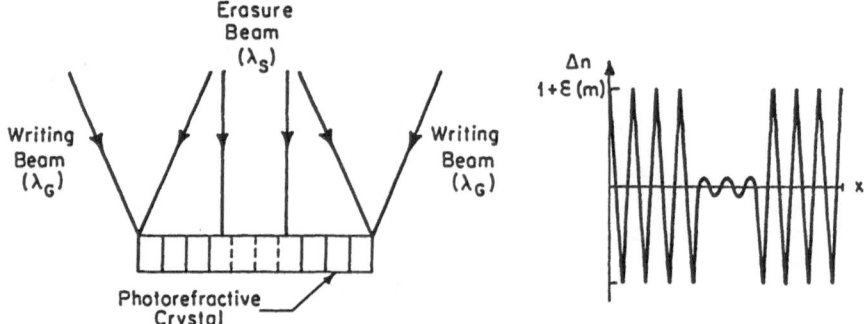

Fig. 7.5. PICOC in the simultaneous erasure/writing mode (SEWM), in which the carrier grating and the incoherent image-bearing signal are recorded simultaneously

process, and the readout function are performed simultaneously. A stable image is transcribed after the space-charge field has reached steady state.

A further distinction can be made in the operating modes between strictly cyclic exposure/readout, with an upper limit on the readout time interval, and operation in which prolonged readout times are required. Cyclic readout can be achieved by any one of the three sequencing modes introduced above, and the sensitometry requirement for achieving good quality images can be expressed in terms of optical exposure (i.e., optical energy per unit area). When prolonged readout is required, degradation of the stored space-charge profile can occur, due either to dark current or erasure induced by the optical readout beam. The most appropriate sequencing mode for prolonged readout is the simultaneous erasure/writing mode (SEWM) because it constantly regenerates the space-charge field profile. However, the sensitometry requirement for this latter mode is better expressed in terms of optical power rather than optical exposure, assuming readout time intervals long compared with the time required to achieve a stable steady state readout image. In exchange for this optical energy penalty, SEWM offers a considerably simplified experimental configuration with no need for temporal sequencing, a much greater tolerance for photorefractive crystals with increased dark conductivity, and readout of essentially unlimited duration. Readout light beams of much shorter wavelength and/or much higher intensities can be accommodated in SEWM without incurring unacceptable erasure of the charge pattern. Because of its experimental convenience and analytical simplicity, SEWM is emphasized in this chapter, with more detailed discussion of GEM and GIM given later in Sect. 7.4.4.

Optical Implementations. The original implementation of PICOC described by *Kamshilin* and *Petrov* [7.6] is a modification of the nondegenerate four-wave mixing geometry to include simultaneous exposure by an incoherent image-bearing beam. This configuration requires a readout wavelength separate and distinct from the coherent grating writing wavelength, which then allows the readout wavelength to be selected for significantly reduced grating erasure rates. Thus, the grating inhibition mode (GIM) and the grating erasure mode (GEM) are best implemented in this configuration. However, the optical alignment is more intricate with this architecture than it is with the degenerate four-wave mixing geometry introduced next, since the Bragg angle of the readout beam will not be the same as the Bragg angle of the coherent writing beams (shown in Fig. 7.6).

An alternative optical implementation is a modification of the conventional degenerate four-wave mixing geometry to include simultaneous exposure by an incoherent image-bearing beam, as shown in Fig. 7.7. This implementation has the advantage of extremely easy optical alignment, as the readout beam is readily Bragg aligned by retroreflecting one of the two coherent grating writing beams. It has the disadvantage that the readout beam, being at the same wavelength as the coherent grating writing beams, must erase the grating structure being probed at rates comparable to the writing process, assuming a readout light intensity comparable to the writing light intensities. Thus this implementation can be utilized for the simultaneous erasure/writing mode (SEWM), and can be adopted for the grating erasure mode (GEM) and the grating inhibition mode (GIM) only by significantly reducing the probe beam intensity, with a correspondingly reduced readout signal intensity.

As a specific example of the PICOC process, consider a degenerate four-wave mixing configuration in which the coherent writing beams and the incoherent image-bearing beam were made to illuminate the same face of a 1.3 mm thick crystal of bismuth silicon oxide, obtained from Crystal Technology, Inc. An electric field of 4 kV/cm was applied along the $\langle \bar{1}10 \rangle$ axis, as shown in

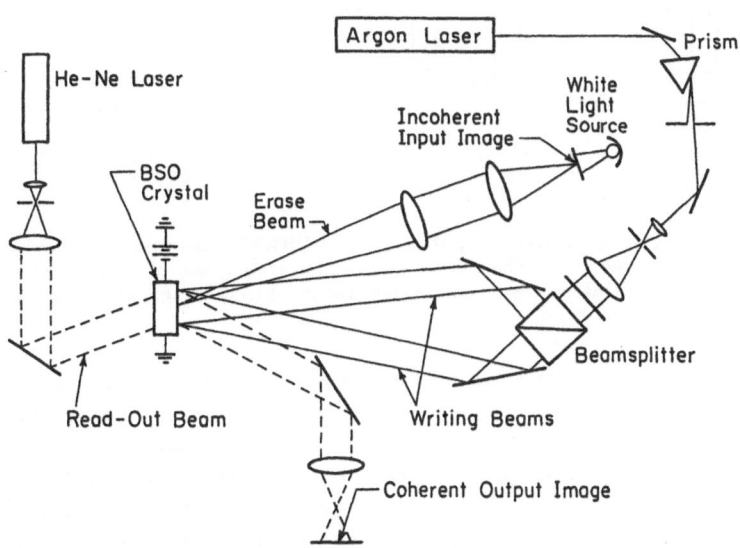

Fig. 7.6. Nondegenerate four-wave mixing architecture to perform the photorefractive incoherent-to-coherent optical conversion

281

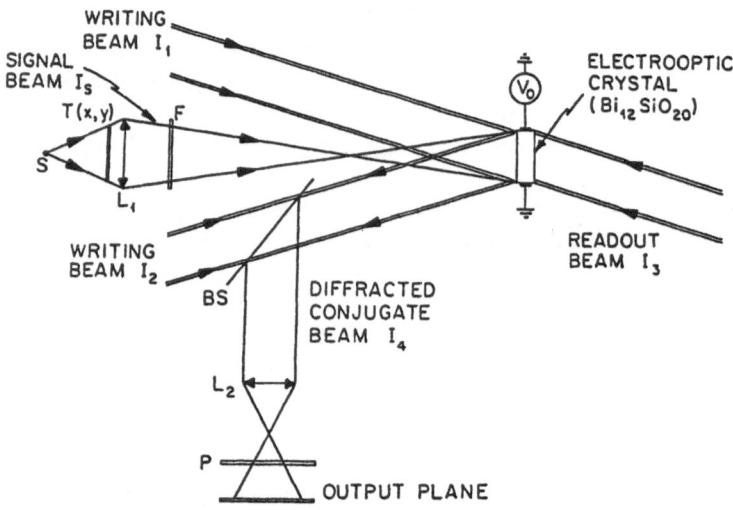

Fig. 7.7. Degenerate four-wave mixing architecture to perform the photorefractive incoherent-to-coherent optical conversion

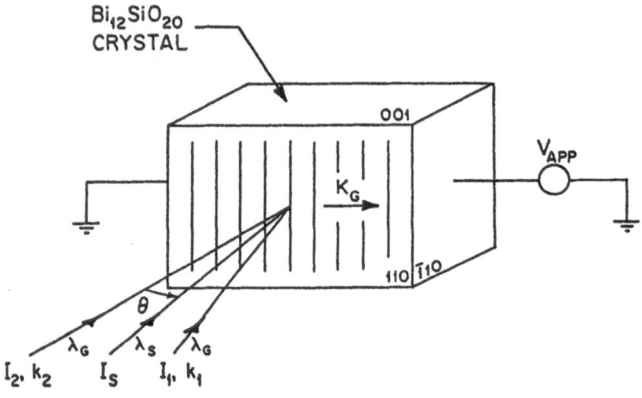

Fig. 7.8. Bismuth silicon oxide crystal orientation for the PICOC transverse electrooptic configuration and recording geometry. The volume holographic grating with wave vector K_G is formed by the coherent writing beams I_1 and I_2, and the incoherent image information is encoded on beam I_S

Fig. 7.8. A 300 cycles/mm grating was written by the 515 nm line of an argon ion laser with the grating wave vector oriented parallel to the applied bias field to maximize the diffraction efficiency. The image-bearing light source was either a xenon arc lamp, a tungsten lamp, or the 488 nm line of the argon ion laser. The average coherent grating intensity was $0.4 \, \mathrm{mW/cm^2}$ and the image-bearing light intensity was typically $8.0 \, \mathrm{mW/cm^2}$. The coherent grating writing beams were polarized orthogonal to the applied electric field. A polarizer was inserted at the output to minimize coherent optical scatter from the crystal [7.15].

Sample converted images obtained from two binary transparencies (a spoke target and an U.S. Air Force resolution target) and from two black-and-white slides with continuous tone gray scale are shown in Fig. 7.9. The original transparency and its converted image have reversed contrast, as shown in this figure and as explained by Fig. 7.2. An approximate resolution of 15 line pairs/mm (determined from the resolution target image) was achieved without optimizing factors such as the Bragg readout condition. Such optimization results in striking enhancements of the resolution to of order 50 line pairs/mm, as discussed in Sect. 7.5 below. Similar images of comparable quality have also been recorded in a bismuth silicon oxide crystal in which the $\langle 001 \rangle$ axis is aligned parallel to the coherent grating wave vector and to the applied bias electric field.

Having reviewed in broadest terms the physical principles and modes of operation of the PICOC process, let us now delineate the scope and limitations

Fig. 7.9a–d. Examples of the conversion of binary and gray-level transparencies: (a) spoke target, (b) U.S. Air Force resolution target, (c) airplane, and (d) an incoherent student

of the analytical model, as given in the next section, in preparation for more detailed studies of the recording process in Sect. 7.4 and the readout process in Sect. 7.5.

7.3 Delineation of the Analytical Model

A reasonably accurate and complete model of the photorefractive incoherent-to-coherent optical conversion process is based upon a set of equations that govern trap and electron balance, electron transport, and the buildup of a space-charge field, as detailed in Sect. 7.4.1. This model exhibits both nonlocal response due to charge transport and striking nonlinearities. No analytical solution has yet been identified that is broadly applicable to the full range of important experiments (e.g., experiments involving large modulation depths and photo-induced variations in the recombination time). Numerical solutions are certainly feasible, but have not yet been fully explored. To help refine physical insight into the conversion process, an approximate model capable of analytic solution needs to be defined, but its interpretation must be tempered with careful attention to its limitations.

Two such approximate solution models are identified herein. The first approach, the "constant recombination time approximation," is based upon the analytical studies of *Moharam* et al. [7.16], and was presented previously by the authors [7.8]. This approach is discussed in sufficient detail in Sect. 7.3.1 to enable comparison with the second approach, the "perturbation series approximation," as introduced in Sect. 7.3.2. These two approaches are compared and contrasted in this section. The perturbation series approximation is then used as the basis for continued discussion of PICOC performance characteristics throughout the remainder of the chapter, and is detailed more fully in Sect. 7.4.1.

7.3.1 Constant Recombination Time Approximation

One such approximate model of the PICOC process [7.8] evolves from analytic solutions of single spatial frequency grating recording as derived by *Moharam* et al. [7.16]. This model approximates quite well the nonlinearity of the conversion process, but it is limited in scope to steady state behavior. Hence the analytical solutions of this model (derived in [7.8]) are suitable only for studying the simultaneous erasure/writing mode (SEWM) in the steady state regime, and cannot cope with the grating erasure mode (GEM) and the grating inhibition mode (GIM) response. A further limitation of this approximation is its assumption of a constant recombination time (which is equivalent to assuming an infinite trap-limited saturation field strength), and so this will hereinafter be referred to as the "constant recombination time" model. A more accurate study of the photorefractive effect, as defined by a set of photoconductivity equations, allows for photoinduced variations of the recombination time from its nominal value. Although these variations may be quite small in amplitude compared with the nominal value, they nevertheless provide a very important mechanism that can significantly modify the space-charge field for certain recording configurations.

A key assumption of the constant recombination time model is the absence of self-diffraction. Self-diffraction is the process by which the grating written at the entrance of the photorefractive crystal diffracts a portion of the coherent writing beams, modifying the interference pattern and hence the grating that is recorded deeper in the crystal [7.13, 17, 37]. This process can be neglected for sufficiently thin crystals and low diffraction efficiencies (e.g., 2 mm of bismuth silicon oxide) and allows a considerable simplification of the analysis. If alternative crystals with significantly higher electrooptic coefficients and/or thicker crystals are used, then the diffraction efficiency would increase, self-diffraction effects would be far more pronounced, and thus the mathematical framework described in this chapter would need to be modified.

7.3.2 Perturbation Series Approximation

By recasting the chosen set of photorefractive equations in Fourier transform space, additional physical insights into the transcription process can be achieved which lead naturally to a perturbation series formulation of the conversion process. Analytic solutions can be derived for the first few terms of this series without restricting such solutions to the steady state regime, and so define an additional approximate model of the PICOC process. Limiting the analysis to the first few terms means that the strong nonlinearities that characterize the incoherent-to-coherent optical conversion process are estimated at best, but the most compelling advantage of this approach is its ability to model the temporal evolution of the space-charge field. Study of the temporal behavior is crucial to the analysis of the grating erasure mode (GEM) and the grating inhibition mode (GIM). In this chapter, we use a perturbation model that predicts the various spatial frequency components of the space-charge field.

Consider for example a coherent grating beam $I_G(x)$ of the form

$$I_G(x) = I_0(1 + m_G \cos K_G x) \quad , \tag{7.1}$$

in which m_G is the modulation depth and K_G is the wave vector associated with the coherent grating, and an incoherent signal beam $I_S(x)$ of similar form

$$I_S(x) = I_1(1 + m_S \cos K_S x) \quad , \tag{7.2}$$

in which m_S is the modulation depth and K_S is the wave vector associated with the image profile. The coordinate system is defined such that x is parallel to the applied bias electric field E_0, which is also parallel to the coherent grating wave vector K_G; z is orthogonal to the entrance face of the photorefractive crystal; y is defined to complete a right-handed coordinate system. Because of the nonlinearities in the recording process, the space-charge field $E(x)$ transcribed by these light beams contains spatial frequencies that do not exist in the original light profiles. These intermodulation terms have the form

$$E(x) = \sum_{m=-\infty}^{+\infty} \sum_{n=-\infty}^{+\infty} E_{mn} \exp[i(mK_G + nK_S)x] \quad . \tag{7.3}$$

If the recording process were perfectly linear, then only terms such as E_{10} (i.e., $m = 1$ and $n = 0$) and E_{01} ($m = 0$ and $n = 1$) would appear in which one of the two subscripts is zero, assuming light profiles as defined by $(7.1, 2)$. The nonlinearity of the recording process induces new spatial harmonics such as E_{11} which describe the modulation of the coherent grating by the incoherent image beam and hence are central to the photorefractive incoherent-to-coherent optical conversion process.

The intermodulation decomposition (7.3) provides a natural framework for a perturbation series analysis of the photorefractive incoherent-to-coherent optical conversion process in powers of the modulation depths m_G and m_S, a technique first demonstrated by *Kukhtarev* et al. as applied to the analysis of single grating transcription [7.18, 19], and extended by *Ochoa* et al. [7.20] and by *Refregier* et al. [7.21]. In practice, a typical image consists of a multiplicity of spatial frequency components, not just the single frequency signal term postulated in (7.2). Such a spectrally rich image will necessarily introduce an additional summation over the spectral harmonics in (7.3). The resultant series, if limited to terms of the form E_{11}, predicts a linear transcription of the image profile for which the response to each spatial frequency component can be evaluated separately, such that the total response is determined by summing up all harmonics. Higher order terms, such as E_{12}, describe nonlinear distortion of the image spectrum in which new image frequencies are generated that do not exist in the original incoherent image profile. These higher order terms prove to be extremely tedious to calculate. If these higher order terms contribute significantly to the conversion process, then alternative methods of analysis such as numerical modeling are recommended. The analysis presented in this chapter concentrates on the E_{11} term.

Analytical expressions defining the temporal evolution of the first few harmonic terms of (7.3) are readily derived. As shown in Appendix 7.A and in [7.22], specific perturbation terms for the simultaneous erasure/writing mode (SEWM) in the saturation limit can be approximated by

$$E_{01} = -\tfrac{1}{2} m_S^{\text{eff}} E_0 \tag{7.4}$$

$$E_{10} = -\tfrac{1}{2} m_G^{\text{eff}} E_0 \quad \text{and} \tag{7.5}$$

$$E_{11} = \tfrac{1}{2} m_S^{\text{eff}} m_G^{\text{eff}} E_0 \quad , \tag{7.6}$$

in which E_0 is the applied bias field shown in Fig. 7.8, and m_G^{eff} and m_S^{eff} are effective modulation depths to be defined in the next section.

If bismuth silicon oxide is chosen as the photorefractive medium of interest, then the refractive index modulation $\Delta n(x)$ induced along a principal electrooptic axis of the crystal by the linear electrooptic effect is proportional to the induced space-charge field $E(x)$ [7.23]. Each of the spatially periodic terms in the space-charge field (7.3) thus induces a distinct volume phase grating through the linear electrooptic effect. Such a superposition of phase gratings

286

diffracts an incident collimated readout laser beam into a multiplicity of discrete beams, as shown in Fig. 7.10.

Another fundamental insight into the PICOC process that evolves naturally from the perturbation series approach is the existence of a one-to-one mapping of E_{mn}, the spatial frequency components of the space-charge field, into I_{mn}, the diffraction orders and suborders of the spatially modulated readout light. This identification presumes weak diffraction efficiencies, as have thus far been typical of most PICOC implementations. In the far field diffraction limit, shown in Fig. 7.10, the spatially modulated readout light profile $I_R(x)$ decomposes into the usual set of discrete diffraction orders associated with the coherent grating, labeled by subscript m, while each of these orders further decomposes into subharmonics associated with the incoherent image signal, labeled by subscript n. This mapping offers a most convenient method to test the behavior of distinct spatial frequency components of the space-charge field E_{mn} during the conversion process.

For typical coherent grating spatial frequencies and typical crystal thicknesses, the volume phase gratings in the photorefractive crystal are recorded deep within the Bragg regime. Therefore the optical readout process can respond at most to one diffraction order such as I_{10} and its immediate subharmonics such as I_{11}, while excluding other diffraction orders and associated sidebands such as I_{01}. Thus for the illumination profiles $I_G(x)$ and $I_S(x)$ of the form (7.2,3), and assuming selective diffraction into only the +1st diffraction order and its immediate sidebands, a typical form of the modulated readout light profile $I_R(x)$ would be

$$I_R(x) = I_{10} + I_{11} \cos (K_S x) + \text{higher image harmonics} \quad . \qquad (7.7)$$

Furthermore, assuming perfect Bragg alignment for the I_{10} term, and assuming image frequencies K_S small enough to avoid Bragg misalignment of the immediately adjacent subharmonics, the diffracted intensity terms I_{10} and I_{11} are proportional to the squares of the corresponding space-charge field components E_{10} and E_{11} for sufficiently low diffraction efficiencies. Combining (7.4–7) and approximating for small modulation depths m_S^{eff} gives

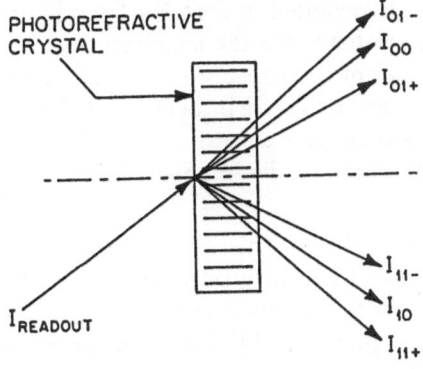

PHOTOREFRACTIVE
CRYSTAL

I_{01-}
I_{00}
I_{01+}

I_{11-}
I_{10}
I_{11+}

I_{READOUT}

Fig. 7.10. Far-field diffraction pattern created by the conversion of an incoherent grating into its coherent replica, showing the mapping of spatial harmonics of the space-charge field into diffraction orders of the modulated readout light beam

$$I_R(x) \propto (m_G^{\text{eff}})^2 [1 - 4m_S^{\text{eff}} \cos{(K_S x)}] \tag{7.8}$$

in which higher order terms have been neglected. Thus we can see from (7.8) that the incoherent image has been transcribed onto the coherent readout beam with a reversal of image contrast. The mapping of E_{mn} into I_{mn} also enables the possibility of spatial filtering of the image beam to change the negative image into a positive image (by a schlieren technique), and to improve the contrast ratio without suffering an associated reduction in image intensity.

The perturbation series approach that leads to (7.8) is most effective whenever the higher spatial harmonics in (7.7) can be neglected in favor of the lowest harmonics, and this applies whenever the modulation depths m_G^{eff} and m_S^{eff} are sufficiently small (or the spatial frequencies are sufficiently large). Such small modulation depths occur only for very restrictive conditions which seldom match the experimental conditions. Therefore the scope of the perturbation solution is limited, which reflects as much a fundamental difficulty with the PICOC process as it does a difficulty with the analytic technique. The PICOC process requires a nonlinear response to generate the modulation of the coherent grating by the incoherent image field, and hence strong levels of nonlinear distortion inextricably occur with significant levels of modulated light intensity. Restricting the attention to the lowest order spatial harmonics underestimates the nonlinearities, but leads to simple and powerful analytical expressions for the temporal response that contain considerable physical insight into the PICOC process and are unobtainable by any other analytical technique.

With these comments as backdrop, we can now proceed with a more detailed study of the recording process in Sect. 7.4 and of the readout process in Sect. 7.5.

7.4 The Recording Process

During the recording process, a space-charge electric field $E(x)$ is formed in the photorefractive crystal in response to the combined illumination by a coherent grating light beam $I_G(x)$ and an incoherent image beam $I_S(x)$. A physical model describing this transcription process is reviewed in Sect. 7.4.1, and particular numerical and analytical solutions are listed for the simultaneous erasure/writing mode (SEWM). The implications of these results on the nonlinear transfer function of the conversion process are explored in Sect. 7.4.2. Issues affecting the resolution of the recording process are discussed in Sect. 7.4.3. Finally, the temporal evolution characteristics are studied in Sect. 7.4.4.

7.4.1 Physical Model and Sample Solutions

The model of photoconductivity that is most frequently selected to describe holographic grating formation in photorefractive crystals assumes a single mobile charge species (electrons) and a single trapping level [7.13], although more intricate models involving hole transport [7.24, 25] and multiple trapping levels [7.26] have occasionally been proposed to achieve a better fit with particular experiments. The simple single trapping level/single mobile charge species model which has been chosen to describe the PICOC recording process consists of the following set of equations:

$$\frac{\partial N_D^+}{\partial t} = [S_G I_G(x,t) + S_S I_S(x,t) + \beta](N_D - N_D^+) - \gamma_R N_D^+ n \tag{7.9}$$

$$\frac{\partial n}{\partial t} = \frac{\partial N_D^+}{\partial t} + \frac{1}{e}\frac{\partial j}{\partial x} \tag{7.10}$$

$$\frac{\partial E}{\partial x} = \frac{e}{\varepsilon\varepsilon_0}(N_D^+ - n - N_A^-) \tag{7.11}$$

$$j = en\mu E + kT\mu\frac{\partial n}{\partial x} \tag{7.12}$$

in which

N_D is the total concentration of donor-like trapping centers,

$N_{D\ eq}^+$ is the concentration of ionized donor-like trapping centers in quasi-equilibrium under dark conditions,

$N_D^+(x,t)$ is the concentration of ionized donor-like trapping centers,

N_A^- is the concentration of negatively charged acceptor-like centers that compensate for the charge $N_{D\ eq}^+$ under dark thermal quasi-equilibrium conditions (N_A^- is a constant of the crystal),

$n(x,t)$ is the concentration of electrons in the conduction band,

$E(x,t)$ is the internal space-charge electric field,

e is a positive number with magnitude equal to the electronic charge,

S_G is the cross section of photo-ionization for the coherent grating beams with wavelength λ_G divided by the photon energy, hereinafter referred to as a photo-ionization cross section,

S_S is the cross section of photo-ionization for the incoherent image beam with wavelength λ_S divided by the photon energy, hereinafter referred to as a photo-ionization cross section,

β is the thermal generation rate of electrons into the conduction band,

γ_R is the carrier recombination constant,

$I_G(x,t)$ is the optical intensity profile for the coherent grating,

$I_S(x,t)$ is the optical intensity profile for the incoherent image,

$j(x,t)$ is the current density in the crystal,

μ	is a positive number with magnitude equal to the charge carrier mobility,
k	is Boltzmann's constant,
T	is the absolute temperature of the crystal,
ε	is the static dielectric constant of the crystal, and
ε_0	is the free space electric permeability.

Rationalized MKS units are assumed. The x coordinate is transverse to the nominal light propagation direction, and parallel to the applied bias field direction; t denotes time. Equation (7.9) is the rate equation describing the excitation of electrons into the conduction band and subsequent recombination into traps. Equation (7.10) states the conservation of electric charge. Equation (7.11) is Maxwell's first equation for the electric field. Equation (7.12) defines the current density in terms of drift and diffusion components. The material parameters for bismuth silicon oxide assumed in the numerical calculations are listed in Table 7.1 taken from the works of *Tanguay* [7.27] and *Valley* and *Klein* [7.28]. We wish to solve the equations for the space-charge electric field $E(x)$ which is induced by exposure to the input optical intensities $I_G(x)$ and $I_S(x)$.

Single Grating Response. Consider first the case of a single spatial frequency grating $I_G(x)$, as defined by (7.1) of the previous section, recorded in the absence of an incoherent image beam $I_S(x)$. Because of the nonlinearity of the recording process, the key variables of the photoconductivity model consist of a superposition of harmonics of the incident light beam, i.e.,

$$n(x,t) = \sum_{m=-\infty}^{+\infty} n_m(t) \exp{(imK_Gx)} \quad , \tag{7.13}$$

$$N_D^+(x,t) = \sum_{m=-\infty}^{+\infty} N_{Dm}^+(t) \exp{(imK_Gx)} \quad , \tag{7.14}$$

$$E(x,t) = \sum_{m=-\infty}^{+\infty} E_m(t) \exp{(imK_Gx)} \quad , \qquad \text{and} \tag{7.15}$$

$$j(x,t) = \sum_{m=-\infty}^{+\infty} j_m(t) \exp{(imK_Gx)} \quad . \tag{7.16}$$

These harmonic decompositions can be substituted into (7.9–12), resulting in a set of coupled differential equations defining the temporal evolution of each harmonic component. This coupled set of equations can either be integrated numerically, as demonstrated by *Moharam* et al. [7.16] on a reduced subset of the equations, or else be solved approximately by perturbation series methods.

Consider first a perturbation series expansion in powers of the modulation depth m_G. Analytical expressions for the first order terms have been derived

Table 7.1. Material parameters of bismuth silicon oxide ($B_{12}SiO_{20}$)

Parameter	Symbol	Value	Reference
Mobility	μ	$0.03\,\mathrm{cm^2/Vs}$	[7.28]
Carrier lifetime	τ	$5 \times 10^{-6}\,\mathrm{s}$	[7.28]
Donor-like trap density	N_D	$10^{19}\,\mathrm{cm^{-3}}$	[7.28]
Dark ionized trap density	$N_{D\,eq}^+$	$10^{16}\,\mathrm{cm^{-3}}$	[7.28]
Recombination coefficient	γ_R	$2 \times 10^{11}\,\mathrm{cm^3/s}$	[7.28]
Dielectric constant	ε	56	[7.28]

	Symbol	488 nm	515 nm	633 nm	Units	Reference
Index of refraction	n_0	2.650	2.615	2.530		[7.27]
Optical absorption	α	7.0	2.8	0.6	$\mathrm{cm^{-1}}$	[7.27]
Electrooptic coefficient	r_{41}	4.52	4.51	4.41	pm/V	[7.27]
Photo-ionization cross section (divided by photon energy)	S_G	–	0.42	–	$\mathrm{cm^2/Joule}$	[7.28]

Note: A photo-ionization cross section S_S at 488 nm has been estimated to be of order $1\,\mathrm{cm^2/Joule}$ by $S_S = (\alpha_S \lambda_S / \alpha_G \lambda_G)S_G$, assuming identical quantum efficiency at 488 and 515 nm. However see the discussion in Sect. 7.4.2 and *Sprague* [7.30].

by *Kukhtarev* et al. [7.18, 19], such that in the steady state regime the first spatial harmonic component E_1 of the space-charge field is given by

$$E_1 = -\frac{1}{2}m_G \frac{E_0 + iE_D(K_G)}{D(K_G)} \tag{7.17}$$

to first order in the modulation depth m_G, in which E_0 is the applied bias electric field strength. In (7.17), the denominator function $D(K_G)$ is defined by

$$D(K_G) = 1 - i\frac{E_0 + iE_D(K_G)}{E_q(K_G)} \quad , \tag{7.18}$$

and the diffusion field $E_D(K_G)$ and the trap-limited saturation field strength $E_q(K_G)$ are defined by

$$E_D = \frac{kTK_G}{e} \quad , \tag{7.19}$$

$$E_q = \frac{eN_{D\,eq}^+}{\varepsilon\varepsilon_0 K_G} \quad . \tag{7.20}$$

The trap-limited saturation field strength E_q defines the highest electric field that can be generated by a sinusoidal charge distribution with maximum charge density of $N_{D\,eq}^+ = N_A^-$. For reduced values of N_A^- (and hence of $N_{D\,eq}^+$), the amount of space charge and resulting space-charge field strength is limited as described by (7.17, 18). Considerable variation has been reported in the literature in estimates of the equilibrium donor-like trap density $N_{D\,eq}^+$ for bismuth silicon oxide [7.29], which directly affects the estimate of the saturation field strength E_q, and hence the upper limit on the space-charge field strength.

The first order perturbation analysis is accurate only in the limit of low modulation depths m_G, whereas many gratings are written with the highest possible modulation depths. Solutions accurate at higher modulation depths can be obtained by numerical methods, with sample solutions presented in Fig. 7.11 showing the space-charge field profiles induced by a single grating frequency in the steady state regime for various applied bias fields and various grating frequencies. A modulation depth of $m_G = 0.99$ and the material parameters listed in Table 7.1 are assumed for all curves. Note that even when the writing light profile is cosinusoidal, the resulting space-charge field profiles exhibit significant distortion because of the nonlinearity of the recording process.

In practice, only one of the multiple spatial harmonics of the coherent grating wave vector K_G can dominate the optical readout process because the grating is typically recorded very deeply into the Bragg regime. Figure 7.12 therefore shows numerical solutions for the strength of just the first spatial harmonic component of the space-charge field E_1 as a function of the modulation depth m_G (labeled in the figure as m_G^{eff} in anticipation of the two grating transcription discussion that follows, but to be interpreted for now as m_G). The family of curves shown in Fig. 7.12 is parametrized by the ratio E_q/E_0, which scales the relative magnitudes of the trap-limited saturation field to the applied bias field. Those portions of the response curves in Fig. 7.12 that can be approximated by a straight line, namely for low levels of the modulation depth m_G up to about 0.5 or so, are the regions in which the linear approximation of *Kukhtarev* et al. [7.18, 19] most accurately describes the transcription process. In anticipation of the discussion of nonlinear PICOC response given in Sect. 7.4.2, we find that high effective modulation depths m_G^{eff} (defined by (7.23) below) for which the linear approximation breaks down are associated with weak levels of average incoherent image intensity I_1 compared with the

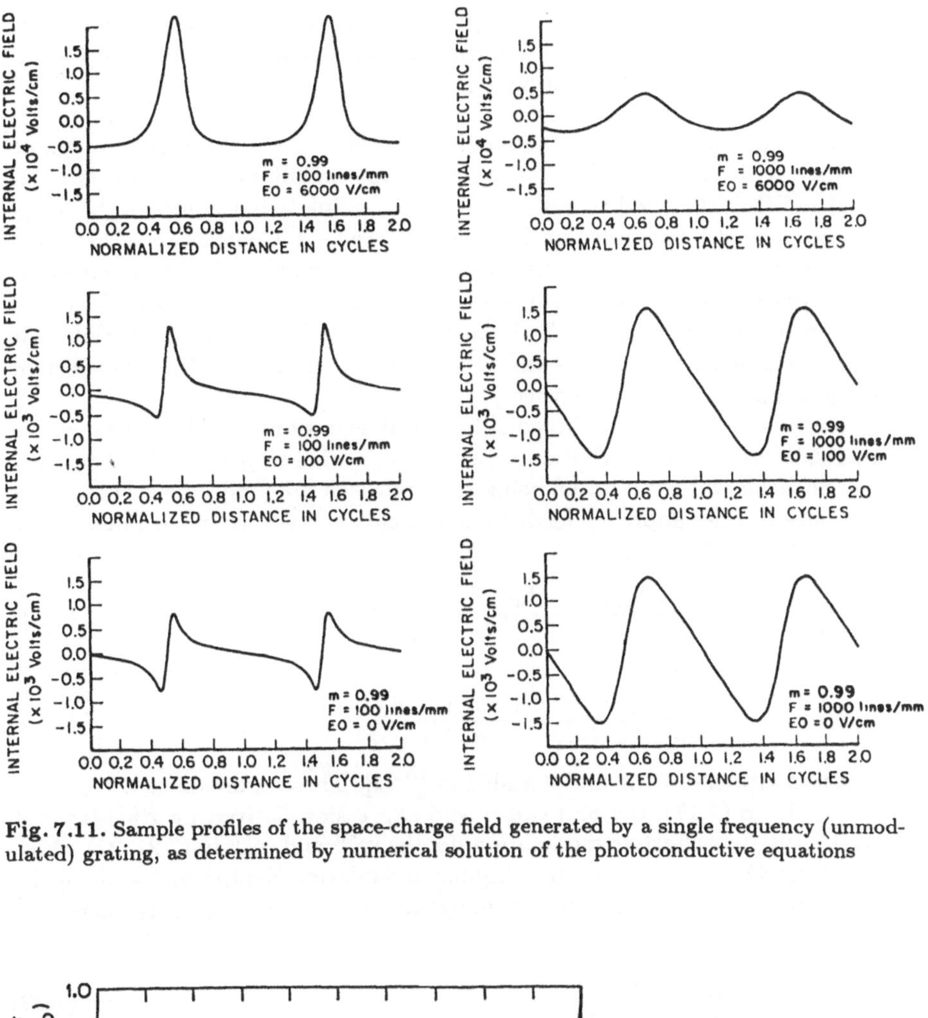

INTERNAL ELECTRIC FIELD (×10⁴ Volts/cm)

m = 0.99
F = 100 lines/mm
E0 = 6000 V/cm

NORMALIZED DISTANCE IN CYCLES

INTERNAL ELECTRIC FIELD (×10⁴ Volts/cm)

m = 0.99
F = 1000 lines/mm
E0 = 6000 V/cm

NORMALIZED DISTANCE IN CYCLES

INTERNAL ELECTRIC FIELD (×10³ Volts/cm)

m = 0.99
F = 100 lines/mm
E0 = 100 V/cm

NORMALIZED DISTANCE IN CYCLES

INTERNAL ELECTRIC FIELD (×10³ Volts/cm)

m = 0.99
F = 1000 lines/mm
E0 = 100 V/cm

NORMALIZED DISTANCE IN CYCLES

INTERNAL ELECTRIC FIELD (×10³ Volts/cm)

m = 0.99
F = 100 lines/mm
E0 = 0 V/cm

NORMALIZED DISTANCE IN CYCLES

INTERNAL ELECTRIC FIELD (×10³ Volts/cm)

m = 0.99
F = 1000 lines/mm
E0 = 0 V/cm

NORMALIZED DISTANCE IN CYCLES

Fig. 7.11. Sample profiles of the space-charge field generated by a single frequency (unmodulated) grating, as determined by numerical solution of the photoconductive equations

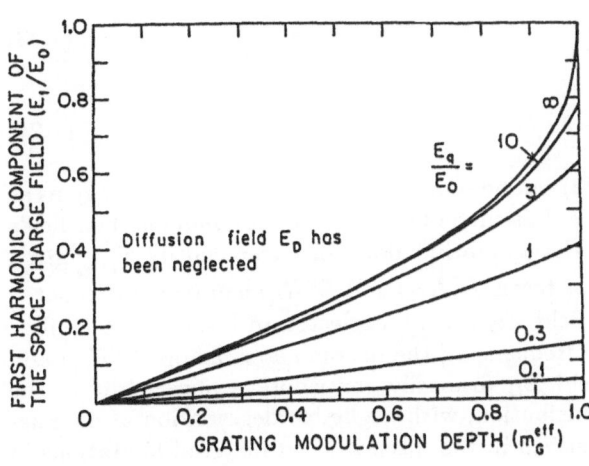

FIRST HARMONIC COMPONENT OF THE SPACE CHARGE FIELD (E_1/E_0)

$\frac{E_q}{E_0} =$

Diffusion field E_D has been neglected

GRATING MODULATION DEPTH (m_G^{eff})

Fig. 7.12. First spatial harmonic components of the space-charge field profile as a function of the modulation depth of the coherent carrier grating for various levels of the equilibrium donor-like trap density $N_{D\ eq}^+$

293

average coherent grating light intensity I_0, while low modulation depths for which the linear approximation is most accurate occur for strong levels of average incoherent light illumination I_1.

Two Grating Transcription. When an incoherent image beam is present simultaneously with the coherent grating, as in the simultaneous erasure/writing mode (SEWM), then additional terms corresponding to harmonics of the image frequencies must be included in the space-charge field. Consider for example an image consisting of just one sinusoidal component, as given by (7.2), with a resulting space-charge field decomposition of the form of (7.3). Double harmonic decompositions analogous to (7.3) can be assumed for each of the variables in the photoconductive equations (i.e., the ionized trap density N_D^+, the electron density n, and the current density j).

The resulting set of coupled differential equations can be solved either numerically or by a perturbation series analysis with respect to this harmonic decomposition, with the latter being the approach adopted herein [7.22]. The perturbation series approach leads to analytical expressions for the linear terms E_{10} and E_{01} of the form

$$E_{10} = -\frac{1}{2}m_G \frac{S_G I_0}{S_G I_0 + S_S I_1} \frac{E_0 + iE_D(K_G)}{D(K_G)} \tag{7.21}$$

$$E_{01} = -\frac{1}{2}m_S \frac{S_S I_1}{S_G I_0 + S_S I_1} \frac{E_0 + iE_D(K_S)}{D(K_S)} , \tag{7.22}$$

as given in Appendix 7.A and detailed in [7.22]. These expressions correspond very closely to (7.17), the first order field expression derived by *Kukhtarev* et al. [7.18, 19]. The effective modulation depths m_G^{eff} and m_S^{eff} in (7.21, 22) which account for the reduction of the original modulation depths m_G and m_S by the presence of both the incoherent image and the coherent grating beams are defined by

$$m_G^{\text{eff}} = m_G \frac{S_G I_0}{S_G I_0 + S_S I_1} \tag{7.23}$$

$$m_S^{\text{eff}} = m_S \frac{S_S I_1}{S_G I_0 + S_S I_1} . \tag{7.24}$$

The expressions (7.21, 22) can be simplified over broad operating regions as follows. For spatial frequencies less than of order 200 cycles/mm and bias fields over $2\,\text{kV/cm}$, and assuming the equilibrium donor-like trap density $N_{D\,\text{eq}}^+$ given in Table 7.1, the denominator terms $D(K_G)$ and $D(K_S)$ can be approximated by unity, and the diffusion field E_D can be neglected in favor of the applied bias field. For higher spatial frequencies, the denominator factors $D(K_G)$ and $D(K_S)$ and the diffusion field E_D primarily contribute a phase shift to the coherent grating's charge distribution, with negligible degradation of its magnitude. (This assertion is justified in the discussion on material limitations in

Sect. 7.4.3.) Thus the lowest harmonics of the space-charge field can be approximated by

$$E_{10} = -\tfrac{1}{2}m_G^{\text{eff}}E_0 \tag{7.25}$$

$$E_{01} = -\tfrac{1}{2}m_S^{\text{eff}}E_0 \quad . \tag{7.26}$$

By invoking similar approximations, the magnitude of the intermodulation term E_{11}, which is crucial to the incoherent-to-coherent optical conversion process, is well approximated by

$$E_{11} = \tfrac{1}{2}m_G^{\text{eff}}m_S^{\text{eff}}E_0 \quad , \tag{7.27}$$

as discussed in Sect. 7.4.3 and in [7.22]. An identical expression for E_{11} was derived by *Marrakchi* et al. [7.8], starting from the constant recombination time approximation discussed in Sect. 7.3.1. This derivation involves an additional linearization of the response in the limit of low modulation depths m_G^{eff} and m_S^{eff}.

In addition to the magnitude expressed by (7.27), the perturbation series analysis predicts that the E_{11} field is phase shifted with respect to the incident coherent grating. For steady state response in SEWM, this phase shift does not significantly impact the performance of the conversion process, assuming that it remains reasonably constant over the recording bandwidth; for temporal response it can significantly degrade the usefulness of PICOC for particular optical processing architectures.

Equations (7.25–27) indicate that the SEWM response in the steady state limit is predominantly governed by the reduced modulation depths m_G^{eff} and m_S^{eff}. The consequent impact on the overall readout image light intensity and on the modulation transfer characteristic is discussed in the nonlinear transfer response analysis, presented next.

7.4.2 Nonlinear Transfer Response

The image transfer for PICOC involves a performance trade-off between two competing mechanisms: the contrast ratio (which involves the ratio of I_{11} and I_{10}) improves steadily with increasing intensities of incoherent image-bearing light (in the absence of spatial filtering), but at the same time the *average* intensity (which involves I_{10} alone) steadily declines with increasing incoherent intensity because the uniform background in the incoherent light erases the carrier grating pattern in the photorefractive crystal.

In many optical information processing applications, optimization of the image contrast ratio is desirable within the image intensity constraints implied by the performance trade-off described above. In other types of signal processing applications such as correlation with a Vander Lugt filter, the dc image content contained in the I_{10} diffraction component does not contribute to the processing, or can be readily modified by spatial filtering, whereas maximizing the intensity of nonzero spatial frequency components such as I_{11} is critical to good conversion performance. For these cases, an optimum level of incoherent image-bearing beam intensity exists for maximizing the I_{11} component, beyond which the I_{11} component decays because of space-charge erasure. Therefore the following study of the nonlinear transfer response includes explicit consideration of the I_{11} diffraction order term.

To obtain a quantitative estimate of these effects, consider as a representative example the simultaneous erasure/writing mode (SEWM) in the steady state regime, with a single spatial frequency coherent grating given by (7.1), and a single spatial frequency incoherent image profile given by (7.2). One possible method of assessing the image transfer response is to determine the modulation depth of the readout image as a function of the input (incoherent and coherent) light characteristics. A convenient parameter that describes these characteristics is the product $B\,R$, in which $R = I_1/I_0$ is the ratio of the average incoherent image light to the average coherent grating light intensity levels, and $B = S_{\mathrm{S}}/S_{\mathrm{G}}$ is the ratio of the photoconductive sensitivity of the incoherent image beam with respect to the coherent grating beam. With these definitions, the effective modulation depths $m_{\mathrm{G}}^{\mathrm{eff}}$ and $m_{\mathrm{S}}^{\mathrm{eff}}$ defined by (7.25, 26) can be expressed as

$$m_{\mathrm{G}}^{\mathrm{eff}} = \frac{m_{\mathrm{G}}}{1 + B\,R} \quad \text{and} \tag{7.28}$$

$$m_{\mathrm{S}}^{\mathrm{eff}} = \frac{m_{\mathrm{S}}B\,R}{1 + B\,R}$$

The diffracted intensity I_{10} is directly proportional to the square of $m_{\mathrm{G}}^{\mathrm{eff}}$, while the modulation depth of the readout image is similarly proportional to $m_{\mathrm{S}}^{\mathrm{eff}}$. The image transfer response is plotted in Fig. 7.13, which shows the improve-

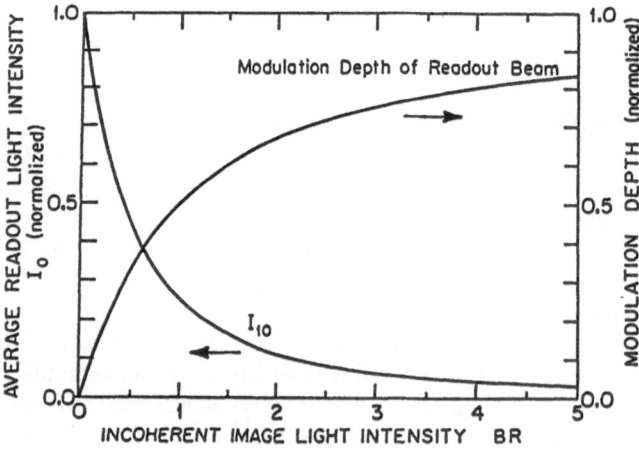

Fig. 7.13. Normalized diffraction efficiency of the I_{10} beam (uniform erasure), and the modulation transfer function for modulated erasure, as a function of the intensity ratio R

ment in modulation depth of the readout light beam with increasing levels of incoherent image illumination, but with concomitant decay of the average transferred image intensity I_{10}.

An unusual but intuitive feature of the photorefractive incoherent-to-coherent conversion process (as opposed to more typical linear spatial light modulation techniques) is that the image quality is primarily determined by the *ratio* of the incoherent image intensity to the coherent grating intensity, not by the *sum* of these intensities (assuming negligible dark conductivity β in the photogeneration rate equation (7.9)). Remember, however, that this analysis assumes a recording process that has already reached steady state, and hence does not address the question of the time required to reach saturation, which is indeed a function of the total light intensity. Temporal response issues are considered in Sect. 7.4.4 below.

To test the predictions of this nonlinear transfer model, the erasure of the readout beam's first diffraction order I_{10} in response to spatially uniform incoherent illumination has been measured by the nondegenerate four wave mixing configuration discussed in Sect. 7.2, with results as shown in Fig. 7.14. The grating was written with the 515 nm line of an argon ion laser, while the 488 nm line was used to simulate the spatially uniform image beam. For comparison, predictions of two different models of the conversion process are included in Fig. 7.14. The curve labeled "linear approximation" corresponds to the perturbation analysis term E_{10} presented in this chapter, while the second curve labeled "constant recombination time approximation" derives from (7.9) of the previously published model [7.8], which in turn is based upon the saturation regime model of *Moharam* et al. [7.16]. The linear approximation model has been scaled in intensity to match the experimental points at $R = 0.6$, and the constant recombination time approximation (Moharam's analytical solution) has been scaled to converge with the linear approximation for very large beam ratios R.

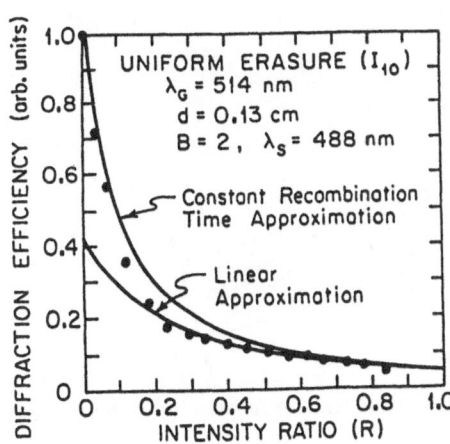

Fig. 7.14. Experimental diffraction efficiency of the I_{10} beam (uniform erasure) as a function of the intensity ratio R. Corresponding theoretical predictions for the linearized model and the constant recombination time model are also shown for comparison

297

Note that the experimental data points and both models converge reasonably well for high levels of incoherent image illumination, which correspond to large intensity ratios R that by (7.28) imply small effective modulation depths m_G^{eff}. Recall from Fig. 7.12 that small modulation depths correspond to the most accurate region of the linearized models of the recording process, as assumed in the lowest order terms of the perturbation analysis (as well as in the linearized constant recombination time approximation discussed following (7.27), which was shown to yield expressions identical to the perturbation analysis). In contrast, for small intensity ratios R with concomitantly large effective modulation depths m_G^{eff}, the full nonlinear constant recombination time model as developed in [7.8] offers a significantly better recording approximation than the linearized models for the simultaneous erasure/writing mode (SEWM) in saturation, as shown clearly in Fig. 7.14 by the curve marked "constant recombination time approximation". Note in addition that the choice of scaling of the constant recombination time approximation does not automatically guarantee good agreement with the data near $R = 0$.

A more challenging test of the theory is to predict accurately the conversion response to a *sinusoidally* modulated image beam. Such a test can be performed with the nondegenerate four wave mixing geometry described previously in which a 488 nm argon ion laser line passes through a Michelson interferometer to generate a sinusoidal spatial modulation, as shown in Fig. 7.15. The sinusoidal image modulation introduces an additional diffraction order I_{11} not observed in the uniform erasure case. The diffracted light sideband intensity I_{11} has been measured as a function of the intensity ratio R, with results as shown in Fig. 7.16, and as compared against the theoretical prediction of the perturbation expansion method.

The general features of Fig. 7.16 can be understood with reference to Fig. 7.13. For low levels of image intensity, corresponding to low intensity ratios R, the coherent grating's charge pattern is not significantly erased by the

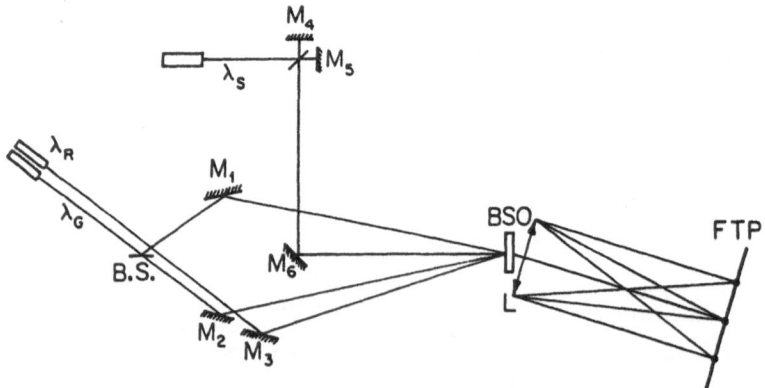

Fig. 7.15. Experimental arrangement for sensitivity and transfer function measurements, as described in detail in the text

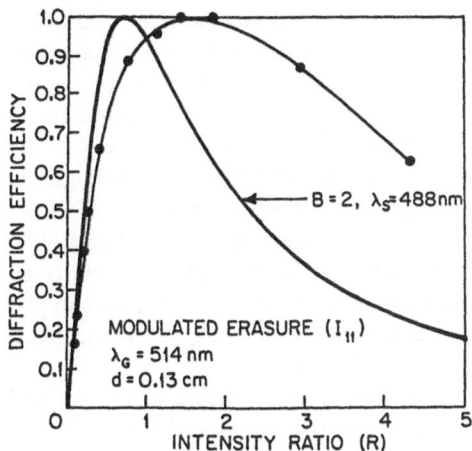

Fig. 7.16. Normalized experimental diffraction efficiency of the I_{11} beam (strongly modulated erasure) as a function of the intensity ratio R. The corresponding theoretical prediction is also shown for comparison

image beam, as indicated by the high level of I_{10} in Fig. 7.13. In this regime, increasing the incoherent image-bearing light intensity increases the transfer of image modulation onto the space-charge grating profile without destroying that profile. For high levels of incoherent light intensity, corresponding to larger intensity ratios R, the transfer of image modulation onto the coherent grating's space-charge field profile is very high as shown by the readout beam's modulation depth curve in Fig. 7.13, but the high intensity of the uniform incoherent image light I_1 strongly erases the coherent grating's space-charge profile, as shown by the I_{10} curve in Fig. 7.13. Hence the I_{11} diffracted light component, which is derived from the combination of the modulation depth and the average grating intensity I_{10}, exhibits the peaking behavior shown in Fig. 7.16.

The match between the theoretical curve in Fig. 7.16 and the experiment is not expected to be perfect because the perturbation theory requires small modulation depths m_G and m_S for reasonable accuracy, whereas the experiment is configured with the highest possible modulation depths. To impose small modulation depths in the experiment would make the diffracted I_{11} light intensity too weak to measure reliably, so a more exacting test of the theory must await numerical modeling of high modulation depth transcriptions.

An additional difficulty in achieving agreement between the theoretical models and the experiment concerns the appropriate value for the photoconductive sensitivity ratio B. For the experiment described in Fig. 7.14, in which the coherent writing wavelength is 515 nm, and the incoherent image wavelength is 488 nm, an estimate of $B = 2$ can be derived from a single photon absorption model that assumes the quantum efficiencies of the photoconductive processes for both the coherent grating beam and the incoherent image beam to be identical [7.8]. However, measurements of the photoconductive quantum efficiency reported by *Sprague* [7.30] show a strong dependence on the light wavelength, such that an increase in the sensitivity ratio B by a factor of 1.5 can reasonably be argued due to this dispersion of quantum efficiency. Furthermore, this factor can be expected to vary from one crystal sample to another, depending upon the detailed growth conditions. On the other hand, a reduction of the effective sensitivity ratio B of as much as 1.7 can also be argued for the crystal thickness used in this experiment because of the dispersion of the optical absorption, i.e., the ratio of coherent grating to incoherent image beam intensities changes continuously as both beams propagate through the crystal [7.8]. Because of these conflicting arguments, the nominal ratio of $B = 2$ in Figs. 7.14 and 7.16 has been assumed for the wavelengths considered. This gives a good fit for the uniform erasure I_{10} experimental points, especially for the constant recombination time approximation, but this ratio causes the predicted modulated erasure I_{11} curve to peak at a significantly lower intensity ratio R than is indicated by experiment.

In summary, the broad features of the recording nonlinearities are well understood and successfully modeled by the perturbation series approach, e.g., the decay of the coherent grating pattern with increasing amounts of image intensity, and the dependence of the I_{11} diffracted light intensity component on exposure parameters. Detailed agreement will require both further analysis to model the nonlinearities more accurately, and better information about the wavelength dispersion of the photoconductive sensitivities of the coherent grating and the image-bearing beams.

7.4.3 Spatial Resolution Issues for the Recording Process

A number of distinct factors influence the ultimate resolution achievable with the PICOC spatial light modulator. These factors can be classified as geometric, configurational, and materials related in nature. The geometric and materials related factors influence the recording of the image, and hence are reviewed in this section. The configurational factors influence the readout of the volume gratings, and as such are reviewed in the subsequent section.

Geometrical Limitations. Geometric resolution limitations derive principally from the incorporation of an incoherent imaging system in the four-wave mixing geometry, and from the finite crystal thickness d required to create a volume holographic grating. These effects are illustrated in Figs. 7.17 and 7.18. Distinctly different resolution performance is expected, depending upon the optical absorption coefficients α_G of the coherent grating light and α_S of the incoherent image light.

Figure 7.17 describes the case for low optical absorption ($\alpha_G d \ll 1$ and $\alpha_S d \ll 1$), such that the induced holographic grating has essentially uniform amplitude throughout the volume of the crystal. As can be expected from physical considerations, the optimum focal point occurs in the center of the crystal, and is not localized on the front surface of the crystal. The spatial frequency response will then be proportional to $(W/2)^{-1}$, which in turn is equal to $4n_0 F\#/d$ for the case of 1:1 imaging, in which W is the diameter of the incoherent image beam at the front surface of the crystal (as shown in Fig. 7.17), n_0 is the refractive index of the electrooptic crystal, $F\#$ is the F-

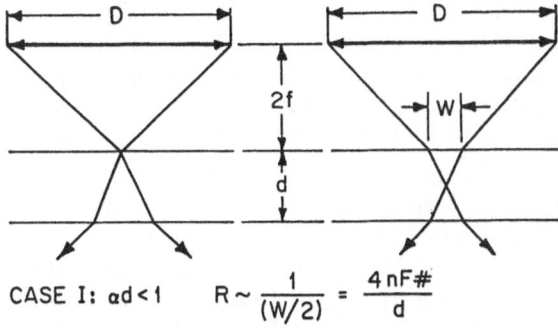

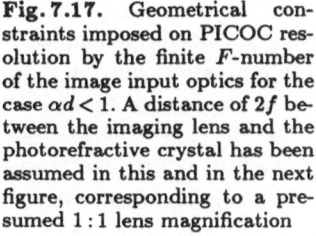

Fig. 7.17. Geometrical constraints imposed on PICOC resolution by the finite F-number of the image input optics for the case $\alpha d < 1$. A distance of $2f$ between the imaging lens and the photorefractive crystal has been assumed in this and in the next figure, corresponding to a presumed $1:1$ lens magnification

CASE I: $\alpha d < 1$ $\qquad R \sim \dfrac{1}{(W/2)} = \dfrac{4nF\#}{d}$

EXAMPLE: FOR $n=2.5$, $d=1\,\text{mm}$, $F\#=5$,

$$R = 50 \text{ line pairs/mm}$$

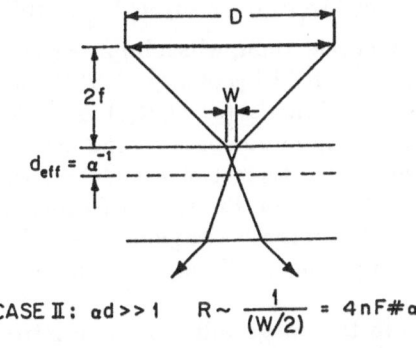

CASE II: $\alpha d \gg 1$ $\qquad R \sim \dfrac{1}{(W/2)} = 4nF\#\alpha$

EXAMPLE: FOR $n=2.5$, $d=1\,\text{mm}$, $F\#=5$, $\alpha=100\,\text{cm}^{-1}$

$$R = 500 \text{ line pairs/mm}$$

Fig. 7.18. Geometrical constraints imposed on PICOC resolution by the finite F-number of the image input optics for the case $\alpha d \gg 1$

number of the incoherent imaging system, and d is the crystal thickness. For example, for $n_0 = 2.5$, $d = 1\,\text{mm}$, and an F-number of 5, the resolution limit is approximately 50 cycles/mm.

In contrast, Fig. 7.18 describes the case for which $\alpha_G d \gg 1$, such that the induced holographic grating has significant amplitude only within a thin layer of thickness $d_{\text{eff}} = \alpha_G^{-1}$. In this case, the resolution is given by $4n_0F\#\alpha_G$. For $n_0 = 2.5$, $d = 1\,\text{mm}$, $\alpha_G = 100\,\text{cm}^{-1}$, and an F-number of 5, the resolution limit is approximately 500 cycles/mm, a factor of 10 improvement in resolution. However, the diffraction efficiency is reduced by a factor of order 100 because of the reduction in effective thickness of the grating. If the absorption coefficient α_S is chosen to be significantly larger than α_G, then the resolution will be constrained by α_S instead of α_G.

Material Limitations. An additional resolution limitation stems from material-dependent parameter constraints which influence the physics of grating formation, in particular the finite supply of compensating traps N_A^-, which is equal to the equilibrium donor-like trap density $N_{D\,\text{eq}}^+$. If the trap density is limited, then the space-charge field that can be recorded is similarly limited because sufficient space charge cannot be generated to establish any higher field strengths. This limitation becomes progressively more severe at higher spatial frequencies as expressed by the $D(K_G)$ factor in (7.17), in which the trap-limited saturation space-charge field E_q is a function of the spatial frequency and the equilibrium donor-like trap density $N_{D\,eq}^+$, as shown in (7.20). The reduction in space-charge field with an increase in spatial frequency (as exhibited by a corresponding decrease in E_q) is shown in Fig. 7.12. If an unlimited supply of compensating traps were available for establishing the space-charge field, then the saturation field E_q would be infinite and the factor $D(K_G)$ would converge to unity, indicating negligible degradation in the space-charge field. In practice, the finite level of compensating traps leads both to a reduction of the space-charge field and to a phase shift as consequences of the finite saturation field. These two effects are discussed in more detail below.

Consider for example an equilibrium donor-like trap density $N_{D\,eq}^+$ of 10^{16} cm^{-3} and a coherent grating frequency of 300 cycles/mm, which implies a saturation field E_q of order $19\,kV/cm$ and a diffusion field E_D of order $0.5\,kV/cm$. For an applied bias field of $6\,kV/cm$, the effect of the factor $D(K_G)$ in (7.17) is only a 5% reduction in the magnitude of the induced space-charge field in the linear approximation. As these parameters are typical of PICOC operation, the resultant effect induced by this mechanism on the spatial frequency response is therefore negligible compared with alternative response degradation mechanisms such as Bragg detuning on readout, as discussed in the next section.

The addition of harmonic components to the image will in general introduce an additional amplitude variation in the space-charge field through the denominator factor $D(K)$ in (7.17) (as shown explicitly in (7.A4) of Appendix 7.A). This variation is negligible compared with the effect induced by the coherent grating spatial frequency described above.

To verify the predicted high bandwidth of the recording process and to eliminate depth of focus issues as discussed previously, the Michelson interferometer configuration shown in Fig. 7.15 was used to record sinusoidal image patterns onto a bismuth silicon oxide crystal. The ratio R of the image-bearing light intensity I_1 to the coherent grating light intensity I_0 was adjusted experimentally to maximize the intensity of the I_{11} diffraction order. As the image grating frequency was varied, the angular alignment of the readout beam was also varied to maintain optimum Bragg angle alignment, thereby removing Bragg detuning effects on the readout resolution. Thus the measured diffracted intensity I_{11} corresponds to the strength of the space-charge field stored in the crystal, with results as shown in Fig. 7.19. A slight decrease in diffraction efficiency with increasing spatial frequency is observed, but the effective bandwidth for grating writing far exceeds the expected bandwidths for Bragg detuning on readout, as discussed in the readout section below.

A far more serious implication of (7.17) for the space-charge field component E_1 is a shift in the phase of the recorded space-charge field profile with

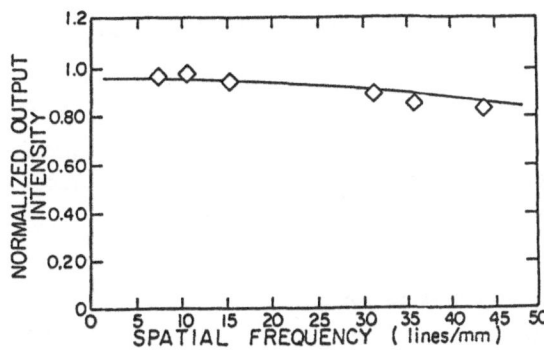

Fig. 7.19. Measured resolution response of the photorefractive recording process, limited by material dependent constraints. These measurements were performed by continuously Bragg matching the I_{11} order, thereby removing any dependence of bandwidth on Bragg detuning, as discussed in Sect. 7.5

respect to the incident coherent grating illumination profile. Considering again an equilibrium donor-like trap density $N_{D\,eq}^{+}$ of $10^{16}\,\mathrm{cm}^{-3}$, a coherent grating frequency of 300 cycles/mm, and an applied bias field of $6\,\mathrm{kV/cm}$, the resultant phase shift is of order $18°$. This phase shift poses no problems for PICOC performance so long as it remains uniform over the full aperture of the photorefractive crystal and the full bandwidth, as it in fact varies weakly with K_S and m_S. On the other hand, an image-induced differential phase shift can prove to be problematic for particular optical processing architectures.

Resolution Anisotropy. A final resolution issue that has been predicted by the perturbation series is a moderate anisotropy in the recording of image structures parallel as opposed to perpendicular to the applied bias field for the simultaneous erasure/writing mode (SEWM), and a severe resolution anisotropy for the grating inhibition mode (GIM). Features of this analysis are quite intriguing and hence are briefly reviewed here.

The perturbation series analysis is conducive to a physically intuitive interpretation of the perturbation terms as a sequence of discrete recording events. For the simultaneous erasure/writing mode (SEWM), two recording paths contribute to the I_{11} diffraction order. In one recording path, the incoherent image writes a space-charge field E_{01}, independent of the coherent grating profile. This space-charge field E_{01} then modulates the recording of the coherent grating light beam. This transcription path is analogous to the grating inhibition mode (GIM). In the second SEWM transcription path, the coherent grating profile writes a space-charge field E_{10}, which then modulates the recording of the incoherent image-bearing beam. This second path is analogous to the grating erasure mode (GEM). The GEM-like path exhibits nearly perfect isotropy of response to an arbitrary image, but the GIM-like path exhibits a very strong anisotropy, with image structures oriented perpendicular to the applied bias field generating much weaker space-charge fields than structures oriented parallel to the bias field.

To elaborate, consider an image profile consisting of a single spatial frequency, similar to (7.2), but oriented such that the wave vector K_S is orthogonal

303

to the applied bias field E_0. Thus the incident image intensity $I_S(y)$ is given by

$$I_S(y) = I_1 \left(1 + m_S \cos K_S y\right) , \tag{7.30}$$

in which the y axis is orthogonal to the applied bias field E_0. This image profile, in combination with the coherent grating profile as given by (7.1), induces a space-charge field $E(x, y)$ of the form

$$\begin{aligned}
E(x, y) = {} & \hat{x} E_{10} \cos K_G x + \hat{y} E_{01} \cos K_S y \\
& + (\hat{x} E_{11X} + \hat{y} E_{11Y}) \cos K_G x \cos K_S y \\
& + \text{higher order terms} \tag{7.31}
\end{aligned}$$

in which $\hat{x}$ and $\hat{y}$ are unit vectors parallel and perpendicular to the applied bias field respectively.

Terms such as E_{01} and E_{11Y} involve recording a charge pattern along a directional orthogonal to the applied bias field, and hence do not benefit from the enhancement of the photoconductivity induced by this applied field. In practice, these terms are very much smaller than terms such as E_{10} and E_{11X} which involve recording a charge pattern along a direction parallel to the applied bias field. Analysis using the perturbation method shows that the E_{11X} term is reduced by a factor of two compared with the E_{11} term given by (7.5) that results when the image wave vector K_S is parallel to the applied bias field E_0.

Thus we find that the material limitations on recording resolution are negligible compared with the geometrical limitations, which in turn are of comparable magnitude to the Bragg sensitivities on readout to be discussed in the Sect. 7.5.

7.4.4 Temporal Response

A temporal response analysis is necessary for the study of the grating inhibition mode (GIM) and the grating erasure mode (GEM) since these involve temporal sequencing, and the issue of timing is crucial to the optimization of their performance. The temporal analysis is also of interest for the simultaneous erasure/writing mode (SEWM) because it clarifies the duration and nature of the transient writing period before a stable response is achieved, and also because it leads to a very important reciprocity law between the incident light power and the response time of the converter (assuming that the dark conductivity of the photorefractive crystal can be neglected). A final and very important issue that arises from the temporal response is an image-induced phase modulation of the coherent grating's charge profile that can degrade PICOC performance in some coherent processing architectures.

The analysis reported here is restricted to small modulation regimes for simplicity. The results are modified substantially when operating with large modulation depths. In particular, the transient period increases significantly in such a regime.

To introduce some of the key concepts, consider the temporal evolution of the space-charge field in response to coherent grating illumination in the absence of an incoherent image. *Kukhtarev*'s study of this problem using first order perturbation analysis [7.19] (in the absence of self-diffraction effects) has shown that if the photorefractive crystal is illuminated with an intensity

$$I(x,t) = \begin{cases} 0 & \text{for} \quad t < 0 \\ I_0(1 + m_{\mathrm{G}} \cos K_{\mathrm{G}} x) & \text{for} \quad t > 0 \end{cases} \tag{7.32}$$

then the space-charge field component $E_1(t)$ has the form

$$E_1(t) = -\tfrac{1}{2} m_{\mathrm{G}} [E_0 + \mathrm{i} E_{\mathrm{D}}(K_{\mathrm{G}})] D(K_{\mathrm{G}})^{-1} \{1 - \exp[-t/T(K_{\mathrm{G}})]\} \tag{7.33}$$

in which $D(K_{\mathrm{G}})$, $E_{\mathrm{D}}(K_{\mathrm{G}})$, and $E_q(K_{\mathrm{G}})$ have been defined by (7.18–20). The time constant $T(K)$ is defined by

$$T(K) = T_0 \frac{\{1 - \mathrm{i}[E_0 + \mathrm{i} E_{\mathrm{D}}(K)]/E_{\mathrm{M}}(K)\}}{\{1 - \mathrm{i}[E_0 + \mathrm{i} E_{\mathrm{D}}(K)]/E_q(K)\}} \tag{7.34}$$

in which T_0 is the dielectric relaxation time, defined by

$$T_0 = \frac{\varepsilon \varepsilon_0}{\mu e n_0} \quad, \tag{7.35}$$

and n_0 is the zeroth order estimate of the electron density given by

$$n_0 = S_{\mathrm{G}} I_0 N_{\mathrm{D}} \tau \quad . \tag{7.36}$$

In (7.36), τ is the free carrier lifetime, as given by

$$\tau = (\gamma_{\mathrm{R}} N_{\mathrm{D\,eq}}^+)^{-1} \quad . \tag{7.37}$$

A group of parameters occurs in (7.34) having the dimensions of an electric field. This field parameter is assigned the symbol $E_{\mathrm{M}}(K_{\mathrm{G}})$, and is defined by

$$E_{\mathrm{M}}(K_{\mathrm{G}}) = (\mu K_{\mathrm{G}} \tau)^{-1} \quad . \tag{7.38}$$

The physical interpretation of this field parameter can be best understood by considering the average transport length of the mobile charges. When the drift contribution to the current density in (7.12) dominates over the diffusion contribution, the mobile charges during their limited lifetime τ travel an average distance $L = \mu \tau E$ while under influence of an electric field E. The drift transport length is equal to the grating period when the field strength $E = 2\pi E_{\mathrm{M}}$.

The space-charge field in (7.33) exponentially approaches its saturation limit, generally with a complex time constant $T(K_{\mathrm{G}})$ which denotes oscillatory as well as decaying behavior. At low spatial frequencies, the drift-induced transport length is very short compared with the grating period (equivalent to having a field E_{M} large compared to the applied bias field E_0). In this case, the decay predicted by (7.33) is governed essentially by the dielectric relaxation time T_0 and exhibits negligible oscillatory behavior. This gives the fastest possible response time. For higher spatial frequencies and high applied bias fields, for which the drift-driven charge transport length greatly exceeds the grating period, the response time increases substantially beyond the dielectric relaxation time and in addition the response exhibits oscillatory behavior. An intuitive interpretation of this phenomenon is that the finite transport length blurs the charge pattern being transcribed, forcing a longer recording time to achieve a given level of charge profile modulation. Furthermore, the blur pattern is one-sided because the applied bias field forces the mobile charges always in one direction, inducing a phase shift of the charge pattern being transcribed. In all cases, the response time is inversely proportional to the incident light intensity, so that doubling the incident intensity reduces the response time by a factor of two.

By similar analysis, one finds that the erasure of the resulting space-charge field by a spatially uniform incident light beam also exhibits an exponential response with a time constant that is inversely proportional to the erasure light intensity. In the absence of an applied bias field, this response is described by a simple exponential function with no shift in the phase of the coherent grating's charge profile. In the presence of an applied bias field, the uniform erasure light beam induces a drift of the charge pattern that was originally recorded by the coherent grating beams, resulting in a phase modulation as well as an amplitude modulation.

Let us now review the analysis of one particular version of the simultaneous erasure/writing mode (SEWM) for the photorefractive incoherent-to-coherent optical conversion (PICOC) process. Consider a crystal in which a coherent grating of wave vector K_G has been written and has reached steady state. At time $t = 0$, an incoherent image grating with wave vector K_S is turned on. The intensity incident upon the crystal is then described by

$$I(x,t) = \begin{cases} I_0(1 + m_G \cos K_G x) & \text{for} \quad t < 0 \\ I_0(1 + m_G \cos K_G x) + I_1(1 + m_S \cos K_S x) & \text{for} \quad t > 0 \end{cases} .$$

(7.39)

The temporal evolution of the various components of the space-charge field can be solved by perturbation techniques, valid for small levels of the modulation depth m_G, with explicit expressions for the lowest order terms as presented in Appendix 7.A. In particular, the temporal response of the E_{11} component has the general form

$$E_{11}(t) = M_0 + M_1 \exp\left[-t/T_1(K_G)\right] + M_2 \exp\left[-t/T_2(K_S)\right] \\ + M_3 \exp\left[-t/T_3(K_G + K_S)\right] + M_4 \exp\left[-t/T_4(K_G, K_S)\right] ,$$

(7.40)

in which the M's are complex coefficients that depend upon the applied field, material parameters, and the incident light intensities, with explicit expressions as given in Appendix 7.A and in [7.22]. Time constants T_1, T_2, and T_3 are given by (7.34) for spatial frequencies corresponding to K_G, K_S, and $(K_G + K_S)$, respectively. The fourth time constant T_4 is given by

$$T_4(K_G, K_S) = [1/T(K_G) + 1/T(K_S)]^{-1} .$$

(7.41)

306

The first term M_1 corresponds to the recording of the coherent grating wave vector K_G in the absence of the image profile. The second term M_2 corresponds to the recording of the image profile wave vector K_S in the absence of the coherent grating. The third term M_3 corresponds to the recording of the intermodulation frequency $K_G + K_S$, and represents a resonant response of the system of photorefractive equations. The fourth term M_4 corresponds to the nonresonant response driven by the product of the coherent grating and the incoherent recording processes, which arises through the nonlinearity of the recording process. The steady state value of the E_{11} field component is governed by the term M_0.

The longest response time to reach steady state is comparable in value to the pure coherent grating response as discussed above in (7.34–38) and in [7.19]. This overall response time t_{sys} obeys the following reciprocity law

$$t_{sys} = \frac{G}{S_G I_0 + S_S I_1} \tag{7.42}$$

in the absence of appreciable dark conductivity β in the photorefractive crystal, in which the proportionality constant G involves only material parameters and the applied bias field E_0. As a result, the rate at which new information can be recorded is determined by the total available intensity incident on the crystal.

Similar temporal response analyses have been performed for both the grating inhibition mode (GIM) and the grating erasure mode (GEM) [7.22]. In the GEM mode, the response time constant of the system is inversely proportional to the incoherent erasing intensity I_1, rather than the sum of the incoherent and coherent intensities. Conversely, in the GIM mode, the response time constant of the system is inversely proportional to the coherent grating intensity I_0.

Sample perturbation analysis solutions for GEM, GIM, and SEWM are shown in Figs. 7.20, 7.21, and 7.22 respectively for a $1\,\mathrm{mW/cm^2}$ average intensity coherent grating beam at $515\,\mathrm{nm}$ wavelength with a grating spatial frequency of 300 cycles/mm and a small modulation depth m_G, a $1\,\mathrm{mW/cm^2}$ average intensity incoherent image-bearing beam at $488\,\mathrm{nm}$ wavelength with an image spatial frequency of 10 cycles/mm and also with a small modulation depth m_S, an applied bias field of $6\,\mathrm{kV/cm}$, and with the material parameters for bismuth silicon oxide as given in Table 7.1.

Figure 7.20 shows the GEM response for three diffracted light components: the direct recording of the unmodulated coherent grating component I_{10}, the direct recording of the incoherent image-bearing beam I_{01}, and the intermodulation component I_{11} (the image-modulated grating component). The coherent grating has been recorded to saturation and then turned off before the time interval shown in the figure. The time $t = 0$ is defined when the image-bearing beam is turned on. Thus the coherent grating frequency component I_{10} starts at its saturated level and decays for times $t > 0$ because of erasure by the image-bearing beam. The direct recording of the image-bearing beam I_{01} grows from an initial value of zero to its saturation level. Note that the response time for

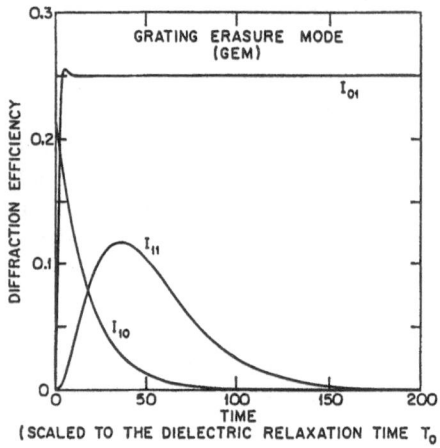

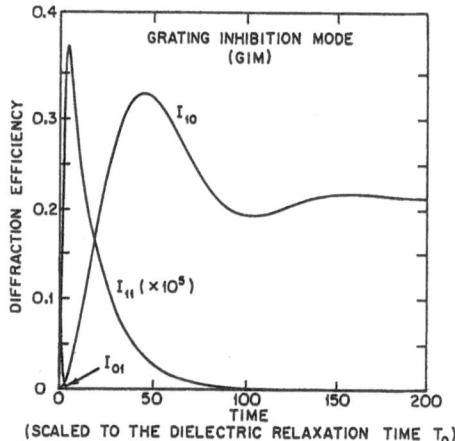

Fig. 7.20   Fig. 7.21

Fig. 7.20. Temporal evolution for the grating erasure mode (GEM) as predicted by the perturbation series analysis. Time $t = 0$ starts after the coherent grating has been recorded and the incoherent image-bearing beam has just been turned on. In general the I_{11} diffraction component would be much smaller than either the I_{10} or the I_{01} components in this and in the next two figures, corresponding to small modulation depths m_G^{eff} and m_S^{eff}, but for simplicity all three curves are shown as if these modulation depths were unity

Fig. 7.21. Temporal evolution for the grating inhibition mode (GIM) as predicted by the perturbation series analysis. Time $t = 0$ starts after the incoherent image-bearing beam has been recorded and the coherent grating beams have just been turned on

the I_{01} image component, with its much lower spatial frequency, is significantly faster than that for the I_{10} coherent grating component, with its order of magnitude higher spatial frequency. The image-modulated grating component I_{11} exhibits a temporal response which is derived from a combination of the I_{10} and I_{01} response, eventually evolving into a slow decay in time when the incoherent image-bearing light beam erases the coherent grating. The I_{11} intensity is eventually erased by the image-bearing beam for very long recording times, so that the image-bearing light exposure time must be truncated.

Figure 7.21 shows a similar set of temporal response curves for the grating inhibition mode (GIM). The image-bearing light has been recorded to saturation, then turned off before the time interval shown in this figure. Time $t = 0$ is defined when the coherent grating light is turned on. In this mode, the pure coherent grating component I_{10} starts with zero intensity and gradually grows to saturation, whereas the directly recorded incoherent image-bearing beam I_{01} starts from its saturation level and is quite rapidly erased by the uniform component of the coherent grating light. The image-modulated grating component I_{11} shows initially a very rapid rise, followed by a much slower erasure by the coherent grating beam, eventually decaying to zero for very long recording times. In GIM, the image-modulated grating I_{11} component generally falls far short of its levels for GEM and SEWM, at least as predicted by the pertur-

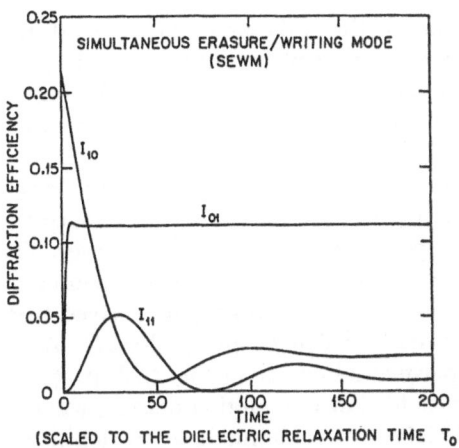

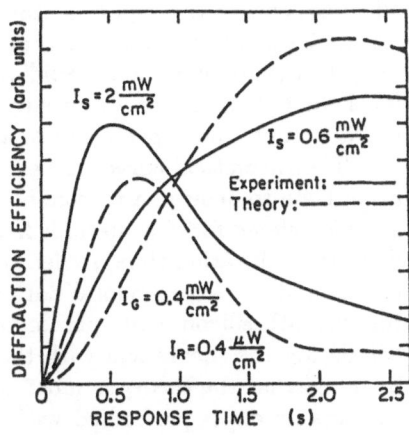

Fig. 7.22   Fig. 7.23

Fig. 7.22. Temporal evolution for one version of the simultaneous erasure/writing mode (SEWM) as predicted by the perturbation series analysis. Time $t = 0$ starts after the coherent grating has been recorded and the incoherent image-bearing beam has just been turned on

Fig. 7.23. Measured diffraction efficiency of the I_{11} beam (strongly modulated erasure) as a function of time for two values of the signal intensity for the simultaneous erasure/writing mode, shown for comparison with the corresponding theoretically predicted response curves from the perturbation series analysis. In this figure, I_S is the intensity of the signal beam(s), I_G is the intensity of the grating recording beams, and I_R is the intensity of the readout beam

bation series analysis, because the direct image recording is erased before the coherent grating recording has a chance to grow.

Figure 7.22 presents the simultaneous erasure/writing mode response, assuming the temporal sequencing given by (7.39). The pure coherent grating recording I_{10} starts with its initial saturation level at time $t = 0$, and slowly drops to a reduced saturation level. The direct recording of the incoherent image-bearing beam I_{01} rapidly builds from zero intensity to saturation. The image-modulated grating component I_{11} exhibits strong oscillations, eventually settling at its saturation level, with a response time much longer than that of the I_{01} component.

Measurements of the temporal response of the I_{11} diffraction order are compared in Fig. 7.23 with the temporal response solutions generated by the perturbation series analysis for the cases of intensity ratios $R = 1.5$ and $R = 5.0$ and for a grating written in the nondegenerate simultaneous erasure/writing mode (SEWM). The experiments are shown as solid lines, the theoretical predictions as dashed lines. In this figure, the coherent grating is established in the saturation regime at time $t = 0$, at which point the incoherent erasure beam is allowed to expose the crystal, as given by (7.39). For a small R ratio ($I_1 = I_S = 0.6\,\text{mW/cm}^2$ in Fig. 7.23), the experimental diffraction efficiency increases monotonically, at least within the time interval of this figure. For a strong incoherent beam ($I_1 = I_S = 2\,\text{mW/cm}^2$ in Fig. 7.23), a transient effect appears in the experimental response within this same time interval. Initially, the incident beam diffracts from the composite grating at wave vector $(K_G + K_S)$ to generate a rapid rise in the amplitude of I_{11}, but the strong incoherent illumination eventually erases the coherent grating and hence the diffraction efficiency decreases to a small steady-state value.

The theoretical curves are scaled in peak intensity and in dielectric relaxation time to achieve a reasonable match with the experiments, but once the scale is defined for one curve, it fixes the scale for both theoretical curves. The dielectric relaxation time needed to achieve agreement between the theoretical curves and the experimental data is a factor 3 slower than that derived from the bismuth silicon oxide parameters given in Table 7.1 per (7.35–37). When comparing the theoretical with the experimental curves in Fig. 7.23, it should be noted that the theory is most accurate for low modulation depths, whereas the experiment is performed with the highest possible modulation depths for both the coherent grating and the incoherent image beams. Even so, the match between the theoretical response and the measured response is quite striking.

The time constant for this particular set of experimental parameters is in the range 0.5 to 1.5 s. To achieve conversion rates of 30 frames per second in bismuth silicon oxide, a total light intensity of order 35 to $45\,\mathrm{mW/cm^2}$ is extrapolated, based upon the reciprocity law given in (7.42), and assuming that the ratio of incoherent image-bearing light to coherent grating light is kept constant.

7.5 The Readout Process

The readout process consists of the optical modulation of the coherent readout beam by the space-charge field. This modulation occurs in PICOC through the linear electrooptic effect which modifies the refractive index within the photorefractive crystal, thus establishing a volume phase grating. The phase hologram is then read out by a coherent auxiliary beam to achieve the conversion.

The diffraction characteristics of such volume phase gratings have been studied using many different analytical techniques, including coupled wave analysis by *Kogelnik* [7.31] with extensive numerical studies by *Klein* and *Cook* [7.32], a Born approximation to a scattering integral by *Gordon* [7.33], and the optical beam propagation method by *Yevick* and *Thylen* [7.34] and *Johnson* and *Tanguay* [7.35]. Methods for studying the polarization properties of light diffraction in electrooptic crystals such as bismuth silicon oxide include anisotropic versions of the coupled wave formalism [7.23, 36] and the optical beam propagation method [7.35].

This section examines the readout of the phase gratings and its consequences for the performance of PICOC as a spatial light modulator. Because of the high spatial frequencies typically used in PICOC, the grating exhibits pronounced Bragg diffraction characteristics with rapid degradation of the readout quality whenever the Bragg condition is detuned, whether by increasing the spatial frequency of the incoherent image, by slight angular misalignment, or by sub-optimum alignment of the incoherent image beam. The Bragg sensitivity is discussed in the section on the isotropic phase grating properties. In addition, because of the optical activity exhibited by bismuth silicon oxide, the polarization states of both the transmitted probe light and the diffracted signal light will in general be elliptical, with implications for the optimum readout conditions, as discussed in the section on polarization properties.

7.5.1 Isotropic Phase Grating Model

The space-charge field induces a small perturbation in the index of refraction through the linear electrooptic effect, such that

$$\Delta n(x) = \tfrac{1}{2} n_0^3 r_{41} E_1 \cos\left(K_G x\right) \tag{7.43}$$

in which n_0 is the nominal index of refraction and r_{41} is the electrooptic coefficient appropriate for bismuth silicon oxide. The index perturbation in turn modulates the optical phase fronts of an incident light beam. Thus the sinusoidal space-charge field induces a volume phase grating in the crystal. To gain some feeling for the readout performance issues involved in PICOC, consider a 2 mm thick piece of bismuth silicon oxide with a simple unmodulated sinusoidal grating with spatial frequency of 300 cycles/mm, space-charge field of 5 kV/cm, and probed by a 633 nm laser beam. A space-charge field E_1 of order 5 kV/cm induces an index perturbation Δn of order 2×10^{-5}, assuming the material parameters for bismuth silicon oxide listed in Table 7.1.

Dimensionless parameters that characterize the diffraction characteristics of such a phase grating include the grating thickness parameter Q and the grating strength v, defined by

$$Q = \frac{\lambda_R d K_G^2}{2\pi n_0} \quad \text{and} \tag{7.44}$$

$$v = \frac{2\pi \Delta n d}{\lambda_R} \tag{7.45}$$

in which λ_R is the wavelength of the readout light *in vacuo* and d is the thickness of the grating [7.32]. For a 300 cycles/mm grating in a 2 mm thick crystal, read out by a 633 nm laser beam, the associated thickness parameter Q is almost 300, which is considered to be a very thick grating [7.30]. A space-charge field of the order of 5 kV/cm induces a grating strength of the order of 0.35 radians, which gives fairly weak diffraction efficiency (on the order of a few per cent). The combination of large thickness parameter Q and weak grating strength v places the grating deep in the Bragg regime with an optical diffraction efficiency η given approximately by

$$\eta = [(v/2\sigma) \sin \sigma]^2 \quad , \quad \text{with} \tag{7.46}$$

$$\sigma = \left(\xi^2 + \frac{v^2}{4}\right)^{\frac{1}{2}} \quad \text{and} \tag{7.47}$$

$$\xi = \tfrac{1}{2} K_G d \sin(\theta_{in} - \theta_B) \qquad (7.48)$$

(according to [Ref. 7.31, Eqs. 17, 42, and 43] or [Ref. 7.32, Eqs. 6–8, 35, and 36]), in which θ_{in} is the entrance angle of the incident probe light measured with respect to the constant phase lines of the coherent grating, and $\theta_B(K_G)$ is the Bragg angle associated with wave vector K_G, defined by

$$\sin \theta_B = \frac{\lambda_R K_G}{4\pi n_0} . \qquad (7.49)$$

The parameter ξ in (7.47, 48) is a measure of the Bragg misalignment; it is equal to zero when the readout beam is perfectly Bragg aligned with respect to the volume hologram.

For perfect Bragg alignment in which $\theta_{in} = \theta_B$, and assuming a grating strength $v = 0.35$ radians, the peak diffraction efficiency is estimated to be of order 3 %. Doubling the thickness of the grating would increase the diffraction efficiency to 12 % (ignoring polarization issues), but also increases the Bragg detuning sensitivity, as described in the following paragraphs.

Bragg detuning impacts PICOC performance in two ways: the angular alignment sensitivity of the hologram to the coherent readout beam, and the spatial frequency response of the hologram readout with its concomitant angular alignment dependence on the incoherent image-bearing beam. Consider first the alignment sensitivity to the coherent readout beam. An angular misalignment $\Delta\theta$ from the optimum Bragg alignment introduces a Bragg mismatch ξ of

$$\xi = \tfrac{1}{2} K_G d \Delta\theta \quad , \qquad (7.50)$$

as seen from (7.48). Assuming a grating strength $v = 0.35$ radians, one finds from (7.46, 47) that the angular alignment sensitivity of the diffraction efficiency has essentially a sinc^2 profile. The angular misalignment $\Delta\theta$ needed to reach the first null of this profile occurs when $\xi = \pi$, i.e., for

$$\Delta\theta = \frac{2\pi}{K_G d} . \qquad (7.51)$$

Hence the angular alignment needed to achieve optimum diffraction efficiency is extremely sensitive, on the order of 0.1° for a 2 mm thick photorefractive crystal with a 300 cycles/mm grating frequency.

One possible alignment of the PICOC system is to orient the readout beam to be Bragg matched precisely to the coherent grating, thereby maximizing the intensity of the I_{10} diffracted order. However, when an incoherent grating is also incident upon the crystal, the nonlinear recording process creates a new grating with wave vector $(K_G + K_S)$ that in general does not satisfy the same Bragg condition as the coherent grating wave vector K_G. The resulting I_{11} intensity component is then attenuated by an amount dependent upon the image wave vector $K_S = 2\pi f_S$, in which f_S is the spatial frequency of the image profile. The dependence of the attenuation on the image frequency f_S is a function of the orientation of the image-bearing beam with respect to the coherent grating.

The importance of the image-bearing beam orientation on the spatial frequency response of the converter is strikingly illustrated by Fig. 7.24, in which are shown converted images of two orthogonal orientations of a 5 line pair/mm Ronchi ruling and the associated coherent Fourier transforms. As can be seen from the figure, a significant difference in resolution exists between cases in which the wave vector of the ruling (incoherent grating) is parallel or perpendicular to the coherently written (holographic) grating. This difference derives principally from the fact that a different wave vector matching condition exists for these two cases.

Consider the alternative wave vector matching conditions shown in Figs. 7.25 and 7.26. In Fig. 7.25, the incoherent grating wave vector K_S is parallel to the coherent grating wave vector K_G, a condition achieved by symmetrically disposing the incident coherent beams about the normal to the crystal while simultaneously arranging the incoherent imaging system such that its optical axis is parallel to the crystal normal. In this case, significant Bragg detuning occurs for even small incoherent grating wave vectors.

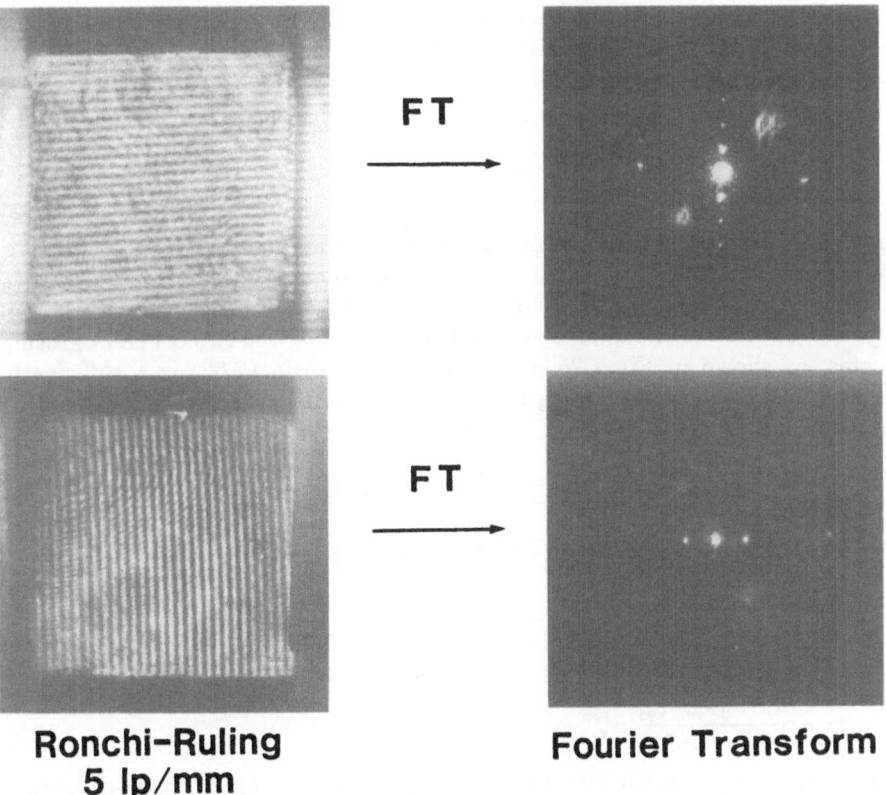

Ronchi-Ruling   **Fourier Transform**
5 lp/mm

Fig. 7.24. Photograph of the Fourier transform of a Ronchi ruling recorded in the PICOC configuration, showing a strong anisotropy in the resolution performance when the image-bearing light beam bisects the two coherent grating writing beams

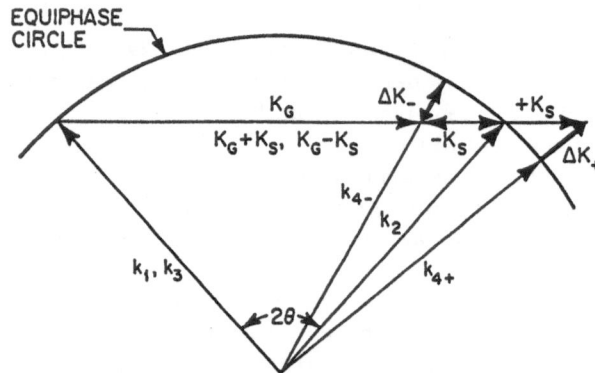

EQUIPHASE
CIRCLE

Fig. 7.25. Wave vector mismatch diagram for the case in which the image-bearing beam bisects the two coherent grating writing beams

Let the spatial bandwidth of the readout process be defined as that spatial frequency f_S for which the magnitude of the I_{11} diffracted intensity component degrades to 25 % of its peak value. This occurs when $\xi = 1.9$, as determined from (7.46,47), assuming a small grating strength v on the order of 0.35 radians. The Bragg mismatch parameter ξ is a function of the image spatial frequency f_S, and can be approximated by the first term or two in a Taylor series expansion of the form

$$\xi(f_S) = \xi_0 + f_S \frac{d\xi}{df_S} + \frac{1}{2} f_S^2 \frac{d^2\xi}{df_S^2}$$
$$+ \text{ higher order terms} \tag{7.52}$$

in which ξ_0 represents the Bragg mismatch associated with the coherent grating. Standard alignment procedure is to set this zeroth order mismatch term ξ_0 to zero, i.e., to Bragg match the incident readout beam to the coherent grating.

For the case of the image-bearing beam bisecting the two coherent grating beams, as shown in Fig. 7.25, the first order term in (7.52) is a sufficiently accurate estimate of the Bragg mismatch parameter ξ, and this term can be derived from (7.48) to give

$$\frac{d\xi}{df_S} = -\frac{1}{2} K_G d \frac{d\theta_B}{df_S} \quad . \tag{7.53}$$

By (7.49), the term $d\theta_B/df_S$ is equivalent to

$$\frac{d\theta_B}{df_S} = \frac{\lambda_R}{2 n_0 \cos \theta_B} \quad . \tag{7.54}$$

Hence the spatial bandwidth f_S is

$$f_S = \frac{7.6 n_0 \cos \theta_B}{\lambda_R K_G d} \quad , \tag{7.55}$$

which can be expressed in terms of the fringe spacing Λ_G of the coherent grating wavelength as

$$f_S = \frac{1.2 n_0 \Lambda_G \cos \theta_B}{\lambda_R d} \quad . \tag{7.56}$$

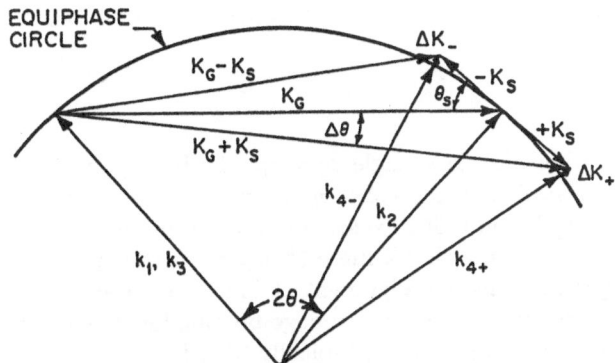

Fig. 7.26. Optimum wave vector alignment diagram in which the image-bearing light is tilted with respect to the bisector of the two coherent grating writing beams by the Bragg angle associated with the probe beam

For the parameters used in our experiments, this bandwidth is estimated to be on the order of 8 cycles/mm. Note that doubling the crystal thickness d to increase the diffraction efficiency by almost a factor of four has the adverse effect of reducing the spatial bandwidth f_S by a factor of two for this particular alignment configuration.

Compare now the resolution performance associated with the alignment of Fig. 7.25 with that for Fig. 7.26. In Fig. 7.26, the incoherent grating wave vector is arranged to lie tangentially to the circle defined by the readout beam wave vector, such that a significantly increased angular deviation of the diffracted beam is allowed before serious Bragg detuning effects occur. Such a wave vector tangency condition is automatically satisfied when the incoherent image wave vector is normal to the coherent grating wave vector (as it is in the y orientation normal to the plane of incidence). Alternatively, the wave vector tangency condition is satisfied when the central ray of the incoherent image beam is parallel to the diffracted probe light, for which the horizontal and vertical resolutions become degenerate. This is not the case for the situation depicted in Fig. 7.25, which explains the observations apparent in Fig. 7.24. An equivalent condition has been described by *Kamshilin* and *Petrov* [7.6] for the nondegenerate four-wave mixing optical architecture.

In the geometry of Fig. 7.26, the addition of an image frequency f_S shifts the pointing direction of the combined wave vector $(K_G + K_S)$ such that the incident readout light remains almost perfectly Bragg matched over a much broader range of image frequencies. Mathematically, this condition is equivalent to setting the first order term $d\xi/df_S$ of the Taylor series expansion in (7.52) for the Bragg mismatch parameter ξ equal to zero. To demonstrate this, consider the more general case in which the image profile wave vector K_S is oriented at a small angle θ_S with respect to the coherent grating wave vector K_G. The combination $(K_G + K_S)$ rotates through a small angle $\Delta\theta$ compared with wave vector K_G alone, with $\Delta\theta$ given approximately by

$$\Delta\theta = \frac{K_S \sin\theta_S}{K_G} = \frac{2\pi f_S \sin\theta_S}{K_G} \quad . \tag{7.57}$$

315

The Bragg mismatch parameter ξ induced by the image wave vector K_S is to first order given by

$$\xi = \frac{d\xi}{d\Delta\theta}\Delta\theta + \frac{d\xi}{df_S}f_S \quad , \tag{7.58}$$

and these two terms cancel when the skew angle $\theta_S = \theta_B$, the Bragg angle for the readout light beam associated with the coherent grating frequency. For a 633 nm readout beam wavelength and a 300 cycles/mm grating frequency, the Bragg angle θ_B is of order 5°, implying that a mere 5° separates the optimum resolution alignment of Fig. 7.26 from the alignment of Fig. 7.25. In addition, the angular alignment tolerance on the incoherent image-bearing beam is much tighter than the nominal alignment angle of 5°, typically of order 0.7°.

With the alignment of Fig. 7.26, the degradation in diffraction efficiency then becomes a second order function of the image spatial frequency f_S. The second derivative term $d^2\xi/df_S^2$ in (7.52) is

$$\frac{d^2\xi}{df_S^2} = \frac{\pi d\lambda_R}{n_0} \tag{7.59}$$

and the resulting spatial bandwidth of the tangential configuration is

$$f_S = \left(\frac{1.2\,n_0}{\lambda_R d}\right)^{1/2} \quad . \tag{7.60}$$

This expression was first published by *Kamshilin* and *Petrov* [7.6]. For our experimental parameters, a tangential geometry increases the converter's bandwidth from 8 to 48 cycles/mm. It is interesting to note that, in addition to the increased bandwidth of the tangential geometry, the spatial resolution is independent of the coherent grating frequency, and that doubling the thickness of the crystal d does not halve the converter's bandwidth, as it would for the alignment configuration shown in Fig. 7.25, but only reduces it by a factor of $\sqrt{2}$.

Further experimental tests of the Bragg detuning hypotheses are presented in Figs. 7.27–29, in which the image source was a Michelson interferometer (shown in Fig. 7.15) to alleviate the depth of focus issues discussed previously. In these experiments, the intensity of the diffracted component I_{11} was measured as a function of the spatial frequency of the image source grating. Figure 7.27 shows the response for the geometry in which the incoherent image beam bisects the coherent grating writing beams to achieve the wave vector mismatch condition diagrammed in Fig. 7.25. The predicted frequency response rolloff with bandwidth of 8 cycles/mm, shown by the solid line, compares reasonably well with the experimental data points which indicate a bandwidth of 6 cycles/mm. Figure 7.28 shows the rolloff when the signal beam wave vector is tangent to the equiphase circle as shown in Fig. 7.26, giving much improved frequency response of 45 cycles/mm which is in excellent agreement with the theoretically predicted 48 cycles/mm. Figure 7.29 shows the rolloff when the signal light wave vector is nominally aligned to be perpendicular to the applied

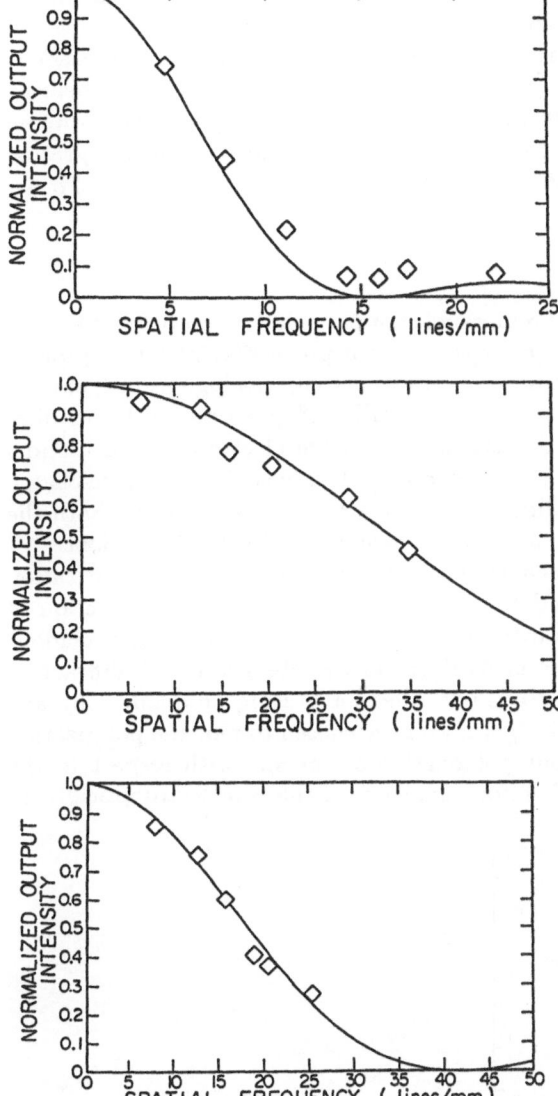

Fig. 7.27. Experimental measurement of the diffraction efficiency as a function of image spatial frequency associated with the alignment shown in Fig. 7.25

Fig. 7.28. Experimental measurement of the diffraction efficiency as a function of image spatial frequency associated with the optimum alignment shown in Fig. 7.26

Fig. 7.29. Experimental measurement of the diffraction efficiency as a function of image spatial frequency for the alignment in which the image-bearing beam's wave vector is normal to the applied bias electric field

bias field and to the coherent grating wave vector. The theoretical bandwidth for this configuration should be identical to that shown in Fig. 7.28, but the measurements indicate a bandwidth of only 25 cycles/mm. This discrepancy may be attributable to the very high alignment sensitivity of the incoherent image beam relative to the plane defined by the two coherent writing beams, e.g., a deviation of approximately 0.7° would cause a deterioration of the bandwidth from 48 cycles/mm to 25 cycles/mm.

To summarize, the presence of the coherent grating in PICOC defines a volume hologram that is typically operated quite deep into the Bragg regime, with very strong misalignment sensitivities. As a consequence, the readout beam must be aligned typically within 0.1° of optimum, the image-bearing beam must be aligned typically within 0.7° of its optimum, and the optimum alignment for the image-bearing beam is not to bisect the two coherent grating beams (Fig. 7.25), but rather offset from this by a small angle typically of order 5° (as shown in Fig. 7.26).

7.5.2 Polarization Issues

Bismuth silicon oxide has quite remarkable optical polarization properties that strongly influence the optimization of the readout process [7.23]. These properties include significant levels of natural optical activity (as high as 46°/mm for the 488 nm argon ion laser wavelength – see Fig. 7.30), a uniform linear birefringence induced by the applied bias field through the electrooptic effect, and a spatially varying linear birefringence induced by the image-defined space-charge field. In almost all readout configurations, these properties will cause both the readout and the diffracted signal beams to exhibit elliptical polarizations.

Consider Fig. 7.31, in which is shown the evolution of the polarization states for both the readout and the diffracted signal beams as a function of depth into the bismuth silicon oxide crystal for a 633 nm wavelength readout beam and an applied bias field of 6 kV/cm in the absence of self-diffraction effects. A hologram induced by a single coherent grating unmodified by any image profile is assumed. In this figure, we have chosen to plot the polarization states that result from an input polarization set at 45° with respect to the grating wave vector, which is therefore along one of the two electrooptically in-

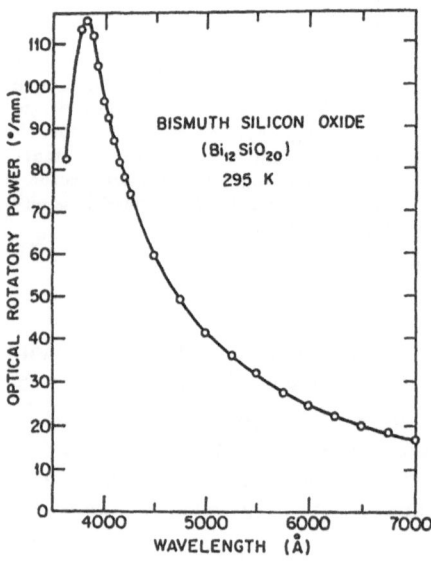

Fig. 7.30. Optical rotatory power of bismuth silicon oxide as a function of the wavelength of the readout light (after [7.27])

duced principal axes of the crystal's index ellipsoid. As such, the polarizations of the readout and signal beams are nearly parallel for very shallow depths but quickly evolve toward a 90° major axis separation with increasing depth because of the influence of the optical activity. This separation of polarization states enables the use of polarization analyzer techniques to suppress the scattered readout beam light in favor of the diffracted signal light. The intricacy of the polarization state evolution can be appreciated from Fig. 7.31.

One potential application of the polarization properties in bismuth silicon oxide has been demonstrated by *Herriau* et al. [7.15] for obtaining optimum holographic readout. They attained excellent suppression of scattered readout light noise for a nearly on-axis recording configuration by placing a polarization analyzer into the diffracted signal beam path. The analyzer is then adjusted to eliminate the scattered readout beam, and since the diffracted signal beam generally has a different polarization state from that of the transmitted readout beam, a significant fraction of the signal beam will pass through the analyzer. This technique greatly improves the signal-to-noise ratio of the holographic reconstruction process.

The presence of spatial modulation further complicates the polarization properties of the diffracted light, especially when the incoherent image beam is misaligned from the optimum wave vector matching configuration shown in Fig. 7.26, which introduces a strong polarization state dispersion. That is, the

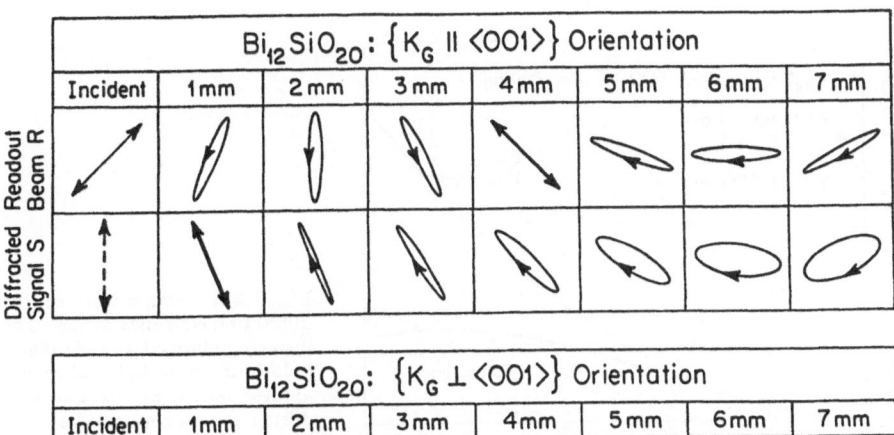

Fig. 7.31. Sample evolution of the polarization states of the incident readout beam and the diffracted signal beam for a simple sinusoidal grating

polarization states of the modulated readout beam's diffraction orders (I_{mn} in Fig. 7.10) are strongly dependent on the image spatial frequency when the optimum alignment of Fig. 7.26 is not achieved. Such dispersion further degrades the resolution if a polarization analyzer is used to separate the scattered readout light from the diffracted signal light. In contrast, when the optimum alignment of Fig. 7.26 is met, the polarization dispersion is negligible. This is one more reason why the alignment of Fig. 7.26 is crucial to achieving the best performance from the PICOC device.

The most serious issue concerning the polarization properties of volume holograms in bismuth silicon oxide is the degradation of the light diffraction efficiency that is imposed by the optical rotatory power [7.23]. However, one readout configuration has been identified in extensive polarization analyses that does not degrade the diffraction efficiency, namely the crystal geometry of Fig. 7.8 with no applied bias field and with a circularly polarized readout light beam. The diffraction efficiency for a linearly polarized readout beam is compared with that of a circularly polarized readout beam in Fig. 7.32, showing the marked improvement in diffraction efficiency that can be achieved by using circularly polarized light at the correct alignment.

In conclusion, the polarization properties of light diffraction in bismuth silicon oxide significantly affect the performance of the PICOC modulator, and can be exploited to improve this performance in terms of signal-to-noise ratio and to attain the highest possible diffraction efficiency.

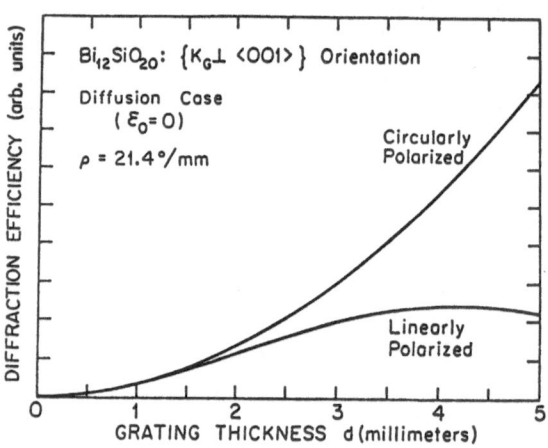

Fig. 7.32. Comparison of the diffraction efficiency achieved by linearly polarized and circularly polarized probe light when the applied bias field is set to zero during the readout process

7.6 Conclusion

Through the use of the photorefractive incoherent-to-coherent optical conversion (PICOC) process, we have successfully converted incoherent images into their negative coherent replicas. The PICOC system is inexpensive, easily implementable, and compares favorably in its performance with other photorefractive spatial light modulators. In addition, we have developed a mathematical framework within which to analyze the performance of the converter. Using this formalism, we have derived the conditions necessary to achieve high linearity, good contrast ratio, and fast temporal response. Experiments have been conducted which verified the essential features of these predictions, and which have underscored important materials issues. It is of considerable importance to extend the model to account more accurately for these materials issues, to predict the response of the converter at higher modulation depths, and to explore the optical phase modulation induced in the output image and its impact on various image processing architectures. In addition, testing of the photorefractive incoherent-to-coherent optical converter in a representative optical image processing system will undoubtedly highlight significant features worthy of further study and advanced development.

7.A Appendix. Steady State and Temporal Behavior of the Space-Charge Field Components in PICOC (Simultaneous Erasure/Writing Mode)

In this Appendix, the expressions for the lowest order components of the space-charge field E_{mn} are presented, both in the steady state and with full temporal evolution. Details of the derivation of these expressions can be found in [7.22].

7.A.1 Steady State Behavior

If the light incident on the photorefractive crystal is given by

$$I(x) = I_0(1 + m_G \cos K_G x) + I_1(1 + m_S \cos K_S x) \tag{7.A1}$$

then to first order the steady state response of the E_{01}, E_{10}, and E_{11} components are

$$E_{10} = -\frac{1}{2} m_G^{\text{eff}} \frac{E_0 + iE_D(K_G)}{D(K_G)} \tag{7.A2}$$

$$E_{01} = -\frac{1}{2} m_S^{\text{eff}} \frac{E_0 + iE_D(K_S)}{D(K_S)} \tag{7.A3}$$

$$E_{11} = \frac{1}{4} m_S^{\text{eff}} m_G^{\text{eff}} [D(K_G) D(K_S) D(K_G + K_S)]^{-1}$$
$$\times \{F_1[E_0 + iE_D(K_G)] + F_2[E_0 + iE_D(K_S)]\} \tag{7.A4}$$

in which

$$m_G^{\text{eff}} = m_G \frac{S_G I_0}{S_G I_0 + S_S I_1} \tag{7.A5}$$

$$m_S^{\text{eff}} = m_S \frac{S_S I_1}{S_G I_0 + S_S I_1} \tag{7.A6}$$

$$D(K) = 1 - i\frac{E_0 + iE_D(K)}{E_q(K)} \tag{7.A7}$$

$$F_1 = D(K_G) + \frac{E_D(K_S)}{E_q(K_G)} \tag{7.A8}$$

$$F_2 = D(K_S) + \frac{E_D(K_G)}{E_q(K_S)} \quad \text{and} \tag{7.A9}$$

$$E_q(K) = \frac{eN_{D\,eq}^+}{\varepsilon\varepsilon_0 K} \tag{7.A10}$$

$$E_D(K) = \frac{kTK}{e} \tag{7.A11}$$

for which K can assume the values K_G, K_S, and $(K_G + K_S)$.

7.A.2 Temporal Response

The temporal behavior of the space-charge field components for the simultaneous erasure/writing mode (SEWM) is discussed in this section. Consider the following sequencing of light intensity profiles:

$$I(x,t) = \begin{cases} I_0(1 + m_G \cos K_G x) & \text{for } t < 0 \\ I_0(1 + m_G \cos K_G x) + I_1(1 + m_S \cos K_S x) & \text{for } t > 0 \end{cases} \tag{7.A12}$$

The temporal evolution of the various components of the space-charge field is given by

$$E_{10} = -\{(m_G - m_G^{\text{eff}})\exp[-t/T(K_G)] + m_G^{\text{eff}}\}$$
$$\times [E_0 + iE_D(K_G)]D^{-1}(K_G) \tag{7.A13}$$

$$E_{01} = -m_S\{1 - \exp[-t/T(K_S)]\}[E_0 + iE_D(K_S)]D^{-1}(K_S) \tag{7.A14}$$

$$E_{11} = M_0 + M_1 e^{-t/T_1} + M_2 e^{-t/T_2} + M_3 e^{-t/T_3} + M_4 e^{-t/T_4} \tag{7.A15}$$

in which

$$M_0 = \tfrac{1}{4}m_G^{\text{eff}}m_S^{\text{eff}}[D(K_G)D(K_S)D(K_G + K_S)]^{-1}$$
$$\times \{F_1[E_0 + iE_D(K_G)] + F_2[E_0 + iE_D(K_S)]\} \tag{7.A16}$$

$$M_1 = \frac{1}{4}(m_G - m_G^{\text{eff}})m_S^{\text{eff}}[D(K_G)D(K_S)D(K_G + K_S)]^{-1}\frac{T_1}{T_1 - T_3}$$
$$\times \left\{F_1[E_0 + iE_D(K_G)] + F_2[E_0 + iE_D(K_S)]\left(1 - \frac{T_0}{T_1}\right)\right\} \tag{7.A17}$$

$$M_2 = -\frac{1}{4}m_{\mathrm{G}}^{\mathrm{eff}}m_{\mathrm{S}}^{\mathrm{eff}}[D(K_{\mathrm{G}})D(K_{\mathrm{S}})D(K_{\mathrm{G}}+K_{\mathrm{S}})]^{-1}\frac{T_2}{T_2-T_3}$$
$$\times \left\{ F_1[E_0 + \mathrm{i}E_{\mathrm{D}}(K_{\mathrm{G}})]\left(1-\frac{T_0}{T_2}\right) + F_2[E_0 + \mathrm{i}E_{\mathrm{D}}(K_{\mathrm{S}})]\right\} \quad \text{(7.A18)}$$

$$M_4 = -\frac{1}{4}(m_{\mathrm{G}} - m_{\mathrm{G}}^{\mathrm{eff}})m_{\mathrm{S}}^{\mathrm{eff}}[D(K_{\mathrm{G}})D(K_{\mathrm{S}})D(K_{\mathrm{G}}+K_{\mathrm{S}})]^{-1}\frac{T_4}{T_4-T_3}$$
$$\times \left\{ F_1[E_0 + \mathrm{i}E_{\mathrm{D}}(K_{\mathrm{G}})]\left(1-\frac{T_0}{T_2}\right)\right.$$
$$\left. + F_2[E_0 + \mathrm{i}E_{\mathrm{D}}(K_{\mathrm{S}})]\left(1-\frac{T_0}{T_1}\right)\right\} \quad \text{(7.A19)}$$

$$M_3 = -(M_0 + M_1 + M_2 + M_4) \quad . \quad \text{(7.A20)}$$

The dielectric relaxation time constant T_0 is defined by

$$T_0 = \frac{\varepsilon\varepsilon_0}{\mu e n_0} \quad ; \quad \text{(7.A21)}$$

the time constants T_1, T_2, and T_3 are defined by

$$T(K) = \frac{T_0 C(K)}{D(K)} \quad \text{(7.A22)}$$

for $K = K_{\mathrm{G}}$, K_{S}, and $(K_{\mathrm{G}} + K_{\mathrm{S}})$ respectively; and the time constant T_4 is defined by

$$T_4 = \left(\frac{1}{T_1} + \frac{1}{T_2}\right)^{-1} \quad . \quad \text{(7.A23)}$$

In (7.A22), the factor $C(K)$ is defined by

$$C(K) = 1 - \mathrm{i}\left(\frac{E_0 + \mathrm{i}E_{\mathrm{D}}(K)}{E_{\mathrm{M}}(K)}\right) \quad . \quad \text{(7.A24)}$$

Note that the GEM temporal response can be derived from the SEWM expression by setting the average coherent grating intensity $I_0 = 0$ for $t > 0$. Thus the M_0 and M_2 terms disappear for GEM, leaving the M_1, M_4, and $M_3 = -(M_1 + M_4)$ terms. The GIM response can in turn be derived from the GEM response.

Acknowledgements. The authors thank F. Lum, D. Seery, M. Garrett, and Y. Shi for their technical assistance. This research was supported in part at the University of Southern California by the Air Force Systems Command (RADC) under Contract No. F19628-83-C-0031, the Defense Advanced Research Projects Agency (Office of Naval Research), the Joint Services Electronics Program, and the Army Research Office; and at the California Institute of Technology by the Air Force Office of Scientific Research and the Army Research Office.

References

7.1   D. Casasent: Proc. IEEE **65**, 143–157 (1977)
7.2   A.R. Tanguay, Jr.: Opt. Eng. **24**, 2–18 (1985)
7.3   B.A. Horwitz, F.J. Corbett: Opt. Eng. **17**, 353–364 (1978)
7.4   M.P. Petrov, A.V. Khomenko, M.V. Krasin'kova, V.I. Marakhonov, M.G. Shlyagin: Sov. Phys. Tech. Phys. **26**, 816–821 (1981)
7.5   Y. Owechko, A.R. Tanguay, Jr.: J. Opt. Soc. Am. **A1**, 635–652 (1984)
7.6   A.A. Kamshilin, M.P. Petrov: Sov. Tech. Phys. Lett. **6**, 144–145 (1980)
7.7   Y. Shi, D. Psaltis, A. Marrakchi, A.R. Tanguay, Jr.: Appl. Opt. **22**, 3665–3667 (1983)
7.8   A. Marrakchi, A.R. Tanguay, Jr., J. Yu, D. Psaltis: Opt. Eng. **24**, 124–131 (1985)
7.9   M.W. McCall, C.R. Petts: Opt. Commun. **53**, 7–12 (1985)
7.10   L.M. Bernardo, O.D.D. Soares: Appl. Opt. **25**, 592–593 (1986)
7.11   M.B. Klein, G.J. Dunning, G.C. Valley, R.C. Lind, T.R. O'Meara: Opt. Lett. **11**, 575–577 (1986)
7.12   M. Peltier, F. Micheron: J. Appl. Phys. **48**, 3683–3690 (1977)
7.13   D.L. Staebler, J.J. Amodei: J. Appl. Phys. **43**, 1042–1049 (1972)
7.14   R. Grousson, S. Mallick: Appl. Opt. **19**, 1762–1767 (1980)
7.15   J.P. Herriau, J.P. Huignard, P. Aubourg: Appl. Opt. **17**, 1851–1852 (1978)
7.16   M.G. Moharam, T.K. Gaylord, R. Magnusson, L. Young: J. Appl. Phys. **50**, 5642–5651 (1979)
7.17   V. Kondilenko, V. Markov, S. Odulov, M. Soskin: Opt. Acta **26**, 238–251 (1979)
7.18   N.V. Kukhtarev, V.B. Markov, S.G. Odulov, M.S. Soskin, V.L. Vinetskii: Ferroelectrics **22**, 949–960 (1979)
7.19   N.V. Kukhtarev: Sov. Tech. Phys. Lett. **2**, 438–460 (1976)
7.20   E. Ochoa, F. Vachss, L. Hesselink: J. Opt. Soc. Am. **A3**, 181–187 (1986)
7.21   Ph. Refregier, L. Solymar, H. Rajbenbach, J.P. Huignard: J. Appl. Phys. **58**, 45–57 (1985)
7.22   J. Yu, D. Psaltis, R.V. Johnson, A.R. Tanguay, Jr.: "Temporal evolution of multiple gratings in photorefractive crystals", to be published
7.23   A. Marrakchi, R.V. Johnson, A.R. Tanguay, Jr.: J. Opt. Soc. Am. **B3**, 321–336 (1986)
7.24   R. Orlowski, E. Kratzig: Solid State Commun. **27**, 1351–1354 (1978)
7.25   M.B. Klein, G.C. Valley: J. Appl. Phys. **57**, 4901–4905 (1985)
7.26   G.C. Valley: Appl. Opt. **22**, 3160–3164 (1983)
7.27   A.R. Tanguay, Jr.: "The Czochralski growth and optical properties of bismuth silicon oxide," Ph. D. dissertation (Yale University, New Haven, Conn., 1977)
7.28   G.C. Valley, M.B. Klein: Opt. Eng. **22**, 704–711 (1983)
7.29   G. Lesaux, J.C. Launay, A. Brun: Opt. Commun. **57**, 166–170 (1986)
7.30   R.A. Sprague: J. Appl. Phys. **46**, 1673–1678 (1975)
7.31   H. Kogelnik: Bell Syst. Tech. J. **48**, 2909–2947 (1969)
7.32   W.R. Klein, B.D. Cook: IEEE Trans. Sonics and Ultrason. **SU14**, 123–134 (1967)
7.33   E.I. Gordon: Appl. Opt. **5**, 1629–1639 (1966)
7.34   D. Yevick, L. Thylen: J. Opt. Soc. Am. **72**, 1084–1089 (1982)
7.35   R.V. Johnson, A.R. Tanguay, Jr.: Opt. Eng. **25**, 235–249 (1986)
7.36   F. Vachss, L. Hesselink: J. Opt. Soc. Am. **A1**, 1221 (1984)
7.37   A. Marrakchi, R.V. Johnson, A.R. Tanguay, Jr.: IEEE J. Quant. Electron. **QE-23**, 2142–2151 (1987)

8. Photorefractive Crystals in PRIZ Spatial Light Modulators

Mikhail P. Petrov and Anatolii V. Khomenko

With 12 Figures

Cubic photorefractive crystals are used as active elements in optically addressed two-dimensional spatial light modulators (SLMs). The major devices utilizing these materials are the PROM[1] [8.1–3] and PRIZ[2] [8.1, 4], which employ mainly the crystals $Bi_{12}SiO_{20}$ [8.2–4] , $Bi_{12}GeO_{20}$ [8.5, 6], and $Bi_{12}TiO_{20}$ [8.7]. In this chapter we shall discuss the PRIZ device, which exhibits a number of unique features and has high performance parameters.

8.1 Background

Optically addressed SLMs are essentially reversible photosensitive media that are used as input devices in real-time coherent optical processing systems. On these devices the 2-D information to be processed is recorded as an image in parallel (the whole image at once) or in succession (point by point). The image can be written with both coherent and incoherent light, then it is read out with coherent light to be introduced into a coherent optical processor. The SLM can thus perform information storage and incoherent-to-coherent conversion of images. The readout is followed by image erasure and recording of a new image. A fast cycle rate (> 10 cycles/s), along with a high information capacity ($> 10^5$ pixels for every image), provides coherent optical information processing systems with a high throughput.

The readout light in the SLMs is modulated through the electro-optic effect, which involves changes of the refractive indices of the crystal induced by the electric field. All the photorefractive crystals listed above belong to the crystal class 23, which allows the linear electro-optic effect. This means that the refractive indices show a linear dependence on the electric field.

With no applied field, the cubic crystal is optically isotropic. However, if we cut out a sample from the crystal, deposit electrodes on its opposite faces, and apply a voltage to them, the crystal becomes birefringent (optically uniaxial or biaxial). The orientation of the principal optical axes and also the magnitude of the refractive index changes is dependent on the electric field direction in the crystal (to be discussed in more detail in Sect. 8.2). Light with an arbitrary linear polarization can be regarded as two orthogonally polarized

---

[1] PROM is an abbreviated form of Pockels Readout Optical Modulator.

[2] PRIZ is a Russian acronym that translates as Image Transformer.

waves (eigenmodes), which propagate in birefrigent media with different velocities. These beams acquire different incremental phase shifts as they travel through the crystal, and the light beam resulting from their interference behind the crystal has a polarization different from that of the incident beam. By varying the voltage applied to the electrodes, we can therefore modulate the polarization state of light emerging from the crystal. This modulation can then be converted into combined amplitude and phase modulation, if polarization analyzer is placed behind the crystal in the beam path.

Thus electro-optic crystals with a pair of electrodes can perform single-channel temporal modulation of a light beam. So as to accomplish space-time (multichannel) light modulation, it is necessary to induce a nonuniform electric field that can be varied independently in every part of the crystal. For this purpose diverse techniques are used in electro-optic SLMs. For instance, in the TITUS device, an electron-beam charge deposition on the surface is employed to produce a nonuniform electric field within the crystal [8.8]. In other SLMs, photoconductive layers [8.8] that serve as photosensitive media or microchannel electron amplifiers [8.9] are utilized. In the photorefractive SLMs to be discussed here the charge is produced within the electro-optic crystal bulk due to its photoconductivity. The photorefractive crystals in these SLMs therefore play the role of both a photosensitive element and a modulating medium.

Figure 8.1 shows a schematic of a photorefractive SLM PRIZ. In the simplest case, it consists of a photorefractive crystal layer, which is typically 300–700 μm thick, and two transparent electrodes. In another version of the PRIZ, as well as in the PROM, electrodes are isolated from the crystal surface by dielectric layers. Images are written with light from the blue-green region, which produces impurity conductivity in photorefractive crystals by exciting electrons from donor levels into the conduction band. In the process of image writing,

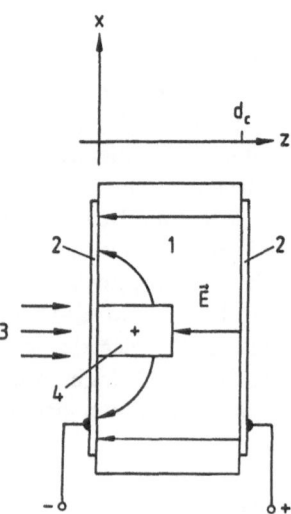

Fig. 8.1. Structure of the PRIZ SLM. *(1)* photorefractive crystal, *(2)* transparent electrodes, *(3)* write light, *(4)* photoinduced charge

326

an external electric field is produced by applying a voltage of 1.5–2.5 kV between electrodes. The excited electrons drift in the applied external field and leave the region near the negative electrode where the uncompensated positive charge remains. The images are read out with polarized light from yellow, red, or infrared regions, since the photorefractive crystals involved have a low sensitivity at these wavelengths. This prevents rapid erasure of the recorded images during readout.

8.2 Basic Parameters of Spatial Light Modulators

In order to provide an adequate framework for further discussion, we consider in this section how the basic performance parameters of SLMs, such as transfer function, sensitivity to write light, speed, inherent noise, and phase distortions, can be defined.

Since the SLM, as any other photosensitive medium, has a limited resolution, its response, i.e., the change of the complex amplitude transmittance T, for a delta function input image $I(x,y) = I_0 \delta(x,y)$ (where $I(x,y)$ is the image intensity) will not be localized in an infinitesimal region. The function $h(x,y)$ that represents the SLM response to a 2-D delta function input image is called the impulse response. If an arbitrary image $I(x,y)$ is written on the modulator and the modulator is treated as linear space invarient system, then its response can be described by

$$T(x,y) = \iint I(x',y')h(x-x',y-y')dx'\,dy' \tag{8.1}$$

where $T(x,y)$ is the amplitude transmittance of the SLM. Generally, it is a complex function. In practice, however, the transfer function $\kappa(\nu,\xi)$, which is the Fourier-transform of $h(x,y)$, is more often used (where ν and ξ are spatial frequencies)

$$\kappa(\nu,\xi) = \iint h(x,y)\exp[2\pi i(\nu x + \xi y)]dx\,dy \quad . \tag{8.2}$$

Using the function $\kappa(\nu,\xi)$ we can calculate the Fourier spectrum of the SLM response

$$\tilde{T}(\nu,\xi) = \kappa(\nu,\xi)\tilde{I}(\nu,\xi) \tag{8.3}$$

where $\tilde{T}(\nu,\xi)$ and $\tilde{I}(\nu,\xi)$ are the Fourier transforms of $T(x,y)$ and $I(x,y)$, respectively.

The transfer function can be defined experimentally using the holographic technique. For this purpose, an image of a sinusoidal grating is produced by an interferometer in coherent write light and written on the modulator

$$I(x,y) = I_0[1 + m \sin 2\pi(\nu x + \xi y)] \tag{8.4}$$

where m is the modulation coefficient of the recorded grating, $m \leq 1$. The readout light amplitude behind a thin modulator or polarization analyzer that is placed just after the modulator can be written as

$$a(x,y) = a_0 T(x,y)$$
$$= a_0 I_0 [\kappa(0,0) + m\kappa(\nu,\xi) \sin 2\pi(\nu x + \xi y)] \tag{8.5}$$

where $a_0 I_0$ is the complex amplitude of the readout light that uniformly illuminates the device. The readout light diffracts from the recorded grating and the first-order intensity $I_1(\nu,\xi)$ is measured experimentally. The SLM can be regarded as a thin hologram [8.10] whose intensity can be calculated using the Fourier expansion of (8.5):

$$I_1(\nu,\xi) = \left| \frac{a_0 I_0 m\kappa(\nu,\xi)}{2} \right|^2 . \tag{8.6}$$

The diffraction efficiency η is determined as the ratio between the diffraction intensity $I_1(\nu,\xi)$ and the intensity of the incident readout light I_R, $\eta = I_1(\nu,\xi)/I_R$. The maximum achievable diffraction efficiency η_{max} for a specific SLM characterizes the readout light utilization. It follows from (8.6) that by measuring η during recording of gratings with different ν and ξ, but with the same amplitude $I_0 m$, we can define the modulus of the transfer function to within a known constant multiplier

$$|\kappa(\nu,\xi)| = \frac{2}{I_0 m} \sqrt{\eta(\nu,\xi)} . \tag{8.7}$$

Using the transfer function, we can determine one of the SLMs vital parameters, namely, the resolution or the reconstructed spatial frequency band. The literature does not give a single definition for this parameter. We assume that the modulator efficiently reproduces the spatial frequency band for which $\eta(\nu,\xi) \geq 0.1\eta_{max}$. This corresponds to an approximately 3-fold decrease of the transfer function amplitude at boundaries of this band.

The SLM sensitivity can be defined in a general case as the magnitude of the control signal at which the output signal attains a given value. Since SLMs are intended primarily for coherent optical information processing systems, the output response should be taken as a change of the complex amplitude transmittance of the device. So as to determine the SLM sensitivity, the control signal for the optically addressed SLMs discussed here is represented by the exposure value, i.e. the density of the write light energy needed to record a test image. Sensitivity is most readily measured in holographic experiments. We shall assume further that the sensitivity S is equal to the exposure per cm^2 at which $\eta = 1\%$. By setting $\sqrt{\eta} = 0.1$ in (8.7) and taking into account that the average exposure is $I_0 t$ (where t is the exposure time), we obtain

$$S = \frac{0.2}{m|\kappa(\nu,\xi)|} . \tag{8.8}$$

Note that the magnitude of S decreases with increasing sensitivity of the SLM to write light and is different at different spatial frequencies. So as to eliminate this uncertainty, we assume that the sensitivity is measured at the spatial frequency where $\kappa(\nu, \xi)$ peaks.

It should be noted that the definition of sensitivity given above is a convenient and frequently used one, but is not the only one possible. According to one of the most fundamental definitions, the SLM sensitivity is the light energy needed to record one bit of information. However, because of difficulties encountered in defining the information capacity of the modulator, the experimental determination of this value is complicated.

Speed of operation is another vital parameter of SLMs. In our case it is defined as the maximum possible number of write-read-erase cycles per second.

The list of SLM parameters, as well as those of any other analogue device, is incomplete if it does not include noise characteristics and dynamic range. In addition, the readout light wavefront phase distortions are essential for such SLMs.

Definition of the information capacity of the modulator employed in a coherent optical system is a fairly involved problem. In an incoherent case, the information capacity of the device can be evaluated using the relation

$$C \simeq \frac{S_{\mathrm{ap}}}{S_{\mathrm{pix}}} \log_2 \left(\frac{I_{\mathrm{ms}}}{I_{\mathrm{n}}} + 1 \right) \qquad (8.9\mathrm{a})$$

where S_{ap} is the area of the modulator, S_{pix} is the area of one pixel, and $I_{\mathrm{ms}}/I_{\mathrm{n}}$ is the maximum ratio of the signal intensity to noise for one pixel.

In a coherent case, phase distortions and additional coherent noise should be taken into account. Phase distortions result in broadening of the diffraction spot in the Fourier plane of a coherent system. Because of difficulties involved in quantitative experimental measurement of noise, these data are not usually available. For a very rough estimate of the information capacity of the modulator we can use the ratio

$$C \simeq \frac{S_{\mathrm{FP}}}{S_{\mathrm{d}}} \qquad (8.9\mathrm{b})$$

where S_{d} is the area of the diffraction spot, and S_{FP} is the area of the Fourier plane where the signal-to-noise ratio is more than 1, when a spread band spectrum signal (but not a sinusoidal grating) is recorded.

Equations (8.9a, b) can both underestimate and overestimate the values of C. However, as a rule, C for a coherent system turns out to be well below that for an incoherent one.

The nonlinear distortions introduced by SLMs arise from the recording and readout mechanisms. For many applications, the modulator's nonlinearity has a deleterious effect. Moreover, parameters such as resolution, sensitivity, and diffraction efficiency can be correctly determined only for the linear case. Real SLMs can be treated as linear systems only within certain limits. Knowledge of nonlinear distortions is therefore important, among other things, for unambiguous evaluation of the parameters listed above. The degree of nonlinear distortion can be measured in holographic experiments involving recording a sinusoidal grating on the modulator. If nonlinear distortions take place, not only the first, but also higher diffraction orders are observed. The ratio between the most intense higher order and the first diffraction order is assumed to characterize the nonlinearity. As a rule, among higher orders the second I_2 is most intense. In this case the corresponding ratio is called the second-harmonic distortion (SHD)

$$\text{SHD} = I_2/I_1 \quad . \tag{8.10}$$

8.3 Anisotropy of the Transfer Function and Diffraction Efficiency

In this section we consider the readout of images from a photorefractive SLM and define the symmetry properties of the transfer function of this device in terms of properties of the electro-optic effect exhibited by cubic crystals.

For discussing the refractive properties of anisotropic media, one may use a so-called index ellipsoid, i.e., a characteristic surface of the dielectric impermeability tensor [8.11]. In a general case, if an electric field is applied to cubic crystals, the index ellipsoid is deformed from a sphere into an ellipsoid. The lengths of the axes of the index ellipsoid cross-section passing through the origin and perpendicular to the light propagation direction are proportional to new refractive indices n_1 and n_2 for the wave components polarized along the principal axes. In the following discussion we shall assume that the readout light is a plane wave incident normal to the active surface of the modulator. The orientation of the index ellipsoid cross-section will therefore coincide with that of the crystal layer.

The index ellipsoid depends on the magnitude and direction of field E and its cross-section of interest depends on the light propagation direction k in the crystal. For these reasons, the electro-optic effect is traditionally classified into two types. The longitudinal electro-optic effect results from the electric field E or its components E_l parallel to k. Correspondingly, the transverse field or the field components E_t that are projections of E on the plane orthogonal to k contribute to the transverse electro-optic effect.

Let us consider three basic orientations of the active surface of the photorefractive layer, namely (100), (110), and (111). The (100) orientation results in the longitudinal effect and is utilized in the PROM [8.2], whereas two latter orientations ensure that the transverse effect is observed and are used in the PRIZ. Since the direction of the electric field longitudinal components is independent of the orientation of the recorded grating wave vector in the crystal layer plane, it is apparent that the transfer function of the PROM is isotropic. For the (100) orientation, the index ellipsoid principal cross-section with coordinates U_1 and U_2 can be described by

$$\left[\frac{1}{n_0^2} - \frac{1}{2}r_{41}E_l(x,y,z)\right]U_1^2 + \left[\frac{1}{n_0^2} + \frac{1}{2}r_{41}E_l(x,y,z)\right]U_2^2 = 1 \tag{8.11}$$

where n_0 is the refractive index of the crystal with no applied field and r_{41} is the electro-optic coefficient of the cubic crystal, in this case the principal axes of the index ellipsoid being along the [010] and [001] axes. Taking into account that $r_{41}E_1(x,y,z) \ll n_0$ in all realizable situations, we can write from (8.11) that

$$n_1(x,y,z) = n_0 + \tfrac{1}{2}n_0^2 r_{41}E(x,y,z) \quad ;$$
$$n_2(x,y,z) = n_0 - \tfrac{1}{2}n_0^3 r_{41}E(x,y,z) \quad . \tag{8.12}$$

When emerging from the crystal, the two light wave components polarized along the principal axes have phase shifts $\varphi_1(x,y)$ and $\varphi_2(x,y)$ induced by the longitudinal electro-optic effect

$$
\begin{aligned}
\varphi_{1,2}(x,y) &= \frac{2\pi}{\lambda}\int\limits_0^{d_c}[n_{1,2}(x,y,z) - n_0]dz \\
&= \pm\frac{\pi}{\lambda}n_0^3 r_{41}\int\limits_0^{d_c}E_1(x,y,z)dz \\
&= \pm\frac{\pi n_0^3 r_{41}U(x,y)}{\lambda d_c}
\end{aligned}
\tag{8.13}
$$

where $U(x,y)$ is the potential difference between the crystal layer surfaces. Interference of these orthogonal waves outside the crystal yields a wave with a modulated state of polarization. We shall consider below how the phase modulation of the readout light eignemodes within the crystal can be related to the polarization modulation outside it (8.22). Equation (8.13) indicates that, if the (100) cut is used, at least one of the electrodes must be isolated from the crystal surface by a dielectric layer, otherwise both surfaces of the crystal turn out to be equipotential, i.e. $U(x,y) = $ const. and spatial light modulation does not occur.

For the (111) cut crystal and an arbitrary orientation of the electric field $E(x,y,z)$ with respect to the crystal axes, the index ellipsoid cross-section can be written in the form

$$\left[\frac{1}{n_0^2} - \sqrt{\frac{2}{3}}r_{41}E_t(x,y,z)\right]U_1^2 + \left[\frac{1}{n_0^2} + \sqrt{\frac{2}{3}}r_{41}E_t(x,y,z)\right]U_2^2 = 1 \tag{8.14}$$

where $E_t(x,y,z) = |E_t(x,y,z)|$ is the length of projection of the electric field vector in the crystal $E(x,y,z)$ on the (111) plane. As follows from (8.14), the index ellipsoid cross-section is independent of the electric field longitudinal components $E_1(x,y,z)$. Since the external field E_{ex} produced by the voltage applied to electrodes of a SLM is longitudinal, it does not affect the crystal birefringence. The transverse components of the total field in the crystal bulk are determined by the photoinduced charge. If a sinusoidal grating given by (8.4) is recorded on the modulator, $E_t(x,y,z) \parallel K$, where K is the wave vector of the recorded grating. It is apparent from (8.14) that the lengths of the principal axes of the index ellipsoid cross-section are independent of the direction of $E_t(x, y, z)$, but the orientation of these axes OU_1 and OU_2 does depend on it. It can be shown [8.12] that the angle between the cross-section axis OU_1 and the crystal axis [110], ψ, and the angle γ between the direction of the vector $E_t(x, y, z)$ and the same axis [110] (see Fig. 8.2) are related by

$$2\psi + \gamma = \frac{\pi}{2} \quad . \tag{8.15}$$

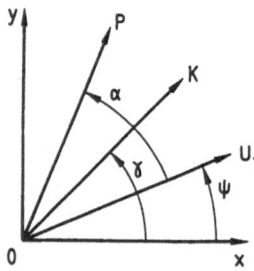

Fig. 8.2. Coordinate system for the transverse electro-optic effect in a cubic crystal. (OU_1) an induced principal axis, (K) wave vector of the recorded grating, (P) polarization plane of the readout light before the crystal. The x-axis coincides with the [110] crystal axis for the (111) orientation, and with the [100] axis for the (110) orientation

Thus, if the vector $E_t(x, y, z)$ is rotated, the index ellipsoid cross-section is "rotated" half as fast without changing its shape. For instance, if the direction of $E_t(x, y, z)$ is reversed, the "fast" and "slow" index ellipsoid axes will interchange their positions. This means that the phase modulation becomes of opposite sign, which is consistent with the linearity of the electro-optic effect.

Similarly, for the (110) orientation, the expression for the index ellipsoid cross-section has the form [8.13]

$$\left[\frac{1}{n_0^2} - E_t(x, y, z)r_{41}(\sin 2\psi \sin \gamma + \sin^2 \psi \cos \gamma)\right]U_1^2$$

$$+ \left[\frac{1}{n_0^2} - E_t(x, y, z)r_{41}(\cos^2 \psi \cos \gamma - \sin 2\psi \sin \gamma)\right]U_2^2 = 1 \tag{8.16}$$

where the angles ψ and γ are read off from the [001] axis (Fig. 8.2) and are related by

$$\psi = -\operatorname{arctg} 2 \operatorname{tg} \gamma \quad . \tag{8.17}$$

Equations (8.16, 17) show that for the (110) crystal orientation, the direction of $E_t(x, y, z)$ affects not only the orientation of the index ellipsoid cross-section principal axes, but also their lengths, i.e. the refractive indices n_1 and n_2.

For the (111) orientation, n_1 and n_2 are affected by both the transverse and longitudinal components of the electric field [8.14]. In a general case, in order to take into account the contribution of the longitudinal components of the electric field, two more terms should be inserted into (8.16):

$$U_1^2\sqrt{\tfrac{1}{3}}r_{41}E_l(x, y, z) + \sqrt{\tfrac{1}{3}}r_{41}E_l(x, y, z)U_2^2 \quad . \tag{8.18}$$

However, here and in the following we omit these terms, since the (111) PRIZ we discuss has no dielectric layers isolating electrodes from the crystal surface.

Therefore, as noted above, the longitudinal field does not contribute to the spatial light modulation and merely leads to a uniform phase shift proportional to the potential difference between the electrodes.

The crystal effect on the readout beam can be represented as phase modulation of the light wave components polarized along the principal axes. Since on recording a sinusoidal grating the direction of the transverse component of the internal electric field $E_t(x,y,z)$ varies with x and y but not with z, the index ellipsoid axes orientation is the same along the z axis. Therefore, if we consider the crystal to be sufficiently thin, the phase shifts induced by the transverse electro-optic effect can be written

$$\varphi_{1,2}(x,y) = \frac{2\pi}{\lambda} \int_0^{d_c} [n_{1,2}(x,y,z) - n_0]dz$$

$$= \frac{2\pi}{\lambda} B_{1,2} \int_0^{d_c} E_t(x,y,z)dz \quad . \tag{8.19}$$

For the (111) cut, using the expression for the index ellipsoid cross-section (8.14) and assuming that the electro-optic effect induced increments of the refractive index are small, we obtain

$$B_1 = \sqrt{\tfrac{2}{3}} n_0^2 r_{41} \quad ; \quad B_2 = -\sqrt{\tfrac{2}{3}} n_0^3 r_{41} \quad . \tag{8.20}$$

Similarly, from (8.16), for the (110) orientation;

$$B_1 = n_0^3 r_{41}(\sin 2\psi \, \sin \gamma + \sin^2 \psi \, \cos \gamma) \quad ;$$
$$B_2 = n_0^3 r_{41}(\cos^2 \psi \, \cos \gamma - \sin 2\psi \, \sin \gamma) \quad . \tag{8.21}$$

Let us denote the amplitude of one of the readout light eigenmodes near the input crystal face by a_1 and the other by a_2. Then the light amplitude represented by a complex 2-D vector immediately at the crystal output face is

$$\begin{vmatrix} a_1'(x,y) \\ a_2'(x,y) \end{vmatrix} = \hat{T}(x,y) \begin{vmatrix} a_1 \\ a_2 \end{vmatrix} = \begin{vmatrix} a_1 \, e^{i\varphi_1(x,y)} \\ a_2 \, e^{i\varphi_2(x,y)} \end{vmatrix} \quad ;$$

$$\hat{T}(x,y) = \begin{vmatrix} e^{i\varphi_1(x,y)} & , & 0 \\ 0 & , & e^{i\varphi_2(x,y)} \end{vmatrix} \tag{8.22}$$

where $\hat{T}$ is a full Jones matrix for the electro-optical crystal [8.15]. If the image recording mechanism is considered to be linear, i.e. if the internal field $E(x,y,z)$ is linearly related to the readout light intensity, then

$$\varphi_{1,2}(x,y) = \varphi_{1,2}(0,0) + \varphi_{1,2}(\nu,\xi) \sin 2\pi(\nu x + \xi y) \quad . \tag{8.23}$$

However, (8.22) indicates a nonlinear relationship between the light amplitude at the SLM output and the phase shifts $\varphi_1(x,y)$ and $\varphi_2(x,y)$. Consequently, not only the first, but also higher diffraction orders of the readout light will be observed, in spite of the linearity of the recording mechanism. The light amplitude in the nth diffraction order can be obtained by expanding (8.22) in a Fourier series after substituting (8.23) into it

$$\begin{vmatrix} a_{1n} \\ a_{2n} \end{vmatrix} = \begin{vmatrix} a_1 J_n(\varphi_1(\nu,\xi)) e^{i\varphi_1(0,0)} \\ a_2 J_n(\varphi_2(\nu,\xi)) e^{i\varphi_2(0,0)} \end{vmatrix} \qquad (8.24)$$

where J_n is the nth order Bessel function.

If the modulator is used in combination with a polarization analyzer, the diffracted light amplitude behind the analyzer is obtainable by multiplying (8.24) by the Jones matrix of the analyzer. For linearly polarized light and a linear analyzer, the first-order diffraction efficiency is

$$\eta = \tfrac{1}{4} \sin^2 2\alpha [J_1^2(\varphi_1(\nu,\xi)) + J_1^2(\varphi_2(\nu,\xi))$$
$$- 2J_1(\varphi_1(\nu,\xi)) J_1(\varphi_2(\nu,\xi)) \cos (\varphi_1(0,0) - \varphi_2(0,0))] \qquad (8.25)$$

where α is the angle between the readout light polarization direction and the polarization of one of the light wave eigenmodes in the crystal. If readout is performed with circularly polarized light and a circular analyzer is placed behind the SLM, we get

$$\eta = \tfrac{1}{4} [J_1^2(\varphi_1(\nu,\xi)) + J_2^2(\varphi_2(\nu,\xi))$$
$$- 2J_1(\varphi_1(\nu,\xi)) J_1(\varphi_2(\nu,\xi)) \cos (\varphi_1 0,0) - \varphi_2(0,0))] \quad . \qquad (8.26)$$

Expressions (8.25, 26) (where the optical activity of photorefractive crystals is ignored) are applicable to both the longitudinal- and transverse-effect SLMs. In the latter case, as will be shown later, $\varphi_1(0,0) = \varphi_2(0,0) = 0$.

Equation (8.25), with allowance made for (8.15), (8.17) and (8.20) predicts that for linearly polarized readout light the diffraction efficiency of the PRIZ depends on the direction of the recorded grating wave vector ($2\pi\nu$ and $2\pi\xi$ are the components of this vector), whereas (8.26) indicates that this is not the case for circularly polarized light. These conclusions are consistent with the experimental data plotted in Fig. 8.3 [8.13]. It is seen from the figure that, if linearly polarized light is used, the anisotropy of $\eta(\nu,\xi)$ results in suppression of Fourier components of the image whose wave vectors are oriented in certain directions, i.e. a so-called directional image filtering occurs. Inasmuch as in the

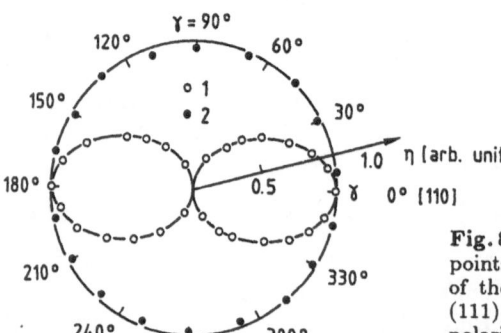

Fig. 8.3. Theoretical curves and experimental points for diffraction efficiency η as a function of the grating wave vector direction for the (111) orientation of the crystal. (1) linearly polarized readout light, (2) circularly polarized light

PRIZ directional filtering is a consequence of the transfer function anisotropy, it can be performed in both coherent and incoherent light.

So far we have discussed only the three main orientations of cubic photorefractive crystals, namely, (100), (110), and (111). For an arbitrarily oriented crystal plate, both the longitudinal and transverse electro-optic effects are observed at the same time. However, as noted above, if electrodes are deposited directly on the crystal surface, spatial light modulation via the longitudinal effect is impossible. In the SLMs with dielectric layers the readout light is modulated through both varieties of the electro-optic effect.

8.4 Image Recording

During image recording in photorefractive SLMs, a photoinduced electric charge is produced within the crystal. It has been shown elsewhere [8.1, 4, 16–18] that vital parameters of SLMs, such as sensitivity, resolution, and diffraction efficiency, depend not only on the charge density, but also on how the charge is distributed within the crystal bulk. In particular, this determines which of the electro-optic effects, i.e. longitudinal or transverse, is the most efficient for achieving the maximum light modulation amplitude. To illustrate this, let us suppose that during recording both the positive and negative charges are formed within the crystal bulk, so that the crystal as a whole is electrically neutral, the positive and negative charge being strictly symmetrically distributed across the crystal thickness (along the z axis in Fig. 8.1) about the crystal center. In this case, the light cannot be modulated through the transverse effect, because the transverse components of the field at positions symmetric about the crystal center (at $d_c/2 + \Delta z$ and $d_c/2 - \Delta z$) are equal in moduli and opposite in sign. Thus, according to (8.19) $\varphi_1(x,y) = \varphi_2(x,y) \equiv 0$. However, the longitudinal components are of the same sign at these positions and, in principle, if the crystal is properly oriented, light can be modulated through the longitudinal electro-optic effect. On the other hand, if charge of one sign only is induced in the crystal and is localized symmetrically across the crystal thickness, modulation proceeds via the transverse effect alone.

References [8.19–22] report on the experimental study of the photoinduced charge formation process. A special electro-optic experimental technique was employed to determine the longitudinal field distributions $E_l(z)$ at different stages of the readout light exposure. The photoinduced charge localization within the crystal bulk was then estimated.

A theoretical treatment of the image recording mechanism in photorefractive SLMs which takes into account the write-light-induced excitation of electrons from donor levels into the conduction band, their drift in an internal electric field within the crystal, and trapping and recombination on donors has been presented in [8.19–23]. The effects of injection from the transparent electrode-crystal contact and of the write light absorption depth in the crystal have been analyzed both theoretically and experimentally [8.20, 21].

It has been shown that the charge formation process can be divided into three stages. In the first stage, a positive charge with an exposure-independent thickness and a density increasing linearly with exposure is formed near the negative electrode (Fig. 8.4). This is caused by the drift of electrons excited by write light towards the positive electrode, leaving positively charged centers at sites of their excitation. Because of a blocking electrode-crystal contact, injection of electrons into the crystal does not compensate for the electrons leaving the region near the negative electrode, and eventually a positive charge accumulates there. It has been shown [8.19] that during the first stage the positive charge is formed in a crystal layer of thickness

$$z_0 = \tau_d(\tau_d\alpha - 1)^{-1}\ln\tau_d\alpha \tag{8.27}$$

where $\tau_d = \mu\tau E_0$, μ and τ are the mobility and lifetime of electrons in the conduction band and α is the write light absorption constant. Analysis of (8.27) shows that z_0 is commensurable with the least of two parameters: either with the write light absorption depth α^{-1}, or with the electron drift length in the external electric field τ_d.

The second stage is characterized by a variation of the positive charge layer thickness with exposure. This is caused by the formation of a so-called "bottle neck" in the first stage, i.e. the crystal region where the external electric field is to a great extent offset by the photoinduced charge field. Because of a small electric field, electrons begin to accumulate in the "bottle neck" to be subsequently trapped or recombined on donors, thereby compensating for the positive charge. Eventually the positive charge thickness decreases, and the "bottle neck" itself is shifted towards the negative electrode. In the second stage, the positive charge density continues to grow linearly with the write light exposure. However, as the layer thickness decreases, the total positive charge in the crystal is not growing linearly with exposure.

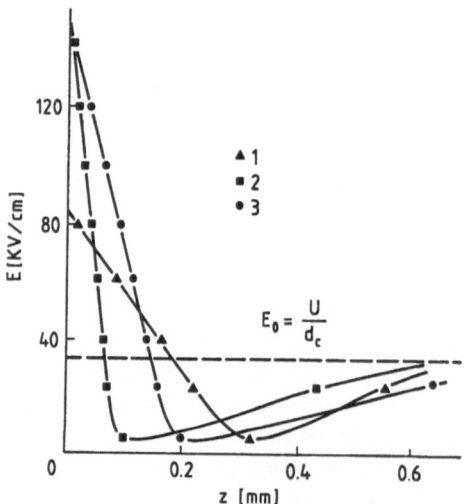

Fig. 8.4. Distribution of the longitudinal electric field at different stages of the photoinduced charge formation. (1) linear regime, (2) contraction of the charge, (3) saturation of the summary charge

The third stage, like the second one, corresponds to a nonlinear regime of image recording. Here the total positive charge ceases to increase with exposure. Hence, the electric field near the negative electrode also ceases to grow with exposure. As evidenced by experiments, this is due to the fact that in the third stage the electric field near the negative electrode becomes so large (300–500 kV/cm) that it causes a considerable increase of electron injection from the electrode into the crystal, which compensates for the photoexcited electrons leaving the region near the electrode. Thus the positive charge ceases to accumulate, which in turn means that the electric field near the electrode no longer grows. The magnitude of the SLM write light exposure at which the process of photoinduced charge formation passes to the third stage is controlled by the properties of the contact. For instance, if during charge formation the crystal is illuminated from the negative electrode side with ultraviolet light which is absorbed in a thin crystal layer several μm thick near the contact, injection will increase. In this case the third stage sets in at smaller exposures [8.23].

8.5 The Transfer Function

As shown above, the transfer function $\kappa(\nu, \xi)$ of a SLM is anisotropic, i.e. it depends on the direction of the recorded grating wave vevtor $\boldsymbol{K}$. The anisotropy arises from the properties of the transverse electro-optic effect and is independent of the wave vector length $K = 2\pi\sqrt{\nu^2 + \xi^2}$. In this case $\kappa(\nu, \xi)$ can be represented in polar coordinates as a function with separable variables

$$\kappa(K, \psi) = \kappa_{\mathrm{I}}(K)\kappa_{\mathrm{A}}(\psi) \tag{8.28}$$

where ψ is the angle governing the wave vector direction, and $\kappa_{\mathrm{I}}(K)$ and $\kappa_{\mathrm{A}}(\psi)$ are the isotropic and anisotropic multipliers of the transfer function, respectively. Here we shall consider the dependence $\kappa_{\mathrm{I}}(K)$ which for a given model of the charge distribution in the SLM structure can be derived from a solution of the electrostatic problem.

Studies of the photoinduced charge distribution showed that image recording in photorefractive SLMs occurs essentially as a volume layer of positive charge. In the linear regime this layer can be several hundred μm thick and is located near the negative electrode. In the remaining part of the crystal, the negative charge can accumulate under certain conditions. However, for the PRIZ in the linear regime of recording, the total positive charge is usually considerably larger than the negative one, and thus we can justifiably neglect the readout light modulation via the negative charge field. In the theoretical calculation of $\kappa_{\mathrm{I}}(K)$ [8.4] only the positive charge was taken into account and it was assumed that the charge layer thickness is independent of the local exposure of SLM to write light and the charge density is unchanged along the z-coordinate up to the layer boundary and is zero outside the layer. This model for charge distribution is consistent with the experimental evidence for the linear regime of image recording corresponding to small values of the write light exposure.

Earlier, this model was successfully used to calculate the transfer function of the photorefractive PROM device [8.16]. For the PRIZ SLM that has a blocking layer between the negative electrode and the crystal, we obtain [8.1]

$$I^{(K)} = \frac{\sigma_K}{K}[1 - C(K)l]$$

$$C(K)$$
$$= \frac{\varepsilon_d[\cosh K d_c + \cosh K z_0 - \cosh K(d_c - z_0) - 1] + \varepsilon_c \tanh K d_d \cosh K z_0}{K z_0(\varepsilon_d \tanh K d_c + \varepsilon_c \tanh K d_d)\cosh K d_c}$$

$$(8.29)$$

where σ_K is the modulation amplitude of the photoinduced surface charge density on recording a sinusoidal grating, ε_d is the relative dielectric permeability of the dielectric layer, d_d and z_0 are thicknesses of the dielectric layer and the positive charge layer, respectively. For the modulator with no dielectric layer, the dependence $\kappa_I(K)$ can be obtained from (8.29) by taking the limit $d_d \to 0$. In this connection it should be noted that, even in the case when the electrode is deposited directly on the crystal surface, the electrode-crystal contact is blocking. This feature is responsible for accumulation of positive charge near the negative electrode during image recording.

The analysis shows that at low spatial frequencies ($K/2\pi < 1$ line/mm) $[1 - C(K)] \sim K^2$. Hence, (8.29) predicts that at $K \to 0$ $\kappa_I(K) \to 0$ and $\kappa(0) = 0$. Thus, the PRIZ suppresses low spatial frequencies in the recorded image spectrum and does not reproduce the zero spectral component, i.e. the information on the average intensity level in the recorded image. The latter is due to the fact that no transverse components of the internal electric field are induced upon uniform illumination of a photorefractive crystal. The function $\kappa_I(K)$ peaks at spatial frequencies $K/2\pi \cong 1/d_c$ and then decreases with increasing K, with $C(K) \to 0$. As a result, at fairly high K (for parameters typical for the PRIZ and $K/2\pi > 20$ lines/mm) $\kappa_I(K) \sim 1/K$. The experimental holographic measurements of $\kappa_I(K)$ at low write light exposures, when the condition of linearity of recording is fulfilled, support these conclusions on the behaviour of the PRIZ transfer function. The transfer function anisotropy was discussed in Sect. 8.3.

8.6 Image Transformation by the PRIZ

Because of specific features exhibited by the PRIZ transfer function, the recorded images are reproduced in a transformed form. This is reflected in the name of the modulator, which is the Russian acronim that translates as "image transformer".

Let us consider at first the 1-D impulse response of a SLM, which is the modulator response to recording a light line described by the $\delta(x)$ function. In this case, a volume positive charge situated along the line orthogonal to the x axis is induced in the crystal. The transverse components of the charge field reverse their direction at $x = 0$. The PRIZ utilizes the linear electro-optic effect and if a polarization analyzer is placed behind the modulator, we can claim that reversal of the field direction leads to a change of sign of the amplitude transmittance. Hence, the 1-D impulse response $h(x)$ of the PRIZ must change its sign at $x = 0$, i.e., it is represented by an off function [8.26].

The impulse response is known to be related to the transfer function $\kappa(K)$ by the Fourier transform. Figure 8.5 shows the experimental $\kappa(K)$ curve for the PRIZ. In the same figure is also plotted the function $\mathrm{sgn}(K)$, which is, to within a phase factor, the transfer function of an ideal optical system performing a Hilbert transformation. The integral Hilbert transformation of the function $f(x)$ is determined as

$$\chi(x) = f(x) \otimes g(x) \tag{8.30}$$

where $\otimes$ denotes the convolution operation with a kernel $g(x) = 1/\pi x$. The functions $\kappa(K)$ and $\mathrm{sgn}(K)$ share an important property in that they are odd functions and $\kappa(K)$ can be considered as an approximation of the function $\mathrm{sgn}(K)$ in a certain, though very limited, spatial frequency interval. Thus we can say that the PRIZ automatically implements transformations of the Hilbert type. The physical origins of this feature are as follows:

1)   The transverse field of the charge whose density is described by a 1-D δ-function (charged thread) is $E_t \sim 1/x$, i.e., it coincides with the kernel of the Hilbert transform $g(x)$, to within a phase factor.
2)   The linearity of the electro-optic effect.

Thus, not only the PRIZ device, but also any other SLM using the linear transverse electro-optic effect will have an odd transfer function and produce image transforms of the same type.

In a 2-D case, when we consider the modulator response to a 2-D $\delta(x,y)$ function, i.e. a point, we use the polar coordinate system. The transverse component of the photoinduced charge field $E_t(r,\varphi)$ is directed in this case along radius-vectors and its length is independent of φ in a cubic photorefractive crystal. As before, when the transverse component reverses its direction, i.e.

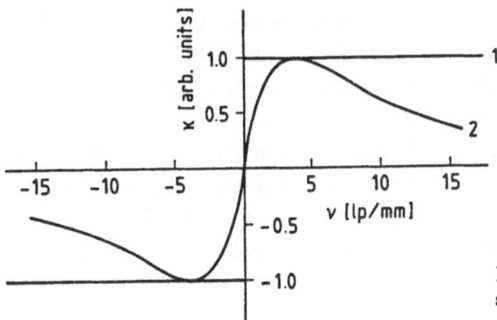

Fig. 8.5. Transfer functions of an ideal system performing the Hilbert transform (1) and of the PRIZ SLM (2)

when φ changes by π, the impulse response of the PRIZ must change its sign. Therefore, the following relation must be valid:

$$h(r,\varphi) = -h(r,\varphi + \pi) \qquad (8.31)$$

for all values of the variables r and φ. The analysis of the electro-optic properties of cubic crystals indicates that there are two major types of the PRIZ impulse response. The first occurs when linearly polarized light is used for readout and a linear polarization analyzer is placed behind the modulator. In this case, for two values of angle φ differing by π, $h(r,\varphi) = 0$ and the impulse response changes sign at these values of φ. An alternative version of the impulse response is found when readout is with circularly polarized light and a circular polarization analyzer is used. Here $h(r,\varphi)$ can be represented as $h(r,\varphi) = h(r)\,e^{i\varphi}$, if the origin for φ is properly chosen. Both versions of the impulse response meet the condition (8.31).

In order to show specific features of the transforms of 2-D images produced by the PRIZ, we shall use an analogy with the point charge field. In the 2-D case, the transverse components of the point charge field are $E_t \sim 1/r^2$. Now, for circularly polarized light and an ideal modulator $h(r,\varphi) = e^{i\varphi}/r^2$. Thus the modulator produces the integral transform [8.24]

$$
\begin{aligned}
\chi(x,y) &= I(x,y) \otimes h(x,y) \\
&= \iint \frac{I(x',y')\exp\{i\,\mathrm{arctg}\,[(x'-x)/(y'-y)]\}}{(x'-x)^2 + (y'-y)^2}\,dx'\,dy'
\end{aligned}
\qquad (8.32)
$$

of a 2-D signal $I(x,y)$ represented by the recorded image intensity.

The 2-D transfer function is related to $h(r,\varphi)$ by the Hankel transform and has the form $\kappa(K,\psi) = 2\pi i e^{-i\psi}$ in polar coordinates, i.e. it is dependent only on the polar angle ψ characterizing the direction of the recorded grating wave vector K. In any cross-section passing through the origin, the 2-D transfer function $2\pi i e^{-i\psi}$ is equal to $\mathrm{sgn}\,(K)$ to within a phase factor. In this sense, $e^{-i\psi}$ is a 2-D analog of the 1-D function $\mathrm{sgn}\,(K)$. Hence, the curves presented in Fig. 8.9 can be interpreted as cross sections of 2-D transfer functions of the PRIZ and an ideal modulator with the impulse response proportional to the point charge field. As in the 1-D case, the PRIZ transfer function differs from that of an ideal device (8.32).

It should be noted that the PRIZ implements transformation of the Hilbert type, both in coherent and incoherent write and readout light. Until recently, such transformations have been performed in optical systems through spatial filtering alone, which, as is well known, can be accomplished only in coherent light. However, the major distinguishing feature of the PRIZ is that, unlike in spatial image coherent filtering, it utilizes the input signal in the form of intensity of the image being transformed and not its amplitude. This prevents its use for visualization of phase images, whereas this problem is often solved in optics by using the Hilbert transform.

8.7 Linearity of Image Recording

It should be recalled that in optically addressed SLMs used in coherent optical information processing systems, the input signal is the write light intensity and the output signal is the amplitude transmittance. In many SLM applications it is desirable to have a linear relation between the input and output signals. In practice, however, this is fulfilled only to a certain degree of accuracy and, as a rule, nonlinear distortions of one type or other can be observed. The nonlinear relation between the input and output signals in the electro-optic modulator can arise from two factors. First, in the electro-optic effect, including the linear electro-optic effect, the readout light amplitude is nonlinearly related to the electric field induced within the crystal (Sect. 8.3). Second, the nonlinearity of the modulator can be caused by the photoinduced charge formation mechanism. The nonlinearities associated with the first factor are considered theoretically elsewhere [8.25]. From the experimental data for the PRIZ device, the authors show that the major reason for the nonlinearity of the photorefractive response of the SLM is the photoinduced charge formation mechanism [8.25].

The PRIZ nonlinearity was stuided in [8.25] using a holographic technique. A high-contrast grating $I_c(1+\cos 2\pi\nu x)$ was projected onto the modulator with light from an argon laser using an interferometer. At the same time the device was illuminated with incoherent light at the same wavelength with intensity I_n. The overall recorded image intensity was

$$I(x) = I_c(1 + m \cos 2\pi\nu x) + I_n = I_0(1 + m \cos 2\pi\nu x) \qquad (8.33)$$

where $I_0 = I_c + I_n$ and $m = I_c/(I_c + I_n)$ is the depth of modulation, with $I_0 m$ the modulation amplitude of the recorded grating intensity. Linearly polarized light from a He-Ne laser was used for readout and a polarization analyzer was placed behind the modulator. The intensity of different diffraction orders was measured as a function of the write light exposure $W = I_0 t$.

When the grating $I(x)$ was recorded, higher diffraction orders, both odd and even, were observed. The presence of higher diffraction orders indicates that the modulator's response is nonlinear. In all experiments, the second order was the most intense. Therefore, the second-harmonic nonlinear distortion factor SHD was determined as the ratio of the second to the first-order diffraction intensity. Figure 8.6 shows the dependence SHD(W) obtained for $m = 1$ at different spatial frequencies. It is apparent from the figure that the nonlinear distortion grows with increasing exposure, and at a fixed exposure, the smaller the spatial frequency, the larger the distortions.

The dependences $\eta(W)$ for the PRIZ are curves with maxima, and the smaller the frequency ν, the smaller the exposure where η peaks. Because $\eta(W)$ varies with spatial frequencies, $\eta(\nu)$ are different for different exposures W, as shown in Fig. 8.7. It is seen that, as W increases, the spatial frequency transmission band becomes broader. At the same time, according to Fig. 8.6, nonlinear distortions grow. The data listed in Table 8.1 show how the resolution ν_R determined from the curves in Fig. 8.7 and the maximum diffraction efficiency

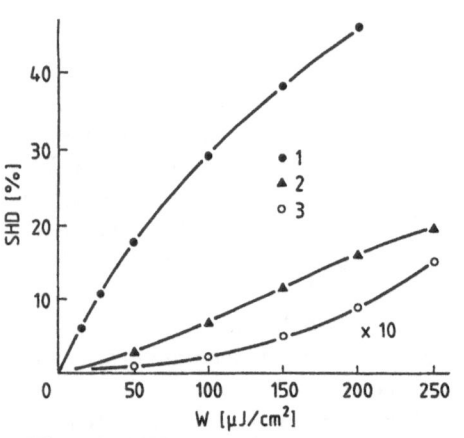

Fig. 8.6. SHD as a function of exposure energy W for different spatial frequencies. *(1)* 2 lines/mm; *(2)* 4 lines/mm; *(3)* 16 lines/mm

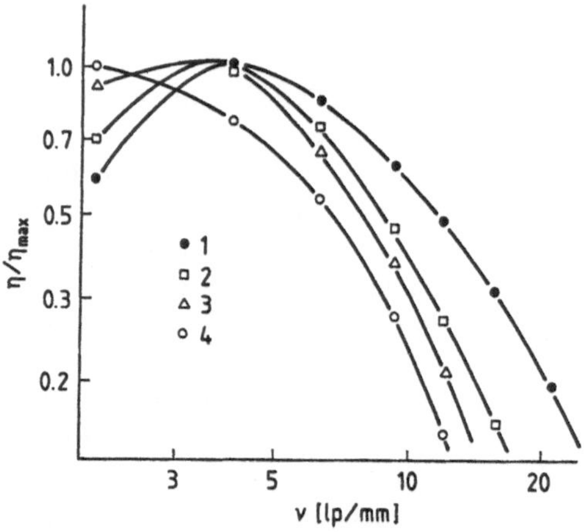

Fig. 8.7. η as function of ν for different exposure energies W. *(1)* 6, *(2)* 13, *(3)* 25 $\mu J/cm^2$

Table 8.1. Resolution and nonlinear distortions of the PRIZ SLM for different exposures

W [$\mu J/cm^2$]	η_{max}	SHD $\nu_1 = 2\,lp/mm$	SHD $\nu_2 = 4\,lp/mm$	ν_R
6	0.12	3.7	1	14
13	0.24	5.7	1.0	16
25	0.27	11.0	1.7	20
50	0.30	18.0	3.0	30

η_{max} grow with increasing nonlinear distortions (SHD was estimated for two spatial frequencies $\nu_1 = 2$ lines/mm and $\nu_2 = 4$ lines/mm).

In [8.25] SHD was calculated using the model for the photoinduced charge distribution, which suggests that not only the density but also the thickness of the photoinduced charge layer depends on the value of exposure (according to results of the charge formation studies – see Sect. 8.4). As a result, when recording is over, the layer thickness turns out to be modulated. It was shown that the transverse field of such a charge contains not only the first, but also higher harmonics, thus leading to appearance of higher diffraction orders. Furthermore, it was shown that on recording a sinusoidal grating, the maxima of the first-, second- and subsequent-order intensities are achieved one after another and SHD increases monotonically with exposure. These conclusions are supported by the experimental evidence. Figure 8.6 shows the experimental points and the theoretical curves for $\eta(W)$ for different depths m of modulation of the recorded grating and a fixed modulation amplitude mI_0. As m decreases, the diffraction efficiency of the modulator also decreases. This is due to the fact that, as the uniform background illumination I_n increases, the photoinduced charge region contracts towards the electrode faster. As a consequence, two phenomena are obseved at a fixed exposure: first, the total charge in the crystal associated with the effect of the modulated part of write light $[I_0(1 + \cos Kx)]$ decreases and second, a narrower layer enhances screening of the transverse field components by the electrode.

Variation in the shape of the $\eta(\nu)$ curves with exposure shown in Fig. 8.7 can be also attributed to changes in the space charge layer thickness. Indeed, it was shown in [8.19], where the authors studied the charge layer thickness in the PRIZ as a function of voltage applied to electrodes, that the dependence $\eta(\nu)$ changes with varying layer thickness: the thinner the layer, the higher the resolution of the modulator determined in accordance with Sect. 8.2. This is consistent with the data shown in Fig. 8.7.

The facts listed above confirm the conclusion that the major reason for the PRIZ nonlinearity is the nonlinearity of the photoinduced charge formation process during image recording. A major specific feature of the process that is responsible for the observed nonlinearities is a nonuniform narrowing of the space-charge region during write light exposure. At small exposures, deviations from linearity of the device are small. However, since not only nonlinear distortions, but also resolution and diffraction efficiency grow with increasing exposure, it is appropriate to choose the maximum possible exposure at a given level of nonlinear distortions.

It should be noted that the nonlinearity of charge formation associated with formation of a "bottle neck" modulating the charge thickness can cause some differences in parameters of the modulator depending on whether the image is recorded simultaneously or successively, point by point, using scanning device, for instance, a CRT.

8.8 Noise, Phase Distortions, Dynamic Range

When information is written and then read out from SLMs the signal-to-noise ratio inevitably becomes poorer due to noise introduced by the SLM and any other element of the optical system. SLM noise can be caused by a number of factors. Among them are such modulator defects as scratches and dust on the active surfaces and mechanical stress in the device's components, which induce birefringence via the elasto-optic effect. Moreover, the poorer signal-to-noise ratio during image recording on photorefractive SLMs can arise from an inhomogeneity of the photoelectric parameters throughout the crystal, which affect the photoinduced charge formation process. The inhomogeneities result in differing sensitivities of various parts of the SLM active area.

The PRIZ has a very low inherent noise level compared to other presently available modulators. This is due mainly to the simplicity of the PRIZ structure. It is also essential that the photorefractive crystals of the BSO type used in the modulator can be fabricated with a high optical and photoelectric quality.

Thus, the inherent noise of the PRIZ is generally lower than that of other elements of the optical systems where it is employed.

In addition to low inherent noise, the PRIZ offers a further advantage allowing one to obtain minimum noise at the optical system output. As mentioned in Sect. 8.3, the polarization of the light diffracted from the modulator is orthogonal to the initial polarization of the readout beam. The polarization analyzer transmits only the diffracted light and absorbs the light with the initial polarization, thus allowing considerable suppression of noise from the optical elements situated in the readout beam path before the SLM. In fact, typical defects of optical elements such as dust and scratches scatter light mostly without depolarization. The scattered light creates a halo around the zero order in the focal plane of a coherent optical system that decreases the signal-to-noise ratio. The amplitude of the light producing the halo is proportional to the average amplitude transmittance of a SLM, i.e. to the zero-order amplitude. If the PRIZ is used in combination with an analyzer, its average transmittance is zero on recording any image inasmuch as the PRIZ transfer function $\kappa(0,0) = 0$. In this case, the light scattered by defects is absorbed by the analyzer, thereby increasing the signal-to-noise ratio in the readout image. Fig. 8.8 demonstrates an improved signal-to-noise ratio in the optical system involving the PRIZ. It is seen that the zero spectral component in the image is suppressed and image contouring results. The photograph of the spectrum (Fig. 8.8d) shows that the signal-to-noise ratio increased, especially in the low frequency part. It should be emphasized that suppression of the zero order in the readout image can be performed by other SLMs that utilize the electro-optic effect for readout, for instance, the PROM [8.27] and PHOTOTITUS [8.8]. However, the transfer function of these devices $\kappa(0) \neq 0$, and therefore the zero component is not automatically suppressed as in the PRIZ and requires compensation of the average birefringence of the crystal, which depends on the recorded image.

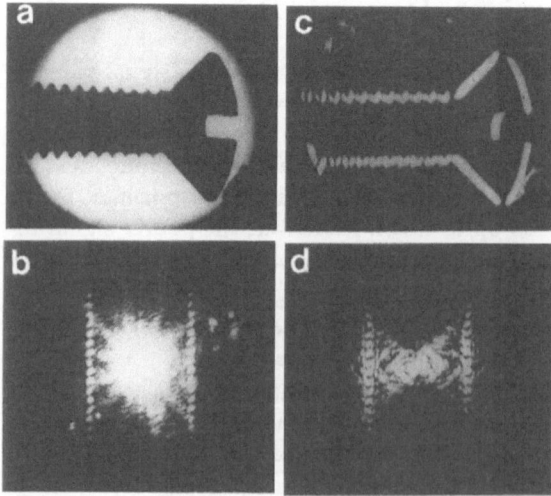

Fig. 8.8a–d. Image transformation by the PRIZ. (a) Image of the initial object; (b) power of the Fourier spectrum of the object; (c) the image read out by the PRIZ; (d) power of the spectrum that is read out from the PRIZ

The low noise level of the PRIZ and also the possibility of suppressing noise from other optical elements provides a large dynamic range at the output. For instance, the signal-to-noise ratio in the focal plane reaches 60 dB.

So as to eliminate the readout light wavefront phase distortions in the PRIZ, a photorefractive crystal plate with flat faces of large area and fairly small thickness ($\simeq 0.5$ mm) should be fabricated. Plates with an active surface up to 30 mm in diameter with phase distortions of not more than $\lambda/4$ were fabricated from the BSO crystal [8.29].

8.9 Speed of Operation

The minimum cycle time of the PRIZ is the sum of the minimum times for write light exposure, image development, readout, and erasure. If a high-power write light is used, the exposure can be as short as nanoseconds. With these short write light pulses, the image appears in the readout light with a delay, because time is needed for the charge excited within the crystal bulk to separate. This time is commensurable with the electron transit time through the crystal plate and amounts to about 10^{-4} s for BSO. The image readout time is determined by factors external with respect to the modulator, such as readout light intensity, sensitivity, and speed of image receiver. The erase light pulse duration can be short, and its energy must be an order of magnitude higher than that of the write pulse. Therefore, after the erase pulse and compensation of the internal field in the crystal take place, some time is needed for the free charge to relax, before voltage is applied to the modulator to record a new image. As evidenced by experiments, this time varies from one crystal sample to another. For the crystals studied it was in the range of one to some hundreds of milliseconds. It was shown that the modulators made from crystals with short relaxation times can be operated at a rate of at least 30 cycles/s.

8.10 Dynamic Image Selection

So far the PRIZ operation has been discussed for the cyclic operating mode, in which recording, readout, and erasure of images are performed successively. However, the PRIZ that has no insulating layers between electrodes and the photorefractive crystal can be operated in a continuous mode as well. In this case the operation is not divided into cycles, the device is operated with a fixed voltage across its electrodes, recording and readout are accomplished continuously and no erasure is performed. It has been shown that in this mode the PRIZ responds only to time-varying images or their parts [8.28, 29]. For instance, it can select the image of a moving object, while suppressing images of immobile objects. For this reason, this mode of operation is called dynamic image selection (DIS). It was first observed in studies of the PRIZ device. Later, the SLM using liquid crystals was proposed; this also exhibited the DIS feature [8.30].

Figure 8.9 gives the temporal dependence of the PRIZ response to the write light switching on and off in the DIS mode. It is seen that, when the write light goes on, the device response peaks and then decays to zero, i.e. the image disappears. However, when the light goes off, the image appears again. The readout light amplitude is of opposite sign when the write light is switched on or off, as shown in Fig. 8.9. If the write light intensity is fixed, but the image or its parts move, the modulator will respond only to the moving object, since the write light intensity will vary only in the part of the device's active surface where the image of a moving object is found at a given moment. It has been shown [8.29] that the PRIZ in the DIS mode can increase the ratio between the intensity of the moving image and that of the immobile one by a factor of 20. It was found at the same time that the PRIZ in the DIS mode should be treated as a nonlinear element, because, for instance, the magnitude of the response to a moving image depends on the intensity of a fixed background.

Let us consider briefly the major features of the physical mechanism responsible for the DIS effect. After the write light is switched on, a nonuniform positive charge is induced within the crystal. This is due to the fact that the electrode-crystal contact is a blocking one and during some time after switching on the write light the contact supplies an insufficient number of electrons into the crystal to compensate for the write-light excited electrons leaving the illuminated regions of the crystal under the action of the external electric field

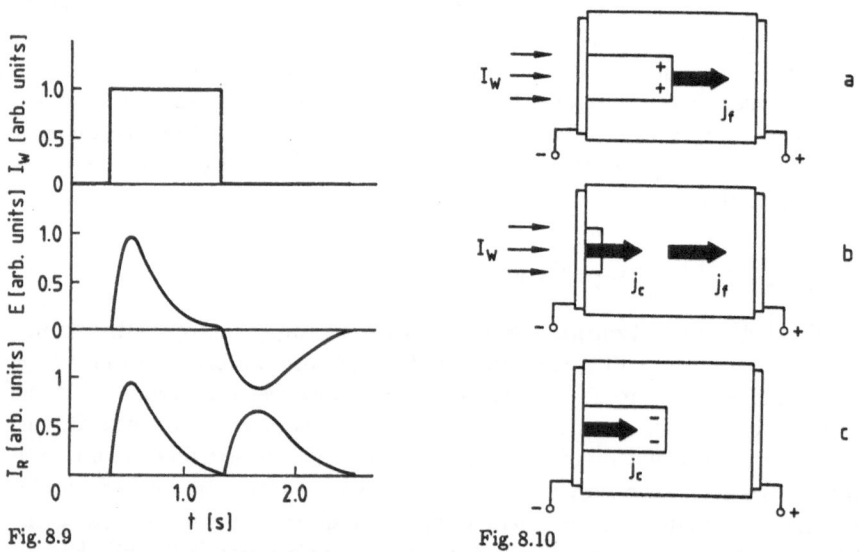

Fig. 8.9

Fig. 8.10

Fig. 8.9. PRIZ response to switching on and switching off the write light. (I_W) write light intensity, (I_R) readout light intensity, (E) readout light amplitude

Fig. 8.10a–c. Distribution of internal fields and charge in the photorefractive crytal in the DIS mode. (**a**) Response to switching on; (**b**) after a prolonged illumination with write light; (**c**) response to switching off. (j_c) electron current through the contact, (j_f) photocurrent, (I_W) write light

(Fig. 8.10a). The transverse components of the internal electric field induced by the positive charge in the crystal determine the modulator's response to the write light switching on. The positive charge layer thickness decreases and the field strength in the region near the electrode increases in the course of time (Fig. 8.10b). As noted earlier, this corresponds to the nonlinear regime of operation. The field growth results in an increase of the current through the contact, which slows down and then stops the positive charge accumulation in the crystal. Contraction of the charge layer towards the electrode accompanied by slowing down of the charge density growth causes a decrease of the readout light modulation amplitude (Fig. 8.9) by virtue of the fact that the transverse components of the charge field near the electrode turn out to be weakened to a great extent by the field of the "mirror" charge produced by the electrode. This effect is particularly pronounced on recording images with low spatial frequencies.

When the write light is switched off, generation of free electrons in the crystal stops. However, the current through the electrode-crystal contact, which is determined by the field strength in the region near the contact, does not cease on switching off the light (Fig. 8.10c). In this case, the field strength and, hence, the current through the contact prove to be stronger in the regions where the write ligth intensity was higher. Injection of electrons into the crystal leads to field strength growth within the crystal bulk, thus increasing the electron drift length in the crystal. As a result, a nonuniform negative charge is produced within the crystal bulk rather far from the electrode, and the transverse components of its field govern the readout light modulation on switching off the write light. As a negative charge accumulates, the field near the electrode decreases, thereby leading to a reduction of the current through the contact. As a consequence, accumulation of the negative charge stops. The negative charge then relaxes with time due, for instance, to the excitation of electrons from traps into the conduction band by the readout light. Experiments indicated that the response amplitude to switching off the write light appears to be different for the PRIZ devices fabricated from different BSO crystals. This may be caused by visualization of "latent" images, whose mechanism will be discussed in the next section.

It should be noted that the temporal and spatial properties of the PRIZ in the DIS mode are interrelated. For instance, the transmitted spatial frequency band is dependent on the rate with which the image to be recorded changes. Therefore, even in the linear approximation, one should introduce the transfer function $\kappa(\nu, \xi, f)$ which depends on three parameters: the spatial frequencies ν, ξ and the temporal frequency f [8.31–33]. Here f describes not the light frequency, but the frequency of the intensity variation in the recorded image. The nonstationary image must be presented in this case by superposition of running waves with differing amplitudes, frequencies and propagation directions. Then each running wave of light intensity produces a wave of crystal birefringence variation modulating the readout light.

8.11 Recording of "Latent" Images

The PRIZ offers one more way of image recording, which is unusual for SLMs and which can be called recording of "latent" images. In this mode, the image induced by the write light exposure cannot be read out immediately and requires an additional development process.

So as to realize this mode, the modulator is pre-illuminated by a uniform red light. Then the input image is focused onto the device in blue light. Voltage is not applied to electrodes during recording. It has been shown [8.34] that, though the image cannot be read out after the write light exposure is over, a "latent" pattern is stored in the crystal. For its visualization, a voltage should be applied to electrodes and, what is more, in some cases a uniform illumination of the crystal is needed. Figure 8.11 shows how the diffraction efficiency of the PRIZ utilizing BSO grows on visualization of the sinusoidal grating image. The point $t = 0$ corresponds to the moment when the write light is switched off and voltage switched on. Visualization begins from the moment the crystal is uniformly illuminated with red light ($\lambda = 633$ nm). This light is used as a readout light as well. It was shown that visualization can be stimulated not only by red, but also, e.g., by green light ($\lambda = 530$ nm).

The mechanism of latent recording can be briefly described as follows. When the crystal is illuminated with write light, electrons are excited from donor levels into the conduction band. Some are then trapped by the traps that are emptied before recording by pre-illuminating the modulator with red light. Since the external field is not applied during the write light exposure, the spatial redistribution of electrons in the conduction band can occur through diffusion alone. The efficiency of the diffusion mechanism of charge formation during image recording is, however, well below that of the drift mechanism. The diffusion-induced internal fields are small and cannot produce a notable modulation of the readout light. Thus, it can be assumed that there is no spatial charge redistribution during write light exposure. However, the latent image recording which involves the spatially nonuniform electron redistribution in energy states in the forbidden gap does occur. During visualization, the electrons that are in higher energy states are excited by light and drift in the

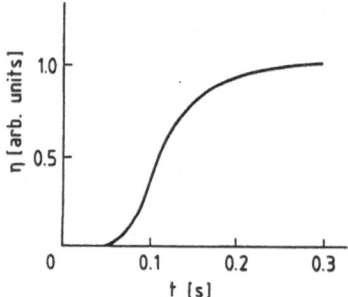

Fig. 8.11. η as a function of t on visualization of a "latent" image

external field. As a result, an uncompensated positive charge of donors whose field spatially modulates the readout light is induced.

Studies of the latent image recording revealed that the SLM samples made from different BSO crystals can have different latent image storage times, sensitivities to visualization with red and green light differing by orders of magnitude, and some samples do not require illumination for image visualization. In such samples, visualization is likely to occur through thermal excitation of electrons. This is the evidence for the appreciably differing trapping level structures in various samples of BSO.

In Sect. 8.6 it was demonstrated that nonlinear distortions during image recording in the PRIZ arise from the nonlinearity of the photoinduced charge formation process. In the latent image recording mode, charge is formed after recording is finished and can be controlled. This allows recording and reconstruction of images without nonlinear distortions over a wide range of exposures. Moreover, in certain PRIZ samples image visualization occurred more than 10 minutes after the write light exposure. This allows an appreciable increase in the information storage and recording times of the PRIZ as compared with a usual operating mode.

8.12 Photoinduced Piezoelectric Phase Modulation

Crystals of the BSO type typically employ the electro-optic effect to modulate the readout light. However, along with electro-optic properties, these crystals exhibit pronounced piezoelectric properties which can also be used to spatially modulate light. Indeed, it has been shown in [8.35] that the internal field of the photoinduced charge gives rise to deformation of the BSO crystal plate. Displacements of the plate surface associated with deformations produce a marked phase modulation of the readout light reflected from the crystal surface.

This effect was studied in experiments using the PRIZ with round crystal plates of (110) orientation. Unlike the usual technique of image readout with light passing through the crystal, in this case we observed diffraction of the light reflected from the crystal surface on recording a sinusoidal grating. Diffraction was observed only when a negative potential was applied to the electrode deposited on the reflecting surface.

It was shown that the diffraction efficiency depends on the spatial frequency ν and on the direction of the recorded grating wave vector with respect to the crystal axes. The dependences $\eta(\nu)$ are curves with peaks and with $\eta(0) = 0$ typical for the PRIZ. The position of the peak is defined by the conditions of recording in a range of 1–10 lines/mm. The maximum of $\eta(\nu)$ does not exceed 0.1 %. We can infer from this that the periodic displacement of the crystal surface is not more than 70 Å in these experiments. Figure 8.12 presents η as a function of the recorded grating wave vector direction. The symmetry of this dependence points to the fact that the crystal surface deformation is

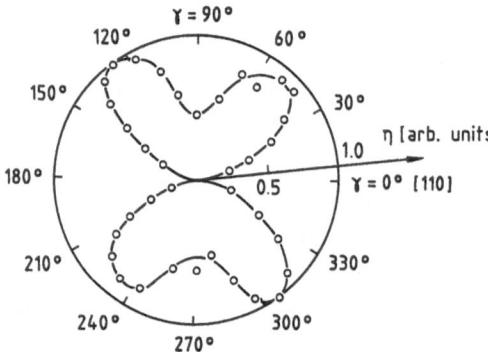

Fig. 8.12. Diffraction efficiency as a function of the grating wave vector orientation for the phase modulation through the photoinduced piezoeffect

mainly due to the transverse components of the internal electric field, since the longitudinal field would give isotropic orientational dependences.

The basic difference between the photoinduced piezoeffect and the electro-optic effect in the PRIZ is that the magnitude of displacement of a given point of the surface depends in a general case on the mechanical deformation field throughout the crystal and on conditions at its boundaries. But changes in birefringence resulting from the electro-optic effect are local, i.e. they are essentially independent of the field magnitude at other points. That is why, for instance, the orientational dependences in the two cases differ appreciably, in spite of identical tensors describing symmetry of the piezoelectric and electro-optic tensors.

8.13 Conclusions

Table 8.2 summarizes the major parameters of the PRIZ SLM made from the photorefractive BSO crystal measured in the cyclic operating mode. Besides $Bi_{12}SiO_{20}$, the PRIZ devices also utilize $Bi_{12}GeO_{20}$ and $Bi_{12}TiO_{20}$.

Table 8.2. PRIZ parameters

Write light wave lengths	400–500 nm
Sensitivity (for 1 % η at 3 lines/mm)	5 μJ/cm^2
Readout light wave lengths	550–700 nm
Resolution at SHD=1 % at SHD=10 %	15 lines/mm 30 lines/mm
Maximum diffraction efficiency	1 %
Active area (diam.) at phase distortions not more that $\lambda/4$	30 mm
Speed of operation	30 cycles/s

The estimates indicate that the parameters listed in Table 8.2 are consistent with the theoretical limiting parameters for a modulator based on crystals with a given value of the electro-optic effect and dielectric permeability. This is due to a high efficiency of the photoinduced charge formation in the BSO crystal caused by a high quantum efficiency and long drift length of electrons. Moreover, it is vital that the PRIZ utilizes the transverse electro-optic effect which provides a most efficient readout light modulation, with the volume photoinduced charge. A low noise level of the modulator is to a large extent associated with the technological effectiveness of BSO crystal growth and simplicity of the PRIZ SLM structure. For this reason, we believe that the PRIZ device is one of the most promising SLMs for use in coherent optical information processing systems.

Acknowledgements. The authors would like to thank Dr. L.I. Korovin, Dr. V.V. Bryksin, V.I. Marakhonov, M.G. Shlyagin, A.M. Bliznetsov, and Dr. M.V. Krasinkova for their participation in studies of the PRIZ SLM and Mrs. N. Nazina for her invaluable help in translating and preparing the manuscript.

References

8.1 M.P.Petrov, S.I. Stepanov, A.V. Khomenko: *Photosensitive Electrooptic Media in Holography and Optical Information Processing* (Nauka, Leningrad 1983) in Russian
8.2 J. Feinleib, D.S. Oliver: Appl. Opt. **11**, 2752–2759 (1972)
8.3 B.A. Horwitz, F.J. Corbett: Opt. Engineer. **17**, 353–364 (1978)
8.4 M.P. Petrov, A.V. Khomenko, M.V. Krasinkova, V.I. Marakhonov, M.G. Shlyagin: Zh. Tkh. Fiz. **51**, 1422–1432 (1981)
8.5 A.T. Klipko, P.E. Kotliar, E.C. Nezhevenko: Avtometria, No. 4, 34–43 (1976)
8.6 A.N. Oparin, O.I. Potaturkin, V.I. Feldbush, P.M. Shilov: Avtometria, No. 3, 57–61 (1984)
8.7 M.P. Petrov, V.I. Marakhonov, V.I. Berezkin, M.V. Krasinkova, A.V. Khomenko: Pisma v Zh. Tekh. Fiz. **11**, 260–263 (1985)
8.8 G. Marie: Ferroelectrics **10**, 9–14 (1976)
8.9 C. Warde, A.M. Weiss, A.D. Fisher: Proc. SPIE **218**, 59–66 (1980)
8.10 R.J. Collier, C.B. Burkchardt, L.H. Lin: *Optical Holography* (Academic Press, New York 1971)
8.11 A.S. Sonin, A.S. Vasilevskaya: *Electrooptic Crystals* (Atomizdat, Moscow 1971)
8.12 E.R. Mustel, V.N. Parygin: *Methods for Modulation and Scanning of Light* (Nauka, Moscow, 1970)
8.13 M.P. Petrov, A.V. Khomenko: Fiz. Tverd. Tela **23**, 1350–1356 (1981)
8.14 Y. Owechko, A.R. Tanguay, Jr.: J. Opt. Soc. Am. A1, 644–652 (1984)
8.15 F. Sherklif: *Polarized Light* (Mir, Moscow 1965)
8.16 M.P. Petrov, A.V. Khomenko, V.I. Berezkin, M.V. Krasinkova: Mikroelektronika **8**, 20–23 (1979)
8.17 M.P. Petrov, V.I. Marakhonov, A.V. Khomenko: Zh. Tekh. Fiz. **53**, 1347–1351 (1983)
8.18 Y. Owechko, A.R. Tanguay, Jr.: J. Opt. Soc. Am. A1, 635–643 (1984)
8.19 V.V. Bryksin, L.I. Korovin, V.I. Marakhonov, A.V. Khomenko: Fiz. Tverd. Tela **24**, 2978–2984 (1982)
8.20 V.V. Bryksin, L.I. Korovin, V.I. Marakhonov, A.V. Khomenko: Pisma v Zh. Tekh. Fiz. **9**, 385–390 (1983)
8.21 V.V. Bryksin, L.I. Korovin, V.I. Marakhonov, M.P. Petrov, A.V. Khomenko: Pisma v Zh. Tekh. Fiz. **9**, 1011–1015 (1983)
8.22 V.V. Bryksin, L.I. Korovin, V.I. Marakhonov: Zh. Tkh. Fiz. **53**, 1133–1138 (1983)
8.23 V.V. Bryksin, L.I. Korovin: Fiz. Tverd. Tela **25**, 55–61 (1983)
8.24 M.P. Petrov. A.V. Khomenko, V.I. Marakhonov: Opt. Commun. **50**, 296–299 (1984)

8.25 M.G. Shlyagin, A.V. Khomenko, V.V. Bryksin, L.I. Korovin, M.P. Petrov: Zh. Tekh. Fiz. **55**, 119–126 (1985)
8.26 M.P. Petrov, A.V. Khomenko: Optik **67**, 247–256 (1984)
8.27 S. Lipson, P. Nisenson: Appl. Opt. **13**, 2052–2060 (1974)
8.28 M.P. Petrov, A.V. Khomenko, V.I. Marakhonov, M.G. Shlyagin: Pisma v Zh. Tekh. Fiz. **6**, 1265–1268 (1980)
8.29 D. Casasent, F. Caimi, M.P. Petrov. A.V. Khomenko: Appl. Opt. **21**, 3846–3854 (1982)
8.30 A.A. Vasiliev, I.N. Kompanetz, A.V. Parfenov: Optik **67**, 223–236 (1984)
8.31 M.P. Petrov, A.V. Khomenko: Opt. Commun. **37**, 253–255 (1981)
8.32 M.P. Petrov: "Electrooptic Photosensitive Media for Image Recording and Processing" in *Current Trends in Optics*, Proc. ICO-12 Conf., Graz, Austria 1981, pp. 161–172
8.33 M.P. Petrov. V.A. Kapustin, A.V. Khomenko: "Processing of time-varying images" in *Advances in Scientific Photography*, **23** (Nauka, Leningrad 1985) pp. 126–131
8.34 M.P. Petrov, M.G. Shlyagin, N.O. Shalaevskii, V.M. Petrov, A.V. Khomenko: Zh. Tekh. Fiz. **55**, 2247–2250 (1985)
8.35 A.M. Bliznetsov, M.P. Petrov, A.V. Khomenko: Pisma v Zh. Tekh. Fiz. **10**, 1094–1098 (1984)

Additional References with Titles

Anderson, D.Z., Saxena, R.: "Theory of multimode operation of a unidirectional ring oscillator having photorefractive gain: weak field limit". JOSA B **4**, 164 (1987)

Arvidsson, G., Laurell, F.: "Second harmonic generation in channel waveguides fabricated by titanium diffusion in magnesium doped lithium niobate". Proc. 4th Europ. Conf. Integrated Optics (SETG, Ltd., Glasgow 1987), p. 198

Ashley, P.R., Chang, W.S.C.: "Transfer curve linearization of Bragg modulator using harmonic correction". Appl. Opt. **25**, 1368 (1986)

Astratov, V.N., Il'inskii, A.V.: "The evolution of the photoinduced space charge and electric field distribution in photorefractive sillenite ($Bi_{12}GeO_{20}$, $Bi_{12}SiO_{20}$) crystals". Ferroelectrics **75**, 251 (1987)

Au, L.B., Solymar, L.: "Space-charge field in photorefractive materials at large modulation". Opt. Lett. **13**, 660 (1988)

Belabaev, K.G., Kiseleva, I.N., Obukhovski, V.V., Odoulov, S.G., Taratouta, R.A.: "New parametric holographic type scattering of light in lithium tantalate crystals". Sov. Phys. Solid State **28**, 321 (1986)

Belic, M.R., Lax, M.: "Exact solution to the stationary holographic four-wave mixing in photorefractive crystals". Opt. Commun. **56**, 197 (1985)

Betts, G.E., Johnson, L.M.: "Experimental evaluation of drift and nonlinearities in lithium niobate interferometric modulators". Proc. SPIE **835**, 152 (1987)

Bledowski, A., Krolikowski, W., Kujawski, A.: "Multistability in reflection grating real-time holography". IEEE J. QE-**22**, 1547 (1986)

Bledowski, A., Krolikowski, W.: "Exact solution of degenerate four-wave mixing in photorefractive media". Opt. Lett. **13**, 146 (1988)

Bliznetsov, A.M., Khomenko, A.V., Kuzmin, Yu.I.: "Studies of the reciprocity law for the PRIZ SLM". Zh. Tek. Fiz. **58**, 596–600 (1988)

Bliznetsov, A.M., Bryksin, V.V., Korovin, L.I., Miridonov, S.V., Khomenko, A.V.: "Injection mechanism of the dynamic image selection for the PRIZ SLM". Zh. Tek. Fiz. **57**, 1268–1275 (1987)

de Bougrenet de la Tocnaye, J.L., Pellat-Finet, P., Huignard, J.P.: "Effect of using a $Bi_{12}SiO_{20}$ light amplifier on the formation and competition of modes in optical resonators". JOSA B **3**, 315 (1986)

Bristow, J.P.G., Wey, A.C.T., Sriram, S., Ott, D.M.: "Depolarization of single mode channel waveguides on lithium niobate". Proc. SPIE **835**, 233 (1987)

Bublyaev, R.A., Levin, V.V., Marasin, L.E., Popov, Yu.V., Kharberger, L.Yu.: "Photorefraction in lithium niobate planar waveguides". Opt. i Spektrosk. **61**, 185 (1986) [English transl.: Opt. Spectrosc. (USSR) **61**, 119 (1986)]

Buchal, Ch., Ashley, P.R., Appleton, B.R.: "Solid-phase epitaxy of ion-implanted $LiNbO_3$ for optical waveguide fabrication". J. Mater. Res. **2**, 222 (1987)

Bylsma, R.B., Bridenbaugh, P.M., Olson, D.H., Glass, A.M.: "Photorefractive properties of doped cadmium telluride". Appl. Phys. Lett. **51**, 889 (1987)

Bylsma, R.B., Olson, D.H., Glass, A.M.: "Photorefractive imaging of semiconductor wafers". Appl. Phys. Lett. **52**, 1083 (1988)

Bylsma, R.B., Glass, A.M., Olson, D.H.: "Optical signal amplification at $1.3\,\mu$m by two-wave mixing in InP:Fe". Electron Lett. **24**, 360 (1988)

Carnera, A.: "Optical waveguides in LiNbO$_3$ produced by Ti in-diffusion, ion exchange and ion implantation". *Electro-optic and Photorefractive Materials*, ed. by P. Günter, Springer Proc. Phys., Vol. 18 (Springer, Berlin, Heidelberg 1987) p. 179

Carrascosa, M., Agullo-Lopez, F.: "Erasure of holographic gratings in photorefractive materials with two active species". Appl. Opt. **27**, 2851 (1988)

Chen, R., Chang, W.S.C.: "Anomalous attenuation and depolarization scattering in y-cut LiNbO$_3$ proton exchanged waveguides". IEEE J. QE-**22**, 880 (1986)

Chen, R., Tsai, C.S.: "Thermally annealed single-mode proton exchanged channel-waveguide cutoff modulator". Opt. Lett. **11**, 546 (1986)

Cheng, L.J., Yeh, P.: "Cross-polarization beam coupling in photorefractive GaAs crystals". Opt. Lett. **13**, 50 (1988)

Choquette, K.D., McCaughan, L., Smith, W.K.: "Improved optical switching extinction in three-electrode Ti:LiNbO$_3$ directional couplers". Appl. Phys. Lett. **51**, 2097 (1987)

Cronin-Golomb, M., Fischer, B., Kwong, S.K., White, J.O., Yariv, A.: "Nondegenerate optical oscillation in a resonator formed by two phase-conjugate mirrors". Opt. Lett. **10**, 353 (1985)

Cronin-Golomb, M., Paslaski, J., Yariv, A.: "Vibration resistance, short coherence length operation and mode-locked pumping in passive phase conjugate mirrors". Appl. Phys. Lett. **47**, 1131 (1985)

Cronin-Golomb, M., Yariv, A.: "Plane wave theory of nondegenerate oscillation in the linear photorefractive passive phase conjugate mirror". Opt. Lett. **11**, 242 (1986)

Cronin-Golomb, M., Yariv, A.: "Self-induced frequency scanning and distributed Bragg reflection in semiconductor laser with phase conjugate feedback". Opt. Lett. **11**, 455 (1986)

Cronin-Golomb, M., Yariv, A.: "Semi-self-pumped phase conjugate mirrors". Opt. Lett. **12**, 714 (1987)

Donaldson, A., Wong, K.K.: "Phase-matched mode convertor in LiNbO$_3$ using near-Z-axis propagation". Electron. Lett. **23**, 1378 (1987)

Duffy, J.F., Al-Shukri, S.M., De La Rue, R.M.: "Guided-wave acousto-optic interaction on proton-exchanged lithium tantalate". Electron. Lett. **23**, 849 (1987)

Eason, R.W., Smout, A.M.C.: "Bistability and noncommutative behavior of multiple-beam self-pulsing and self-pumping in BaTiO$_3$". Opt. Lett. **12**, 51 (1987)

Eknoyan, O., Yoon, D.W., Taylor, H.F.: "Low-loss optical waveguides in lithium tantalate by vapor diffusion". Appl. Phys. Lett. **51**, 384 (1987)

Ewbank, M.D., Yeh, P.: "Frequency shift and cavity length in photorefractive resonators". Opt. Lett. **10**, 496 (1985)

Ewbank, M.D., Yeh, P., Khoshnevisan, M., Feinberg, J.: "Time reversal by an interferometer with coupled phase conjugate reflectors". Opt. Lett. **10**, 282 (1985)

Ewbank, M.D., Yeh, P., Feinberg, J.: "Photorefractive conical diffraction in BaTiO$_3$". Opt. Commun. **59**, 423 (1986)

Fabre, J.C., Jonathan, J.M.C., Roosen, G.: "43m photorefractive materials in energy transfer experiments". Opt. Commun. **65**, 257 (1988)

Fabre J.C., Jonathan, J.M.C., Roosen, G.: "Analysis of energy transfer in semi-insulating InP-Fe generated by nanosecond light pulses." Topical Meeting on Photorefractive materials effects and devices, Technical digest **17**, 101 (1987)

Fainman, A., Klancnik, E., Lee, S.H.: "Optimal coherent image amplification by two-wave coupling in photorefractive BaTiO$_3$". Opt. Eng. **25**, 228 (1986)

Fischer, B., Weiss, S.: "Solvable optimized four-wave mixing configuration with cubic photorefractive crystals". Appl. Phys. **53**, 257 (1988)

Fischer, B.: "Theory of self-frequency detuning of oscillations by wave mixing in photorefractive crystals". Opt. Lett. **11**, 236 (1986)

Fischer, B., White, J.O., Cronin-Golomb, M., Yariv, A.: "Nonlinear vectorial two-beam coupling and forward four-wave mixing in photorefractive materials". Opt. Lett. **11**, 239 (1986)

Fischer, B., Sternklar, S.: "New optical gyroscope based on the ring passive phase conjugator". Appl. Phys. Lett. **47**, 1 (1985)

Fischer, B., Sternklar, S.: "Self Bragg-matched beam steering using the double color pumped photorefractive oscillator". Appl. Phys. Lett. **51**, 74 (1987)

Foote, P.D., Hall, T.J.: "Influence of optical activity on two beam coupling constants in photorefractive $Bi_{12}SiO_{20}$". Opt. Commun. **57**, 201 (1986)

Ford, J.E., Fainman, Y., Lee, S.H.: "Time integrating interferometry using photorefractive fanout". Opt. Lett. **13**, 856 (1988)

Gan'shin, V.A., Ivanov, V.Sh., Korkishko, Yu.N., Petrova, V.Z.: "Some characteristics of ion exchange in lithium niobate crystals". Zh. Tekh. Fiz. **56**, 1354 (1986) [English transl.: Sov. Phys. Tech. Phys. **31**, 894 (1986)]

Gan, X., Ye, S., Sun, Y.: "Alternating electric field enhancement of two-wave mixing gain in photorefractive BSO". Opt. Commun. **66**, 155 (1988)

Gauthier, D.J., Narum, P., Boyd, R.W.: "Observation of deterministic chaos in a phase conjugate mirror". Phys. Rev. Lett. **58**, 1640 (1987)

Gericke, V., Hertel, P., Krätzig, E., Nisius, J.P., Sommerfeldt, R.: "Light-induced refractive index changes in $LiNbO_3$:Ti waveguides". Appl. Phys. B **44**, 155 (1987)

Gheen, G., Cheng, L.J.: "Image processing by four-wave mixing in photorefractive GaAs". Appl. Phys. Lett. **51**, 1481 (1987)

Glavas, E., Townsend, P.D., Droungas, G., Dorey, M., Wong, K.K., Allen, L.: "Optical damage resistance of ion-implanted $LiNbO_3$ waveguides". Electron. Lett. **23**, 73 (1987)

Goltz, J., Tschudi, T.: "Angular selectivity of volume holograms recorded in photorefractive crystals; an analytical treatment". Opt. Commun. **67**, 164 (1988)

Goltz, J., Laeri, F., Tschudi, T.: "Nearly degenerate four-wave mixing in photorefractive crystals, an analytical treatment". Opt. Commun. **64**, 63 (1987)

Grant, M.F., Donaldson, A., Gibson, D.R., Wale, M.: "Recent progress in lithium niobate integrated optics technology under a collaborative Joint Opto-Electronics Research Scheme (JOERS) programme". Opt. Eng. **27**, 2 (1988)

Guibelalde, E., Calvo, M.L.: "Coupled wave analysis for a reflection dephased mixed hologram grating". Opt. Quant. Electron. **18**, 213 (1986)

Günter, P.: "Electro-optical effects in ferroelectrics". Ferroelectrics **74**, 305–307 (1987)

Günter, P., Voit, E.: "Anisotropic Bragg diffraction in photorefractive crystals". Ferroelectrics **78**, 51–60 (1988)

Günter, P., Voit, E., Zha, M.Z., Albers, J.: "Self-pulsation and optical chaos in self-pumped photorefractive $BaTiO_3$". Opt. Commun. **55**, 210 (1985)

Hamel de Montchenault, G., Loiseaux, B., Huignard, J.P.: "Moving grating during erasure in photorefractive $Bi_{12}SiO_{20}$ crystals". Electron. Lett. **22**, 1030 (1986)

Hamel de Montchenault, G., Loiseaux, B., Huignard, J.P.: "Amplification of high bandwidth signals through two-wave mixing in photorefractive $Bi_{12}SiO_{20}$ crystals". Appl. Phys. Lett. **50**, 1794 (1987)

Harvey, G.T.: "The photorefractive effect in directional coupler and Mach-Zehnder $LiNbO_3$ optical modulators at a wavelength of $1.4\,\mu m$". J. Lightwave Technol. **6**, 872 (1988)

Heaton, J.M., Solymar, L.: "Transient effects during dynamic hologram formation in BSO crystals: theory and experiment". IEEE J. **QE-24**, 558 (1988)

Heismann, F., Alferness, R.C., Buhl, L.L., Eisenstein, G., Korotky, S.K., Veselka, J.J., Stulz, L.W., Burrus, C.A.: "Narrow-linewidth, electro-optically tunable $InGaAsP$-Ti:$LiNbO_3$ extended cavity laser". Appl. Phys. Lett. **51**, 164 (1987)

Herriau, J.P., Rojas, D., Huignard, J.P., Bassat, J.M., Launay, J.C.: "Highly efficient diffraction in photorefractive BSO-BGO crystals at large applied fields". Ferroelectrics **75**, 271 (1987)

Herriau, J.P., Huignard, J.P.: "Hologram fixing process at room temperature in photorefractive $Bi_{12}SiO_{20}$ crystals". Appl. Phys. Lett. **49**, 1140 (1986)

Hesselink, L., Redfield, S.: "Photorefractive holographic recording in strontium barium niobate fibers". Opt. Lett. **13**, 877 (1988)

Imbert, B., Rajbenbach, H., Mallick, S., Herriau, J.P., Huignard, J.P.: "High photorefractive gain in two-beam coupling with moving fringes in GaAs:Cr crystals". Opt. Lett. **13**, 327 (1988)

Ja, Y.H.: "High-order finite-element method to solve the nonlinear coupled-wave equations for degenerate two-wave and four-wave mixing". J. Mod. Opt. **35**, 253 (1988)

Jagannath, H., Venkateswarlu, P., George, M.C.: "Optical bistability in self-pumped phase conjugate ring resonators". Opt. Lett. **12**, 1032 (1987)

Jani, M.G., Halliburton, L.E.: "Light-induced migration of charge in photorefractive $Bi_{12}SiO_{20}$ and $Bi_{12}GeO_{20}$ crystals". J. Appl. Phys. **64**, 2022 (1988)

Jiang, J.P., Feinberg, J.: "Dancing modes and frequency shifts in a phase conjugator". Opt. Lett. **12**, 266 (1987)

Jonathan, J.M.C., Hellwarth R.W., Roosen G.: "Effect of applied electric field on buildup and decay of photorefractive gratings." IEEE J. QE-**22**, 1936 (1986)

Jonathan, J.M.C., Roosen, G., Roussignol, Ph.: "Time-resolved buildup of a photorefractive grating induced in $Bi_{12}SiO_{20}$ by picosecond light pulses". Opt. Lett. **13**, 224 (1988)

Juguryan, L.A.: "Polarization characteristics of phase conjugation at four-wave mixing in uniaxial crystals". Optica Acta **33**, 811 (1986)

Kanata, T., Kobayashi, Y., Kubota, K.: "Epitaxial growth of $LiNbO_3$-$LiTaO_3$ thin films on Al_2O_3". J. Appl. Phys. **62**, 2989 (1987)

Kandidova, O.V., Lemanov, V.V., Sukharev, B.V.: "Asymmetric photoinduced scattering of light in $LiNbO_3$:Fe crystals". Sov. Phys. Solid State **28**, 424 (1986)

Khachaturyan, O.A.: "Liquid-phase electroepitaxy of lithium niobate". Pis'ma Zh. Tekh. Fiz. **13**, 55 (1987) [English transl.: Sov. Tech. Phys. Lett. **13**, 23 (1987)]

Klein, M.B., Dunning, G.J., Valley, G.C., Lind, R.C., O'Meara, T.R.: "Imaging threshold detector using a phase conjugate resonator in $BaTiO_3$". Opt. Lett. **11**, 575 (1986)

Kong, H., Lin, C., Biernacki, A.M., Cronin-Golomb, M.: "Photorefractive phase conjugation with orthogonally polarized pumping beams". Opt. Lett. **13**, 324 (1988)

Krolikowski, W.: "Multi-grating phase conjugation in photorefractive media". Opt. Commun. **60**, 319 (1986)

Krolikowski, W., Belic, M.R., Bledowski, A.: "Phase transfer in optical phase conjugation". Phys. Rev. A **37**, 2224 (1988)

Krolikowski, W., Belic, M.R.: "Multigrating phase conjugation: exact results". Opt. Lett. **13**, 149 (1988)

Krumins, A.E., Kuzminov, Yu.S., Odulov, S.G., Polozkov, N.M., Seglinsh, Ya.A.: "Optical oscillator based on frequency-degenerate pumping of a cerium-activated barium strontium niobate crystal". Sov. J. Quantum Electron. **16**, 679 (1986)

Kukhtarev, N., Pavlik, B., Semenets, T.: "Selfdiffraction and phase conjugation of laser beams in electrooptic crystals". Phys. Status Solidi A **94**, 623 (1986)

Kulich, H.C., Rupp, R.A., Hesse, H., Kratzig, E.: "Anisotropic self-diffraction in $KNbO_3$". Opt. Quant. Electron. **19**, 93 (1987)

Kumar, J., Albanese, G., Steier, W.H., Ziari, M.: "Enhanced two-beam mixing gain in photorefractive GaAs using alternating electric fields". Opt. Lett. **12**, 120 (1987)

Kuzminov, Yu.S., Mamaev, A.V., Orazov, K., Polozkov, N.M., Shkunov, V.V.: "Self-conjugation in nominally pure SBN crystals". Sov. Phys. Lebedev Inst. Rep. No. 2, 35 (1988)

Kwong, S.K., Cronin-Golomb, M., Yariv, A.: "Optical bistability and hysteresis with a photorefractive self-pumped phase-conjugate mirror". Appl. Phys. Lett. **45**, 1016 (1984)

Kwong, S.-K., Cronin-Golomb, M., Yariv, A.: "Oscillation with photorefractive gain". IEEE J. QE-**22**, 1508 (1986)

Laurell, F., Arvidsson, G.: "Frequency doubling in Ti:MgO:$LiNbO_3$ channel waveguides". J. Opt. Soc. Am. B **5**, 292 (1988)

Le Saux, G., Brun, A.: "Photorefractive material response to short pulse illuminations". IEEE J. QE-**23**, 1680 (1987)

Liu, W.-H., Qiu, Y.-S., Zhang, H.-J., Dai., J.-H., Wang, P.-Y., Xu, L.-Y.: "Energy transfer in $Sr_{0.56}Ba_{0.44}Nb_2O_6$:Ce at 633 nm". Opt. Commun. **64**, 81 (1987)

Lowry, M., Jander, D., Kidd, B., Kwiat, P., Peterson, R., McWright, G., Roeske, F.: "A study of the photorefractive effect in $LiNbO_3$:Ti waveguides by using pulsed illumination at 810 nm". Proc. SPIE **720**, 105 (1986)

Luh, Y.S., Fejer, M.M., Byer, R.L., Feigelson, R.S.: "Stoichiometric $LiNbO_3$ single-crystal fibers for nonlinear optical applications". J. Cryst. Growth **85**, 264 (1987)

Mack, R.: "BT&D enters production with MOVPE and automation". Laser Focus/Electro-Optics **24**, No. 1, 131 (Jan. 1988)

McCaughan, L., Choquette, K.D.: "Ti-concentration inhomogeneities in Ti:LiNbO$_3$ waveguides". Opt. Lett. **12**, 567 (1987)

Mainguet, B.: "Characterization of the photorefractive effect in InP:Fe by using two-wave mixing under electric fields". Opt. Lett. **13**, 657 (1988)

Mallick, S., Imbert, B., Ducollet, H., Herriau, J.P., Huignard, J.P.: "Generation of spatial subharmonics by two-wave mixing in a nonlinear photorefractive medium". J. Appl. Phys. **63**, 5660 (1988)

Marotz, J., Ringhofer, K.H., Rupp, R.A., Treichel, S.: "Light-induced scattering in photorefractive crystals". IEEE J. Quant. Electron. QE-**22**, 1376 (1986)

Marrakchi, A.: "Photorefractive spatial light modulation based on enhanced self-diffraction in sillenite crystals". Opt. Lett. **13**, 654–656 (1988)

Marrakchi, A., Johnson, R.V., Tanguay, A.R.: "Polarization properties of enhanced self-diffraction in sillenite crystals". IEEE J. QE-**23**, 2142 (1987)

Marrakchi, A.: "Two-beam coupling photorefractive spatial light modulation with reversible contrast". Appl. Phys. Lett. **53**, 634 (1988)

Medrano, C., Voit, E., Amrhein, P., Günter, P.: "Optimization of the photorefractive properties of KNbO$_3$ crystals". J. Appl. Phys. **64**, 4668 (1988)

Miteva, M., Nikolova, L.: "Oscillating behaviour of diffracted light on uniform illumination of holograms in photorefractive Bi$_{12}$TiO$_{20}$ crystals". Opt. Commun. **67**, 192 (1988)

Motes, A., Kim, J.J.: "Beam coupling in photorefractive BaTiO$_3$ crystals". Opt. Lett. **12**, 199 (1987)

Nightingale, J.L., Becker, R.A., Willis, P.C., Vrhel, J.S.: "Characterization of frequency dispersion in Ti-diffused lithium niobate optical devices". Proc. SPIE **835**, 108 (1987)

Novikov, A.D., Odulov, S.G., Slyusarenko, S.S., Soskin, M.S.: "Threshold conditions for oscillations with the aid of shifted and unshifted dynamic gratings". Sov. J. Quantum Electron. **16**, 573 (1986)

Nozawa, T.: "Growth and properties of an LiNbO$_3$ film on sapphire with an LiNbO$_3$ buffer layer". Electron. Lett. **23**, 1321 (1987)

Obukhovski, V.V., Stoyanov, A.V.: "Photoinduced light scattering in crystals with local response". Sov. Phys. Solid State **28**, 225 (1986)

Odoulov, S.G.: "Anisotropic scattering in photorefractive crystals". J. Opt. Soc. Am. B **4**, 1333 (1987)

Odoulov, S.G.: "Optical oscillation due to vectorial four-wave mixing in photorefractive crystals without photovoltaic effect". Ukrainian Phys. J. **31**, 1645 (1986) (in Russian)

Partlow, D.P., Greggi, J.: "Properties and microstructure of thin LiNbO$_3$ films prepared by a sol-gel process". J. Mater. Res. **2**, 595 (1987)

Partovi, A., Garmire, E.M., Cheng, L.J.: "Enhanced beam coupling modulation using the polarization properties of photorefractive GaAs". Appl. Phys. Lett. **51**, 299 (1987)

Pauliat, G., Günter, P.: "Coherent light oscillators with photorefractive KNbO$_3$ crystals". Opt. Commun. **66**, 329 (1988)

Petersen, P.M., Johansen, P.M.: "Simple theory for degenerate four-wave mixing in photorefractive media". Opt. Lett. **13**, 45 (1988)

Petrov, M.P., Shalaevskii, N.O., Khomenko, A.V., Shlyagin, M.G., Petrov, V.M., Bryksin, V.V., Korovin, L.I.: "Pulsed image readout from the PRIZ SLM". Pis'ma v Zh. Tek. Fiz. **12**, 695–699 (1986)

Psaltis, D., Brady, D., Wagner, K.: "Adaptive optical networks using photorefractive crystals". Appl. Opt. **27**, 1752 (1988)

Ramsey, J.M., Whitten, W.B.: "Controlled scanning of a continuous-wave dye laser with an intracavity photorefractive element". Opt. Lett. **12**, 915 (1987)

Redfield, S., Hesselink, L.: "Enhanced nondestructive holographic readout in strontium barium niobate". Opt. Lett. **13**, 880 (1988)

Reed, G.T., Weiss, B.L.: "Low-loss optical stripe waveguides in LiNbO$_3$ formed by He[+] implantation". Electron. Lett. **23**, 792 (1987)

Reed, G.T., Weiss, B.L.: "Characteristics of He[+] implanted stripe waveguides in LiNbO$_3$". Proc. SPIE **835**, 262 (1987)

Rice, C.E., Holmes, R.J.: "A new rutile structure solid-solution phase in the $LiNb_3O_8$-TiO_2 system, and its role in Ti diffusion into $LiNbO_3$". J. Appl. Phys. 60, 3836 (1986)

Rupp, R.A., Marotz, J., Ringhofer, K.H., Treichel, S., Feng Shi, Kratzig, E.: "Four-wave interaction phenomena contributing holographic scattering". IEEE J. Quant. Electron. QE-23, 2136 (1987)

Rupp, R.A., Drees, F.W.: "Light-induced scattering in photorefractive crystals". Appl. Phys. B 39, 223 (1986)

Rytz, D., Klein, M.B., Mullen, R.A., Schwartz, R.N., Valley, G.C., Wechsler, B.A.: "High-efficiency fast response in photorefractive $BaTiO_3$ at 120°C". Appl. Phys. Lett. 52, 1759 (1988)

Sanford, N.A., Robinson, W.C.: "Direct-current bias stable $Ti:LiNbO_3$ TE-TM mode converters produced by magnesium postdiffusion". Opt. Lett. 12, 531 (1987)

Saxena, R., Anderson, D.Z.: "Effects of an applied field on the steady state characteristics of a unidirectional photorefractive ring oscillator". Opt. Commun. 66, 172 (1988)

Sayano, K., Rakuljic, G.A., Yariv, A.: "Thresholding semilinear phase-conjugate mirror". Opt. Lett. 13, 143 (1988)

Segev, M., Weiss, S., Fischer, B.: "Coupling of diode laser arrays with photorefractive passive phase conjugate mirrors". Appl. Phys. Lett. 50, 1397 (1986)

Seglins, Y., Krumins, A., Ozols, A., Odoulov, S.: "Coherent light oscillator by self-induced holographic gratings in SBN:Ce crystals". Ferroelectrics 75, 317 (1987)

Serdyuk, V.M., Khapalyuk, A.P.: "Diffraction of light by phase holograms in photorefractive ferroelectric crystals". Sov. J. Quantum Electron. 16, 91 (1986)

Serdyuk, V.M.: "Polarization reversal of the wavefront of light beams in photorefractive crystals". Sov. J. Quantum Electron. 16, 555 (1986)

Shandarov, V.M., Shandarov, S.M.: "Recording holograms in $LiNbO_3$:Fe planar optical waveguides". Pis'ma Zh. Tekh. Fiz. 12, 48 (1986) [English transl.: Sov. Tech. Phys. Lett. 12, 20 (1986)]

Shaw, K.D., Cronin-Golomb, M.: "Optical bistability in photorefractive four-wave mixing". Opt. Commun. 65, 301 (1988)

Shepelevich, V.V.: "Effect of optical activity on hologram diffraction efficiency and diffracted light polarization in photorefractive cubic crystals". Sov. Phys. Tech. Phys. 31, 375 (1986)

Shlyagin, M.G., Khomenko, A.V.: "Studies of noises in the PRIZ SLM". Zh. Tek. Fiz. 57, 2101–2104 (1987)

Skeath, P., Burns, W.K., Elam, W.T.: "Relationship of the concentration-dependent Ti center to the $LiNbO_3$ ordinary optical index". Proc. SPIE 836, 32 (1987)

Smout, A.M.C., Eason, R.W.: "Analysis of mutually incoherent beam coupling in $BaTiO_3$". Opt. Lett. 12, 498 (1987)

Stepanov, S.I., Sochava, S.L.: "Effective energy transfer in a two-wave interaction in $Bi_{12}TiO_{20}$". Sov. Phys. Tech. Phys. 32, 1054 (1987)

Sternklar, S., Fischer, B.: "Double-color-pumped photorefractive oscillator and image color conversion". Opt. Lett. 12, 711 (1987)

Sternklar, S., Weiss, S., Fischer, B.: "Tunable frequency shift of photorefractive oscillators". Opt. Lett. 11, 165 (1986)

Sternklar, S., Weiss, S., Segev, M., Fischer, B.: "Beam coupling and locking of lasers using photorefractive four-wave mixing". Opt. Lett. 11, 528 (1986)

Stone, J.: "Photorefractivity in GeO_2-doped silica fibers". J. Appl. Phys. 62, 4371 (1987)

Strohkendl, F.P., Hellwarth, R.W.: "Contribution of holes to the photorefractive effect in n-type $Bi_{12}SiO_{20}$". J. Appl. Phys. 62, 2450 (1987)

Suchoski, P.G., Findakly, T.K., Leonberger, F.J.: "Depolarisation in $Ti:LiNbO_3$ waveguides and its effect on circuit design". Electron. Lett. 23, 1357 (1987)

Temple, D.A., Warde, C.: "Anisotropic scattering in photorefractive crystals". J. Opt. Soc. Am. B 3, 337 (1986); 4, 1335 (1987)

Temple, D.A., Warde, C.: "High order anisotropic diffraction in photorefractive crystals". J. Opt. Soc. Am. B 5, 1800 (1988)

Toda, H., Haruna, M., Nishihara, N.: "Integrated-optic device for a fibre laser Doppler velocimeter". Electron. Lett. 22, 982 (1986)

Tomita, Y., Yahalom, R., Yariv, A.: "Real-time image substraction with the use of wave polarization and phase conjugation". Appl. Phys. Lett. **52**, 425 (1988)

Tschudi, T., Herden, A., Goltz, J., Klumb, H., Laeri, F., Albers, J.: "Image amplification by two- and four-wave mixing in $BaTiO_3$ photorefractive crystals". IEEE J. Quant. Electron. QE-**22**, 1493 (1986)

Twigg, M.E., Maher, D.M., Nakahara, S., Sheng, T.T., Holmes, R.J.: "Study of structural faults in Ti-diffused lithium niobate". Appl. Phys. Lett. **50**, 501 (1987)

Vachss, F., Hesselink, L.: "Holographic beam coupling in anisotropic photorefractive media". JOSA A **4**, 325 (1987)

Vachss, F., Hesselink, L.: "Nonlinear photorefractive response at high modulation depths". J. Opt. Soc. Am. A **5**, 690 (1988)

Valley, G.C.: "Competition between forward- and backward-stimulated photorefractive scattering in $BaTiO_3$". J. Opt. Soc. Am. B **4**, 14 (1987)

Voit, E.: "Anisotropic Bragg Diffraction in Photorefractive Crystals" in *Electro-optic and Photorefractive Materials, Proc. of the Internat. School on Materials Science and Technol.*, ed. by P. Günter (Springer, Berlin, Heidelberg 1986) p. 246–265

Voit, E., Günter, E.: "Optically induced variable light deflection by anisotropic Bragg diffraction in photorefractive $KNbO_3$". SPIE, Vol. 701, ECOOSA '86 (Florence 1986), p. 301–306

Voit, E., Günter, P.: "Photorefractive spatial light modulation by anisotropic self-diffraction in $KnbO_3$ crystals". Opt. Lett. **12**, 769 (1988)

Voit, E., Zha, M.Z., Amrhein, P., Günter, P.: "Reduced $KNbO_3$ crystals for fast photorefractive nonlinear optics". Appl. Phys. Lett. **51**, 2079 (1987)

Volk, R., Sohler, W.: "A highly sensitive excite and probe technique to measure optically induced refractive index changes in $Ti:LiNbO_3$-waveguide resonators". Digest Tech. Papers, Topical Mtg. Integr. Guided Wave Opt. (Opt. Soc. Amer., 1988 Tech. Digest Series, Vol. 5), p. MF6-1

Walsh, K., Hall, T.J., Burge, R.E.: "Influence of polarization state and absorption gratings on photorefractive two-wave mixing in GaAs". Opt. Lett. **12**, 1026 (1987)

Walther, C., Günter, P.: "Photoinduced TE-TM mode conversion in $Ti-LiNbO_3$ waveguides, in *Electro-optic and Photorefractive Materials*, ed. by P. Günter, Proc. in Phys., Vol. 18 (Springer, Berlin, Heidelberg 1987) p. 381

Weiss, S., Sternklar, S., Fischer, B.: "Double phase-conjugate mirror: Analysis, demonstration and application". Opt. Lett. **12**, 114 (1987)

Weiss, S., Segev, M., Sternklar, S., Fischer, B.: "Photorefractive dynamic optical interconnects". Appl. Opt. **27**, 3422 (1988)

White, J.O., Valley, G.C., McFarlane, R.A.: "Coherent coupling of pulsed dye oscillators using nonlinear phase conjugation". Appl. Phys. Lett. **50**, 880 (1987)

Whitten, W.B., Ramsey, J.M.: "Mode selection in a continuous-wave dye laser with an intracavity photorefractive element". Opt. Lett. **12**, 117 (1987)

Yahalom, R., Yariv, A.: "Optical thresholding and switching using a fiber coupled phase conjugate mirror". Opt. Lett. **13**, 889 (1988)

Yariv, A., Kwong, S.K.: "Associative memories based on message bearing optical modes in phase conjugate resonators". Opt. Lett. **11**, 186 (1986)

Yeh, P.: "Theory of unidirectional photorefractive ring oscillators". J. Opt. Soc. Am. B **2**, 1924 (1985)

Yeh, P.: "Photorefractive two-beam coupling in cubic crystals". J. Opt. Soc. Am. B **4**, 1382 (1987)

Yeh, P., Chiou, A.E.T., Hong, J.: "Optical interconnection using photorefractive dynamic holograms". Appl. Opt. **27**, 2093 (1988)

Zhang, G., Li, Q.X., Ho, P.P., Alfano, R.R., Liu, S., Wu, Z.: "Degenerate stimulated parametric scattering in $LiNbO_3:Fe$". J. Opt. Soc. Am. B **4**, 882 (1987)

Zhang, G., Li, Q.X., Ho., P.P., Liu, S., Wu, Z.K., Alfano, R.: "Dependence of specklon size on the laser beam size via photoinduced light scattering in $LiNbO_3:Fe$". Appl. Opt. **25**, 2955 (1986)

Papers on photorefractive phenomena published in the Proceedings of the First

European Conference on Applications of Polar Dielectrics Zürich
(ECAPD-1/ISAF '88)

To be published in Ferroelectrics, April 1989 (Ed. P. Günter)

Agullo-Lopez, F.: "Point defects in electrooptic oxides"

Ammann, F.E., Herdden, A., Tschudi, T., Taranenko, V.B.: "Coupling channels in a photorefractive ring-resonator containing gain by two-wave mixing"

Amrhein, P., Voit, E., Günter, P.: "Photorefractive incoherent to coherent optical conversion using electro-chemically reduced KNbO₃-crystals"

Ditman, L.S., Jr.: "Material parameters determination in barium titanate using a laser probe technique"

Donnerberg, H., Catlow, C.R.A., Tomlinson, S.M.: "Computer-simulation of defect-structures in LiNbO₃"

Feng, X., Ying, J., Wang, J., Liu, J.: "The OH⁻ absorption spectrum as a probe of defect structure of LiNbO₃ crystal"

Godefroy, G.: "Photorefractive properties of monodomain single crystals doped barium titanate"

Ingold, M., Günter, P.: "Linear longitudinal electro-optic effect in oxygen octahedra ferroelectrics"

Ingold, M., Keller, J., Pauliat, G., Günter, P.: "Optical bistability of a nematic liquid crystal with photorefractively amplified feedback"

Klose, F., Wöhlecke, M., Kapphan, S.: "UV-excited luminescence of LiNbO₃ and LiNbO₃:Mg"

Kobialka, T., Herden, A., Tschudi, T.: "Spatial chaos in a phase-conjugate ring-resonator during self-oscillation"

Koppitz, J., Kuznetsov, A.I., Schirmer, O.F., Wöhlecke, M., Grabmeier, B.C.: "Threshold effects in LiNbO₃:Mg caused by change of electron-lattice coupling"

Krumins, A.E., Rupp, R.A., Seglins, Y.A., Knite, M.E.: "Hologram recording at diffused phase transition in transparent PLZT ferroelectric ceramics"

Kulagin, N.: "Optical, dielectric properties and defects in perovskite crystals"

Maillard, A., Rupp, R.: "Effect of applied electric field on diffraction efficiency of BaTiO₃ undoped and iron doped samples"

Medrano, C., Voit, E., Amrhein, P., Günter, P.: "Optimization of the photorefractive properties of KNbO₃ crystals"

Montemezzani, G., Ingold, M., Günter, P.: "Multiple photorefractive gratings in Ce-doped LiNbO₃ and KNbO₃ crystals"

Oates, D.E., Pan, J.Y.: "Bulk-acoustic-wave resonators using holographic reflection gratings in photorefractive materials"

Odoulov, S.G.: "Vectorial interaction in photovoltaic media"

Odoulov, S., Oleinik, O.: "Photorefractive barium sodium niobate"

Pauliat, G., Ingold, M., Günter, P.: "Self-induced coherent light resonators with photorefractive KNbO₃ crystals"

Possenriede, E., Schirmer, O.F., Donnerberg, H.J., Godefroy, G., Maillard, A.: ESR identification of Fe containing defects in BaTiO₃"

Rytz, D., Klein, M.B., Mullen, R.A., Schwartz, R.N., Valley, G.C., Wechsler, B.A.: "Temperature dependence of the photorefractive properties of BaTiO₃"

Sommerfeldt, R., Grabmaier, B.C., Holtmann, L., Krätzig, E.: "The light-induced charge transport in LiNbO₃:Mg, Fe crystals"

Stace, C., Powell, A.K., Hall, T.J.: "Coupling modulation in BSO with applied ac fields"

Stepanov, S.I., Petrov, M.P., Sochava, S.L.: "Optical oscillators and phase conjugators using photorefractive Bi₁₂TiO₂₀"

Voit, E., Amrhein, P., Günter, P.: "Optically addressable fiber interconnection by anisotropic Bragg diffraction in photorefractive KNbO₃"

Wechsler, B.A., Klein, M.B.: "Effects of crystal growth and processing conditions on the photorefractive properties of BaTiO₃: A theoretical model"

Xiao, D.Q., Wang, X., Fu, B.L., Lü, M.K.: "The anomalous photovoltaic effect in potassium iodate crystals"

Xu, Y., Huang, Z., Li, W., Wang, H., Zhu, D.: "Photorefractive effects in KNSBN crystals"

Zha, M.Z., Zhang, Z., Günter, P.: "Effects of absorption and interior reflection on antiparallel beam coupling"

Subject Index